Hydrogen Materials Science and Chemistry of Metal Hydrides

Hydrogen Materials Science and Chemistry of Metal Hydrides

Editor:

Mulmudi Hemant Kumar, Ph.D.

ArclerPress

www.arclerpress.com

Hydrogen Materials Science and Chemistry of Metal Hydrides

Editor: Mulmudi Hemant Kumar, Ph.D.

708 3rd Avenue, 6th Floor
New York
NY 10017
United States of America
www.arclerpress.com

ISBN: 978-1-68094-427-3

Library of Congress Control Number: 2016951120

About the Editor

Mulmudi Hemant Kumar, Ph.D.

Hemant obtained his PhD from Nanyang Technological University, Singapore in 2014. His interests span from material synthesis via solution processed methods to it application in energy harvesting devices such as photovoltaics and water splitting. He is currently working as a research fellow at the Australian national university on silicon/perovskite solar cells.

Contents

List of Contributors

D. Minic
Physical Chemistry Faculty, University of Belgrade, Studentski Trg 12-16, Belgrade, Serbia, Yugoslavia

M. v. Susic
Serbian Academy of Sciences and Arts, Knez-Mihailova 35, Belgrade, Yugoslavia

Robert A. Varin
Department of Mechanical and Mechatronics Engineering, University of Waterloo, Waterloo, Ontario, N2L 3G1, Canada

Leszek Zbroniec
Department of Mechanical and Mechatronics Engineering, University of Waterloo, Waterloo, Ontario, N2L 3G1, Canada

Marek Polanski
Faculty of Advanced Technology and Chemistry, Military University of Technology, 2 Kaliskiego Str., 00-908 Warsaw, Poland

Jerzy Bystrzycki
Faculty of Advanced Technology and Chemistry, Military University of Technology, 2 Kaliskiego Str., 00-908 Warsaw, Poland

Jason Graetz
Department of Sustainable Energy Technologies, Brookhaven National Laboratory, Upton, NY 11973, USA

M. Kunowsky
Group Carbon Materials and Environment, Department of Inorganic Chemistry, University of Alicante, Apartado 99, 03080 Alicante, Spain

J. P. Marco-Lózar
Group Carbon Materials and Environment, Department of Inorganic Chemistry, University of Alicante, Apartado 99, 03080 Alicante, Spain

A. Linares-Solano
Group Carbon Materials and Environment, Department of Inorganic Chemistry, University of Alicante, Apartado 99, 03080 Alicante, Spain

Morten B. Ley
Max Planck Institute for Coal Research, Kaiser-Wilhelm-Platz 1, 45470 Mülheim an der Ruhr, Germany

Mariem Meggouh
Max Planck Institute for Coal Research, Kaiser-Wilhelm-Platz 1, 45470 Mülheim an der Ruhr, Germany

Romain Moury
Max Planck Institute for Coal Research, Kaiser-Wilhelm-Platz 1, 45470 Mülheim an der Ruhr, Germany

Kateryna Peinecke
Max Planck Institute for Coal Research, Kaiser-Wilhelm-Platz 1, 45470 Mülheim an der Ruhr, Germany

Michael Felderhoff
Max Planck Institute for Coal Research, Kaiser-Wilhelm-Platz 1, 45470 Mülheim an der Ruhr, Germany

Hai-Wen Li
Institute for Materials Research (IMR), Tohoku University, Sendai 980-8577, Japan

Yigang Yan
Institute for Materials Research (IMR), Tohoku University, Sendai 980-8577, Japan

Shin-ichi Orimo
Institute for Materials Research (IMR), Tohoku University, Sendai 980-8577, Japan

Andreas Züttel
Department of the Environment, Energy and Mobility (EMPA), Abt. 138 "Hydrogen & Energy", Überlandstrasse 129, 8600 Dübendorf, Switzerland

Craig M. Jensen
Department of Chemistry, University of Hawaii, Honolulu, HI 96822, USA

Kwo-hsiung Young
BASF/Battery Materials-Ovonic, 2983 Waterview Drive, Rochester Hills, MI 48309, USA

Jean Nei
BASF/Battery Materials-Ovonic, 2983 Waterview Drive, Rochester Hills, MI 48309, USA

Thirugnasambandam G. Manivasagam
Department of Chemical Engineering and Chemistry, Eindhoven University of Technology, MB Eindhoven 5600, The Netherlands

Kamil Kiraz
Department of Chemical Engineering and Chemistry, Eindhoven University of Technology, MB Eindhoven 5600, The Netherlands

Peter H. L. Notten
Department of Chemical Engineering and Chemistry, Eindhoven University of Technology, MB Eindhoven 5600, The Netherlands
Department of Electrical Engineering, Eindhoven University of Technology, MB Eindhoven 5600, The Netherlands

Michael Felderhoff
Max Planck Institute for Coal Research, Kaiser-Wilhelm-Platz 1, 45470 Mülheim / Ruhr, Germany

Borislav Bogdanović
Max Planck Institute for Coal Research, Kaiser-Wilhelm-Platz 1, 45470 Mülheim / Ruhr, Germany

Kwo-hsiung Young
BASF/Battery Materials-Ovonic, 2983 Waterview Drive, Rochester Hills, MI 48309, USA
Department of Chemical Engineering and Materials Science, Wayne State University, Detroit, MI 48202, USA

Diana F. Wong
BASF/Battery Materials-Ovonic, 2983 Waterview Drive, Rochester Hills, MI 48309, USA
Department of Chemical Engineering and Materials Science, Wayne State University, Detroit, MI 48202, USA

Jean Nei
BASF/Battery Materials-Ovonic, 2983 Waterview Drive, Rochester Hills, MI 48309, USA

Pragya Jain
Hydrogen Energy Centre, Department of Physics, Banaras Hindu University, Varanasi 221 005, India
Hydrogen Research Institute, University of Quebec at Trois-Rivieres, Trois-Rivières, QC G9A 5H7, Canada

Viney Dixit
Hydrogen Energy Centre, Department of Physics, Banaras Hindu University, Varanasi 221 005, India

Ankur Jain
Institute for Advanced Materials Research, Hiroshima University, Higashi-Hiroshima 739-8530, Japan

Onkar N. Srivastava
Hydrogen Energy Centre, Department of Physics, Banaras Hindu University, Varanasi 221 005, India

Jacques Huot
Hydrogen Research Institute, University of Quebec at Trois-Rivieres, Trois-Rivières, QC G9A 5H7, Canada

Ivaldete da Silva Dupim
Engineering Center, Modeling and Applied Social Sciences, Federal University of ABC (UFABC), Av. United 5001, Santo Andre, SP 09210-580, Brazil

Sydney Ferreira Santos
Engineering Center, Modeling and Applied Social Sciences, Federal University of ABC (UFABC), Av. United 5001, Santo Andre, SP 09210-580, Brazil

Jacques Huot
Department of Chemistry, Biochemistry and Physics, University of Quebec at Trois-Rivieres (UQTR), 3351 des Forges, Trois-Rivières, QC G9A 5H7, Canada

Pascal Schouwink
Department of Quantum Matter Physics, Laboratory of Crystallography, University of Geneva, Quai Ernest-Ansermet 24, CH-1211 Geneva, Switzerland

Fabrice Morelle
Institute of Condensed Matter and Nanosciences, Université Catholique de Louvain, Place L. Pasteur 1, 1348 Louvain-la-Neuve, Belgium

Yolanda Sadikin
Department of Quantum Matter Physics, Laboratory of Crystallography, University of Geneva, Quai Ernest-Ansermet 24, CH-1211 Geneva, Switzerland

Yaroslav Filinchuk
Institute of Condensed Matter and Nanosciences, Catholic University of Louvain, Place L. Pasteur 1, 1348 Louvain-la-Neuve, Belgium

Radovan Černý
Department of Quantum Matter Physics, Laboratory of Crystallography, University of Geneva, Quai Ernest-Ansermet 24, CH-1211 Geneva, Switzerland

Luc Aymard
Reactivity Laboratory and Chemistry of Solids - LRCS, UMR CNRS-UPJV 7314, 33 rue Saint-Leu, 80039 Amiens, France

Yassine Oumellal
Institute of Chemistry and Materials Paris-Est - ICMPE, UMR CNRS-UPEC 7182, 2-8 Rue Henri Dunant, 94320 Thiais, France

Jean-Pierre Bonnet
Reactivity Laboratory and Chemistry of Solids - LRCS, UMR CNRS-UPJV 7314, 33 rue Saint-Leu, 80039 Amiens, France

Kwo-hsiung Young
Department of Chemical Engineering and Materials Science, Wayne State University, Detroit, MI 48202, USA
BASF/Battery Materials-Ovonic, 2983 Waterview Drive, Rochester Hills, MI 48309, USA

Shigekazu Yasuoka
FDK Corporation, 307-2 Koyagimachi, Takasaki 370-0071, Gunma, Japan

Ying Zhu
Center for Microfibrous Materials, Department of Chemical Engineering, Auburn University, 212 Ross Hall, Auburn, AL 36849-5127, USA

Wenhua H. Zhu
Center for Microfibrous Materials, Department of Chemical Engineering, Auburn University, 212 Ross Hall, Auburn, AL 36849-5127, USA

Zenda Davis
Center for Microfibrous Materials, Department of Chemical Engineering, Auburn University, 212 Ross Hall, Auburn, AL 36849-5127, USA

Bruce J. Tatarchuk
Center for Microfibrous Materials, Department of Chemical Engineering, Auburn University, 212 Ross Hall, Auburn, AL 36849-5127, USA

Huan Yang
School of Chemistry and Chemical Engineering, Huazhong University of Science and Technology, Wuhan 430074, China
Key laboratory of Material Chemistry for Energy Conversion and Storage, Huazhong University of Science and Technology, Ministry of Education, Wuhan 430074, China

Yubing Qiu
School of Chemistry and Chemical Engineering, Huazhong University of Science and Technology, Wuhan 430074, China

Xingpeng Guo
School of Chemistry and Chemical Engineering, Huazhong University of Science and Technology, Wuhan 430074, China
Key laboratory of Material Chemistry for Energy Conversion and Storage, Huazhong University of Science and Technology, Ministry of Education, Wuhan 430074, China

Yalian Yang
State Key Laboratory of Mechanical Transmission (SKLMT), Chongqing University, Chongqing 400030, China

Xiaosong Hu
National Engineering Laboratory for Electric Vehicles, Beijing Institute of Technology, Beijing 100081, China
Department of Signals and Systems, Chalmers University of Technology, 41296 Gothenburg, Sweden

Datong Qing
State Key Laboratory of Mechanical Transmission (SKLMT), Chongqing University, Chongqing 400030, China

Fangyuan Chen
State Key Laboratory of Mechanical Transmission (SKLMT), Chongqing University, Chongqing 400030, China

Yajuan Yu
Beijing Key Laboratory of Environmental Science and Engineering, School of Chemical Engineering and Environment, Beijing Institute of Technology, Beijing 100081, China

Bo Chen
Beijing Key Laboratory of Environmental Science and Engineering, School of Chemical Engineering and Environment, Beijing Institute of Technology, Beijing 100081, China

Kai Huang
College of Environmental Science and Engineering, Beijing Forestry University, Beijing 100083, China

Xiang Wang
Beijing Key Laboratory of Environmental Science and Engineering, School of Chemical Engineering and Environment, Beijing Institute of Technology, Beijing 100081, China

Dong Wang
Beijing Key Laboratory of Environmental Science and Engineering, School of Chemical Engineering and Environment, Beijing Institute of Technology, Beijing 100081, China

Pattaraporn Sridechprasat
The Petroleum and Petrochemical College, Chulalongkorn University, Bangkok 10330, Thailand

Labhatrada Phuirot
The Petroleum and Petrochemical College, Chulalongkorn University, Bangkok 10330, Thailand

Pramoch Rangsunvigit
The Petroleum and Petrochemical College, Chulalongkorn University, Bangkok 10330, Thailand
Center of Excellence for Petrochemical and Materials Technology, Chulalongkorn University, Bangkok 10330, Thailand

Boonyarach Kitiyanan
The Petroleum and Petrochemical College, Chulalongkorn University, Bangkok 10330, Thailand
Center of Excellence for Petrochemical and Materials Technology, Chulalongkorn University, Bangkok 10330, Thailand

Santi Kulprathipanja
UOP, A Honeywell Company, 50 East Algonquin Rd., Des Plaines, IL 60017, USA

Preface

Hydrides are compounds containing hydrogen having reducing property. These compounds have wide ranging applications in hydrogen storage, electrochemical, heat storage, other solar related applications and in batteries. This book is a compilation of reviews, research articles on hydrogen material science and application of metal hydrides. Chapters 1-6 deal with the hydrogen economy, modern concepts of energy conversion and storage, effects of structural changes in metal hydrides on hydrogen storage, metal borohydrides and development of hydrogen storage tanks. Electrochemical applications, optical properties and related heat storage materials of solar are discussed in chapters 7-9. Chapters 10-13 focus on additive effects on the properties of metal hydrides and also the effect of hydrogen desorption on cold rolled metal hydrides. Chapters 14-19 highlight the role of metal hydrides in the application of lithium-ion, Nickel/metal hydride batteries. Simulation, prediction of state of health and cell health assessment test is also discussed briefly. These articles give us in-depth details of end-of-life treatment Policy Analysis for Lithium-Ion Batteries and Nickel-Metal hydride batteries. This book is useful to students, chemists and researchers working in the field of batteries and hydrogen storage. The compilation of articles is very broad and up to date. I believe it will enhance the knowledge bank of the personnel pertaining to this subject.

Editor
Mulmudi Hemant Kuma

INTRODUCTION

Hydrogen is a chemical element with chemical symbol H and atomic number 1. With an atomic weight of 1.00794 u, hydrogen is the lightest element on the periodic table. Its monatomic form (H) is the most abundant chemical substance in the Universe, constituting roughly 75% of all baryonic mass. Non-remnant stars are mainly composed of hydrogen in the plasma state. The most common isotope of hydrogen, termed protium (name rarely used, symbol 1H), has one proton and no neutrons.

The universal emergence of atomic hydrogen first occurred during the recombination epoch. At standard temperature and pressure, hydrogen is a colorless, odorless, tasteless, non-toxic, nonmetallic, highly combustible diatomic gas with the molecular formula H_2. Since hydrogen readily forms covalent compounds with most nonmetallic elements, most of the hydrogen on Earth exists in molecular forms such as water or organic compounds. Hydrogen plays a particularly important role in acid–base reactions because most acid-base reactions involve the exchange of protons between soluble molecules. In ionic compounds, hydrogen can take the form of a negative charge (i.e., anion) when it is known as a hydride, or as a positively charged (i.e., cation) species denoted by the symbol H+. The hydrogen cation is written as though composed of a bare proton, but in reality, hydrogen cations in ionic compounds are always more complex. As the only neutral atom for which the Schrödinger equation can be solved analytically, study of the energetics and bonding of the hydrogen atom has played a key role in the development of quantum mechanics.

Hydrogen gas was first artificially produced in the early 16th century by the reaction of acids on metals. In 1766–81, Henry Cavendish was the first to recognize that hydrogen gas was a discrete substance, and that it produces water when burned, the property for which it was later named: in Greek, hydrogen means "water-former".

Industrial production is mainly from steam reforming natural gas, and less often from more energy-intensive methods such as the electrolysis of water. Most hydrogen is used near the site of its production site, the two largest uses being fossil fuel processing (e.g., hydrocracking) and ammonia production, mostly for the fertilizer market. Hydrogen is a concern in metallurgy as it can embrittle many metals, complicating the design of pipelines and storage tanks.

Hydrogen gas (dihydrogen or molecular hydrogen) is highly flammable and will burn in air at a very wide range of concentrations between 4% and 75% by volume. The enthalpy of combustion is −286 kJ/mol:

$$2\ H_2(g) + O_2(g) \rightarrow 2\ H_2O(l) + 572\ kJ\ (286\ kJ/mol)$$

Hydrogen gas forms explosive mixtures with air in concentrations from 4–74% and with chlorine at 5–95%. The explosive reactions may be triggered by spark, heat, or sunlight. The hydrogen autoignition temperature, the temperature of spontaneous ignition in air, is 500 °C (932 °F). Pure hydrogen-oxygen flames emit ultraviolet light and with high oxygen mix are nearly invisible to the naked eye, as illustrated by the faint plume of the Space Shuttle Main Engine, compared to the highly visible plume of a Space Shuttle Solid Rocket Booster, which uses an ammonium perchlorate composite. The detection of a burning hydrogen leak may require a flame detector; such leaks can be very dangerous. Hydrogen flames in other conditions are blue, resembling blue natural gas flames.

The destruction of the Hindenburg airship was a notorious example of hydrogen combustion and the cause is still debated. The visible orange flames in that incident were the result of a rich mixture of hydrogen to oxygen combined with carbon compounds from the airship skin.

H2 reacts with every oxidizing element. Hydrogen can react spontaneously and violently at room temperature with chlorine and fluorine to form the corresponding hydrogen halides, hydrogen chloride and hydrogen fluoride, which are also potentially dangerous acids.

Energy Materials

Multiple geophysical and social pressures are forcing a shift from fossil fuels to renewable and sustainable energy sources.

To effect this change, we must create the materials that will support emergent energy technologies. Solar energy is a top priority of the department, and we are devoting extensive resources to developing photovoltaic cells that are both more efficient and less costly than current technology. We also have extensive research around next-generation battery technology.

Photovoltaics

Widespread application of photovoltaic power to provide a significant fraction of the world's energy needs will require a dramatic lowering of photovoltaic cell material cost and the use of inexpensive, abundant materials and low-cost fabrication strategies. One candidate material that has the potential to meet these requirements is Cu_2ZnSnS_4 (CZTS). CZTS has a favorable band gap for solar cell applications, is made entirely of abundant materials and has a high absorption coefficient to minimize the quantity of material used in devices.

Energy storage

The frontiers of energy storage research are expanding, thanks to the burgeoning science of nanotechnology. Stanford engineer Yi Cui and his team have manufactured new energy storage devices out of paper and cloth, with a range of potential applications. Their research also has shown that using silicon nanowires to replace carbon anodes in lithium ion batteries can significantly improve their performance.

Hydrogen storage

For the transition to a hydrogen-based economy to become feasible and economically practical, many materials challenges must be addressed. Not the least of these is the engineering of a hydrogen storage material with high storage density (both gravimetric and volumetric), appropriate equilibrium pressure, favorable reaction kinetics, relative safety and low cost. Metal hydrides represent one attractive way to store large amounts of hydrogen due to the very high potential volumetric capacity, which can even exceed that of liquid hydrogen. So far, however, no single material has met all the requirements for a practical, reversible on-board storage material.

We combine the flexibility and control of physical vapor deposition to fabricate thin film samples with precise chemical compositions and microstructures in order to probe metal hydride reactions in a very controlled way. Using a variety of thin film and powder characterization techniques (from X-ray diffraction to quartz crystal microbalance measurements and gas adsorption in a Sievert's type apparatus) we monitor the thermodynamic, kinetic and structural properties of these materials to gain a more fundamental understanding about the processes limiting their practical implementations. By applying the knowledge learned from these highly controlled systems, we can engineer materials to better meet the challenges of a hydrogen-based economy.

What is a Hydride?

Traditional hydrides are simple compounds in which hydrogen bears a negative charge. They often contain one or more positive, metal ions — as in, for example, lithium aluminum hydride (LiA_lH_4). These substances are bases and are powerful reducing agents that can be dangerous to handle. Nevertheless, in the search for suitable replacements for fossil fuels, metal hydrides are considered likely candidates. This may be particularly true for transition metal hydrides.

Some of the more common traditional metal hydrides are those of sodium, calcium and nickel. These substances are categorized respectively as the hydrides of alkali, alkaline earth and transition metals. For an alkali or an alkaline earth metal hydride, chemical bonding is most commonly of the covalent, ionic and mixed ionic varieties. Nickel hydride, used in the manufacture of vehicular batteries, is formed through combining the elements under high pressure. This metallic hydride exhibits a different kind of chemical bond, which is believed to be essential to the hydrogen storage process.

Nickel hydride resembles to some degree the hydride of its fellow transition metal, palladium. These two elements unite with hydrogen through a variety of metallic bond called "interstitial bonding." In this type of bonding, larger atoms have smaller atoms — in this case hydrogen — inserted between them. Not requiring the stringent conditions needed for nickel, palladium hydride forms at room temperature and atmospheric pressure, storing up to 900 times its volume in hydrogen. Although palladium is prohibitively expensive, it could theoretically be used and would present a safer, more efficient means of carrying vehicular hydrogen than pressurized tanks of gas.

Palladium atoms are nearly 5.5 times as large as those of hydrogen. Nickel atoms are 4.6 times larger than hydrogen. This compares to a ratio of 2.1 times for iron and carbon, which bond interstitially to form carbon steel. Whatever relationship atomic size ratio bears to the ease of diffusive insertion, this correlation in bonding to that of carbon steel indicates both nickel and palladium hydrides are alloys of sorts.

If hydrides are to be considered serious contenders for use, some challenges must be met — one example of this can be seen in fuel storage. For one, as hydrogen gas is diffused into a metal, it quickly builds up a back-pressure that slows the further diffusion. Doping the primary metal with another metallic element may lessen this tendency. Another problem is that with each repeated cycle, the hydride metal substrate expands and contracts. Substrate pieces can break down into smaller particles, producing fines that become a source of difficulty unless filtered out. Finally, hydrides must out-perform the contenders, which include possibly liquefied hydrogen and liquid boron-hydrogen complexes.

In Chemistry, What Is the Oxidation Number?

Atoms combine through the giving and accepting of negatively charged electrons, forming an ionic bond, or sharing electrons, forming a covalent bond. Typically, ionic bonds form between metals and non-metals, while covalent bonds form only between non-metals. An atom's oxidation number represents the number of electrons it has given to, accepted from or shared with other atoms, and will be positive when it has lost or shared electrons and negative when it has gained or shared another atom's electrons.

Originally, oxidation meant combining with oxygen; the opposite process, removal of oxygen, was known as reduction. For example, when calcium combines with oxygen, the oxygen removes two electrons from the calcium, acquiring a charge of -2, leaving the calcium with a charge of +2. An ionic bond has formed; calcium has been oxidized and oxygen reduced, giving them oxidation numbers of +2 and -2, respectively. Covalent bonding also involves oxidation and reduction: to form water, two hydrogen atoms combine with one oxygen atom, each sharing an electron with it. The oxidation numbers of the two hydrogen atoms are each +1 and the oxygen atom -2.

Oxidation need not involve oxygen; today, oxidation simply means loss of electrons and reduction, gain of electrons. As a rule, the elements toward the bottom left corner of the periodic table have the greatest tendency to lose electrons, while those toward the top right corner have the greatest tendency to gain electrons. Oxygen has a strong tendency to remove electrons from other elements and is therefore an oxidizing agent; however, other elements can also oxidize. Fluorine, for example, is an even stronger oxidizing agent.

The oxygen oxidation number is usually -2, but many elements can have more than one possible oxidation number. For iron it can be +2 or +3, while for hydrogen it is +1 when combined with non-metals, as in water (H_2O), but -1 when combined with metals, as in sodium hydride (Na^+H^-), where it accepts an electron. The nitrogen oxidation number can vary from -3 to +5. The oxidation numbers of the ions in an ionic compound are equivalent to the charges on the ions, while those of the atoms in a compound always add up to zero.

Elements can combine to form ions that have an overall oxidation number. For example, an atom of sulfur can combine with four oxygen atoms to form a sulfate group (SO_4^{2-}) with an oxidation number of -2. Ammonia, a compound of nitrogen and hydrogen with the formula NH_3, can combine with an additional hydrogen atom to form the ammonium group NH_4+, with an oxidation number of +1. This behaves like a metal and forms ionic compounds such as ammonium sulfate ($(NH_4^+)_2SO_4^{2-}$), where two ammonium groups combine with one sulfate group.

Metal Hydrides

Metal-hydrogen bonds, also known (misleadingly) as metal hydrides, are ubiquitous X-type ligands in organometallic chemistry. There is much more than meets the eye to most M-H bonds: although they're simple to draw, they vary enormously in polarization and pKa. They may be acidic or hydridic or both, depending on the nature of the metal center and the reaction conditions. In this post, we'll develop some heuristics for predicting the behavior of M-H bonds and discuss their major modes of reactivity (acidity, radical reactions, migratory insertion, etc.). We will also touch on the most widely used synthetic methods to form metal hydrides.

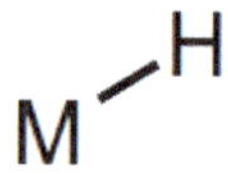

General Properties

Metal hydrides run the gamut from nucleophilic/basic to electrophilic/acidic. Then again, the same can be said of X–H bonds in organic chemistry, which may vary from mildly nucleophilic (consider Hantzsch esters and NADH) to extremely electrophilic (consider triflic acid). As hydrogen is what it is in both cases, it's clear that the nature of the X fragment—more specifically, the stability of the charged fragments X+ and X-—dictate the character of the X–H bond. Compare the four equilibria outlined below—the stabilities of the ions dictate the position of each equilibrium. By now we shouldn't find it surprising that the highly π-acidic $W(CO)_5$ fragment is good at stabilizing negative charge; for a similar reason, the $ZrCp_2Cl$ fragment can stabilize positive charge.*

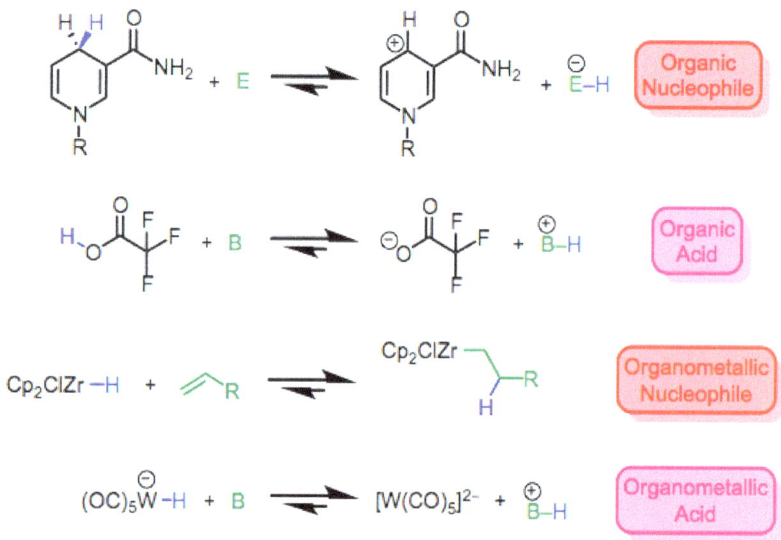

Metal-hydrogen bonds may be either hydridic (nucleophilic) or acidic (electrophilic). The nature of other ligands and the reaction conditions are keys to making predictions.

Let's turn our attention now to homolytic M–H bond strength. A convenient thermodynamic cycle allows us to use the acidity of M–H and the oxidation potential of its conjugate base in order to determine bond strength. This clever method, employed by Tilset and inspired by the inimitable Bordwell, uses the cycle in the figure below. BDE values for some complexes are provided as well. From the examples provided, we can see that bond strength increases down a group in the periodic table. This trend, and the idea that bridging hydrides have larger BDEs than terminal M–H bonds, are just about the only observable trends in M–H BDE.

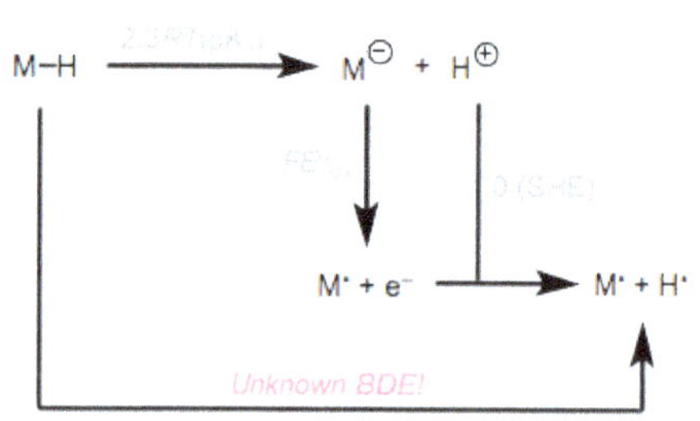

Complex	BDE (kcal/mol)
$CpCr(CO)_3H$	61
$CpMo(CO)_3H$	69
$CpW(CO)_3H$	72
$Mn(CO)_5H$	68
$Re(CO)_5H$	75

Why is knowing M-H BDEs useful? For one thing, the relative BDEs of M-C and M-H bonds determine the thermodynamics of β-hydride elimination, which results in the replacement of a covalent M-C bond with an M-H bond. Secondly, complexes containing weak M-H bonds are often good hydrogen transfer agents and may react with organic radicals and double bonds, channeling stannane and silane reductants from organic chemistry.

Hydricity refers to the tendency of a hydride ligand to depart as H–. A similar thermodynamic cycle relates the energetics of losing H– to the oxidation potentials of the conjugate base and the oxidized conjugate base; however, this method is complicated by the fact that hydride loss establishes an open coordination site. For five-coordinate hydrides that form 16-electron, square planar complexes upon loss of hydride, the bite angle of chelating phosphines plays an interesting role. As bite angle increases, hydricity does as well.

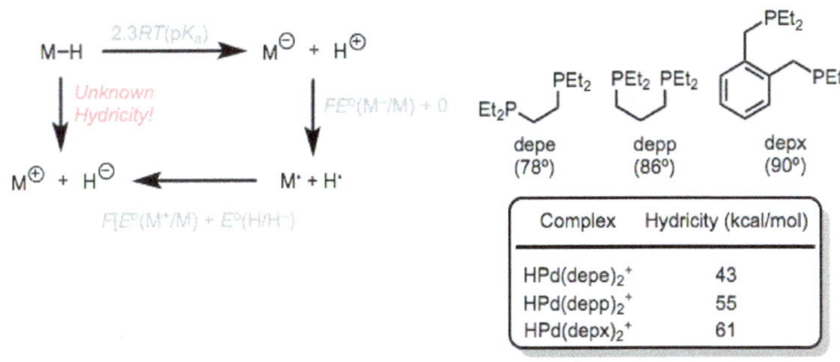

A thermodynamic cycle for hydricity, with some examples. Hydricity and bite angle are well correlated in five-coordinate palladium hydrides.

Bridging hydrides are an intriguing class of ligands. A question to ponder: how can a ligand associated with only two electrons possibly bridge two metal centers? How can two electrons hold three atoms together? Enter the magic of three-center, two-electron bonding. We can envision the M–H sigma bond as an electron donor itself! With this in mind, we can imagine that hydrides are able to bind end-on to one metal (like a standard X-type ligand) and side-on to another (like an L-type π system ligand, but using sigma electrons instead). Slick, no? We'll see more side-on bonding of sigma electrons in a future post on sigma complexes.

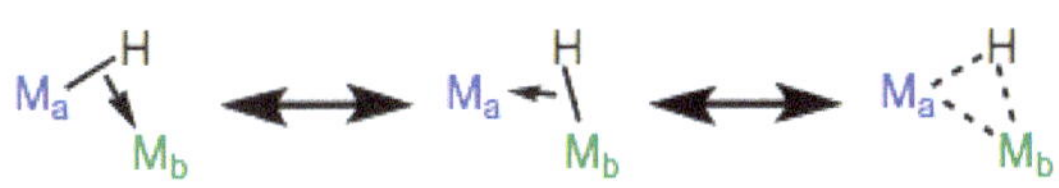

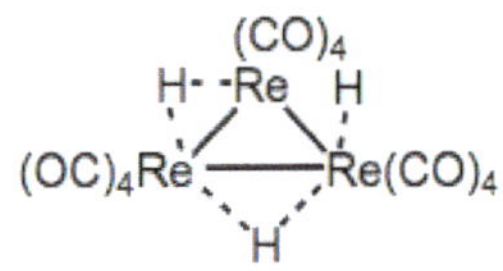

Consistent with the idea that bridging is the result of "end-on + side-on" bonding, bond angles of bridging hydrides are never 180°.

Synthesis

Here we'll discuss four ways to make hydrides: metal protonation, oxidative addition of H2, addition of nucleophilic main-group hydrides (borohydrides, aluminum hydrides, and silanes), and β-hydride elimination.

Just as in organic chemistry, the basicity of an organometallic complex is inversely related to the acidity of its conjugate base. Furthermore, charges have a predictable effect on the basicity of organometallic complexes: negatively charged complexes lacking π-acidic ligands are highly basic. Even neutral complexes containing strong donor ligands, like the tungsten complex below, can be protonated effectively. Notice that protonation is actually a kind of oxidative addition—the oxidation state of Fe in the first reaction below goes from -2 to 0 to +2! All coordination events of isolated electrophiles can also be viewed in this light. Reactions of this type are sometimes called oxidative ligations to distinguish them from oxidative addition reactions, which involve the addition of two ligands to the metal center with oxidation.

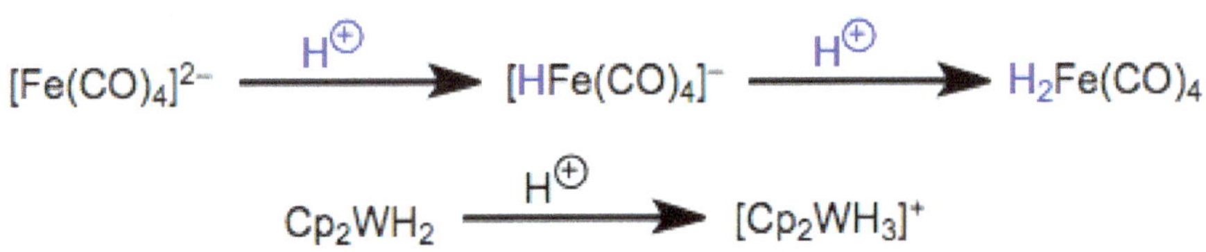

Contrast oxidative ligation with the oxidative addition of dihydrogen (H2), a second method for the synthesis of hydride complexes. A key requirement here is that the starting metal center is at least d2—two electrons are formally lost from the metal center, and metals can't possess a negative number of d electrons. An open coordination site on the starting material must also be present (or possible through ligand dissociation). The reaction below is a standard example of the addition of H2 to Vaska's complex, but there are some funky variations on this theme. These riffs include third-order homolytic cleavage of H2 by two metal radicals, and oxidative addition followed by deprotonation by the starting complex (apparent heterolytic cleavage).

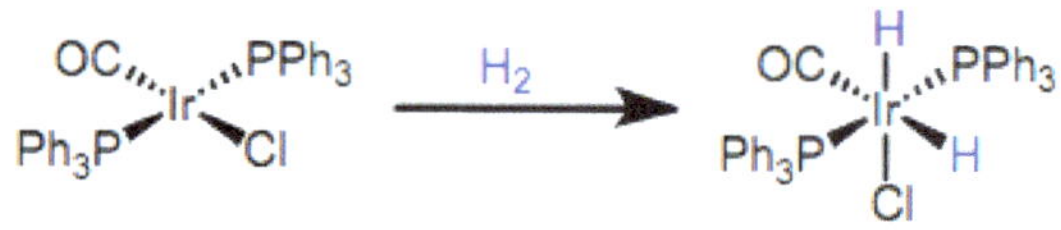

Main-group hydrides like borohydrides and aluminum hydrides are great sources of H– for organometallic complexes. These reactions seem more natural than metal protonations, since we often think of metals as electropositive or electrophilic species. Indeed, the combination of main-group sources of nucleophilic hydride with complexes containing metal–leaving group bonds is a very general method for the synthesis of metal hydride complexes. Check out the reaction below—what's the most likely mechanism? Is associative or dissocative substitution more likely? Hint: count electrons!

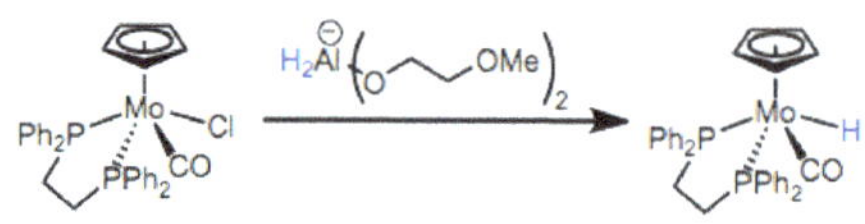

β-hydride elimination forms metal hydride complexes and double bonds within organic ligands. Alkoxide ligands are commonly used for this purpose—elimination to form the hydride complex and aldehyde is more favorable than the reverse, migratory insertion of C=O into the M–H bond. Since the unsaturated byproduct is thrown away, it's desirable to make it something small, cheap, and gaseous. Hydroxycarbonyl, formate, and tert-butyl ligands have been applied successfully with this goal in mind…what are the byproducts?

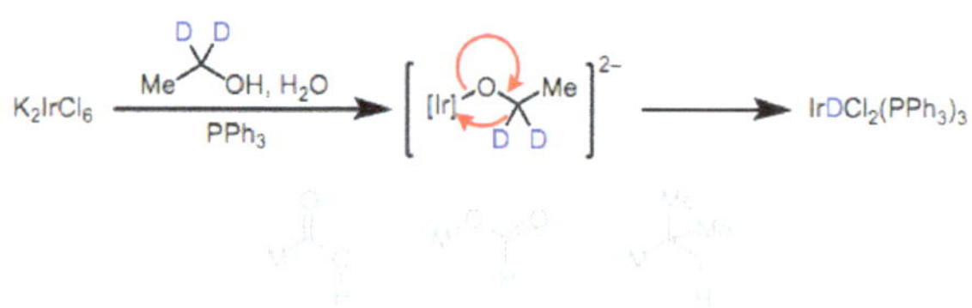

Reactions

Metal hydrides are characterized by nucleophilic, electrophilic, and radical behavior. The exact behavior of a given metal hydride complex depends on its electronic properties, its M–H bond dissociation energy, and the nature of the reacting partner. Basic metal hydrides react with acids to free up a coordination site on the metal center (file this reaction alongside photolytic M–CO cleavage and M–CO cleavage with amine oxides).

Migratory insertion reactions involving M–H bonds are extremely important in a practical sense (see hydrogenation and hydroformylation), and are conceptually related to nucleophilic hydride transfers from the metal center. We can think of insertions of π bonds into M–H as internal nucleophilic attack by the hydride ligand at one end of the π bond, with coordination of the metal center to the other. The figure below depicts the transition state for migratory insertion of an olefin into M–H and an example reaction. Notice the coplanarity of M, H, and the C=C bond in the transition state for insertion, which determines the cis configuration of the product. The same is required for the microscopic reverse (β-hydride elimination) but this essential geometry is easily overlooked for β-hydride elimination.

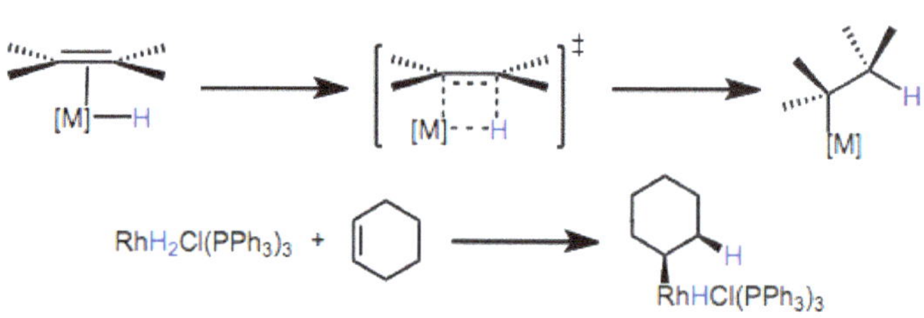

Migratory insertion of an olefin into M–H. Note the relative configuration of M–C and C–H!

It is important to note that insertions of CO into M–H bonds are rare, because such insertions are usually unfavorable thermodynamically. Finally, the radical behavior of certain metal hydrides in the presence of olefins deserves a nod. The behavior of the cobalt hydride complex in the reaction below is typical of these types of reactions.

And there you have it: the properties, synthesis, and reactivity of metal hydrides in a nutshell. Like other ligands we have seen so far, the behavior of hydrides is controlled by the nature of the metal center and its accompanying ligands. However, it's interesting to note that the observed behavior of hydrides often depends on the nature of the reaction conditions, as well. Some complexes display "schizophrenic" behavior, so to speak, putting on their nucleophile hat in the presence of an electrophile and their electrophile hat in the presence of a nucleophile.

CHAPTER 1

MODERN CONCEPTS OF CONVERSION AND STORAGE OF ENERGY BY DISPERSED MATERIALS ABSORPTION

D. Minic and M. v. Susic

Keywords: Hydrogen, Conversion, Storage, Metal hydrides, Carbon nanotubes, Graphite nanotubes

INTRODUCTION

The largest amount of sun's energy and other renewable forms of energy are available only in certain weather conditions, and require storage for later use. The most important technologies used for storage of this type of energy use modem batteries (lithium-ion, nickel metal hydride, air metal hydride, etc.), hydrogen storage and fuel cells.

Hydrogen fuel is, because of its great availability (also its regeneration) in nature and the fact that it releases a great amount of energy while producing no environmental pollution, considered by many to be the *"energy carrier"* of the future. It is called an energy carrier and not an energy source because it takes a primary energy source to produce it. The energy is released on demand by combustion or by electrochemical conversion directly to electricity in the fuel cell. The product of the process is in all cases water and therefore hydrogen is the ultimate "clean fuel". Such fuel might solve the world's growing need for thermal and electrical power along with energy transportation. This is why today hydrogen production and storage are primary research interests in the search for new energy sources.

Hydrogen is the most abundant element in nature (the universe consists of 93% of hydrogen), and it is commonly found in the shape of its most common compounds from which it is extracted through various techniques. For hydrogen production, where it is bound with other elements in nature, various photobiological, chemical and electro-chemical techniques exist. All of the previous techniques require certain consumption of energy. One of the well-known techniques for hydrogen production is photolysis where energy of the sun is used to split water molecules into hydrogen and oxygen molecules.

STORAGE OF ENERGY

Once produced hydrogen can be stored using various techniques which have a number of advantages but also many disadvantages. The prime criterion for choosing which storage technique to use is to provide a safe and easy way to store hydrogen. These techniques deal with physical means of storing energy, where hydrogen might be compressed or transferred to liquid state, as well as chemical storage and gas-solid interaction (absorption) [1-5].

Among all these techniques, today, the most practical and most economical techniques are compression of hydrogen in its gas state, extraction and maintenance of liquid hydrogen and utilization of chemical compounds for hydrogen storage. The newest and most promising hydrogen storage research is focused on creation of new dispersed materials (power balls, hydrides and carbon clusters), which are suitable for hydrogen absorption at room temperature and normal pressure [11-34].

Compression of Hydrogen

Nowadays, commercially available hydrogen in its gas state can be compressed at room temperature and pressure of 24.8 MPa into glass fiber strengthened aluminum cylinders with density 12 kg/m^3 and gravimetric density of 2 wt.%. Polymeric cylinders, strengthened with carbon fibers, provide higher density (15 kg/m^3, 5 wt.%), but it is much lower than the density required by vehicles with fuel cells (62 kg/m^3 and 6.5 wt.%). Light composite cylinders secure hydrogen under gravimetric density of 12 wt.% and pressure 33.8 MPa, when they are completely filled [6-7]. Compared to petrol, thus compressed

Citation: D. Minic and M. v. Susic, Modern concepts of conversion and storage of energy by dispersed materials absorption, DOI 10.2298/SOS0203247M.

hydrogen volume is larger and, consequently, has lower energy density. More precisely, the energy density of petrol equivalent is around 3000 times higher than the one related to compressed hydrogen. On the other hand, compression requires energy consumption to a certain degree, so that 10 % of complete stored energy is consumed, by compression of hydrogen. These 225 liters containers, and larger ones, are used for motoring cars and buses with fuel cells, respectively (Los Angeles, Chicago- USA).

Hydrogen can be compressed in containers at pressure up to 41.22 MPa, while every new 28 liters of compressed gas, added in the same container, increase the pressure for additional 0.101 MPa. Because of extremely high pressure, these containers require permanent control and occasional testing.

Liquid Hydrogen

Liquid hydrogen can exist only at extremely low temperatures (20 K or -253°C) and such a state is hard to reach and maintain [7-9]. Density of thus stored hydrogen is 845 times higher, compared to the one of gaseous hydrogen at normal conditions. It has a low mass (one liter of liquid hydrogen has 0.07 kg in weight while the same volume of petrol has 0.84 kg in weight), but 4 times lower energy density compared to petrol (4 liters of hydrogen are energetically equivalent to a liter of petrol), and has a 3 times higher energy value per weight unit. Conversion of hydrogen into its liquid state by cooling and compressing requires 30 % of stored energy. On the other hand, keeping hydrogen liquid apart from temperature maintaining needs application of special highly isolated containers equipped for liquid hydrogen storage under pressure. Even under these conditions, 1-2 % of stored hydrogen evaporates every day, and it has to be safely stored. Considering all this, such storage of hydrogen as a method is expensive and unsuitable for common application. However, due to its low weight, liquid hydrogen is the fuel of choice in spacecrafts, where conditions for attaining low temperature are much more suitable than the ones on Earth. The purpose of research in this area is construction of light containers made of composite materials, strengthened in order to keep liquid hydrogen and improvement of preparation of hydrogen in this state.

Cryogenic Activated Carbon

Carbon activated by cooling (cryogenic carbon) can be used for hydrogen storage, too [1-2, 10]. This technique has developed rapidly. The best results were achieved with carbon cooled at 77 K (-196°C). Thus stored hydrogen density approaches the density of liquid hydrogen.

Chemically Stored Hydrogen

As the most abundant element in the universe, hydrogen appears in the form of its compounds. These compounds can be used for hydrogen production, and some, for its storage [11]. A stable hydrogen compound, produced in the first stage, can release hydrogen in a chemical reaction, which can be collected and used for energy production in fuel cells. Some of these reactions include ammonia cracking, methanol cracking, electrolysis of water, etc. In this case, there is no need to store released hydrogen, because it is extracted when needed. For example, methanol is introduced directly into the cell. Methanol and water convey into a gas state and produce hydrogen, CO, CO_2 at 280°C. After catalytic oxidation, in order to remove CO (there is a quantity of 10-20 ppm of CO), the gas transfers directly onto the positive terminal of the fuel cell.

Hydrogen can be stored in the form of fossil fuel (petrol, gas, etc.), which are widely used hydrogen sources. In a technique, known as Storage of liquid carrier, fossil fuel undergoes the process of "reforming". Such hydrogen can hold additional CO that originates from fossil fuel and may pollute certain types of fuel cells. Therefore, produced hydrogen has to be purified in order to remove excess CO.

A more current technology of hydrogen chemical storage, which doesn't require energy for releasing hydrogen, is the so called *Power ball technology.* This new technology refers to its compounds, which in an exothermic reaction, reacts with water to extract hydrogen. Alkaline metal hydrides (Na, Li or K) are most suitable for this purpose. A hydride pellet, as needed, reacts with water to release hydrogen. This hydrogen is used as fuel and forms a suitable hydroxide solution that can be recycled. Power balls available today include balls or pellets with water resistant coating (polyethylene). Polyethylene coating prevents the reaction of sodium hydride with water and uncontrolled loss of energy. Thus protected balls can be stored directly in water and as needed, after breaking of the coating, a sodium hydride ball (one inch in diameter) completely reacts with water within 10 seconds, releasing hydrogen, which burns with oxygen. Both of these reactions are exothermic, so the hydrogen production doesn't require energy:

$NaH + H_2O \rightarrow NaOH + H_2$ (pellet reaction)

$H_2 + 0.5\ O_2 \rightarrow 7\ H_2O$ (fuel combustion reaction)

NaH + 0.5 02 --7 NaOH (whole reaction)

The reaction scheme shows that 12 grams of NaH yield 1 gram of hydrogen. In a stored form, power balls (pellets) consume only 65 % of the total volume of its container. Pellet coats and spaces between them, filled with water, fill the rest of the volume. One liter of power balls (NaH density is 1.396 g/cm^3 produces 76 grams of hydrogen or 845.45 liters (measured at sea level and room temperature). It has to be emphasized, for comparison, that one-liter of liquid hydrogen and methanol yield 778 and 1001 liters of hydrogen in a gas state, respectively (measured at sea level and room temperature).

According to the same reaction cheme 1 liter of power balls produce 38.02 moles or 1520.29 grams of NaOH with a density of 2.130 g/cm^3 and volume of 12.7 liters. Thus produced sodium hydroxide can be stored in a container, which previously held power balls and recycled later, for sodium hydride production, using electric or solar energy.

Glass Micro Spheres

These hollow glass balls, 25-500 11m in diameter with 1 11m thick walls, can be used for safe hydrogen storage [12]. Empty glass balls are heated to a temperature range 200-400°C, making its walls more permeable. Balls are' filled with hydrogen by putting it into gaseous hydrogen under pressure. Then, balls are cooled retaining hydrogen. The balls release the contained hydrogen with repeated heating. Such balls are inert, resistant to contamination and require small amount of heat to release hydrogen.

Dispersed Systems

A new concept of hydrogen storage is aiming on the use of dispersed systems, i.e. metal hydrides, polymer-dispersed metal hydrides and carbon clusters. It is known that many dispersed systems, metals and its alloys, absorb hydrogen at relatively low pressures. Because of this property, they are used as hydride containers for the storage of hydrogen. More precisely, metals have a unique ability to absorb hydrogen and release it later, both at room or higher temperatures. The whole amount of hydrogen absorbed in this way, is usually 1-2% of the whole weight of hydride, even though some hydrides absorb 5-7 % at higher temperatures (250°C) [9].

Metal Hydrides

Metal hydrides are solid compounds of hydrogen and metal or metal alloy [13-24]. Certain alloys absorb easily large amounts of gaseous hydrogen reversibly at pressure around ambient pressure as shown in Fig. 1.

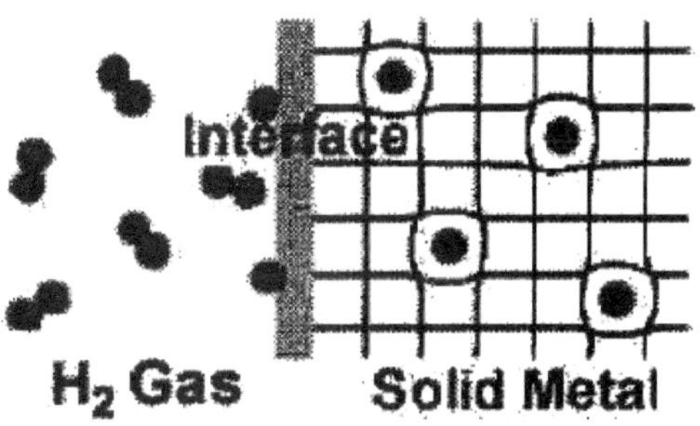

Figure. 1 Simplified model of hydrogen-metal interaction.

Metal-hydrogen systems consist of a metal phase, gas phase (hydrogen in gas state) and interphase surface, Fig. 1. Hz molecules are adsorbed at the interphase surface and divided to give atoms. The yielded hydrogen atoms might be absorbed or dissolved in a metal phase, releasing a large amount of heat. By suitable tailoring of the metal alloy it is possible to obtain almost any desired bond strength resulting in the desired equilibrium pressure. This is the storage pressure at which hydrogen is either absorbed or desorbed. Thus yielded and disordered solution of hydrogen in a metal phase is known as a-hydride phase. In a metal phase, hydrogen atoms might consolidate and form specific configuration with metal atoms, specific for the type of metal, known as α -hydride phase. The way nucleation develops, its location and β - phase growth, depend on characteristics of the metal. The reaction of hydrogen with the metal can be presented like this

$$M + x/2\, H_2 \Leftrightarrow MH_x + \text{heat}$$

The reaction is reversible - it reaches equilibrium, and then changes direction with the change of conditions·. According to the right side of the reaction, heat is released, when hydride is produced and it is consumed when hydrogen escapes from metal hydride. The amount of reaction enthalpy is the strength indicator of the metal-hydrogen bond in metal hydride.

Thermodynamic aspect of hydride forming

The best way to describe equilibrium between metal and hydrogen is by a pressure composition isotherm, also called a *PCT curve.* An ideal isotherm, three temperatures PCT curves is, shown in Fig. 2a

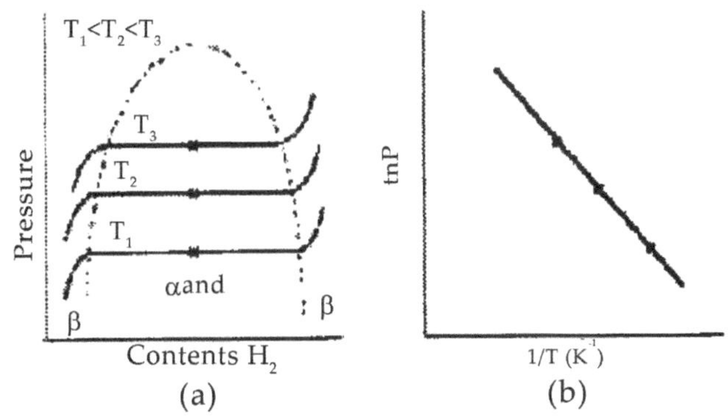

Figure. 2 Three isotherm PCT curves (a); change of equilibrium pressure with temperature (b).

The slope on the left side of PCT curves indicates that with the presence of an α -hydride phase, big change of hydrogen pressure causes a slight change of dissolved hydrogen concentration in the metal phase. Meanwhile, when the β -hydride phase starts to form, more exactly, when both hydride phases are present, a slight change of hydrogen pressure causes a big change of absorbed hydrogen concentrations in the metal phase. Furthermore, when the material saturates with the β -hydride phase (right side of PCT curves) than again, a big change of hydrogen pressure causes a slight change of absorbed concentration in the metal phase. A plato, which indicates the existence of both phases in equilibrium, at the same time, is the so called *equilibrium pressure plato,* and it is a characteristic of that material. Temperature growth also increases the content of hydrogen in the metal phase.

Van't Hoffs equation describes the proportion between pressure and temperature:

$$\ln p = \Delta H/RT - \Delta S/R,$$

where *p* is the equilibrium pressure of hydrogen, H and S are enthalpy and entropy of the reaction of hydrogen absorption, respectively, R is the universal gas constant and T is the absolute temperature.

Van·t Hoffs equation shows that the temperature raise increases the pressure of hydrogen, so it is possible to absorb hydrogen at low pressure and temperature and then to heat the hydride to incite desorption of hydrogen at much higher pressure.

By measuring the equilibrium pressure at various temperatures (Fig. 2b) utilization of this equation enables determination of the enthalpy and entropy of hydrogen absorption in the metal phase.

Application of metal hydrides for hydrogen storage

The ability of metal and its alloys to absorb great amounts of hydrogen at relatively low pressures made possible their utilization for hydrogen storage (Tab. I). The most exploited systems for this purpose are MgH_2, $FeTiH_2$, $LaNiH_6$, Mg_2Nil_4 [4-5]. Hydrogen atoms intensely react with electron gas of metal atoms (they are being absorbed), releasing a great amount of heat. A reduction of rotational degrees of freedom, caused by absorption, brings about a decrease of bonded hydrogen atom diameter, so it is possible to draw them near each other, much closer, than in liquid hydrogen. However, the metal phase weight is still to a great extent higher than that of stored hydrogen. Consequently, metal hydrides are unsuitable for economical application in transport means. Still, hydrogen density in metal hydrides is much higher than the one for gaseous or liquid hydrogen.

Hydrogen, produced in this way, is of a great purity, and the life of such a hydride container depends on the presence of impurities. The alloy acts like a sponge, soaking up water and absorbs hydrogen as well as the impurities present. After the release of hydrogen, these impurities permanently bind to the metal and decrease its ability for further absorption of hydrogen.

In order to obtain alloys suitable for different applications, the. properties of hydrides can be controlled by changing the composition and microscopic structure or by treating the surface, e.g., substituting or adding other metal elements.

Particular progress in this area is in the application of amorphous and disordered metal hydrides with required hydrogen storage qualities. The presence of platinum and palladium in amount less than 0.1 % has a favorable influence on this process as presented in the work of professor Susie et al. In this case, comparing to the disperse pure substance, Pt and Pd are very active catalysts and activators of hydrogen reactivity so that absorption of H_2 carries on much faster and at considerably lower temperatures, due to the known hydrogen spillover effect (HSE) into the crystal lattice. During the HSE at the surface of metal particles (Pt, Pd), which are dispersed at absorbent, easily carries on dissociation of hydrogen molecules according to reaction $H_2 \rightarrow {}_2H$. Thus produced, very reactive hydrogen is spilled into the absorbent crystal lattice and easily yields hydride, where further dissolving take place.

Nanocrystalline Metal Hydrides

This a new class of hydrogen absorbing alloys [24]. These are polycrystals with a typical grain size < 20 nm. Their advantages are to maintain their integrity upon cycling and to show enhanced absorption/desorption kinetics. Examples of these materials, which can be synthesized easily by mechanical alloying, are FeTi and Mg_2Ti. A major drawback is the reduced maximum capacity.

Although, the percentage of absorbed gas is sometimes low, in spite of the high price of these materials and their big weight, hydrides obtain safe hydrogen storage and its delivery at constant pressure.

Metal hydrides, dispersed in a polymetric matrix (PDMH) are the newest approach in this field [26-28]. The goal is to advantageously combine the volumetric storage capacity of metal hydrides, with the low density of polymeric materials, in an effort to improve the gravimetric storage capacity by factors of 2-5 relative to non-dispersed metal hydrides. In addition, it is conceivable that dehydrating temperatures may be reduced, potentially enabling automotive waste heat to effectively dissociate the hydride. Dispersion of the metal hydride in a polymer support phase may also increase the stability of the metal hydride, possible through reduction of hydride decrepitating.

It was shown that relatively low gravimetric capacity of metal hydrides could be slightly enhanced by incorporating a low-density polymer that can both interact with the hydride on a molecular level, and store additional hydrogen within the polymer structure, Fig. 3.

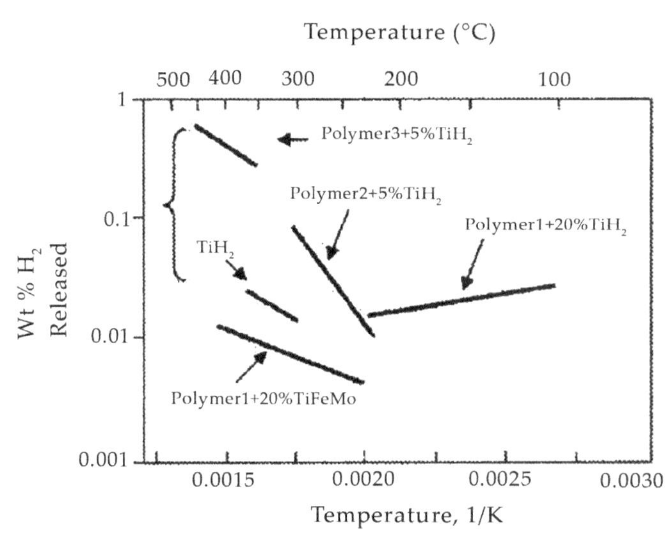

Figure. 3 Hydrogen released from various PDMH.

The exact role that a polymer plays in modifying the hydrogen storage capacity of a metal hydride is unknown at this time. It is plausible that a polymer structure could interact with molecular hydrogen. Given that each polymer contains a high concentration of hydrogen covalently bonded to the polymer structure, it is conceivable that hydrogen species could interact on a molecular level. For example, depending on the charge of hydrogen species, a combination of covalent, ionic, metallic, H-bonding and Van der Waals interactions could exist within and polymer chains and/or with the metal hydride. Free radicals present within the polymer structure could further participate in hydrogen movement through the polymer network. Although the exact location of the free radical sites has yet to be determined, they can be represented as dangling bonds. As a hydrogen molecule diffuses near a dangling bond, it can dissociate into two hydrogen atoms, which can further interact with the polymer free radical to form a covalent bond, or diffuse into the metal hydride lattice.

To date, there is little experimental evidence to support the notation that polymers alone provide significant hydrogen storage capacity, Fig. 3. Investigations continue to examine the inter-relationship between the dispersed metal hydride and the polymer support phase, and the resulting hydrogen storage behavior of the composite.

Carbon Clusters

This is a new approach in the storage of hydrogen [29-34]. The phase diagram of carbon, showed at Fig. 4, illustrates phase transitions between known forms of carbon (graphite and diamond) and it is of a great historical importance, related to the first production of diamond from graphite by synthesis in 1960. Among stable forms of carbon, hexagonal graphite and cubic diamond, hexagonal diamond, hexagonal carbines and liquid carbon forms exist at higher temperatures and pressures. Recent attention was paid to small carbon clusters, fullerene (ball shaped, consists of 12 pentagon and 20 hexagon rings with 60 atoms in total, Fig. Sa) and nanotubes (cylinders of hexagonal graphite rings that from both ends, capped with half fullerene at each end, Fig. 5b; S. Lijima 1991) of very stable structures [33]. Additional system energy decreases with creation of these· structures caused by the presence of side graphite atoms with an unutilized binding potential. These forms of carbon have many unique characteristics, great mechanical strength, high electric conductivity, great specific surface, great ability for gas absorption and unique capillary properties. They have a complicated electronic structure, and depending on their chirality, they may have metal or semiconductor properties.

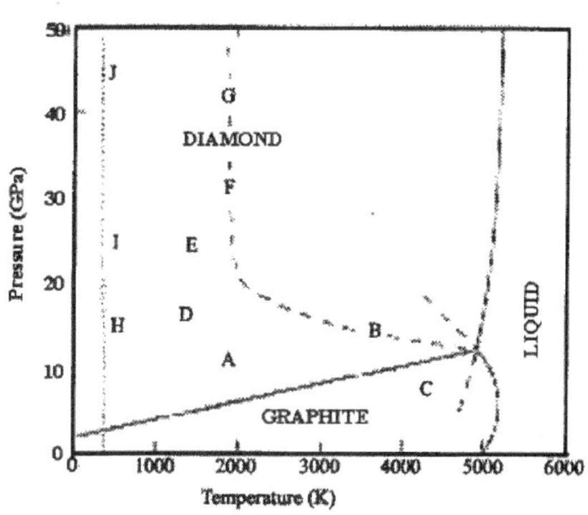

Figure. 4 Phase transitions of various forms of carbon; A: commercial catalytic production of diamond from graphite by synthesis, B: rapid synthesis of diamond from graphite in a solid phase, C: rapid transformation from diamond to graphite, D: synthesis of hexagonal diamond from hexagonal graphite, E: synthesis of hexagonal diamond by compression of graphite, F: synthesis of cubic diamond by compression of graphite, B, F, G: synthesis of cubic diamond from graphite or hexagonal diamond, H, I, J: compression of graphite at high pressure to have diamond like properties.

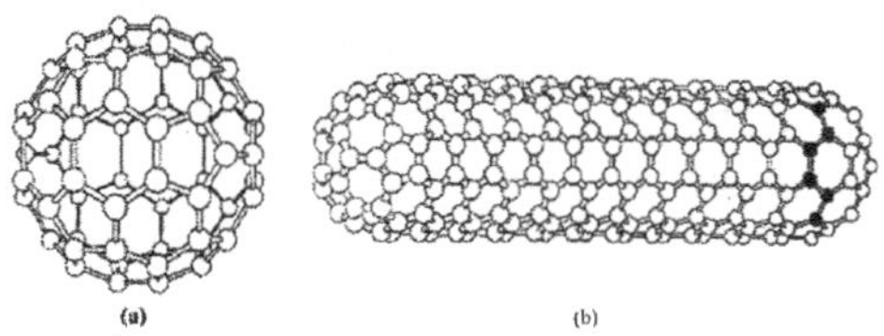

Figure. 5 Structure of stable forms of carbon clusters: (a) fullerene C 60; (b) carbon nanotube.

Carbon nanotubes

Carbon nanotubes (visible under a microscope) are small light cylinders of a graphite structure, a few molecules in diameter (usually 2 nm) that store hydrogen in microscopic pores on the tubes and inside the structure, in a similar way to hydrides. According to literature, carbon nanotubes can store 4.2 to 65 mass% of hydrogen, depending on the purity and presence of a small quantity of some metal components [29-34]. On the other side, for car industry application, it is enough to be thus stored 6.5 % of hydrogen. This big quantity of hydrogen, thus safely stored, presents the main advantage of this approach and at the same time, the reason why, the development of these materials of a great capacity for hydrogen or some hydrocarbon storage is the latest research topic. The most recent research, in this line of work, is pointed to the production of centered clusters of carbon nanotubes, called carbon nanofibers (usually 5-100 nm in diameter). There are two kinds of nanotubes available today: single atom wall nanotubes (SWNTs) and many atom wall nanotubes (MWNTs). Connected to SWNTs, there is distribution of tube diameter and geometry. Three basic conformations are showed in Fig. 6 (armchair, zigzag and chiral single wall carbon nanotubes). Metal properties are provided by first of these three and with 113 of the other two conformations.

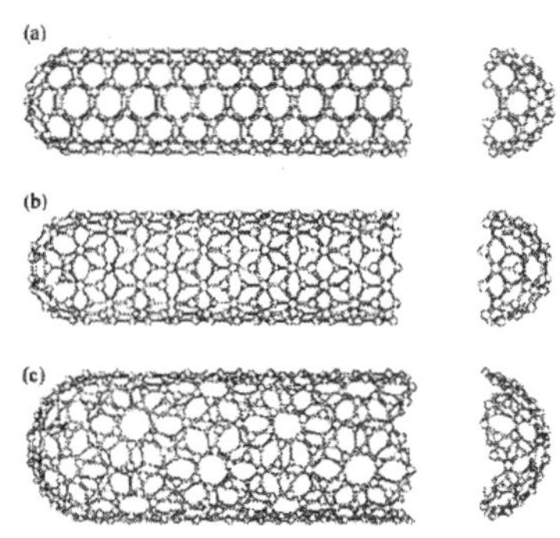

Figure. 6 Schematic models for (a) armchair, (b) zigzag and (c) chiral single-wall carbon nanotubes.

Three methods are used for the preparation of carbon nanotubes. The first method includes application of an arc light on a carbon electrode impregnated with metal in order to transit into a gaseous state. The second method uses a laser to transit the carbon target treated with metal (Ni, Co, Fe) into vapor.

The third method is known as catalytic chemical vapor deposition (CVD). It is the most suitable method, because it doesn't require a high temperature. Heated metal decomposes hydrocarbon gas (methane, ethylene, acetylene) to carbon and hydrogen. Hydrogen extracts in a vapor form, while carbon draws out in nanotubes.

Graphite nanofibers

Graphite nanofibers (GNF) are new materials in this area with a different morphology that is produced from the dissociation of carbon-containing gases over selected metal surfaces. The solid consists of very small graphite platelets, 3-50 nm in width, which are stacked in a perfectly arranged conformation. These materials are capable of sorbing and retaining in excess 20 liters of hydrogen per gram of carbon when the nanofibers are exposed to gas at pressures of 11.35 MPa at 298 K, a value that is over an order of magnitude higher than the one found with conventional hydrogen storage systems. It is probable that the presence of delocalized 7t-electrons on the graphite layers is a major contributory factor to the enhancement of the hydrogen absorption observed in the presented system.

Application of carbon clusters for hydrogen storage

Pores, the size of a· molecule, can absorb big amounts of gas thanks to their great attracting potential (presence of three electrons) [31]. Hence, 6.2 liters of hydrogen per gram of carbon can be absorbed, covering the surface with a single layer of hydrogen molecules.

Considerably bigger quantities of hydrogen can be stored, condensing more hydrogen molecule layers between graphite layers of nanofibers. According to measurements, the distance between graphite layers is 0.34 μm, while the hydrogen molecule diameter is 0.26 μm. Hydrogen loses its energy (rotational), being absorbed on carbon (heavy interaction with electrons from a graphite structure), which leads to a decrease of its effective diameter to 0.064 μm. In this way, 5 hydrogen layers can be packed between graphite layers. Nanofibres are rinsed out of catalytic impurities with acid for introduction of hydrogen into them and then heated to 900°C in vacuum to remove absorbed gases. Afterwards, hydrogen is introduced at a pressure of 11.35 atm during 4 to 24 hours. A comparison of results of different hydrogen storage technologies is shown in Fig. 6 with signification reference to applications in cars with fuel cells.

Recent tendencies in this area introduce a modification of graphite nanotubes or nanofibres, replacing carbon atoms with other atoms (boron, nitrogen, alkaline metal), which can change their electric properties. In this way, it is possible to achieve densities of absorbed hydrogen of up to 67 mass%, as shown in Tab. I

Table. I Gravimetric density, temperature and pressure of various forms of stored hydrogen [3-24].

Material	H_2 (wt.%)	T (K)	P (MPa)
SWNT (low purity)	5-10	133	0.040
SWNT (high purity)	8.25	80	7.18
SWNT (50 % purity)	4.2	300	10-12
GNFs (tubular)	11.26	298	11.35
GNFs (herringbone)	67.55	298	11.35
GNFs (platelet)	53.68	298	11.35
GNFs (herringbone)	0.4	298~773	0.101
Li-GNFs	20.0	473~673	0.101
K-GNFs	14.0	< 313	0.101
Graphite	4.52	298	11.35
Li-graphite	14.0	473~673	0.101
K-graphite	5.0	< 313	0.101
Graphite + 0.01 % Pd	44	436	0.101
Charcoal + 0.005 % Pd	31	430	0.101
Mg + 1% Pd	7.66	552	0.101
Co + 0.03 % Pd	5,85	463	0.180
Ti	3.98	916	0,101
Mg_2Ni	3.60	528	0.101
$CaNi_5$	1.87	316	0.101
FeTi	1.89	>263	0.101
FeTi	15	>263	11.35
$Ti(Fe_{0.2}Mn_{0.1})$	1.90	280	0.101
$Ti(Fe_{0.8}Mn_{0.2})$	1.30	346	0.101
$LaNi_5$	1.49	295	0.101
$LaNi_{4.7}Al_{0.6}$	1.44	318	0.101
$MnNi_{4.5}Al_{0.5}$	3.33	298	0.101
$LaNi_5H_6$ + 0.03 % Pd	8.8	450	0.101
$LaNi_5H_6$ + 0.01 % Pd	4	420	0.101
PDMH + 5 % Ti	0.36	680	0.101
Ta + 0.004 % Pd	9	480	0.101
Crioadsorption	~5	~77	2.0
Isooctane/benzene	17.3	>233	0.1

(SWNTs =Single wall carbon nanotubes; GNFs =Graphite nanofibers)

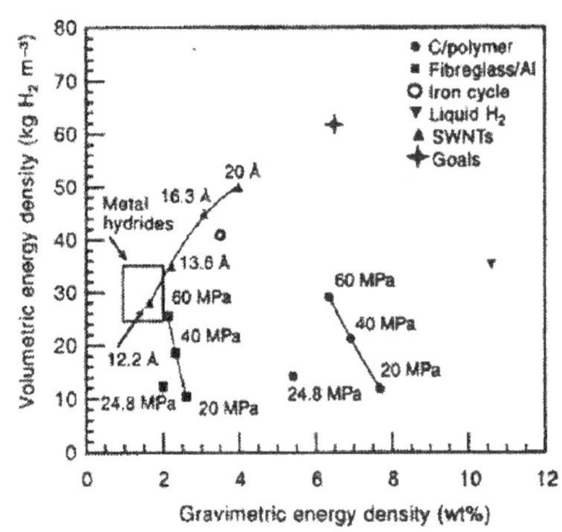

Figure. 7 Installed density of energy for different technologies of hydrogen storage.

REFERENCES

1. J. Nitsch, W. Peshka, W. Schnurnberger, M. Fischer and H. Eirchert in "Hydrogen as an Energy Carrier", eds. C. Winter and J. Nitsch, Springer-Verlag, Berlin, 1988.
2. S. Hynek, W. Fuller and Bentley, J. Int. Hydrogen Energy, 22 (1997) 601.
3. M.V. Susie J. Int. Hydrogen Energy, 22 (1997) 585.
4. C. G. Carter and L. F. Carter, in "Metal-Hydrogen Systems", ed. N.T. Veziroglu, Pergamon, Oxford, 1981.
5. H. Buchner, P. Pelloux-Gervais, M. Muller, F. Grafwallner and P. Luger, in "Hydrogen and Other Alternative Fuels for Air and Ground Transportation", ed. W.H. Pohl, Wiley, Chichester, UG, 1995.
6. G.J. Hansel, W.G. Matern, N.R. Miller, J. Int. Hydrogen Energy, 18 (1993) 783.
7. B.J. Taylor, J. Int. Hydrogen Energy, 11 (1985) 5.
8. J.F. Edsekuty, F.W. Stewart, Safety in the handling of Cryogenic Fluids, The International Cryogenics Monograph Series, Plenum Press, New York, 1996.
9. W. Peschika, Liquid Hydrogen: Fuel of the Future, Springer-Verlag, Wien, New York 1992.
10. R. Chahine, T.K. Bose, Proc. 11th World Hydrogen Energy Conf., Stuttgart, FRG, 1996.
11. M. Epstein, 3'd JAERI Symp. Orai, 1996, Proc. JAERI Conf. 96-010, Japan Atomic Energy Research Institute, 1996, p. 228.
12. E.F. Lynch, Int. Symp., Tokyo, NEDO, 1995, p. 99.
13. R. Schulz, S. Boily, Proc. 7th Canadian Hydrogen Workshop, Quebec City, 1995.
14. Y. Osumi, Int. Sympo. Tokyo, NEDO, 1995, p. 131.
15. G.S. Podkolzin, M.R. Watwe, Q. Yan, J.J. Pablo, A.J. Dumestic, J. Phys. Chern. B, 105 (36) (2001) 8550.
16. C.J. Groen, J. Perez-Ramirez, A.A.L. Peffer, Studies in Surfaces Science and Catalysis, 135 (2001) 2862.
17. L. Liwu, J. Weishen, J. Jifei, X. Zhusheng, T. Zhang, F. Yining, K. Yuan, Sh. Jianyi, Science in China, Series B: Chemistry, 42 (6) (1999) 571
18. S.J. Bradley, W. Busser, Catalysis Letters, 63 (3,4) (1999) 127.
19. Y. Sakka, H. Okuyama, T. Uchikoshi, S. Ohno, Transaction of the Materials Research Society of Japan, 25 (1) (2000) 43.
20. D. Jing-Fa, Current Topics in Catalysis, 2 (1999) 1.
21. C.A. Dillon, M.K. Jones, A.T. Bekkedahl, C.H. Kiang, D.S. Bethune and J.M. Heben, Nature, 386 (1997) 377.
22. D.M. Minh, M.V. Susie, in "Hydrides as hydrogen accumulators", Serbian Academy of Science and Arts, Monographs, Vol. DCXLIV, No.2, Ed. S. Ribnikar, 1999. (in Serbian)
23. M.V. Susie, J. Serb. Chern. Soc., 62 (12) (1997) 1183.
24. M.V. Susie, J. Serb. Chern. Soc., 62 (8) (1997) 631.
25. A. Zaluska, L. Zaluski, J.O. Stro-Olsen, J. of Alloys and Compounds, 288 (1999) 217.
26. D.T. Jarvi, Hydrogen Storage in Polymer Dispersed Metal Hydrides, UTRC Proposal P.300.0024 (1999).
27. S.R. Wayde, Proceedings of the 2001 DOE Hydrogen Program Review, 2001.
28. A.R. Meinzer, U.S. Patent 5360461, Polymeric Storage Bed for Hydrogen.
29. Y. Ye, C.C. Ahn, C. Withman, B. Fultz, J. Liu, G. A. Rinzler, D. Colbert, A.K. Smith and R.E. Smalley, Applied Physics Letters, 74 (1999) 2307.
30. C. Liu, Y.Y. Fan, M. Liu, T.H. Cong, M.H. Cheng and S.M. Dresselhaus, Science, 286 (1999) 1127.
31. A. Chambers, C. Park, K.T.R. Baker and M.N. Rodrigez, J. Phys. Chern. B, 102 (1998) 4253.
32. P. Chen, X. Wu, J. Lin and L.K. Tan, Science, 285 (1999) 91.
33. S. Iijima, Nature, 354 (1991) 56.
34. J. Jagiello, J.T. Bandosz, K. Putyera, A.J. Schwarz, J. Chern. Soc. Faraday Trans., 91 (1995) 2929.

CHAPTER 2

RECENT ADVANCES ON THE EFFECTS OF MICROSTRUCTURAL REFINEMENT AND NANO-CATALYTIC ADDITIVES ON THE HYDROGEN STORAGE PROPERTIES OF METAL AND COMPLEX HYDRIDES

Robert A. Varin, Leszek Zbroniec, Marek Polanski and Jerzy Bystrzycki

Keywords: Solid state hydrogen storage, ball milling, microstructural refinement, particle/grain size, nano-catalytic additives, simple metal and complex hydrides

INTRODUCTION

In principle, solid state hydrogen storage in hydrides as compared to gaseous and liquid storage has many benefits which outweigh its drawbacks. However, the hydrogen storage properties of metal and complex hydrides must be greatly improved before these materials could be applied commercially for supplying Proton Exchange Membrane (PEM) fuel cells particularly in automotive sector. It has been recognized that a refinement of the microstructure (particles and crystallites/grains) of hydrides, most preferably to the nanometric level, accompanied by nanosized catalytic additives can improve greatly hydrogen desorption/absorption rates. This paper is a comprehensive review of the recent advances on the effects of microstructural refinement brought about by ball milling and nanosized catalytic additives on the hydrogen storage behavior of simple MgH_2 metal hydride and selected complex metal hydrides ($LiAlH_4$). The results obtained in the past several years in the allied laboratories at the University of Waterloo (Canada) and Military University of Technology (Warsaw, Poland) are critically reviewed. Most of these results are accessible in the open literature although they are scattered in various journals and it is beneficial to collect them in one place for a quick access.

Citation: Robert A. Varin, Leszek Zbroniec, Marek Polanski and Jerzy Bystrzycki, Recent Advances on the Effects of Microstructural Refinement and Nano-Catalytic Additives on the Hydrogen Storage Properties of Metal and Complex Hydrides, doi:10.3390/en4010001.

THE EFFECTS OF PARTICLE AND GRAIN/ CRYSTALLITE SIZE

MgH_2

One of the most thoroughly investigated simple metal hydrides is magnesium hydride (dihydride), MgH_2. In the 1990's it was investigated as a potential hydrogen storage material for supplying H_2 to a Proton Exchange Membrane (PEM) fuel cell (FC) stack. However, one must realize that the general service conditions of an advanced PEM FC stack, which is a primary candidate for transportation (e.g., automotive) applications, require hydrogen pressure of about 1.1–1.8 bar while the waste-heat generated temperature usually does not exceed 60–70 °C [1]. In view of these practical requirements, a serious thermodynamic constraint for magnesium hydride (MgH_2) is its high enthalpy of formation/decomposition which is within the range of ~71–75 kJ/mol [2] and does not allow dehydrogenation below about 280 °C at 0.1 MPa H_2 (atmospheric pressure). No catalytic additive is able to alter these unfavorable thermodynamics [2]. Taking this into account it has been commonly realized in the last decade that a high temperature MgH_2 hydride is not a suitable hydrogen storage medium for supplying a PEM fuel cell stack although it may be still considered for hydrogen supply to high temperature fuel cells, for instance, Solid Oxide Fuel Cell (SOFC). However, as first reported by Bogdanović et al. [3,4] the disadvantageous thermodynamics of the MgH_2-Mg system for a solid state hydrogen storage can be quite advantageous for heat storage because approximately 0.9 kWh of heat per kg Mg can be stored at a hydrogen-pressure-dependent temperature level between 300 and around 500 °C applicable for solar power generation via Stirling engines or storage of industrial heat in the above temperature ranges. From this viewpoint MgH_2 remains still relatively attractive hydride for this kind of application.

Mechanical (ball) milling technique is the most popular and widespread method of refining the microstructure (particles and grains) with the hope of improving hydrogen storage properties of various hydrides. The group led by Prof. J.O. Ström-Olsen at the McGill University in Montréal, Canada, in which a prominent role was played by L. Zaluski and A. Zaluska [5,6,7], was the first to apply ball milling for the synthesis of MgH_2 from pure Mg with the resulting microstructural refinement. The first attempts to refine microstructure of a commercial MgH_2 powder by ball milling were undertaken by the group of researchers in the Institut de recherche sur l'hydrogène Université du Québec à Trois-Rivières, Quebec, Canada [8,9,10].

However, those early works overemphasized the effect of nanograins (also commonly referred to as "crystallites") formed within the heavily milled powder particles on the hydrogen absorption/desorption properties and somehow marginalized the role of the reduction of particle size. At least, in contrast to the prior work in the Ström-Olsen's group, the authors from the Institut de recherche sur l›hydrogène also pointed out to the importance of the particle size reduction associated with an increase of the specific surface area (SSA). In the context of MgH_2, Dornheim et al. [11] in a comprehensive review paper pointed out that for the optimization of hydrogen absorption rates for MgH_2 it is important to reduce the diffusion path lengths for hydrogen atoms. This can be achieved by both small grain sizes and small particle sizes. On the other hand, hydrogen desorption rates can be enhanced by creating large surface areas as it occurs for small particle sizes and adding catalytic additives as well. However, for ball milled MgH_2 they somehow overlooked the effect exerted by a metastable γ-MgH_2 phase created during ball milling which will be discussed later.

In our laboratory we have investigated the changes in particle and grain/crystallite size (embedded within individual powder particles) as a function of ball milling time and clearly shown that these two events occur simultaneously as can be seen in Figure 1. It is important to emphasize that the particle size reduction occurs within a very similar time frame as does the reduction of grain size.

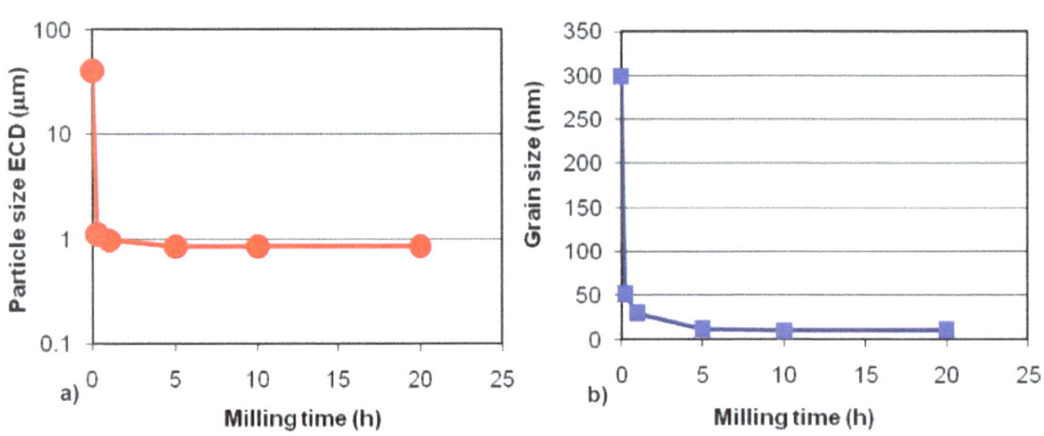

Figure 1. Simultaneous reduction of (a) the particle size and (b) the grain size as a function of ball milling time of MgH_2. Adapted from [2].

This behavior is of a very general nature and practically, does not depend much on the mode of milling and the type of MgH_2 commercial powder used for the experiments [2]. Almost identical time frame for grain and particle size variations as a function of milling makes it quite difficult to identify unambiguously which factor is, indeed, governing hydrogen storage characteristics. This difficulty has led to a common belief that the grain size is mostly responsible for the observed enhancement of hydrogen storage properties which is not necessarily the case as will be discussed later.

Figure 2a shows the DSC peak temperatures for low-temperature (LT) and high-temperature (HT) peak doublets/ shoulders (see Figure 2b for designation of HT and LT) as a function of powder particle size (ECD-Equivalent Circle Diameter) for commercial MgH_2 powder ball milled for various times [12]. The peak temperature decreases slowly and then, after achieving some critical (threshold) ECD range, decreases more rapidly, with decreasing ECD. The total observed drop of LT and HT peak temperatures from the initial peak temperature value for as-received MgH_2 is equal to about 40–50 °C. The critical ECD at which sudden drop of DSC peak temperature commences in Figure 2a is around 2,000 nm (2 μm). Figure 2a also shows the corresponding grain sizes. It seems that there is no clear dependence of DSC peak temperatures on grain size. It must also be pointed out that the microstructure of ball milled MgH_2 always contains a non-equilibriumγ-MgH_2 phase which is a high pressure polymorph of MgH_2 [13] coexisting with equilibrium β-MgH_2 [2,12].Figure 2b shows DSC curves of the MgH_2 sample ball milled for 100h (ECD~621 ± 426 nm; grain size ~3 nm) and the same powder sample after three cycles of desorption/absorption at 350 °C (ECD ~ 625 ± 405 nm; grain size ~ 56 nm) as compared to the DSC curve of as-received commercial MgH_2 [12].

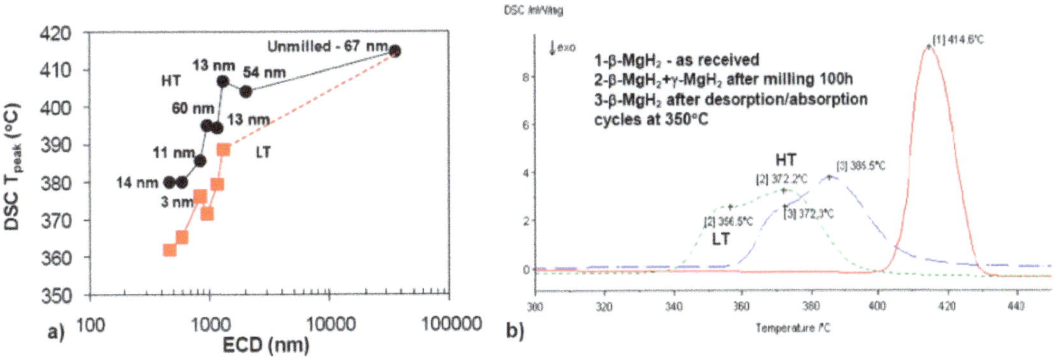

Figure 2. (a) Peak temperature vs. particle size (ECD) for low-temperature (LT) and high-temperature (HT) peak doublets/shoulders for MgH_2 ball milled for various times. A number beside each data point indicates the grain (crystallite) size of the β-MgH_2 phase. (b) DSC curves of the MgH_2 sample ball milled for 100 h (average ECD 621 ± 426 nm; average grain size ~3 nm) and the same after three cycles of desorption/absorption at 350 °C (average ECD 625 ± 405 nm; average grain size ~56 nm) as compared to the DSC curve of as-received commercial MgH_2 (average particle size 36 ± 16.4 μm). DSC heating rate 4 °C/min. ECD-Equivalent Circle Diameter. Adapted from [12].

In Figure 2b the DSC curve for the thermally cycled MgH_2 is shifted back to higher desorption temperatures range as compared to the as-milled sample but its peak maximum temperatures are still lower than those of as-received MgH_2. The XRD pattern of the thermally cycled sample showed no trace of γ-MgH_2. As argued in [12] if the grain size of β-MgH_2 were the principal factor responsible for the reduction of hydrogen desorption temperature then the DSC curve of the thermally cycled sample should have been shifted very close to the original position corresponding to the DSC curve of the as-received MgH_2 powder because the grain size of β-MgH_2 is almost identical in both materials (~ 56 vs. 67 nm, respectively). However, one must keep in mind that the average particle size of the cycled sample (~ 625 nm) is nearly identical to that for ball milled MgH_2 (~ 621 nm) and the DSC curve of the cycled MgH_2 sample is shifted to higher temperatures but not exactly to the position of the peak temperature range of as received MgH_2 (Figure 2b) which has a particle size of ~36 μm [12]. This behavior points towards some mild effect of reduced particle size on the temperature of DSC desorption peaks. However, it seems that the principal factor responsible for lowering the hydrogen desorption peak temperature of ball milled MgH_2 is the presence of γ-MgH_2. It seems that the γ-MgH_2 hydride exhibits lower DSC desorption temperature than β-MgH_2 [14]. Unfortunately, thermodynamic data are not available for γ-MgH_2, so it is unknown if it may have a lower enthalpy of formation than its β-MgH_2 counterpart. The second factor lowering the hydrogen desorption peak temperature seems to be a reduced particle size. In general, as pointed out by Selvam et al. [15] in the dehydriding stage, successively formed fresh Mg at the surface layer of coarse particles may function as the diffusion barrier to the hydrogen escaping from MgH_2 while in small particles the Mg phase may form simultaneously all through the particle and the entire process of desorption is then governed by fast hydrogen diffusion rather than slow Mg–MgH_2 boundary movement. In addition, smaller particles may provide larger total surface area for recombination of hydrogen molecules [16]. It must also be pointed out that a nearly identical dependence of the DSC hydrogen desorption peak maximum temperature on the particle size as in Figure 2a is also observed for MgH_2 which was synthesized by a reactive ball milling under hydrogen [17].

As shown in Figure 3, a beneficial effect of the refinement of particle size and formation of γ-MgH_2 is also reflected in the accelerated rate of hydrogen desorption during isothermal dehydriding for powders with higher specific surface area (SSA) which is roughly inversely proportional to the particle size. The ball milled powders having SSA = 13.7 and 43.9 m^2/g exhibit the highest rate of hydrogen desorption. It also seems that the rate of desorption is independent of the type of ball milled used for processing. It is interesting that the dependence of the rate of hydrogen absorption on SSA (300 °C/10 bar H_2) is not that pronounced as the one observed for desorption [18].

Unfortunately, ball milling and the resulting refinement of microstructure of MgH_2 also bring about a detrimental reduction in the hydrogen storage capacity of the milled powders as shown in Figure 4a. The ball milled MgH_2 powder does not completely desorb even at high temperatures of 375 and 400 °C showing ~6.2–7.0 wt.% desorbed hydrogen which is less than the purity-corrected 7.3 wt.% (at 95% purity) hydrogen capacity for this powder. The full capacity is in practical terms obtained without difficulty by hydrogen desorption from unmilled non-activated commercial powder at the temperature range 350–420 °C [2]. The microstructure of desorbed powders was investigated by XRD as shown in Figure 4b. After each desorption the diffraction peaks of retained MgH_2 are discernible on the XRD pattern. Apparently, for whatever reason, the decomposition process of ball milled MgH_2 hydride does not proceed to completion even at the temperature as high as 400 °C.

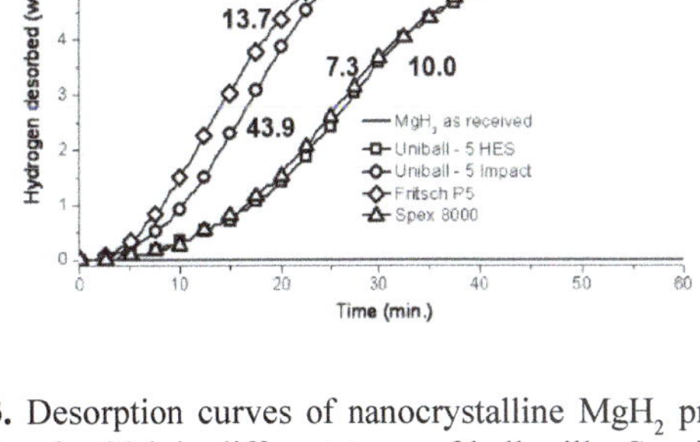

Figure 3. Desorption curves of nanocrystalline MgH_2 prepared by ball milling for 20 h in different types of ball mills. Specific surface area (SSA) of each powder is shown in m^2/g beside each desorption curve. Adapted from [18].

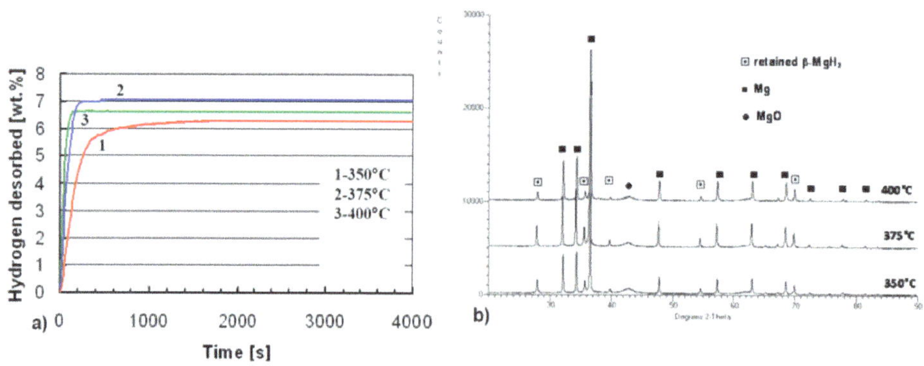

Figure 4. (a) Hydrogen desorption curves under 0.1 MPa H_2 at three different temperatures for MgH_2 milled continuously in argon for 100 h. (b) XRD patterns of the same powders after desorption. Adapted from [2].

The exact mechanism (s) responsible for MgH_2 stabilization during desorption is/are not available at the present time. Nevertheless, there are a few factors that could be relevant to the case. First, the γ-MgH_2 phase is always formed during milling due to the transformation of β-MgH_2 [2]. However, during subsequent high temperature desorption or cycling this orthorhombic hydride phase quickly disappears (e.g., Figure 4b). According to Gennari et al. [14] the initial decomposition of the γ phase stimulates β-MgH_2 decomposition by creating a volume contraction which, in turn, generates stresses acting on β-MgH_2. Conversely, one may argue that if the γ phase decomposes too quickly then the β phase may become too stable and small amounts of it may persist even up to high desorption temperatures. Second, during desorption of a milled MgH_2powder and decomposition of β-MgH_2 there always occurs a simultaneous growth of nanograins of β-MgH_2. For example, for the powder in Figure 4a the nanograin of β-MgH_2 after ball milling for 100 h was ~14 nm in size which increased to ~60–80 nm for the retained β-MgH_2 present after desorption in Figure 4b [2]. Since particle size is not changed during desorption one may hypothesize that the growth of nanograins within β-MgH_2 particles might somehow decelerate the decomposition of β-MgH_2. It should also be pointed out that desorption under low vacuum rather than under 0.1 MPa H_2 pressure removes any stabilization effect of retained MgH_2 and no diffraction peaks of β-MgH_2 are ever observed [19].

Finally, to put the effects of the microstructural refinement by ball milling on the hydrogen desorption properties of MgH_2 into perspective it should be mentioned that the apparent activation energy of desorption can be affected more profoundly by the oxide/hydroxide layer at the surface of powder particles than by microstructural refinement. For example, ball milling under reducing hydrogen atmosphere lowers the apparent activation energy from ~217 kJ/mol for the as received heavily hydrolyzed MgH_2 to ~140 kJ/mol after 25 h of ball milling [19]. In general, there is no dramatic difference between the apparent activation energies for dehydrogenation for unmilled non-activated/activated MgH_2 which are within the range of 120–168 kJ/mol as opposed to the apparent activation energy values for ball milled powders which are within the range of 140–160 kJ/mol for both commercial MgH_2 and the one synthesized from Mg by reactive milling in H_2 [2].

Complex Hydrides

In general, ball milling of complex metal/non metal hydrides such as, for example, $LiAlH_4$(lithium alanate) and $NaAlH_4$ (sodium alanate), is less effective for the microstructural refinement than ball milling of a simple metal hydride MgH_2. For example, high energy impact milling of $LiAlH_4$ for 15 min reduces its average particle size from ~10 to ~3 µm with the average corresponding grain size of ~40–80 nm [20]. For comparison, after about 15 min of high energy ball milling the average particle size of MgH_2 is reduced from the initial ~40 µm to ~1 µm (Figure 1a) and the grain size to ~30–50 nm [2]. Thermal behavior of complex hydrides during DSC tests is not measurably affected by ball milling and resulting microstructural refinement [2,20,21]. However, their dehydriding rates are affected to a modest extent as shown in Figure 5. For long desorption times unmilled $LiAlH_4$ decomposes in only Stage I which is attributed to the decomposition of $LiAlH_4$ into Li_3AlH_6, Al and H_2 as opposed to ball milled alanate which shows clearly two stages of decomposition, Stage I and II. Stage II is attributed to the decomposition of Li_3AlH_6 into LiH, Al and H_2 [2,20,21]. The apparent activation energy estimated from the Johnson–Mehl–Avrami–Kolmogorov (JMAK) and Arrhenius equations [20] for Stage I and II for unmilled $LiAlH_4$ is equal to ~111 kJ/mol (with an excellent coefficient of fit R^2 = 0.998) and ~100 kJ/mol (R^2 = 0.966), respectively. For ball milled $LiAlH_4$ the apparent activation energy for Stage I and II is slightly lower at ~92.6 kJ/mol (R^2 = 0.992) and ~92 kJ/mol (R^2 = 0.979), respectively [20].

THE EFFECTS OF CATALYTIC NANO-ADDITIVES

Elemental Nanometric Metal Catalysts

MgH_2

Nanometric-size transition metals are excellent catalytic additives which increase profoundly the rate of hydrogen desorption/absorption of simple metal hydrides such as MgH_2. One of the best nanometric metallic catalysts we have tested so far is nanometric nickel (n-Ni). Figure 6 shows scanning electron micrographs of two morphologies, spherical and filamentary, of n-Ni with slightly different SSA values produced as experimental batches by Vale Inco Ltd. Canada [22].Figure 7 shows the effect of 5 wt.% of micrometric Ni (m-Ni) and nanometric Ni (n-Ni) additives on DSC hydrogen desorption traces for MgH_2 ball milled for 15 min. Endothermic desorption peak for MgH_2 + 5 wt.% m-Ni is only modestly shifted to lower temperatures exhibiting the onset temperature at ~350 °C and the peak maximum temperature at 392.8 °C as compared to a pure MgH_2 milled for 15 min with the onset at ~380 °C and the maximum at 418.2 °C, respectively. In contrast, the hydrogen desorption peak for the MgH_2 + 5wt.% n-Ni powder is substantially shifted to the lower temperatures with the onset temperature at ~170 °C and the peak maximum temperature at 243.1 °C. However, when the MgH_2 + 5wt.% m-Ni powder was milled for 20 h its desorption properties were improved such that the onset was at ~275 °C and peak maximum temperature at 302.3 °C [2]. Apparently, longer milling time reduces desorption temperature most likely due to a better dispersion of coarser m-Ni particles within the MgH_2 matrix. Nevertheless, the hydrogen desorption peak maximum temperature for the 20 h milled MgH_2 + 5wt.% m-Ni nanocomposite is still higher than that for the 15 min milled MgH_2 + 5wt.% n-Ni nanocomposite. This clearly shows that nanometric Ni (n-Ni) exhibits superb catalytic properties.

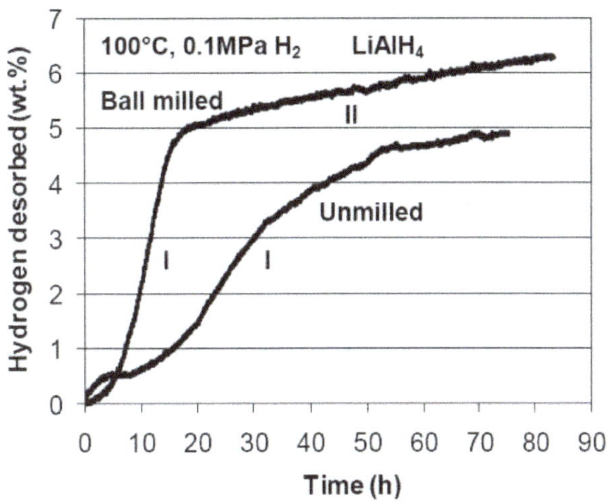

Figure 5. Comparison of volumetric desorption curves at 100 °C under 0.1 MPa H_2 pressure for unmilled (as-received) $LiAlH_4$ and $LiAlH_4$ ball milled under high impact energy. Adapted from [20].

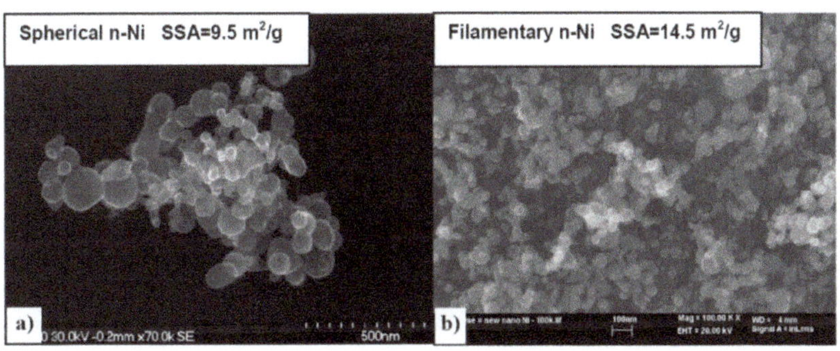

Figure 6. Scanning electron micrographs of the morphology of Vale Inco Ltd. nanometric Ni (n-Ni) with (a) SSA = 9.5 m^2/g and (b) SSA = 14.5 m^2/g used as a catalytic additive. Adapted from [21].

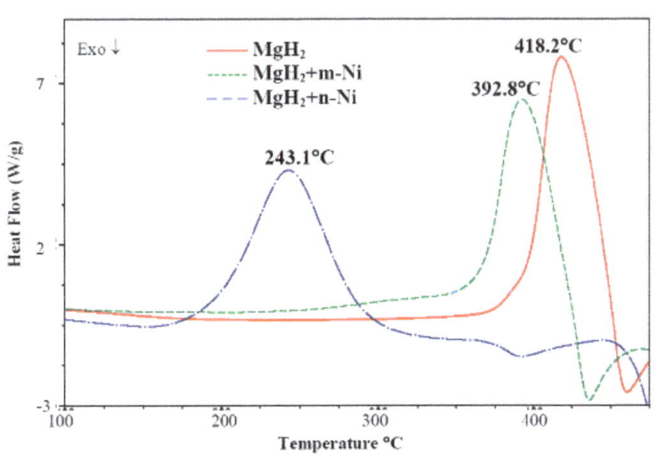

Figure 7. DSC traces of MgH_2 doped with 5 wt.% Vale Inco Ltd. micrometric Ni (m-Ni; SSA = 0.7 m^2/g) and nanometric Ni (n-Ni; SSA = 14.5 m^2/g) subjected to 15 min of ball milling under 700 kPa hydrogen pressure. Heating rate 10 °C/min. Adapted from [2].

Figure 8 shows the improvement of hydrogen desorption rate of MgH_2 ball milled with the addition of 5 wt.% n-Ni. As can be seen in Figure 8a the nanocomposite does not desorb at 200 and 250 °C under 0.1 MPa H_2 due to the high enthalpy change of dehydrogenation for MgH_2 as mentioned earlier in the text. However, at around 275 °C it is able to desorb about 4.5 wt.% H_2. According to the Van't Hoff equation [2] this temperature range is approximately the lowest at which MgH_2 is able to desorb at the atmospheric pressure (0.1 MPa) of hydrogen due to its serious thermodynamic constraint which is its high enthalpy of formation/decomposition being within the range of 71–75 kJ/mol [2]. The desorption kinetics in Figure 8a become faster at higher temperatures although the maximum amount of H_2 desorbed approaches ~6.5 wt.% which is about 0.6 wt.% less than the purity and the 5 wt.% n-Ni additive content corrected theoretical H_2 content [23]. As discussed earlier, MgH_2 is somehow stabilized after ball milling which prevents achieving a full desorption under 0.1 MPa H_2 even in the presence of such a strong catalyst as n-Ni as shown in this work. Figure 8b shows the Arrhenius plot for the estimation of the apparent activation energy for desorption which for the temperature range 275–350 °C is equal to ~94 kJ/mol (note an excellent coefficient of fit $R^2 = 0.982$). This value is much lower than the activation energy for desorption of ball milled MgH_2 which falls within the 140–160 kJ/mol range as mentioned earlier. It is clear that the n-Ni catalytic additive improves profoundly the hydrogen desorption kinetics which is reflected in the reduction of the apparent activation energy for desorption. For the sake of clarity, it must be pointed out that as shown in Figure 9 the reduction of the SSA of a catalytic metal does not necessarily bring about more improvement in the hydrogen desorption kinetics. Figure 9 shows the plot of the rate constant k in the Johnson-Mehl-Avrami-Kolmogorov (JMAK) equation (see Sec. 1.4.1.2 in [2]) as a function of SSA for two desorption temperatures 275 and 300 °C for a variety of n-Ni additives with largely varying SSA [2]. The k increases up to SSA ~10–15 m^2/g and then there is no further dependence with increasing SSA. It is also clear that there is no apparent effect of carbon (0.34–2.97 wt.%) and oxygen (0.05–11.7 wt.%) content in n-Ni on the hydrogen storage properties of MgH_2, neither there is any apparent effect of the n-Ni morphology (spherical or filamentary) on hydrogen storage properties.

Another elemental nanometric metal nano iron (Fe) has been found to improve greatly the hydrogen absorption into Mg. Figure 10a,b show STEM images at various contrast modes of a nanocomposite powder particle showing Fe dispersoids embedded within the Mg matrix. This unusual microstructure was produced via thermal decomposition of the dynamically synthesized Mg_2FeH_6 ternary hydride precursor [24, 25].

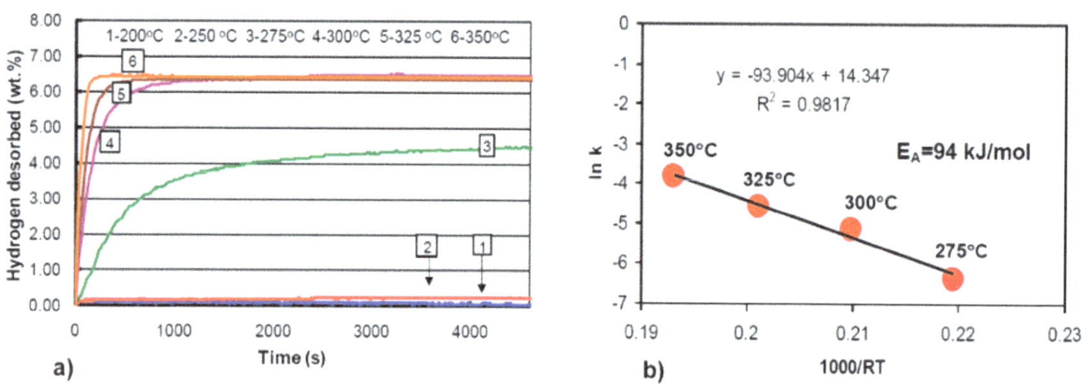

Figure 8. (a) Volumetric hydrogen desorption curves at various temperatures under initial hydrogen pressure of 0.1 MPa for the MgH_2 + 5 wt.% n-Ni nanocomposite directly after milling for 15 min (average ECD = 0.74 ± 0.46 μm). (b) The Arrhenius plot of the desorption rate for the estimation of the apparent activation energy, E_A. Adapted from [23].

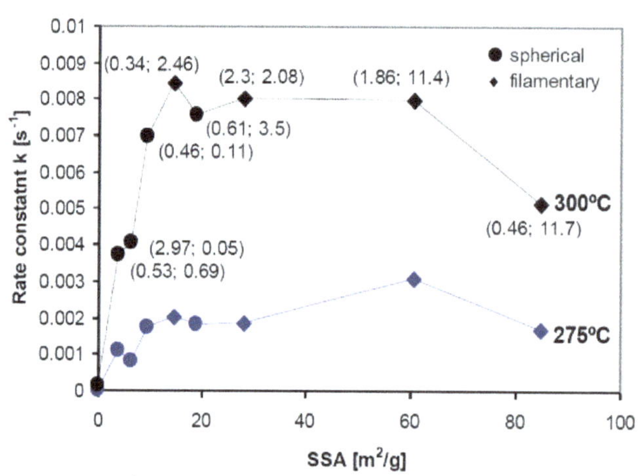

Figure 9. The rate constant, k, in the Johnson-Mehl-Avrami-Kolmogorov (JMAK) equation as a function of the SSA of the Inco n-Ni catalytic additive. The first and second number in parentheses is content of carbon and oxygen in wt.%, respectively. Adapted from [2].

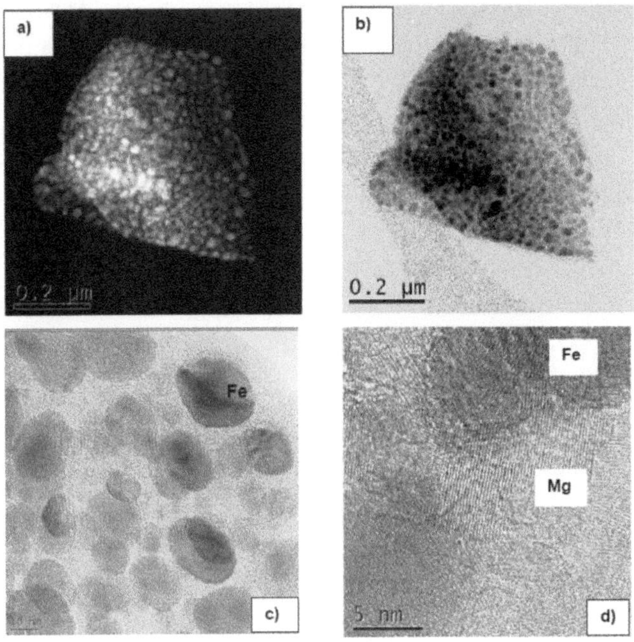

Figure 10. STEM images showing the microstructure of individual 2Mg-Fe powder particle obtained from the decomposition of Mg_2FeH_6. (a) Z-contrast (HAADF-High Angle Annular Dark Field detector) and (b) BF. (c,d) High resolution (HR) STEM bright field (BF) images showing the microstructure of 2Mg-Fe particle at lower and higher (lattice planes contrast) magnification. Adapted from [25].

Figure 10c at a higher magnification shows clearly nearly spherical nanometric Fe dispersoids with an average diameter of ~20–30 nm (notice 10 nm magnification marker) uniformly distributed in the Mg matrix. High resolution high magnification image in Figure 10d shows lattice planes of Mg matrix nanograins about ~5–10 nm in size in which the Fe dispersoids are embedded forming high misorientation angle phase boundaries. This Mg-Fe nanocomposite exhibits quite remarkable hydrogen absorption/desorption properties as shown in Figure 11. Figure 11a shows the total hydrogen absorption curve of the 2Mg-Fe (molar ratio) nanocomposite formed during decomposition of previously synthesized ternary Mg_2FeH_6 precursor [24]. At the temperature of barely 30 °C the sample kept under hydrogen pressure for 10 h shows a significant 1.1 mass % absorption of hydrogen. For the clarity, it is to be pointed out that for the molar ratio $2MgH_2$-Fe used in the synthesis [24] the theoretical maximum hydrogen quantity that could be absorbed is ~3.7 wt.%. It is most likely that prolonging the absorption time much beyond 10 h the 2Mg-Fe nanocomposite could absorb even more H_2 at 30 °C. Figure 11b shows a temperature programmed desorption (TPD) spectrum of the 2Mg-Fe sample hydrogenated for 10h in a. The onset of dehydrogenation occurs at a temperature as low as ~130 °C, with a peak maximum around 250 °C, which is similar to the results reported for MgH_2 catalyzed with Nb_2O_5 [26]. The TPD curve inFigure 11b is unusually shaped (asymmetrical). A TPD peak starts forming at a very low temperature around 130 °C then increases slowly developing a sort of a long "tail" and eventually finishes very rapidly just after reaching its maximum at ~250 °C. A typical shape for this type of peaks is rather symmetrical, as for example, reported in [2,26]. Varin et al. [27] pointed out that a narrow hydrogen TPD peak indicates the presence of small well-crystallized hydride particles. The asymmetry of the TPD peak in the present case might be related to the catalytic effect of Fe particles embedded into the Mg matrix and the grain size distribution for the Mg matrix which is skewed towards very small nanocrystalline grains. Taking this hypothesis into account and assuming that smaller MgH_2 nanograins can desorb hydrogen at lower temperatures due to shorter diffusion paths, we can conclude that most likely, there is a large population of very small nanograins of MgH_2 desorbing hydrogen much below 250 °C. Moreover, it seems that the population of relatively large MgH_2 nanograins is rather limited because of the sharp drop of the TPD peak at around 250 °C in Figure 11b.

Finally, it must be clearly pointed out that the observed effects of nanometric Ni and Fe catalysts on the hydrogen storage behavior of MgH_2 strongly suggest that all the improvement of hydrogen storage properties is owing to the kinetic effect of nanometric elemental metals acting as powerful catalysts rather than to a modification of thermodynamic properties (enthalpy) of the MgH_2 + nanometal catalyst system. As pointed out in [25] He and Zhao [28] reported that a 2.25 at.% V decorated Mg nanoblade (25 nm thick) array absorbed volumetrically at room temperature about 1.7 wt.% H_2 within 10 h under 1 MPa H_2. This result is very comparable to the result shown inFigure 11a in the present work. In addition, they have clearly shown that the absorption enthalpy change ΔH and entropy change ΔS of the Mg-V nanoblade array system is equal to −74.6 ± 6.8 kJ/mol and −131.8 ± 13.0 kJ/mol, respectively, which clearly points toward a purely kinetic effect.

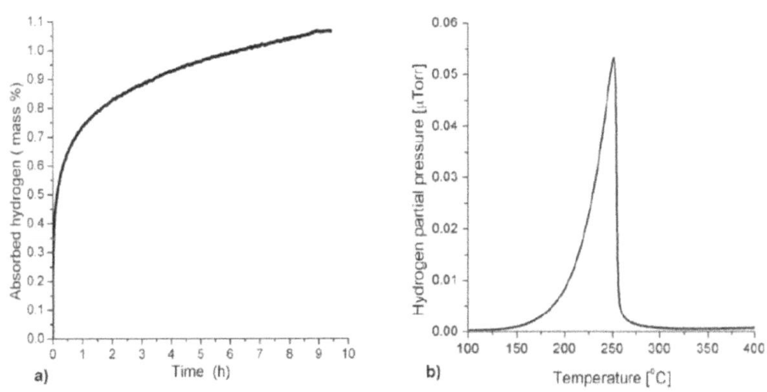

Figure 11. (a) Hydrogen absorption curve for the 2Mg-Fe elemental nanocomposite at 30 °C under 4 MPa of hydrogen pressure up to 10 h. (b) Temperature programmed desorption (TPD) spectrum of 2Mg + Fe nanocomposite sample hydrogenated at 30 °C under 4 MPa of hydrogen pressure up to 10 h (heating rate of 5 °C/min under flowing helium). Adapted from [25].

Complex Hydrides

The addition of barely 5 wt.% of n-Ni to a complex metal hydride $LiAlH_4$ brings about a remarkable changes in its thermal behavior and profound improvement of its dehydriding properties. Figure 12 a,b,c,d show the evolution of DSC curves with increasing ball milling energy for the $LiAlH_4$ + 5wt.% n-Ni nanocomposite as compared in Figure 12a to the DSC behavior of undoped $LiAlH_4$ ball milled under high energy milling mode. It must be mentioned that high energy ball milling does not change the DSC curve shape neither it shifts the peak temperatures. However, ball milling of complex alanates is not very effective. The average ECD particle size with standard deviation (S.D.) was reduced from 9.9 ± 5.2 μm for as received $LiAlH_4$ to 2.8 ± 2.3 μm after 2 h of ball milling [20].

According to the discussion in [21] the first exothermic (peak 1) at 143.8 °C in Figure 12a is most likely attributed to the reaction of the surface aluminum-hydroxyl groups owing to the presence of impurities as first reported by Block and Gray [29]:

$$>Al\text{-}OH + H\text{-}Al< \rightarrow >Al\text{-}O\text{-}Al< + H_2 \quad (1)$$

The endothermic peak 2 at 174.4 °C is attributed to the melting of $LiAlH_4$:

$$LiAlH_4\,(solid) \rightarrow LiAlH_4\,(liquid) \quad (2)$$

An exothermic peak 3 at 201.1 °C is usually attributed to the decomposition of liquid $LiAlH_4$according to the following reaction:

$$LiAlH_4\,(liquid) \rightarrow 1/3Li_3AlH_6\,(solid) + 2/3Al\,(solid) + H_2 \quad (3)$$

However, as reported in [20] two other thermal events such as solidification of Li_3AlH_6 and its decomposition also occur at this peak. Finally, a small and broad endothermic peak 4 at 234.7 °C is due to the final decomposition of the small amount of remaining Li_3AlH_6 [2,29]:

$$1/3Li_3AlH_6\,(solid) \rightarrow LiH\,(solid) + 1/3Al\,(solid) + 0.5H_2\,(final) \quad (4)$$

The last peak 5 at over 400 °C is usually attributed to the decomposition of LiH which was formed as a result of decomposition of Li_3AlH_6. However, decomposition of LiH has no practical value since its temperature is too high.

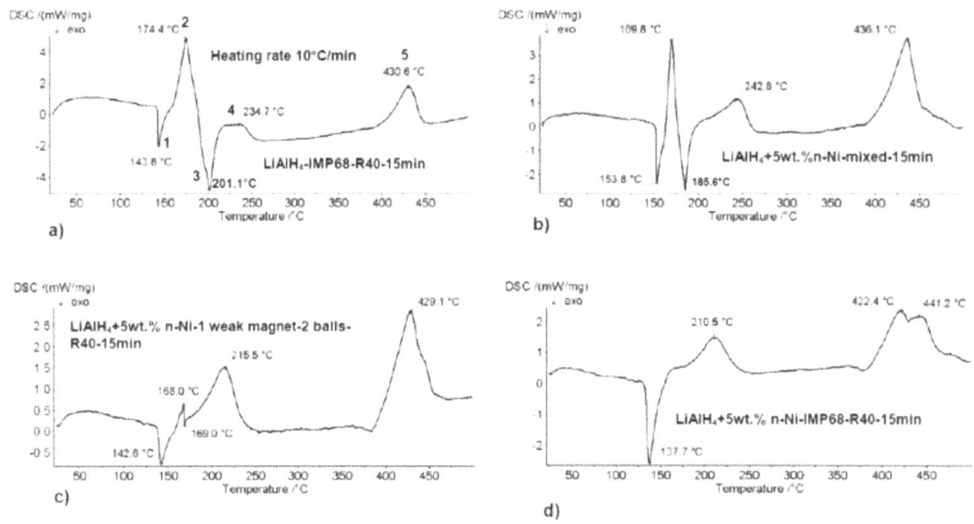

Figure 12. DSC curves for (a) undoped $LiAlH_4$ milled for 15 min, (b) $LiAlH_4$ + 5wt.% n-Ni mixed, (c) $LiAlH_4$ + 5wt.% n-Ni milled for 15 min under low energy shearing mode with 1 weak magnet, 2 balls, (d) $LiAlH_4$ + 5wt.% n-Ni milled for 15 min under high energy milling mode. Heating rate was 10 °C/min. R-ball-to-powder weight ratio; IMP68-high energy impact mode with 2 magnets in a magneto-mill. Adapted from [21].

Figure 12b shows that the addition of 5 wt.% n-Ni to $LiAlH_4$ by simple mixing without ball milling does not change the reference DSC scan in Figure 12a. In contrast, even relatively low energy ball milling of n-Ni doped $LiAlH_4$ using 1 weak magnet and 2 balls changes profoundly a DSC scan as shown in Figure 12c. Both melting peak of $LiAlH_4$ (at 168 °C) and the following exo peak at 169 °C become very small. As can be seen in d high energy ball milling completely eliminates the endothermic melting peak 2 of $LiAlH_4$ and virtually merges the exothermic peak 1 and 3 into one exo peak at 137.7 °C. Now the decomposition of $LiAlH_4$ occurs fully in a solid state:

$$LiAlH_4\,(solid) \rightarrow 1/3Li_3AlH_6\,(solid) + 2/3Al\,(solid) + H_2 \quad (5)$$

This is quite a remarkable sequence which shows that the mere presence of the n-Ni catalyst in a mixture as obtained by just mixing is insufficient and the nano-catalyst must be very intimately embedded in the $LiAlH_4$ particles in order to be able to exert a strong catalytic action on the alanate matrix. That also suggests that it is still insufficient if the n-Ni particles are distributed solely on the surface of hydride particles as it occurs during mixing but they must be also embedded in the bulk of the hydride particles which is only achieved by relatively high energy ball milling. As reported in [21] it is unclear if the reaction of the surface aluminum-hydroxyl groups (reaction (1)) triggers a spontaneous decomposition of ball milled n-Ni doped $LiAlH_4$ at temperatures much lower than its melting point leading to a merger of peak 1 and 3 into one exothermic peak as presented in Figure 12d. This issue requires more studies.

The synthesis of $LiAlH_4$ with 5 wt.% n-Ni by high energy ball milling also brings about some remarkable improvement of isothermal dehydring rate at quite moderate temperature as shown inFigure 13. It is seen that a $LiAlH_4$ + n-Ni nanocomposite system always desorbs hydrogen in a solid state in two stages I and II which corresponds to reaction (5) and (4), respectively. Such a behavior has never been reported in the literature.

The apparent activation energy of Stage I and II as estimated from the Johnson–Mehl–Avrami–Kolmogorov (JMAK) and Arrhenius equations has been reported in [21] as being equal to ~70 ($R^2 = 0.998$) and ~100 kJ/mol ($R^2 = 0.998$), respectively. The one for Stage I is lower than the apparent activation energy for Stage I reported for metal chloride catalytic precursors in $LiAlH_4$ [2,21].

Another effect of nano-catalytic Ni additive is its ability to induce a slow desorption of large quantities of hydrogen at room temperature (RT), 40 and 80 °C for the ball milled nanocomposite of $LiAlH_4$ + 5 wt.% n-Ni as shown in Figure 14.

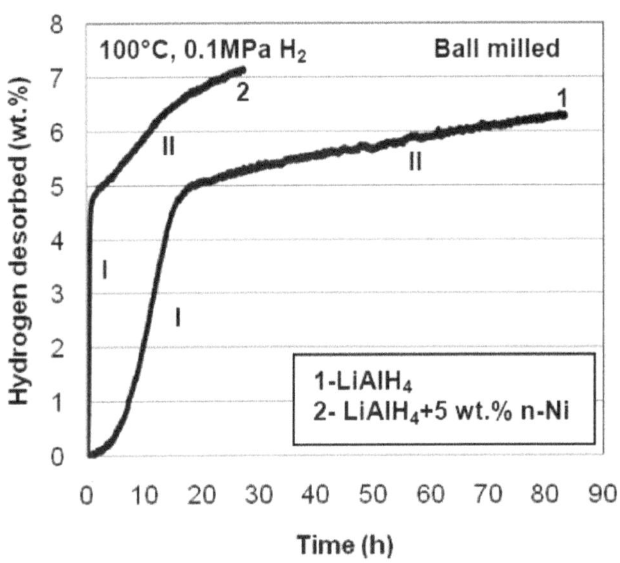

Figure 13. Comparison of volumetric desorption curves at 100 °C under 0.1 MPa H_2 pressure for ball milled undoped $LiAlH_4$(curve 1) and ball milled $LiAlH_4$ + 5 wt.% n-Ni (curve 2). Adapted from [21].

XRD studies have shown that the microstructure after 25 and 65 days of storage at RT consists mainly of a mixture of retained $LiAlH_4$, newly formed Li_3AlH_6 and Al [21]. This kind of microstructure confirms that reaction (5) was incomplete during storage at RT which also agrees well with the release of ~3.4 wt.% H_2 in Figure 14a. At 80 °C after 15 days of storage the microstructure consists mainly of newly formed Li_3AlH_6, Al and possibly some small amount of LiH [21]. This microstructure indicates near-completion of reaction (5) and the beginning of reaction (4). After 30 days at 80 °C the microstructure consists of a minimal amount of retained Li_3AlH_6, mostly Al and some LiH [21]. The rate of H_2 release during storage at RT −80 °C can be easily regulated by increasing or decreasing temperature. These virtues make the $LiAlH_4$ + 5 wt.% n-Ni nanocomposite a potential hydrogen storage material for applications where a continuous supply of hydrogen is required for a prolonged service time.

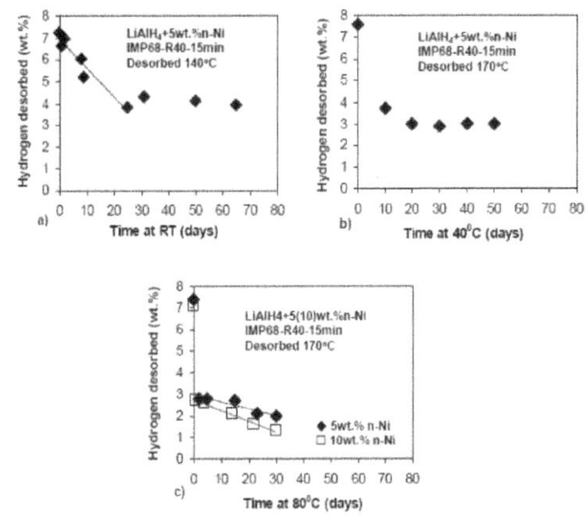

Figure 14. Plots of hydrogen desorbed vs. storage time in days (1 day = 24h) for $LiAlH_4$ + 5(10) wt.% n-Ni mixtures ball milled for 15 min under high energy mode. (a) Stored at room temperature (RT) under Ar; subsequently desorbed at 140 °C under 0.1 MPa H_2 pressure. (b) Stored at 40 °C under Ar; subsequently desorbed at 170 °C under 0.1 MPa H_2 pressure. (c) Stored at 80 °C under Ar; subsequently desorbed at 170 °C under 0.1 MPa H_2 pressure. Adapted from [21].

Nanometric Metal Oxides in MgH_2

Various metal oxides seem to be quite potent catalytic additives to MgH_2. Most of the studies on the effects of metal oxide additives on the dehydriding behavior of MgH_2 until the end of 2008 were critically reviewed in [2]. Most recently, a number of metal oxide nanopowders (size within ~8–25 nm range) were investigated as catalysts for ball milled MgH_2. In order to eliminate the effect of γ-MgH_2 on dehydrogenation behavior (see Sec. 2.1) all the samples were first dehydrogenated at 325 °C under vacuum and then hydrogenated at 300 °C under 10 bar of hydrogen pressure. After the 1h absorption, sample was heated to 325 °C, and the desorption at 0.1 MPa H_2 pressure was conducted [18,30]. Figure 15 shows absorption and subsequent desorption curves for a number of investigated nano-oxides. It is clear that the best catalytic effect on hydrogen desorption rate is exhibited by nano-Cr_2O_3 and nano-TiO_2. The apparent activation energy of dehydrogenation for all additives was also estimated by the Kissinger method [2, 30].

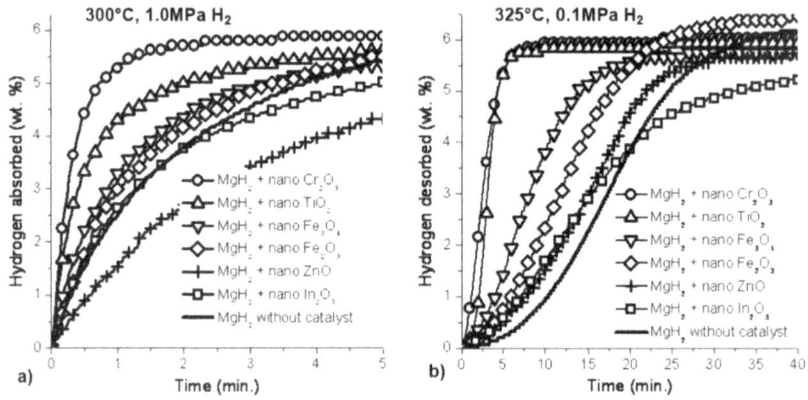

Figure 15. (a) Hydrogen absorption and (b) desorption curves for MgH_2 ball milled with various nano-sized metal oxides. Adapted from [30].

For the samples with nano-Cr_2O_3 and nano-TiO_2, the activation energy was found to be ~84 and ~94 kJ/mol, respectively, as compared to the range of 115–147 kJ/mol for the other nano-oxides inFigure 15. The obtained apparent activation energy range for nano-Cr_2O_3 and nano-TiO_2 is quite comparable to that reported for the nano-Ni doped MgH_2 in Figure 8b which is equal to ~94 kJ/mol. However, the DSC peak temperature for the nano-Cr_2O_3 and nano-TiO_2 doped MgH_2nanocomposite was reported as being equal to 359 and 348 °C, respectively, for the heating rate of 10 °C/min [30]. These are much higher peak temperatures than 243.1 °C shown in Figure 7 for the n-Ni doped MgH_2 nanocomposite registered at the same heating rate (10 °C/min). However, as mentioned in [30] there seems to be no linear dependence between the estimated apparent activation energy by the Kissinger method and the desorption rate measured by volumetric Sieverts' method and DSC. This behavior may be caused by the fact that the DSC peak does not always correspond to the higher reaction rate (exothermic and endothermic peaks can overlap and shift the peak value). Therefore, Kissinger's method, based on slightly shifted peaks, does not give very accurate values. It is also possible that even the correctly estimated activation energy is not the only factor influencing the desorption rate under both atmospheric and dynamic vacuum conditions.

Since nano-Cr_2O_3 is one of the two most efficient catalytic additives as discussed above the cyclic stability of the ball milled MgH_2 + 10 wt.% nano-Cr_2O_3 additive was further investigated in [31]. The desorption step in cycling was conducted under high vacuum of 10^{-6} mbar while absorption was conducted under 10 bar of hydrogen pressure both at 325 °C. To evaluate precisely the amount of hydrogen absorbed by a sample, absorption isotherms were performed before cycling and after each 25 cycles. The total of 150 desorption/absorption cycles were conducted. Figure 16 shows that after only one cycle the total hydrogen capacity is reduced to about 5.2 wt.% from the initial capacity of ~6.7 wt.% and then it further decreases with increasing number of cycles reaching about 4.6 wt.% after 75 cycles. For the next 75 cycles the total hydrogen capacity seems not to be much affected by cycling and remains at a constant level of ~4.6 wt.% which suggests that most of the microstructural changes responsible for the capacity loss occurred within the first 75 cycles. In general, the hydrogen capacity loss with the progressive cycling is quite commonly observed for ball milled MgH_2 both undoped and doped with catalytic additives [2,23]. Microstructural investigations in [31] revealed that a number of microstructural changes occurred during cycling which included powder particle sintering into clusters, segregation of the Cr_2O_3 particles to the interfaces between the sintered Mg particles, which after ball milling were quite uniformly distributed within the individual MgH_2 particles, grain growth of the β-MgH_2 grains/crystallites within the particles from ~30 nm after milling to over 100 nm after 150 cycles, and a gradual degradation of the Cr_2O_3 catalytic additive owing to its reduction by Mg and formation of MgO during an absorption step in cycling [31]. However, it was difficult to establish unambiguously which one of these microstructural alterations was primarily responsible for the peculiar loss of capacity observed in Figure 16.

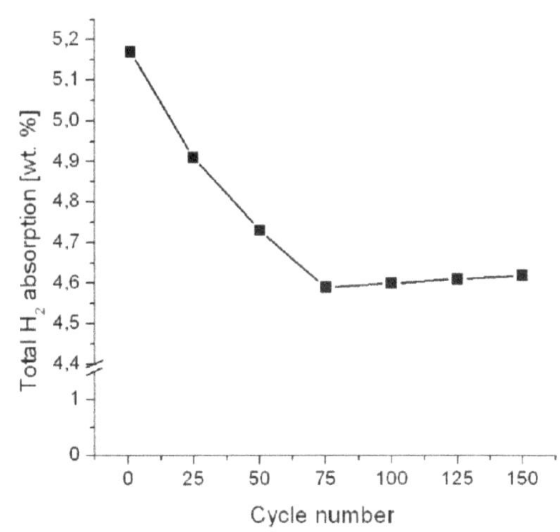

Figure 16. The total hydrogen absorption capacity for MgH_2 + 10 wt.% nano-Cr_2O_3 obtained from PCT curves after a specified number of cycles at 325 °C. Adapted from [31].

Metal Halides Catalytic Precursors

MgH_2

Most recently a number of halides (chlorides&fluoridies) have been evaluated as potential catalysts for MgH_2. The amount of 7 wt.% halides was added to the MgH_2 matrix and ball milled in a planetary mill [32]. The dependence of the dehydrogenation peak temperature of the synthesized nanocomposites on the halide additive was monitored by temperature programmed desorption (TPD) and differential scanning calorimetry (DSC). Figure 17 shows the TPD dehydrogenation peak temperature as a function of the type of a halide additive. A horizontal line delineates all TPD temperatures lower than 300 °C. It is quite clear that for chlorides only the $TiCl_3$additive gives the peak temperature slightly lower than the 300 °C limit. The situation is much better for fluorides because a number of them such as ZrF_4, NbF_5, TaF_5, TiF_3, and FeF_2 have peak temperatures below 300 °C. Apparently the lowest peak temperature is observed for ZrF_4.

The apparent activation energies for dehydrogenation for MgH_2 nanocomposites with $TiCl_3$, ZrF_4, NbF_5, and TaF_5 were calculated by the Kissinger method from TPD/DSC curves and are equal to 85 kJ/mol ($R^2 = 0.998$)/79 kJ/mol ($R^2 = 0.997$), 77 kJ/mol ($R^2 = 0.995$)/82 kJ/mol ($R^2 = 0.999$), 83 kJ/mol ($R^2 = 0.999$)/70 kJ/mol ($R^2 = 0.999$) and 97 kJ/mol ($R^2 = 0.940$)/95 kJ/mol ($R^2 = 0.996$), respectively. Apparently, the apparent activation energy ranges are the lowest ones for NbF_5 and ZrF_4 (note excellent coefficients of fit R^2). The selected ZrF_4, NbF_5, and TaF_5 fluorides have the highest oxidation states compared to other fluorides and chlorides which may be responsible for their enhanced behavior as catalytic species. However, the exact mechanism of the catalytic action exerted by them on MgH_2 is still elusive.

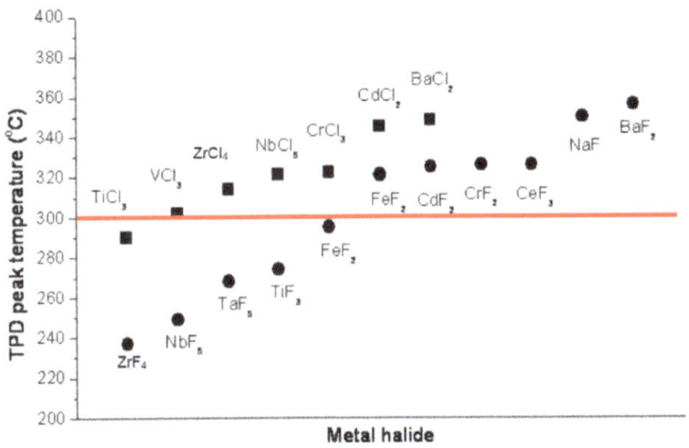

Figure 17. Dehydrogenation TPD peak temperature of ball milled MgH_2 doped with 7 wt.% of various halides as a function of the type of a halide additive. The horizontal line shows the limit of 300 °C for TPD peak temperature. Adapted from [32].

Further studies investigated the cyclic stability of ball milled nanocomposites MgH_2 + 7 wt.% ZrF_4 and NbF_5. For both nanocomposites it has been observed that after 50 absorption/desorption cycles at 325 °C the hydrogen capacity decreased from the initial ~6.0 to ~4.9 wt. % for the MgH_2/ZrF_2 sample and ~4.5 wt.% for the MgH_2/NbF_5 [33]. The causes for the reduced capacity with cycling are not clear but it must be noted that this behavior is very similar to the one presented in Figure 16 for the nano-Cr_2O_3 additive. For the MgH_2/NbF5 sample microstructural studies by STEM, XRD and EDS found that NbF5 was degraded after 50 cycles forming the MgF2 phase and nano-sized Nb-rich precipitates. It seems that the formation of MgF_2 is mainly responsible for the loss of hydrogen storage capacity in this system due to the reduction in the volume fraction of MgH_2 matrix. It is more difficult to explain the cyclic capacity loss for the MgH_2/ZrF_4. Microstructural studies showed no degradation of ZrF_4 after 50 cycles. Possibly, the microstructural changes discussed earlier for the MgH_2/Cr_2O_3 system might be responsible for the capacity loss for the MgH_2/ZrF_4 system. Apparently more research is required to quantitatively explain the observed behavior.

Complex Hydrides

The addition of catalytic metal halide precursors to complex metal hydrides has been very well documented. Some catalytic metal chloride precursors such as $TiCl_3$, $ZrCl_4$, VCl_3, $NiCl_2$ and $ZnCl_2$were added to $LiAlH_4$ which enhanced quite dramatically the rate of desorption and in effect lowered the effective desorption temperature of $LiAlH_4$ [2]. In general, catalytic precursors work in such a way that they react with the hydride matrix forming a metal salt and a free nanometric or amorphous elemental metal/intermetallic catalyst [2].

Most recently, for the first time we have investigated the effect of a manganese chloride ($MnCl_2$) catalytic precursor on the hydrogen storage properties of $LiAlH_4$ [34]. This chloride is less volatile than those of Ti, V and Zr metals. Figure 18a shows a DSC curve for the ball milled $LiAH_4$/$MnCl_2$system. The first exo peak is attributed to the partial decomposition of $LiAlH_4$ to Li_3AlH_6, Al and H_2 in a solid state according to reaction (5) in Sec.3.1.2. However, in contrast to the $LiAH_4$/n-Ni system discussed in Sec.3.1.2 the melting of $LiAlH_4$is not completely eliminated since reaction (5) is incomplete as a small amount of the retained $LiAlH_4$ is melting around the small endo peak at 163.1 °C and decomposes to Li_3AlH_6, Al and H_2 around the exo peak at 171.1 °C. It is to be reminded that the addition of 5 wt.% n-Ni and a similar ball milling energy completely eliminated the melting of $LiAlH_4$ as shown in Figure 12d.

Figure 18b shows that the $MnCl_2$ catalytic precursor is not as effective in enhancing the rate of dehydrogenation as n-Ni discussed in Sec.3.1.2. For the $LiAH_4$/$MnCl_2$ system dehydrogenation occurs in a solid state but only through Stage I (reaction (5) in Sec.3.1.2) whereas for the $LiAH_4$/n-Ni system within a similar time span, dehydrogenation occurs in a solid state through both Stage I and II (reaction (5) and (4), respectively, in Sec.3.1.2). After desorption of ball milled $LiAlH_4$/$MnCl_2$at 100 °C for 140 ks its XRD pattern (not shown here) exhibited the diffraction peaks of LiCl (JCPDS#04-0664), Al and Li_3AlH_6. Apparently, during decomposition besides reaction (5) in Stage I producing Li_3AlH_6 and Al, a second reaction producing LiCl (and again Al) must have occurred with the following general path [34]:

$$nLiAlH_4 + MnCl_2 \rightarrow (n-2)LiAlH_4 + 2LiCl + 2Al + Mn + 4H_2 \quad (6)$$

For 5 wt.% $MnCl_2$, n = 63 so the exact reaction is as follows:

$$63LiAlH_4 + MnCl_2 \rightarrow 61LiAlH_4 + 2LiCl + 2Al + Mn + 4H_2 \quad (7)$$

It is to be pointed out that since no diffraction peaks of Mn were observed after desorption at 100 °C for 140 ks, most likely, elemental Mn catalyzing the reaction is in an amorphous state.

The apparent activation energy of dehydrogenation of the ball milled $LiAlH_4$/$MnCl_2$ system for Stage I (reaction (5) in Sec.3.1.2) is equal to ~80 kJ/mol ($R^2 = 1.0$) as compared to ~70 kJ/mol for the $LiAlH_4$/n-Ni system (Sec.3.1.2).

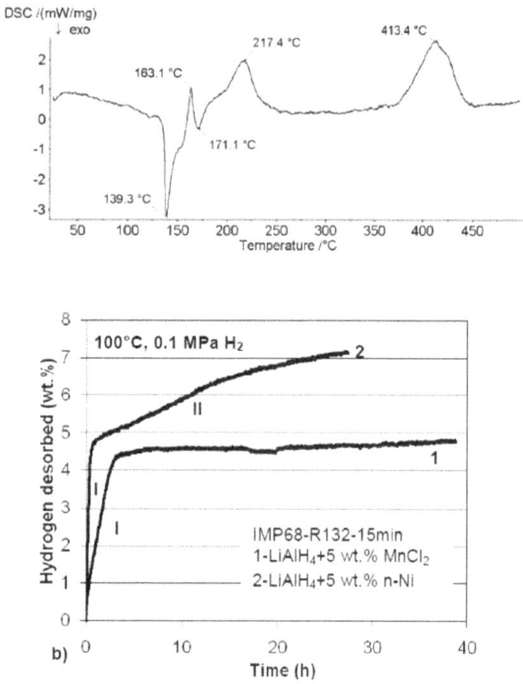

Figure 18. (a) DSC curve for $LiAlH_4$ + 5 wt.% $MnCl_2$ ball milled under high energy impact mode. (b) Comparison of volumetric desorption curves at 100 °C under 0.1 MPa H_2 for catalyzed $LiAlH_4$ ball milled under high energy impact mode. 1 - $LiAlH_4$ + 5 wt.% $MnCl_2$ and 2 - $LiAlH_4$ + 5wt.% n-Ni. R - ball-to-powder weight ratio; IMP68 - high energy impact mode with 2 magnets in a magneto-mill. Adapted from [34].

Figure 19 shows that ball milled $LiAlH_4$ doped with 5 wt.% $MnCl_2$ is able to desorb quite substantial quantities of H_2 at room temperature (RT; ~21 °C), 40 and 80 °C. So far, such a behavior has never been reported for a metal chloride doped $LiAlH_4$. At RT and 80 °C the ball milled $LiAlH_4$+ 5 wt.% $MnCl_2$ nanocomposite within a comparable time span slowly desorbs even slightly greater amount of H_2 than its $LiAlH_4$ + 5 wt.% n-Ni counterpart in Figure 14a,c despite that at elevated temperatures the apparent activation energy of 5 wt.% $MnCl_2$ doped $LiAlH_4$ is slightly greater (~80 kJ/mol vs. ~70 kJ/mol, respectively).

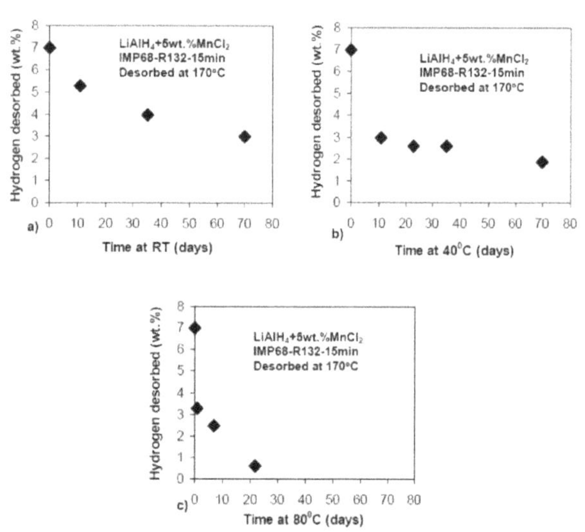

Figure 19. The amounts of hydrogen desorbed during long term storage at (a) room temperature (RT), (b) 40 and (c) 80 °C for ball milled $LiAlH_4$ + 5 wt.% $MnCl_2$. Adapted from [34].

SUMMARY AND CONCLUSIONS

The most recent advances on the effects of microstructural refinement and various nano-catalytic additives on the hydrogen storage properties of metal and complex hydrides have been critically discussed. It is shown that microstructural refinement induced by ball milling is rather of secondary importance to the hydrogen storage properties of metal and complex metal hydrides. In particular, for MgH_2 the presence of the non-equilibrium pressure-induced γ-MgH_2 phase formed during ball milling and the particle refinement to the size level achievable by high energy ball milling can measurably improve the kinetics of dehydrogenation of ball milled MgH_2 but does not affect its thermodynamics. However, after the removal of γ-MgH_2 by annealing the kinetics of dehydrogenation is only very modestly affected by the refined particle size. In this context it seems that the role of the refined grain size within the powder particles is quite minor. It also seems that the reduction of the hydroxide layer on the MgH_2 particles, if ball milling occurs under reducing H_2, is even more important than the particle refinement.

For complex metal hydrides (e.g., $LiAlH_4$) microstructural refinement induced by ball milling still improves the kinetics of dehydrogenation but again the overall effect is at least quite modest. Taking the apparent activation energy of dehydrogenation as a most important kinetic parameter, for example, three-fold refinement by ball milling of the average particle size of $LiAlH_4$ from ~10 to ~3 μm brings about only a very modest decrease of the apparent activation energy by ~10 kJ/mol.

Nano-sized elemental metals (e.g., n-Ni and n-Fe) influence the kinetics of dehydrogenation of ball milled MgH_2 in a major way much more substantially than the microstructural refinement. The observed decrease of the apparent activation energy with respect to the ball milled undoped MgH_2 is within the range of ~50–60 kJ/mol. For complex metal hydrides the reduction in the apparent activation energy approaches ~30 kJ/mol. Quite similar effects on the kinetics of dehydrogenation and particularly on the apparent activation energy of MgH_2 dehydrogenation are obtained by the addition of halide catalytic precursors. Again the reductions of the apparent activation energy within the range of ~50–60 kJ/mol are easily achieved.

For the complex hydrides, like $LiAlH_4$, the addition of metal chloride precursors result in more modest reductions in the apparent activation energy on the order of ~20 kJ/mol as compared to undoped $LiAlH_4$.

A complex metal hydride $LiAlH_4$ after ball milling with nanometric Ni metal catalysts and $MnCl_2$ catalytic precursor is able to desorb large quantities of hydrogen at RT, 40 and 80 °C. This kind of behavior is very encouraging for the future development of solid state hydrogen systems.

ACKNOWLEDGEMENTS

A financial support from the NSERC Hydrogen Canada (H2CAN) Strategic Research Network, NSERC Discovery, NSERC/Vale Inco Inc. Collaborative Research and Development, and the Polish Ministry of Science and Higher Education, Key Project POIG.01.03.01-14-016/08, is gratefully acknowledged.

REFERENCES

1. Homepage of Nuvera Fuel Cells. Available online: http://www.nuvera.com/products/androm.php (Accessed on 20 December 2010).
2. Varin, R.A.; Czujko, T.; Wronski, Z.S. Nanomaterials for Solid State Hydrogen Storage; Springer Science + Business Media: New York, NY, USA, 2009; Chapter 1–3.
3. Bogdanović, B.; Bohmhammel, K.; Christ, B.; A. Reiser, A.; Schlichte, K.; Vehlen, R.; Wolf, U. Thermodynamic investigation of the magnesium–hydrogen system. J. Alloys Compd. **1999**,282, 84–92.
4. Bogdanović, B.; Hofmann, H.; Neuy, A.; Reiser, A.; Schlichte, K.; Spliethoff, B.; Wessel, S. Ni-doped versus undoped Mg–MgH materials for high temperature heat or hydrogen storage.J. Alloys Compd. **1999**, 292, 57–71.
5. Zaluski, L.; Zaluska, A.; Ström-Olsen, J.O. Nanocrystalline metal hydrides. J. Alloys Compd.**1997**, 253-254, 70–79.
6. Zaluska, A.; Zaluski, L.; Ström-Olsen, J.O. Nanocrystalline magnesium for hydrogen storage.J. Alloys Compd. **1999**, 288, 217–225.
7. Zaluska, A.; Zaluski, L.; Ström-Olsen, J.O. Structure, catalysis and atomic reactions on the nano-scale: A systematic approach to metal hydrides for hydrogen storage. Appl. Phys. A**2001**, 72, 157–165.
8. Huot, J.; Liang, G.; Boily, S.; Van Neste, A.; Schulz, R. Structural study and hydrogen sorption kinetics of ball-milled magnesium hydride. J. Alloys Compd. **1999**, 293-295, 495–500.
9. Schulz, R.; Huot, J.; Liang, G.; Boily, S.; Van Neste, A. Structure and hydrogen sorption properties of ball milled Mg dihydride. Mater. Sci. Forum **1999**, 312–314, 615–622.
10. Schulz, R.; Huot, J.; Liang, G.; Boily, S.; Lalande, G.; Denis, M.C.; Dodelet, J.P. Recent developments in the application of nanocrystalline materials to hydrogen technologies.Mater. Sci. Eng. A **1999**, 267, 240–245.
11. Dornheim, M.; Doppiu, S.; Barkhordarian, G.; Boesenberg, U.; Klassen, T.; Gutfleisch, O.; Bormann, R. Hydrogen storage in magnesium-based hydrides and hydride composites. Scr. Mater. **2007**, 56, 841–846.
12. Varin, R.A.; Czujko, T.; Wronski, Z. Particle size, grain size and γ-MgH_2 effects on the desorption properties of nanocrystalline commercial magnesium hydride processed by controlled mechanical milling. Nanotechnology **2006**, 17, 3856–3865.
13. Bastide, J.-P.; Bonnetot, B.; Létoffé, J.-M.; Claudy, P. Polymorphisme de l'hydrure de magnesium sous haute pression. Mat. Res. Bull. **1980**, 15, 1779–1787.
14. Gennari, F.C.; Castro, F.J.; Urretavizcaya, G. Hydrogen desorption behavior from magnesium hydrides synthesized by reactive mechanical alloying. J. Alloys Compd. **2001**, 321, 46–53.
15. Selvam, P.; Viswanathan, B.; Swamy, C.S.; Srinivasan, V. Magnesium and magnesium alloy hydrides. Int. J. Hyd. Energy **1986**, 11, 169–192.
16. Barkhordarian, G.; Klassen, T.; Bormann, R. Kinetic investigation of the effect of milling time on the hydrogen sorption reaction of magnesium catalyzed with different Nb_2O_5 contents. J. Alloys Compd. **2006**, 407, 249–255.
17. Varin, R.A.; Czujko, T.; Chiu, C.; Wronski, Z. Particle size effects on the desorption properties of nanostructured magnesium dihydride (MgH_2) synthesized by controlled reactive mechanical milling (CRMM). J. Alloys Compd. **2006**, 424, 356–364.
18. Polanski, M.; Bystrzycki, J.; Plocinski, T. The effect of milling conditions on microstructure and hydrogen absorption/desorption properties of magnesium hydride (MgH_2) without and with Cr_2O_3 nanoparticles. Int. J. Hyd. Energy **2008**, 33, 1859–1867.
19. Varin, R.A.; Jang, M.; Czujko, T.; Wronski, Z.S. The effect of ball milling under hydrogen and argon on the desorption properties of MgH_2 covered with a layer of $Mg(OH)_2$. J. Alloys Compd. **2010**, 493, L29–L32.
20. Varin, R.A.; Zbroniec, L. Decomposition behavior of unmilled and ball milled lithium alanate ($LiAlH_4$) including long-term storage and moisture effects. J. Alloys Compd. **2010**,504, 89–101.
21. Varin, R.A.; Zbroniec, L. The effects of nanometric nickel (n-Ni) catalyst on the dehydrogenation and rehydrogenation behavior of ball milled lithium alanate ($LiAlH_4$). J. Alloys Compd. **2010**, 506, 928–939.
22. Paserin, V.; Baksa, S.; Zaitsev, A.; Shu, J.; Shojai, F.; Nowosiadly, W. Potential for mass production of nickel-based nanomaterials by carbonyl process. J. Nanosci. Nanotechnol. **2008**,8, 4049–4055.
23. Varin, R.A.; Czujko, T.; Wronski, Z.S. Thermal stability of Vale Inco nanonometric nickel as a catalytic additive for magnesium hydride (MgH_2). Int. J. Hydrogen Ener. **2009**, 34, 8603–8610.
24. Polanski, M.; Plocinski, T.; Kunce, I.; Bystrzycki, J. Dynamic synthesis of ternary Mg_2FeH_6.Int. J. Hydrogen Energy **2010**, 35, 1257–1266.
25. Polanski, M.; Bystrzycki, J.; Varin, R.A.; Plocinski, T. Rapid hydrogenation at 30 °C of magnesium (Mg) and iron (Fe) nanocomposite obtained through a decomposition of Mg_2FeH_6 precursor. Int. J. Hydrogen Energy **2010**, 8.
26. Hanada, N.; Ichikawa, T.; Hino, S.; Fujii, H. Remarkable improvement of hydrogen sorption kinetics in magnesium catalyzed with Nb_2O_5. Int. J. Hydrogen Energy **2006**, 420, 46–49.
27. Varin, R.A.; Li, S.; Wronski, Z.; Morozova, O.; Khomenko, T. The effect of sequential and continuous high-energy impact mode on the mechano-chemical synthesis of nanostructured complex hydride Mg_2FeH_6. J. Alloys Compd. **2005**, 390, 282–96.
28. He, Y.; Zhao, Y. Hydrogen storage and cycling properties of a vanadium decorated Mg nanoblade array on a Ti coated Si substrate. Nanotechnology **2009**, 20, 204008.
29. Block, J.; Gray, A.P. The thermal decomposition of lithium aluminum hydride. Inorg. Chem.**1965**, 4, 304–305.

30. Polanski, M.; Bystrzycki, J. Comparative studies of the influence of different nano-sized metal oxides on the hydrogen sorption properties of magnesium hydride. J. Alloys Compd.**2009**, 486, 697–701.

31. Polanski, M.; Bystrzycki, J.; Robert, A.; Varin, R.A.; Plocinski, T.; Pisarek, M. The effect of chromium (III) oxide (Cr_2O_3) nanopowder on the microstructure and cyclic hydrogen storage behavior of magnesium hydride (MgH_2). J. Alloys Compd. **2010**, 11. in press.

32. Malka, I.E.; Czujko, T.; Bystrzycki, J. Catalytic effect of halide additives ball milled with magnesium hydride. Int. J. Hydrogen Energy **2010**, 35, 1706–1712.

33. Malka, I.E.; Bystrzycki, J.; Plocinski, T. Microstructure and hydrogen storage capacity of magnesium hydride with zirconium and niobium fluoride additives after cyclic loading. J. Alloys Compd. **2010**, 10. in press.

34. Varin, R.A.; Zbroniec, L. Fast and slow dehydrogenation of ball milled lithium alanate ($LiAlH_4$) catalyzed with manganese chloride ($MnCl_2$) as compared to nanometric nickel catalyst. J. Alloys Compd. **2010**, 09. in press.

CHAPTER 3

METASTABLE METAL HYDRIDES FOR HYDROGEN STORAGE

Jason Graetz

INTRODUCTION

The doubling of the world's energy consumption over the next fifty years [1] will require significant changes in the way we produce, distribute, store, and use energy. A crucial step in this process will be reducing our dependence on carbon-based fossil fuels. The rapidly growing global demand for oil will likely outpace production within a few decades (if it has not already). Although petroleum fuels can be produced from coal through a Fischer-Tropsch process, this method is costly and will likely result in greater greenhouse gas (e.g., CO_2) emissions. Hydrogen is a versatile, clean energy carrier that can be produced from natural gas reforming and a variety of carbon free energy sources such as solar, wind, and nuclear energy through water electrolysis. Hydrogen can easily be converted into mechanical or electrical energy through a combustion reaction or fuel cell, respectively. In both cases the hydrogen reacts with oxygen to produce water and energy ($(2H_2^g + O_2^g \rightarrow 2H_2O^g$ with a theoretical energy of theoretical energy density of 120 kJ/g).

Hydrogen storage remains one of the more challenging technological barriers to the advancement of hydrogen fuel cell technologies for mobile applications. Since hydrogen is a gas at standard pressure and temperature (SPT), it has a low volumetric density. Although one kilogram of H_2 can replace about 3 kg or one gallon (3.79 L) of gasoline, on a volumetric scale more than 3500 gallons (1.3×10^4 L $\approx$ volume of a midsize car) of H_2 gas is necessary to replace just one gallon (3.79 L) of gasoline at SPT.

There are a variety of ways to store hydrogen (as shown in Figure 1), and the more conventional methods include compressed gas (typically 350 or 700 bar) and liquefaction, where the hydrogen is cooled to below its boiling point (20 K). However, these options are costly and require extremely high pressures or low temperatures to achieve reasonable hydrogen densities. Solid-state hydrogen storage, where the hydrogen is either attached onto the surface (adsorption) or inserted into a host material, offers improved volumetric energy densities at moderate pressures and temperatures. In fact, many metal hydrides have a hydrogen density greater than that of liquid hydrogen (Figure 1). Most hydrogen storage materials can be divided into two categories, those that bind molecular hydrogen to surfaces via weak dipole (van der Waals) interactions (typically <10 kJ/mol H_2) [2] in a process known as physisorption (e.g., activated carbons and metal organic frameworks (MOFs) and those that trap atomic hydrogen via a strong (typically >10 kJ/mol H_2) chemical bond (e.g., metal hydrides, complex hydrides and chemical hydrides). There has been tremendous progress in the development of new physisorbents for hydrogen storage in recent years [3–8]; however, these materials typically exhibit appreciable storage capacities (e.g., >2 wt.%) only at low temperature (e.g., 77 K).

Citation: Jason Graetz, "Metastable Metal Hydrides for Hydrogen Storage," ISRN Materials Science, vol. 2012, Article ID 863025, 18 pages, 2012. doi:10.5402/2012/863025.

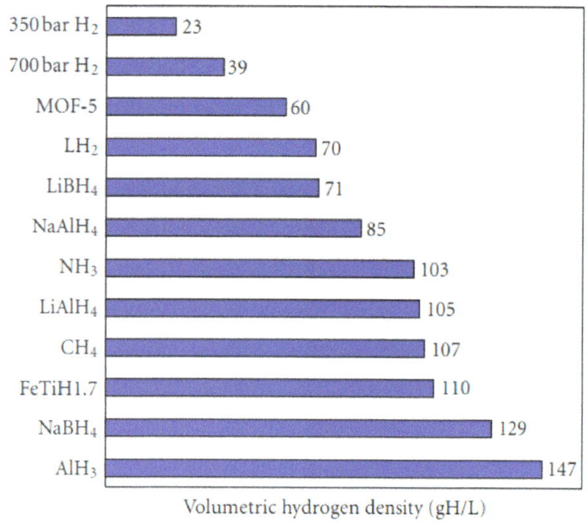

Figure 1: Volumetric hydrogen density (gH/L) of a variety of hydrogen storage methods including compressed gas (350 and 700 bar), physisorption in a metal organic framework (MOF), liquid hydrogen (LH_2), ammonia (NH_3), methane (CH_4), and a variety of metal and complex hydrides. The values present in this graph are only for the storage medium and do not take into account the full system volume (e.g., tank and fuel cell).

METAL HYDRIDES AND COMPLEX HYDRIDES

Hydrogen storage materials that store atomic hydrogen through a chemical bond are often categorized by their "reversibility", or their ability to be rehydrogenated under moderate temperature and pressure conditions. Recently, much attention has been focused on reversible complex metal hydrides, specifically sodium aluminum hydride ($NaAlH_4$). During thermal decomposition of catalyzed $NaAlH_4$ (typically 2–4 mol% Ti) approximately 4 wt.% of hydrogen is released at a temperature of around 160°C. Most importantly, the dehydrogenated material (NaH + Al) can be rehydrogenated under moderate conditions [9]. The demonstration of reversible hydrogen cycling in the complex metal hydrides was a breakthrough in solid-state hydrogen storage. Catalyzed sodium aluminum hydride is capable of cycling approximately twice the hydrogen (on a weight basis) of any of the conventional metal hydrides (e.g., FeTiHx). Since this discovery, hydrogen reversibility has been demonstrated in other complex hydrides including Na_2LiAlH_6, $KAlH_4$, K_3AlH_6, K_2LiAlH_6, and K_2NaAlH_6 [9–11].

Despite the recent progress in the development of new reversible complex hydrides for hydrogen storage, systems based on these materials have an inherent heat problem, which poses a significant engineering challenge for on-board refueling. This problem is due to the large hydrogenation enthalpy (typically greater than 33 kJ/mol H_2) characteristic of metal hydrides reversible at around 100°C. The large enthalpy is necessary to offset the large entropy change (110–130 J/K mol H_2) during hydrogenation and achieve a near-zero free energy change. Based on a generic reversible hydride with a slightly positive free energy (so that $H_2$is released at pressures above 1 atm) and a 5 kg H_2 tank, approximately 83 MJ of heat is released during hydrogenation. Although this is a lot of wasted energy, the real problem arises when we consider the rapid refueling rates necessary to complete the process in approximately 3 minutes (1.67 kg H_2/min). The 83 MJ of heat generated during refueling will need to be dissipated at a rate of ~0.5 MW, which is an extremely challenging engineering problem that is not likely to be solved by moderate improvements in heat exchangers [12]. The large reaction enthalpy also presents a problem for the decomposition reaction since a substantial heat exchange system is required to provide the heat necessary to release the H_2 at a rate to satisfy the fuel cell demand.

CHEMICAL HYDRIDES

Given the challenges associated with adsorbents (low gravimetric capacity at room temperature) and the reversible hydrides (low gravimetric capacity and large heats of reaction), there has been a growing interest in the chemical hydrides, which offer high capacity and rapid hydrogen rates at low temperature. However, these hydrides are not directly reversible under moderate hydrogenation conditions and typically require "off-board" chemical regeneration. Since the barrier to hydrogenation can be thermodynamic, kinetic, or both, these materials can have a range of decomposition enthalpies from weak (e.g., AlH_3 ΔH = 7 kJ/mol H_2) to strong (e.g., $LiBH_4$ ΔH = 67 kJ/mol H_2 for partial decomposition). For the more stable compounds, such as $NaBH_4$, the hydrogen is typically generated through a low temperature hydrolysis reaction:

$$NaBH_4 + 2H_2O \longrightarrow NaBO_2 + 4H_2 \quad (1)$$

In this type of system, a large amount of hydrogen can be generated at rapid rates that easily meet the US Department of Energy's fuel flow rate target 0.02 $gs^{-1}kW^{-1}$. These types of reactions are useful for some applications, especially when water is easily accessible (e.g., marine applications), but there are also a number of limitations, namely, regeneration of the spent fuel. After hydrolysis the reactants are left in deep thermodynamic well that makes the regeneration (refueling) process economically and energetically costly.

On the other end of the spectrum, there is a class of weakly bound chemical hydrides that can release hydrogen at rapid rates through a low temperature thermolysis reaction; these are known as the metastable or kinetically stabilized hydrides. These materials have an equilibrium H_2 pressure (P_{eq}) much greater than 1 bar at room temperature, but do not decompose (or decompose very slowly) due to kinetic limitations. One of the more popular examples is ammonia borane (NH_3BH_3), which has a total gravimetric hydrogen capacity of nearly 20 wt.%. Although there are a number of high capacity hydrogen storage materials where the hydrogen is bound to a light main group element (e.g., C–H, B–H, N–H, or O–H), ammonia borane is somewhat unique in that it contains both protic (N–H) and hydridic (B–H) bonds and a strong B–N bond [13, 14]. Although ammonia borane can be prepared in a metastable state at room temperature, it is thermodynamically unstable and can, in an ideal reaction, decompose exothermically to yield aminoborane and H_2 gas. Despite the strength of the B–N bond, the decomposition is often associated with the formation of a small amount of gaseous species such as ammonia (NH_3) or diborane (B_2H_6). Even small quantities (a few parts-per-million) of these unwanted species is sufficient to poison a low temperature proton exchange membrane fuel cell. In recent years a number of new methods have been identified to limit these side reactions and efforts are underway to design systems equipped with scrubbers to trap these species before they reach the fuel cell. Further information on the latest developments on ammonia borane can be found in a few recent review articles [14–16].

METASTABLE METAL HYDRIDES

Similar to ammonia borane, the metastable metal hydrides offer high capacities and low temperature hydrogen release. However, unlike ammonia borane, thermal decomposition does not result in the formation of gaseous species that can degrade capacity and poison the fuel cell. Of particular interest are the aluminum-based metastable hydrides including aluminum hydride (AlH_3), lithium aluminum hydride ($LiAlH_4$ and Li_3AlH_6), magnesium aluminum hydride ($Mg(AlH_4)_2$), among others. These materials exhibit a low decomposition enthalpy, which reduces the heat required to release the hydrogen at practical pressures. In addition, these materials exhibit rapid H_2 evolution rates at low temperature (80–100°C), due to the large driving force for decomposition.

Many of the metastable aluminum-based hydrides are ideally suited for portable power systems where weight and volume are critical. This is because the energy density of a power source consisting of a metal hydride and a PEM fuel cell is substantially larger than even the best batteries currently available as shown in Table 1. The actual energy density of a system that utilizes AlH_3 as the power source is not known (only a few small prototype systems currently exist), but based on similar systems it is reasonable to estimate a 50% penalty for weight of the tank, fuel cell, and additional hardware and an additional 50% penalty for the fuel cell conversion efficiency. This yields a system energy density that is 25% of the material energy density. Even under this rather conservative scenario, a hydrogen fuel cell using a metal hydride (e.g., AlH_3) as an H_2source yields gravimetric and volumetric energy densities over four times greater than even the best performing lithium polymer batteries.

Table 1: Comparison of hydride—fuel cell system (AlH_3 + 50 kW fuel cell operating at 50% efficiency and assuming an additional 50% energy reduction to account for the fuel cell system weight) with conventional battery systems showing the specific energy and energy density.

Energy source	Specific energy (MJ/kg)	Energy density (MJ/L)
700 bar hydrogen tank [17]	6.2	3.2
350 bar hydrogen tank [17]	6.6	2.1
AlH_3/fuel cell	**3.1**	**4.3**
Li polymer	0.72	1.1
Ni-Metal hydride	0.29	0.36
Ni-Cd	0.18	0.36
Pb-acid	0.11	0.14

This spotlight paper reviews recent research efforts on the metastable metal hydrides and is specifically focused on metastable aluminum-based hydrides (alanes and alanates). This paper covers the synthesis procedures, crystal structures, thermodynamics and decomposition pathways, kinetics of hydrogen release, and regeneration efforts. A more general review on hydrogen storage, metal hydrides, and complex hydrides can be found elsewhere [46–53].

SYNTHESIS

The earliest reports on the synthesis of aluminum hydride date back to the forties when Stecher and Wiberg [54] and later Finholt et al. [55] prepared alane amine and alane etherate adducts. Approximately ten years later Chizinsky et al. [56] prepared the first nonsolvated form of AlH_3 using an ethereal reaction of $AlCl_3$with $LiAlH_4$. In the sixties and seventies Brower et al. prepared a total of seven different polymorphs of AlH_3, $\alpha, \alpha', \beta, \gamma, \delta, \varepsilon, \xi$ using a similar method [57]. The conventional synthesis method for the preparation of α-AlH_3 is given below:

$$4LiAlH_4 + AlCl_3 + Et_2O \longrightarrow 4AlH_3 \cdot nEt_2O + 3LiCl\downarrow + LiAlH_4 + Et_2O \quad (2)$$

$$4AlH_3 \cdot nEt_2O + LiAlH_4 + LiBH_4 + Et_2O \xrightarrow[5\,hrs]{65^{\circ}C} 4AlH_3\downarrow + Et_2O\uparrow + LiAlH_4\downarrow + LiBH_4\downarrow \quad (3)$$

Importantly, excess $LiAlH_4$, and in some cases excess $LiBH_4$, is added to the alane-ether solution to reduce the time and temperature required to remove the ether [58, 59]. The desolvation step (reaction (3)) determines the polymorph type and particle size. In the final step, the excess $LiAlH_4$ and $LiBH_4$ is removed by a wash with diethyl ether (AlH_3 is insoluble in Et_2O) and, in some cases, a wash in dilute hydrochloric acid. Aluminum hydride (etherated or pure) can also be synthesized using any of the binary hydrides or tetrahydroaluminates (alanates) from the 1st and 2nd groups through a reaction with a Brønsted-Lewis acid (e.g., $AlCl_3$) [58, 60–67].

A variety of routes can be used to synthesize the complex aluminum hydrides. The most stable alanates (e.g., $KAlH_4$, $NaAlH_4$, and $NaLiAlH_4$), also known as the reversible alanates, can be synthesized by direct hydrogenation of aluminum and the binary hydride [9, 68–70]:

$$NaH + Al + \frac{3}{2}H_2 \xrightarrow{THF} NaAlH_4 \quad (4)$$

The metastable alanates are not easily formed by direct hydrogenation under reasonable pressures and are typically prepared through solution methods. Probably the most common method for the preparation of these materials is a reaction of a binary hydride with a metal halide to form an alanate, as shown below [32,66, 71, 72]:

$$4LiH + AlCl_3 \xrightarrow{Et_2O} LiAlH_4 + 3LiC \quad (5)$$

$$4CaH_2 + 2AlCl_3 \xrightarrow{THF} Ca(AlH_4)_2 + 3CaCl_2 \quad (6)$$

A similar approach involves a metathesis reaction of a metal halide ($MgCl_2$) and an alanate ($NaAlH_4$) to form the metastable alanate ($Mg(AlH_4)_2$) [35, 69, 73, 74]:

$$NaAlH_4 + LiCl \xrightarrow{Et_2O} LiAlH_4 + 3LiCl$$

$$2NaAlH_4 + MgCl_2 \xrightarrow{DEE} Mg(AlH_4)_2 + 2NaCl \quad (7)$$

However, these procedures are energetically and economically costly since the expensive alkali metals (e.g., Na) form stable salt (e.g., NaCl) byproducts, which are difficult to separate. Interestingly, many alanates can be prepared directly from AlH_3 and a binary metal hydride in a chemical or mechanochemical reaction [75,76], such as:

$$CaH_2 + 2AlH_3 \longrightarrow Ca(AlH_4)_2 \quad (8)$$

CRYSTAL STRUCTURES

The thermodynamic stability of the hydride is determined by its structure and composition. The lightweight aluminum-based hydrides can exist as pure aluminum hydride (AlH_3), a covalently bonded solid, or as a complex hydride consisting of a covalently bonded anion (e.g., $[AlH_6]^{3-}$ or $[AlH_4]^-$) and a cation. Pure aluminum hydride comes in a variety of polymorphs (α, α′, β, γ), all of which are metastable at room temperature [18–28]. The crystal structure of α-AlH_3 is trigonal ($R\bar{3}c$ space group) and consists of corner connected AlH_6 octahedra with each H bridging two octahedra as shown in Figure 2 (left). The complex aluminum hydrides tend to be more stable since the complex anion is stabilized by a cation, typically an alkali (e.g., Li, Na, K) [9, 29–31, 33, 34, 36, 70], or alkaline earth metal (e.g., Mg, Ca) [35, 37, 38, 40, 74], or a transition metal (e.g., Ti, Fe) [32, 39, 77]. The complex aluminum hydrides typically appear in two forms, the tetrahydroaluminates $(M^{n+}(AlH_4)_n)$, where Al is tetrahedrally coordinated by H, and hexahydroaluminates (M_3AlH_6, $MAlH_5$, M_2AlH_7), where Al is octahedrally coordinated by H. The crystal structure of $Mg(AlH_4)_2$, a tetrahydroaluminate, is shown in Figure 2 (right). $Mg(AlH_4)_2$ crystalizes in the $P\bar{3}m1$ space group and consists of AlH_4 tetrahedra forming a close-pack double layer separated by a layer of Mg atoms [44]. A list of some lightweight metastable aluminum hydrides is shown in Table 2 along with their theoretical hydrogen content and decomposition temperature. A more thorough review of these materials and their hydrogen storage properties can be found in a few recent review articles [28, 41, 78].

Table 2: A few examples of unstable and metastable aluminum-based hydrides showing hydrogen weight percent and the initial decomposition temperature (in many cases full hydrogen desorption occurs over multiple steps that occur at higher temperatures). It is also worth noting that the decomposition temperatures listed identify the temperature at which significant H_2 evolution occurs and may be lower than the exothermic or endothermic peaks observed in calorimetric measurements.

Alanes and hexahydride alanates	Wt.%H	Decomposition temperature	Tetrahydride alanates	Wt.%H	Decomposition temperature
α-AlH_3 [18–28]	10.1	100°C	$LiAlH_4$ [29–31]	10.5	120°C
α′-AlH_3 [21]	10.1	90°C	$CuAlH_4$ [32]		−70°C
β-AlH_3 [19, 20, 22]	10.1	80°C	$AgAlH_4$ [32]		−50°C
γ-AlH_3 [19, 20, 23–28]	10.1	80°C	$Be(AlH_4)_2$ [32]		20°C
Li_3AlH_6 [29–31, 33, 34]	11.1	160°C	$Mg(AlH_4)_2$ [35, 36]		110°C
			$Ca(AlH_4)_2$ [37, 38]		150°C
			$Mn(AlH_4)_2$ [32]		25°C
			$Fe(AlH_4)_2$ [32, 39]		< −45°C
			$LiMg(AlH_4)_3$ [37]		100°C
			$CaMg(AlH_4)_3$ [38]		125°C
			$In(AlH_4)_3$ [32]		−40°C
			$Ce(AlH_4)_3$ [32]		25°C
			$Ti(AlH_4)_4$ [32, 40]		−85°C
			$Sn(AlH_4)_4$ [32]		−40°C

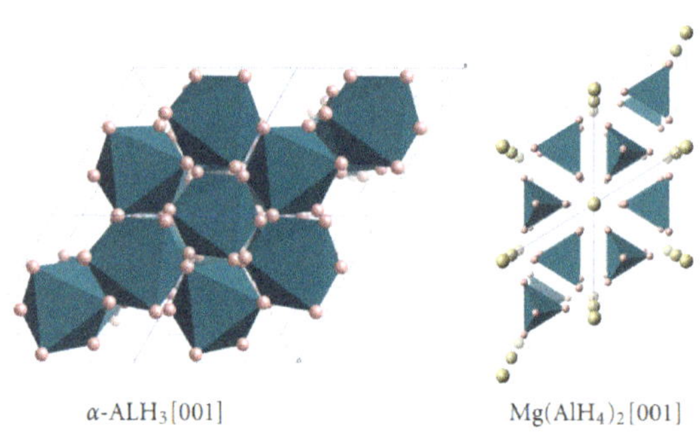

Figure 2: Crystal structures of alpha aluminum hydride (α-AlH_3), showing AlH_6octahedra, and magnesium alanate (Mg $(AlH_4)_2$) showing $[AlH_4]^-$ tetrahedra and Mg^+cation (yellow) viewed along the [001] direction [18, 77].

THERMODYNAMICS AND DECOMPOSITION PATHWAYS

The decomposition thermodynamics, or more specifically, the free energy difference between the reactant and the product, determines the H_2 equilibrium pressure at a given temperature. For binary hydrides (e.g., AlH_3) the decomposition pathway is a single step (to the elements) and the decomposition enthalpy is the negative of the formation enthalpy. Differential scanning calorimetry (DSC) plots from α-, β-, and γ-AlH_3undergoing thermal decomposition (temperature ramp from 35 to 300°C) are shown in Figure 2. The DSC trace from α-AlH_3 exhibits a single endothermic peak at around 170°C and is attributed to the decomposition of α-AlH_3:

$$\alpha\text{-}AlH_3 \longrightarrow Al + \frac{3}{2}H_2 \tag{9}$$

Although the endothermic peak occurs at 170°C, α-AlH_3 will decompose slowly at lower temperatures (complete decomposition in ~10 hrs at 100°C) [19]. Experimentally determined (calorimetry) values for the formation enthalpy of α-AlH_3 range from −6.0 to −7.6 kJ/mol H_2 [20, 45, 79], with a formation entropy of 20 J/K mol H_2 [44]. Based on these values the Gibbs formation energy for α-AlH_3 at 300 K is approximately 31 kJ/mol H_2 [20, 44]. The DSC plots from β-AlH_3 and γ-AlH_3 are also shown in Figure 3. In addition to the endothermic decomposition peak at 170°C, there are distinct exothermic peaks at approximately 100°C for both polymorphs. A similar exothermic peak was observed during decomposition of α′-

AlH_3 [80]. These peaks are due to a polymorphic transition (no H_2 release) from the less stable phase (α′, β, γ) to the more stable α phase. The energies of the polymorph transitions are −1.1 kJ/mol H_2 (α'-AlH_3 → α-AlH_3), −1.0 kJ/mol H_2 (β-AlH_3→ α-AlH_3), and −1.9 kJ/mol H_2 (γ-AlH_3 → α-AlH_3) [20, 79]. Subsequent studies have shown that at higher temperature (>100°C) the decomposition of the less stable polymorphs occurs by a polymorphic transition to the α phase followed by H_2 release [25, 28]. At lower temperatures the decomposition is more complex with a fraction of the material decomposing directly to the elements (e.g., γ-AlH_3 → Al + 3/2H_2) while the remaining material transforms to α-AlH_3 before decomposing (e.g., γ-AlH_3 →α-AlH_3 → Al + 3/2H_2).

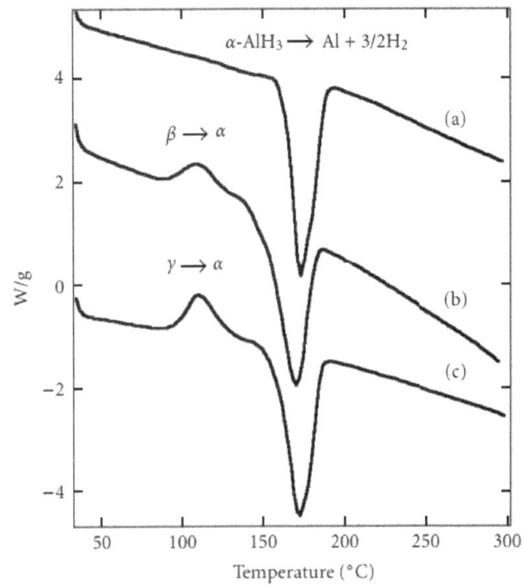

Figure 3: DSC traces (10°C/min) from (a) α-AlH_3, (b) β-AlH_3, and (c) γ-AlH_3 showing the exothermic polymorphic transition at ~100°C and the endothermic decomposition at ~170°C [20].

The decomposition pathways for the complex metal hydrides are not always as simple as those of the binary hydrides. For example, the group 1 tetrahydroaluminates (e.g., $LiAlH_4$) typically decompose to a hexahydridoaluminate first, then to a binary hydride (and Al), and finally to the alloy (LiAl) [31, 43], as shown below:

$$LiAlH_4 \xrightarrow{150°C} \frac{1}{3}(Li_3AlH_6) + \frac{2}{3}Al + H_2$$
$$\xrightarrow{200°C} LiH + Al + \frac{3}{2}H_2 \xrightarrow{450°C} LiAl + 2H_2 \qquad (10)$$

Tomiyasu et al. recently showed that in the case of $LiAlH_4$, the decomposition stems from a lattice instability attributed to the weak of binding of the complex anions in the lattice [81].

For the metastable hydrides, the decomposition can be exothermic or weakly endothermic. Figure 4 shows the DSC traces from three metastable hydrides ($LiAlH_4$, $Ca(AlH_4)_2$, and $Mg(AlH_4)_2$) during thermal decomposition. At each stage of the reaction the material releases some hydrogen and undergoes a structural transition to a more stable phase. The $Mg(AlH_4)_2$ phase undergoes a slightly endothermic transition (ΔH = 1.7–5.5 kJ/mol) at ~110°C directly to the binary hydride (MgH_2) and Al [37, 42, 82]:

$$Mg(AlH_4)_2 \xrightarrow{110°C} MgH_2 + 2Al + 3H_2 \qquad (11)$$

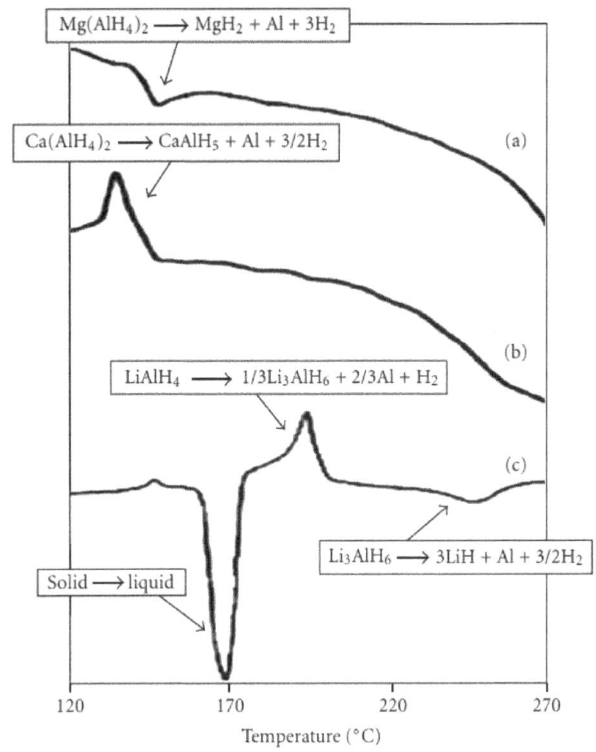

Figure 4: DSC traces (5°C/min) from (a) $Mg(AlH_4)_2$ [33], (b) $Ca(AlH_4)_2$ [33], and (c) $LiAlH_4$ [27] showing exothermic (positive peaks) and endothermic (negative peaks) transitions.

The first decomposition step of $Ca(AlH_4)_2$ is exothermic and occurs at ~150°C [37, 38, 81, 83]:

$$Ca(AlH_4)_2 \xrightarrow{100°C} CaAlH_5 + Al + \frac{3}{2}H_2 \qquad (12)$$

The decomposition enthalpy of this step is −7.4 kJ/mol. Unlike $Mg(AlH_4)_2$, the $Ca(AlH_4)_2$ decomposes to a pentahydride phase, before undergoing a more endothermic transition to the binary hydride at higher temperatures.

The decomposition of $LiAlH_4$ occurs by a three-step reaction first to the hexahydridoaluminate (Li_3AlH_6), then to LiH and Al, and finally to the elements Li and Al (reaction (10)). The DSC trace at the bottom of Figure 4 shows a slight exothermic peak at 150°C, which is attributed to a reaction with hydroxyl groups on the surface. The first endothermic minimum at ~170°C is due to melting [29, 33], which is immediately followed by the exothermic decomposition (175–220°C) of liquid $LiAlH_4$ to the solid Li_3AlH_6. Although considerably more reactive in the liquid state, solid $LiAlH_4$ is metastable and can spontaneously decompose slowly at room temperature [84]. $LiAlH_4$ exhibits a large, negative formation enthalpy (−94 kJ/mol) [42], making it stable against decomposition to the elements; however, $LiAlH_4$ is thermodynamically unstable against decomposition to the hexahydroaluminate phase (step 1 of reaction (10)) [33]. The higher temperature endotherm in the DSC trace (240°C) is attributed to the decomposition of Li_3AlH_6 to LiH and Al (step 2 of reaction (10)) [30, 33]. The final decomposition step is the endothermic hydrogen release from LiH (step 3 of reaction (10)), which occurs at much higher temperatures (>450°C).

A list of decomposition reactions, along with the amount of H_2 released and the reaction enthalpy (ΔH), is shown in Table 3. A plot of the calculated equilibrium pressure as a function of temperature (van't Hoff plot) for several metastable hydrides is shown in Figure 5 [20, 37, 43, 44, 85, 86]. It is important to note that the values used in Figure 5 correspond to H_2 fugacities, not hydrogen pressures. If H_2 is treated strictly as an ideal gas, a reasonable approximation at low pressures (<300 bar), then the hydrogen pressure is roughly equivalent to the hydrogen fugacity. However, at higher pressures the thermodynamic activity of H_2 is considerably higher than what is expected from an ideal gas [87, 88]. The fugacity-to-pressure ratio for H_2gas is ~1 at 1 bar, ~2 at 1 kbar, and ~300 at 10 kbar. Therefore, in some cases, hydrogenation occurs at much lower pressures than what is predicted from the reaction isotherm equation. In the case of α-AlH_3 at 300 K the predicted H_2 fugacity is $f_{H_2} = 5 \times 10^5$ bar, which is equivalent to a hydrogen pressure of 7 kbar [89], which is in reasonable agreement with high pressure hydrogenation experiments [90–93].

Table 3: Decomposition reactions of a few metastable metal hydrides showing H_2content (wt.%) and enthalpy (kJ/mol H_2). Negative enthalpy values indicate an exothermic decomposition reaction.

Reactant	Product	H_2 content (wt.%)	ΔH (kJ/mol H_2)	References
$LiAlH_4$	$1/3Li_3AlH_6 + 2/3Al + H_2$	5.3	(−9.1 −) (−9.6)	[41, 42]
Li_3AlH_6	$3LiH + Al + 3/2H_2$	5.6	27–29	[33, 41]
$Mg(AlH_4)_2$	$MgH_2 + 2Al + 3H_2$	7.0	1.8	[37, 43]
$Ca(AlH_4)_2$	$CaAlH_5 + A + 3/2H_2$	2.9	−7.4	[37]
$LiMg(AlH_4)_3$	$LiMgAlH_6 + 2Al + 3H_2$	4.8	−15	[37]
α-AlH_3	$Al + 3/2H_2$	10	6.0–7.6	[20, 44, 45]
β-AlH_3	$Al + 3/2H_2$	10	5.3	[20]
γ-AlH_3	$Al + 3/2H_2$	10	4.7	[20]

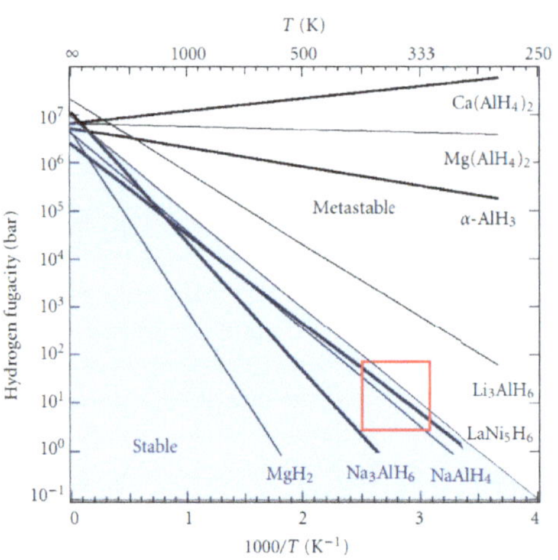

Figure 5: Approximate van't Hoff plots for several reversible and nonreversible hydrides [20, 37, 43, 44, 83, 84]. The compounds in the lower left (shaded) region are stable while the compounds in the upper right are unstable or metastable at ambient pressure and temperature. The box represents the thermodynamic target for "on-board reversible" hydrides. Reproduced by permission of The Royal Society of Chemistry [94].

KINETICS AND H RATES

The rate of H_2 desorption from aluminum hydride and lithium aluminum hydride varies considerably with the properties of the material (e.g., structure, morphology/size, surface composition, and catalyst concentration). The isothermal decomposition curve typically has a similar sigmoidal shape (over a single reaction step) with three distinct regions as shown in Figure 6 for AlH_3: (i) an induction period at the onset of the reaction, corresponding to a break-up of the surface layer and nucleation of aluminum; (ii) an acceleratory period, attributed to growth of the aluminum phase in two and three dimensions; (iii) a decay period, where the concentration of the hydride phase is limited and the growing aluminum particles begin to overlap [28].

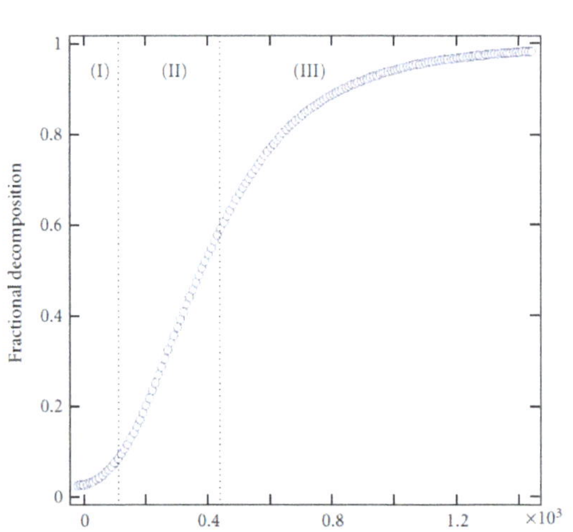

Figure 6: Isothermal decomposition of α-AlH_3 (prepared by Dow Chemical Co.) at 180°C showing (I) induction period, (II) acceleratory period, and (III) decay period [28].

volume for the unreacted fuel (e.g., AlH_3) and the spent fuel (e.g., Al metal) is essential for off-board refueling. With this type of tank configuration only the used portion of the fuel is replaced during off-board refueling.

REGENERATION

Despite their high capacity and rapid low temperature kinetics, the metastable metal hydrides are currently not practical storage media for most applications since they cannot be rehydrogenated at reasonable pressures and temperatures. The primary barrier to using metastable metal hydrides as a hydrogen source for mobile fuel cell applications is the cost associated with making the material. Although these materials are easily synthesized through a variety of chemical routes (see reactions (2)–(8)), the cost of the precursors is generally unacceptably high for large-scale commercial applications. For this reason (along with environmental reasons), there is much interest in identifying a closed-loop regeneration procedure where all of the materials (except for the hydrogen) are reused to reform the hydride. Unfortunately, the byproducts formed during the chemical synthesis of most metastable hydrides are stable compounds (e.g., LiCl) that are energetically costly to separate or recycle. A recent study on AlH_3 revealed that even under optimal conditions, a perfectly efficient recycling reaction for AlH_3 would still require an energy input of 167 kJ/mol H_2, which is equivalent to approximately 70% of the fuel energy contained within AlH_3 [92]. In a more realistic, industrial-scale process this method would likely require a much greater energy input, many time greater than the energy contained within the material.

Given the high cost associated with the direct chemical synthesis, it is reasonable to consider alternative routes of reforming the hydride from the spent (hydrogen depleted) material. Due to their low stability, the metastable metal hydrides are not easily rehydrogenated under moderate pressure and temperature conditions. In recent years there has been considerable effort focused on identifying low energy regeneration pathways to prepare the hydride from the dehydrogenated material (e.g., spent fuel). Baranowski succeeded in forming AlH_3 by direct hydrogenation of Al metal at high pressures (3×10^4 bar at 573 K) [88]. A number of other high pressure [87, 89–91] and thermodynamic studies [20, 44, 45] on the Al–H system have also been performed, which have recently been compiled into a pressure-temperature phase diagram, shown in Figure 11 [28]. The problem is clearly evident in this plot—the direct hydrogenation of aluminum requires pressures >7 kbar (almost 7000 atmospheres) at room temperature, much too high to be practical for large-scale commercial applications. Other metastable hydrides have proven to be even more difficult to rehydrogenate and most of them (e.g., $LiAlH_4$, $Mg(AlH_4)_2$, $Ca(AlH_4)_2$) have never been formed by direct hydrogenation at any pressure.

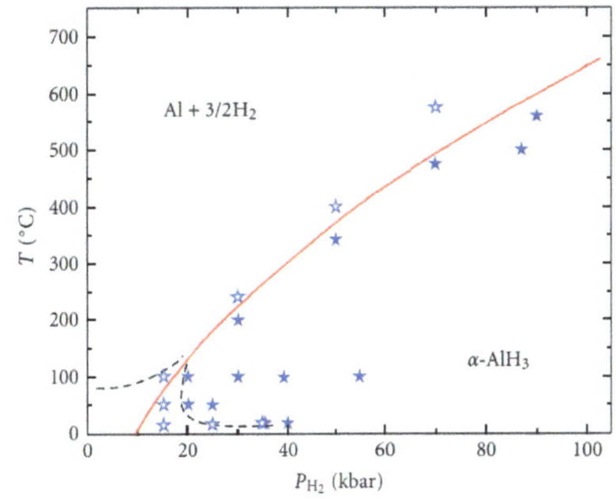

Figure 11: Pressure-temperature phase diagram of the Al–H system showing a solid line representing the calculated equilibrium for AlH_3 Al + 3/2H_2. The markers represent experimental results over a variety of pressure and temperature conditions, with the solid stars indicating the formation of α-AlH_3 and the open stars indicating no reaction [28].

Electrochemical Regeneration

Given the high cost of the direct chemical synthesis and the high pressures required for direct hydrogenation, there are currently a number of efforts focused on regeneration of the metastable metal hydrides. In one method, the hydrogenation is performed electrochemically. In the most direct approach, an electrolytic potential is used to increase the hydrogen activity and drive the hydrogen into the host material. The electrolytic potential (E) is related to the hydrogen pressure (P_{H_2}) through the Nernst equation:

$$E = -\frac{RT}{2F} \ln P_{H_2}, \tag{13}$$

where R is the universal gas constant, T is the temperature, and F is the Faraday constant. Since the potential scales as the natural log of the pressure, this equation indicates that the thermodynamic barrier to hydrogenation can be overcome by a relatively low voltage. Investigations of electrochemical hydrogenation of aluminum date back to the early 1960s when Clasen used an aluminum anode and an iron/mercury cathode with an electrolyte of $NaAlH_4$ in THF [110]. In the late 1990s, Birnbaum et al. used aqueous solutions of H_2SO_4 and HCl along with a recombination poison ($NaAsO_4$) to achieve high hydrogen fugacities for cathodic charging of aluminum [111]. They demonstrated that hydrogen enters vacancies in the Al structure and achieved concentrations of >1000 ppm, but no transformation to α-AlH_3 was observed. More recently, Zidan et al. [112, 113] have investigated the possibility of electrochemically charging Al with H using a cell similar to that of Clasen. In this case the electrochemical cell consisted of a pure Al anode and a platinum (foil or wire) cathode using THF with dissolved $NaAlH_4$ as the electrolyte. In this reaction three AlH_4^- ions (from $NaAlH_4$) combine with one Al^{3+} ion (from anode) to generate four moles of solvated AlH_3 at the anode as shown for the (anode) reaction below:

$$3AlH_4^- + Al^{3+} + nTHF \longrightarrow 4AlH_3 \cdot nTHF \tag{14}$$

The AlH_3-nTHF adduct is a white solid, and it is typically difficult to remove the AlH_3 intact. The desolvation requires high temperatures to break the Al–O bond and once the THF is removed the hydride begins decomposing, leading to a final product consisting of a mixture of Al and AlH_3. Improvements in yield and efficiency were achieved by the use of additives, such as the introduction of $LiAlH_4$ to the THF, along with the addition of LiCl, which acts as an electrocatalytic additive (ECA). The use of LiCl enhances the electrochemical process, leading to a higher cell efficiency and better overall yields [111]. Similar regeneration reactions can be envisioned for other metastable metal hydrides, such as $LiAlH_4$ or $Mg(AlH_4)_2$.

In this particular procedure three-quarters of the AlH_3 come from splitting the alanate ($NaAlH_4$) and one-quarter is due to Al hydrogenation (consumption of the anode). It is worth noting that reaction (14) is similar to the conventional chemical synthesis of AlH_3 (reaction (2)) where the Al^- ions from $AlCl_3$ are replaced by Al^- ions from the anode. Removing the chloride from the reaction is an important step towards achieving a low energy regeneration process since it eliminates the need to break up the salt (e.g., LiCl).

Although the above reaction is an improvement over the conventional synthesis, this procedure still requires a costly precursor, $NaAlH_4$, which is consumed during the reaction. Importantly, however, the $NaAlH_4$ can also be reformed along with the spent fuel and the byproducts of the electrochemical reaction using the well-known hydrogenation process of Bogdanović and Schwickardi [9]. After the alane is formed, sodium from the electrolyte is deposited on the platinum electrode, either in the form a metallic sodium or, if hydrogen gas is present, in the form of sodium hydride. In the latter case, the NaH (recovered from the cathode) can be reacted with the spent material (e.g., Al metal) and H_2 gas to reform $NaAlH_4$ [9]. This last step, the formation of $NaAlH_4$ from NaH and Al, requires a catalyst, typically a transition metal such as titanium. When all steps are combined, electrochemical synthesis of alane along with the reformation of $NaAlH_4$ by direct hydrogenation, this procedure yields a complete, closed-loop regeneration pathway for AlH_3 where the only thing added and consumed during the cycle is hydrogen gas.

Chemical Regeneration

Another promising approach to the regeneration of metastable metal hydrides involves a multistep, chemical process for hydride formation. In its simplest form, this is a two-step process involving hydride stabilization through the formation of a hydride adduct followed by separation and hydride recovery. Although the formation of metastable hydrides by direct hydrogenation typically requires extremely high pressures, this barrier can be substantially reduced by the addition of a stabilizing agent that forms an adduct with the hydride. Since these metastable hydride (e.g., AlH_3) can be generally classified as Lewis acids, the appropriate stabilizing agents are Lewis bases, such as ethers or amines. For example, in the presence of an amine, aluminum powder can be readily hydrogenated at pressures of a few bar, considerably lower than the 30 kbar required to hydrogenate aluminum (at 300°C) in the absence of a stabilizing amine [114–118]. Once the hydride is formed, the second step involves the removal of the stabilizing ligand and the recovery of the pure AlH_3. Generic examples of the regeneration procedure for AlH_3 and $MAlH_4$ are given below:

$$Al + L + \frac{3}{2}H_2 \longrightarrow AlH_3\text{-}L \longrightarrow AlH_3 \downarrow + L \uparrow$$

$$MH + Al + L + \frac{3}{2}H_2 \longrightarrow MAlH_4\text{-}L \longrightarrow MAlH_4 \downarrow + L \uparrow \quad (15)$$

This method can be used to regenerate a variety of metastable metal hydrides (AlH_3, $LiAlH_4$, $Mg(AlH_4)_2$, and $Ca(AlH_4)_2$), which cannot be formed by direct hydrogenation. One of the best examples is $LiAlH_4$, which cannot be reformed from LiH and Al powder directly, but in the presence of tetrahydrofuran will form the adduct $LiAlH_4$-THF at H_2 pressures of ~350 bar [69, 119, 120]. A more recent study demonstrated that the addition of a Ti catalyst lowered the hydrogenation pressure to just a few bar at room temperature [121]. Once the adduct is formed it can decomposed at low temperature (60–70°C) under partial vacuum to remove the THF and recover the AlH_3 [119]. The full regeneration procedure for $LiAlH_4$ is given below:

$$LiH + Al^* + n\text{THF} + \frac{3}{2}H_2 \longrightarrow LiAlH_4\text{–}n\text{THF} \longrightarrow LiAlH_4 \downarrow + n\text{THF} \uparrow \quad (16)$$

where Al* represents titanium-catalyzed aluminum. The hydrogenation reaction (first step in reaction (16)) is reversible and the thermodynamics were investigated by measuring pressure-composition isotherms. The free energy was found to be small and slightly negative ($\Delta G = -1.1$ kJ/mol H_2), suggesting an equilibrium hydrogen pressure of just under 1 bar at room temperature [119]. More recently, a similar approach was adopted using dimethyl ether (DME = Me_2O), rather than THF, as the stabilizing ligand, as shown below [122]:

$$LiH + Al^* + nMe_2O + \frac{3}{2}H_2 \longrightarrow LiAlH_4\text{–}nMe_2O \longrightarrow LiAlH_4 \downarrow + nMe_2O \uparrow \quad (17)$$

In this process the use of DME over THF may be more convenient since the DME is a gas at room temperature and it is easily removed after hydrogenation by simply venting the reactor. In the full regeneration procedure the dehydrogenated product (LiH + Al*) is dissolved in Me_2O at a hydrogen pressure of 100 bar. After 24 hours at room temperature, the gases H_2 and Me_2O are released and the remaining solid material in the reactor is catalyzed $LiAlH_4$. Cycling studies of $LiAlH_4$ (Ti catalyzed) produced by this method (reaction (17)) showed a hydrogen release of 7 wt.% (80–180°C) over 5 regeneration cycles [123].

A similar type of regeneration procedure (consisting of a stabilization step followed by separation) is also being pursued for aluminum hydride. Early studies of Ashby demonstrated hydrogenation of Al powder at moderate pressures in the presence of triethylenediamine (TEDA) [112]. More recently, the hydrogenation pressure required for this reaction was further reduced by the use of a transition metal catalysts (e.g., Ti). The hydrogenation of catalyzed Al* and an amine (NR_3), to form an alane amine adduct, is shown below:

$$Al^* + xNR_3 + \frac{3}{2}H_2 \longleftrightarrow AlH_3\text{–}xNR_3 \quad (18)$$

In this reversible reaction, the catalyzed Al powder (Al*) is stirred into a solution with the amine and a solvent (e.g., tetrahydrofuran). The application of moderate hydrogen pressures (1–100 bar) at low temperature (23–80°C) results in the formation of xNR_3–AlH_3, where x is typically 1 or 2. Figure 12 shows the drop in hydrogen pressure associated with the hydrogenation of Al and triethylenediamine (TEDA) in THF to form TEDA-AlH_3 at room temperature. In this particular case, the TEDA (solid) was dissolved in THF and added to the catalyzed Al* powder in the reactor. There is initially no reaction with the application of hydrogen (~35 bar) until the solution is agitated (stirrer activated), at which point there is an immediate pressure drop due to dissolution of H_2 into the solution. In the case of the Ti-catalyzed Al, the hydrogenation reaction proceeds slowly and is complete after ~3 days. The product, TEDA-AlH_3 is insoluble in THF (and most solvents) and precipitates out of solution as a white solid. It is interesting to note that the same experiment performed with uncatalyzed Al powder does not react, clearly indicating that a small amount of a transition metal catalyst (<2 mol%) is necessary to perform these hydrogenation reactions at low pressures [113].

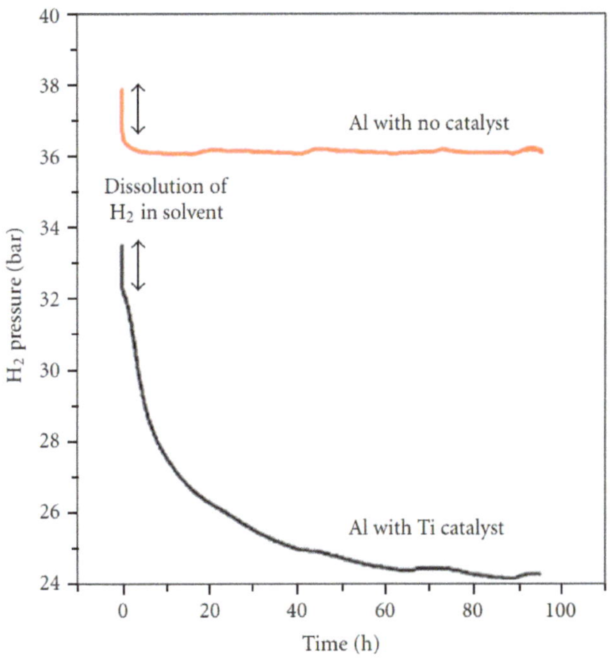

Figure 12: Uptake of hydrogen in a TEDA/THF solution with Al powder in suspension showing the pressure change for Ti-catalyzed (lower trace) and uncatalyzed (upper trace) slurries at room temperature. The rapid drop in pressure within the first few minutes is due to the dissolution of H_2 in the solvent [113].

More recent studies focused on alane adduct formation by direct hydrogenation have demonstrated that a number of other tertiary amines are also effective at stabilizing AlH_3 including trimethylamine (TMA), quinuclidine, hexamine, and dimethylethylamine (DMEA). These compounds typically form 1 : 1 or 2 : 1 adducts (NR_3 : AlH_3) depending upon the concentration. The molecular structures for a few different amine alane adducts are shown in Figure 13 (type of adduct shown (e.g., 2 : 1 or 1 : 1) is not necessarily the most stable).

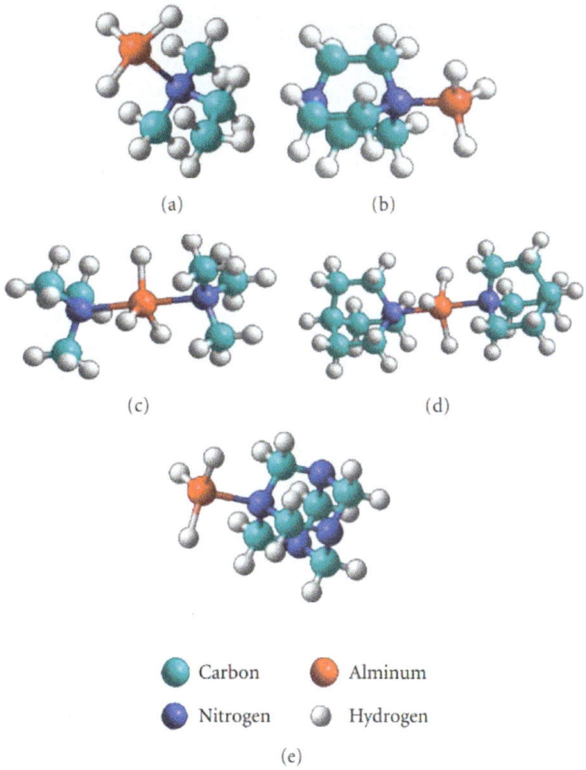

Figure 13: Adducts of AlH_3 formed by direct hydrogenation showing (a) dimethylethylamine (DMEA) alane, (b) triethylenediamine (TEDA) alane, (c) trimethylamine (TMA) alane, (d) quinuclidine alane, and (e) hexamine alane. It is important to note that amine alane adducts can form different complexes (e.g., mono, bis, and dimer) depending upon conditions (e.g., concentration). The structures shown are not necessarily the most stable configurations.

The formation of alane adducts by direct hydrogenation of Al is a relatively simple reaction that works under mild condition with a variety of amines. However, in order to recover AlH_3 from the adduct, the stabilizing molecule (amine) must be removed without decomposing the hydride. This has proven to be the primary obstacle for alane regeneration since all of the amine alane adducts shown in Figure 13 are very stable. Unlike alane etherates, which have a relatively weak Al–O bond and can be separated at low temperature (60–70°C), the Al in the alane amines is more tightly bound to the N, making the direct separation of NR_3–AlH_3 more difficult. Typically, the separation, or desolvation, of most alane amines occurs at temperatures well above 100°C where the AlH_3 is unstable. Once the amine is removed, the alane rapidly releases H_2 and decomposes to Al metal.

The challenges associated with separating alane amines have led to a search for less stable adducts and to date, there is only one known amine that can be separated from AlH_3 at low temperature (70°C), triethylamine (TEA). However, despite numerous attempts, TEA-AH_3 is not easily formed by direct hydrogenation (reaction (18)). In this case, the Al–N bond is weaker than most other alane amines and the large ethyl groups may also sterically hinder adduct formation. Currently, the search is underway for the "Goldilocks" compound, which forms a bond strong enough to stabilize the hydride, but weak enough so that the adduct can be separated at low temperature.

One way of circumventing this issue may be to simply use two different amines for the regeneration, one that forms a strong Al–N bond for the hydrogenation, and a second that forms a weak Al–N bond for the separation. In this case an amine exchange reaction, also known as transamination, is necessary to transform one adduct into the other. Although the transamination is energetically uphill, this type of reaction can be driven by exploiting the law of mass action. In this case, a solution is prepared where the second amine (which forms the less stable adduct) is maintained at a high concentration and the first amine (which forms the more stable adduct) is preferentially removed. This type of reaction places restrictions on the choice of amines, since the first amine must form a more stable adduct and should have a higher vapor pressure than the second amine or the adduct. It was recently demonstrated that the amines in DMEA-AlH_3 and TMA-AlH_3 could be exchanged with TEA to form TEA-AlH_3 under mild conditions, as shown below [114, 116]:

$$NR_3\text{–}AlH_3 + TEA \longrightarrow TEA\text{-}AlH_3 + NR_3 \uparrow \quad (19)$$

The progress of the transamination reaction can be followed using Fourier transform infrared (FTIR) spectroscopy since the positions of the Al–H stretch peaks for TMA-AlH_3 (Al–H peak at 1710 cm^{-1}) and TEA-AlH_3 (Al–H stretch at 1777 cm^{-1}) are separated by 67 cm^{-1}. Figure 14 shows the FTIR spectra from the initial hydrogenation product (TMA-AlH_3) on the bottom and a spectrum from a standard sample of TEA-AlH_3 on the top. Triethylamine was added to the TMA-AlH_3 in diethyl ether and stirred for several hours at 55°C under a sweep of nitrogen gas. The formation of triethylamine alane is indicated by the growth of a new peak at 1777 cm^{-1} (middle spectrum), which is complete after approximately 3 hrs.

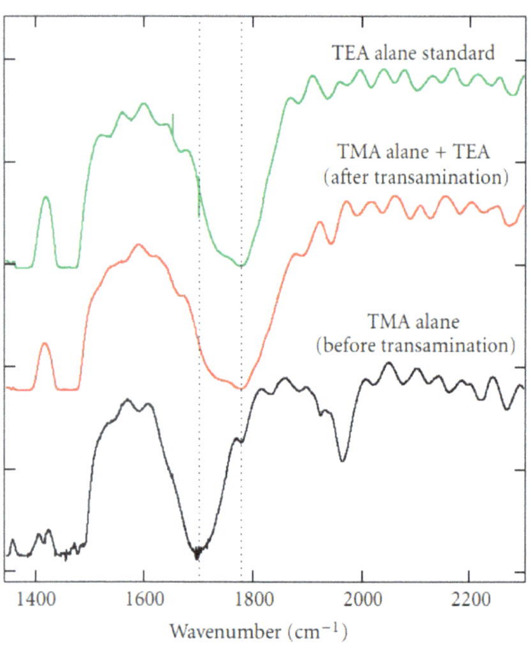

Figure 14: FTIR spectra from the TEA-AlH_3 standard (top), TMA-AlH_3 (bottom), and the reaction product after transamination, identified as TEA-AlH_3 (middle) [116].

The final step in the regeneration process involves the separation of TEA-AlH_3 into TEA and AlH_3. Although this can be accomplished a few different ways, the most common method involves heating the TEA-AlH_3 to 75°C under a nitrogen sweep or partial vacuum for ~4 hours, as shown below:

$$AlH_3\text{-}TEA \longrightarrow AlH_3 \downarrow + TEA \uparrow \quad (20)$$

The TEA is removed as a gas (recovered in a trap) and the remaining solid is crystalline α-AlH_3. X-ray diffraction patterns of the material recovered from TEA-AlH_3 showed primarily α-AlH_3 (~90 mol%), with a small amount of Al (~10 mol%).

The complete low energy regeneration procedure for α-AlH_3 is summarized in Figure 15. In this particular example DMEA is used to stabilize the alane during hydrogenation in step 1 (TMA could also be used for this step). Step 2 involves exchanging TEA for DMEA (transamination) to form TEA-AlH_3, which is separated in step 3 to recover AlH_3. Although each of the steps necessary to regenerate AlH_3 (reactions (18)–(20)) has been independently verified, the integration of all three steps remains a significant challenge. More work is needed to improve the product yields at each stage of the procedure and incorporate and simplify the three steps (hydrogenation, transamination, and separation) into a single process.

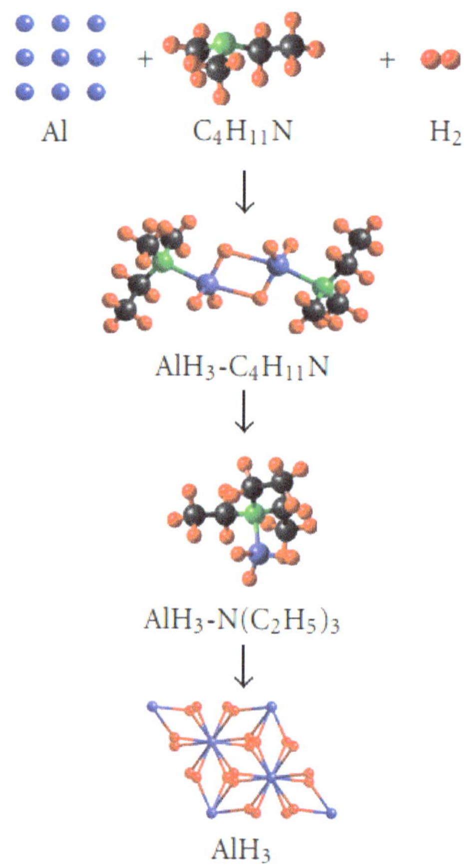

Figure 15: Regeneration pathway for AlH_3 showing hydrogenation of Al and dimethylethylamine ($C_4H_{11}N$) to form DMEA-AlH_3 (step 1), followed by transamination to form TEA-AlH_3 (step 2), and separation to recover AlH_3 (step 3) [114].

One of the primary concerns with the regeneration of metastable metal hydrides is the energy required to execute this process. In 2011 an independent analysis of AlH_3 regeneration described above (using TMA) was conducted by Hua and Ahluwalia assuming 100% conversion during hydrogenation, 75% yields in the transamination step, and 75% in the TEA-AlH_3 separation step [124]. They found that the production of 10 kg of AlH_3 (1 kg of H_2) required ~511 MJ of energy (natural gas and electricity). This energy consumption includes regeneration energy (~341 MJ) and the energy it takes to make the 1 kg of hydrogen in the first place from natural gas (~170 MJ). Given that 10 kg of AlH_3 contains 121 MJ of energy (based on the lower heating value of H_2), the well-to-tank (WTT) efficiency is ~24%. Since the temperatures for all regeneration steps in Figure 15 are typically low (<80°C), improvements in efficiency can be achieved by using industrial waste heat rather than natural gas for the heating requirements. In this case the WTT efficiencies were estimated to be 37% [122]. More recently, the WTT efficiencies for AlH_3 regeneration were raised to 41–54%, based upon improvements in the regeneration procedure [125]. Although these values still fall short of the US Department of Energy target of 60%, the efficiency of alane regeneration is estimated to be greater than other off-board hydrides, such as ammonia borane, which is estimated to have WTT efficiencies of 12–37%.

WHERE DO WE GO FROM HERE?

There is currently a growing need for innovative energy storage systems to supply power at all scales, from small microelectronics to the electrical grid. Although metastable metal hydrides are clearly unsuitable for many of these applications, they exhibit a few important properties that could prove especially useful for mobile applications: (i) low pressure hydrogen storage, (ii) high volumetric and gravimetric hydrogen density, and (iii) low temperature hydrogen release. Recently, two metastable metal hydrides, AlH_3 and $LiAlH_4$, were identified as two of only a few "near term materials" for vehicular hydrogen storage [126]. Despite many of the advantages of using a metastable hydride for automotive applications, two key challenges need to be overcome. The first is regeneration; although a few different pathways have been identified, the cost will need to come down (WTT efficiencies need to increase to 60% or more). The second challenge is the system engineering; current designs are focused on using hydride slurries to move the material into an external reactor, but that introduces a number of new challenges such as keeping the slurry flowing before and after decomposition.

In the near term, the metastable metal hydrides are ideally suited for a variety of other applications. These compounds (e.g., AlH_3, $LiAlH_4$), along with their amine adducts, are valuable chemical reductants that can be used (and is some cases are currently being used) to synthesize a variety of chemical compounds. For hydrogen storage applications, these materials are well suited to high value portable electronics that require lightweight and compact energy storage systems. Increasing demand for power systems built into backpacks, jackets, and clothing for military, industrial, and recreational use has generated much interest in AlH_3 in recent years. In addition, these materials may also be ideally suited to unmanned aerial vehicles (UAV). It is also worth noting that these hydrides can be tailored to release hydrogen on demand, either through low temperature thermolysis, hydrolysis, or both. When reacted with water (hydrolysis), hydrogen is released from the hydride and from the water increasing the total H_2 yield. This type of system may be especially useful for marine application (e.g., unmanned submersibles) where the water is available and does not need to be carried onboard. In many of these applications the amount of hydride used is relatively small so the cost is expected to be low relative to the cost of the device (e.g., UAV). In these cases the hydride would likely be prepared through the conventional synthesis procedure (rather than regenerated), which would allow the spent fuel to be discarded after use.

SUMMARY

The kinetically stabilized metal hydrides offer a new approach to hydrogen storage. The low heat of decomposition and rapid low temperature kinetics make these hydrides promising candidates for high value portable electronics in the near term and possibly automotive applications in the long term. For larger-scale applications a number of key challenges need to be overcome. These materials are not reversible at practical H_2 pressures and the conventional organometallic synthesis routes are costly. The utility of these materials will require new cost-effective and energetically efficient methods to regenerate or recycle the hydride from the spent fuel and reaction products. In addition, new hydrogen storage systems will need to be designed, specifically for the metastable metal hydrides, that enable a controllable hydrogen release within a narrow temperature window.

ACKNOWLEDGMENTS

This paper has been authored by Brookhaven Science Associates, LLC under Contract no. DE-AC02-98CH1-886 with the US Department of Energy.

REFERENCES

1. "Basic research needs for electrical energy storage," Report of the Basic Energy Sciences Workshop on Electrical Energy Storage, US Department of Energy, 2007.
2. M. Hirscher and B. Panella, "Hydrogen storage in metal-organic frameworks," Scripta Materialia, vol. 56, no. 10, pp. 809–812, 2007.
3. A. G. Wong-Foy, A. J. Matzger, and O. M. Yaghi, "Exceptional H_2 saturation uptake in microporous metal-organic frameworks," Journal of the American Chemical Society, vol. 128, no. 11, pp. 3494–3495, 2006.
4. A. Dailly, J. J. Vajo, and C. C. Ahn, "Saturation of hydrogen sorption in Zn benzenedicarboxylate and Zn naphthalenedicarboxylate," Journal of Physical Chemistry B, vol. 110, no. 3, pp. 1099–1101, 2006.
5. B. Panella, M. Hirscher, H. Pütter, and U. Müller, "Hydrogen adsorption in metal-organic frameworks: Cu-MOFs and Zn-MOFs compared," Advanced Functional Materials, vol. 16, no. 4, pp. 520–524, 2006.
6. R. Chahine and T. K. Bose, "Low-pressure adsorption storage of hydrogen," International Journal of Hydrogen Energy, vol. 19, no. 2, pp. 161–164, 1994.
7. B. Panella, M. Hirscher, and S. Roth, "Hydrogen adsorption in different carbon nanostructures,"Carbon, vol. 43, no. 10, pp. 2209–2214, 2005.
8. X. Lin, J. Jia, X. Zhao et al., "High H_2 adsorption by coordination-framework materials," Angewandte Chemie, vol. 45, no. 44, pp. 7358–7364, 2006.
9. B. Bogdanović and M. Schwickardi, "Ti-doped alkali metal aluminium hydrides as potential novel reversible

hydrogen storage materials," Journal of Alloys and Compounds, vol. 253-254, pp. 1–9, 1997.

10. J. Graetz, Y. Lee, J. J. Reilly, S. Park, and T. Vogt, "Structures and thermodynamics of the mixed alkali alanates," Physical Review B, vol. 71, no. 18, Article ID 184115, 2005.
11. L. M. Arnbjerg and T. R. Jensen, "New compounds in the potassium-aluminium-hydrogen system observed during release and uptake of hydrogen," International Journal of Hydrogen Energy, vol. 37, no. 1, pp. 345–356, 2012.
12. G. Sandrock, J. Reilly, J. Graetz, W. M. Zhou, J. Johnson, and J. Wegrzyn, "Accelerated thermal decomposition of AlH_3 for hydrogen-fueled vehicles," Applied Physics A, vol. 80, no. 4, pp. 687–690, 2005.
13. E. Fakioğlu, Y. Yürüm, and T. N. Veziroğlu, "A review of hydrogen storage systems based on boron and its compounds," International Journal of Hydrogen Energy, vol. 29, no. 13, pp. 1371–1376, 2004.
14. F. H. Stephens, V. Pons, and R. T. Baker, "Ammonia-borane: the hydrogen source par excellence?"Dalton Transactions, no. 25, pp. 2613–2626, 2007.
15. A. Staubitz, A. P. M. Robertson, and I. Manners, "Ammonia-Borane and related compounds as dihydrogen sources," Chemical Reviews, vol. 110, no. 7, pp. 4079–4124, 2010.
16. N. C. Smythe and J. C. Gordon, "Ammonia borane as a hydrogen carrier: dehydrogenation and regeneration," European Journal of Inorganic Chemistry, no. 4, pp. 509–521, 2010. ·
17. T. Hua, R. Ahluwalia, J. -K Peng et al., "Technical assessment of compressed hydrogen storage tank systems for automotive applications," U.S. Department of Energy Report prepare by Argonne National Laboratory, 2010, http://www1.eere.energy.gov/hydrogenandfuelcells/pdfs/compressedtank_storage.pdf/.
18. J. W. Turley and H. W. Rinn, "The crystal structure of aluminum hydride," Inorganic Chemistry, vol. 8, no. 1, pp. 18–22, 1969.
19. J. Graetz and J. J. Reilly, "Decomposition kinetics of the AlH_3 polymorphs," Journal of Physical Chemistry B, vol. 109, no. 47, pp. 22181–22185, 2005. ·
20. J. Graetz and J. J. Reilly, "Thermodynamics of the α, β and γ polymorphs of AlH_3," Journal of Alloys and Compounds, vol. 424, no. 1-2, pp. 262–265, 2006. ·
21. H. W. Brinks, A. Istad-Lem, and B. C. Hauback, "Mechanochemical synthesis and crystal structure ofα'-AlD_3 and α-AlD_3," Journal of Physical Chemistry B, vol. 110, no. 51, pp. 25833–25837, 2006.
22. H. W. Brinks, W. Langley, C. M. Jensen, J. Graetz, J. J. Reilly, and B. C. Hauback, "Synthesis and crystal structure of β-AlD_3," Journal of Alloys and Compounds, vol. 433, no. 1-2, pp. 180–183, 2007.
23. V. A. Yartys, R. V. Denys, J. P. Maehlen et al., "Double-bridge bonding of aluminium and hydrogen in the crystal structure of γ-AlH_3," Inorganic Chemistry, vol. 46, no. 4, pp. 1051–1055, 2007.
24. H. W. Brinks, C. Brown, C. M. Jensen, J. Graetz, J. J. Reilly, and B. C. Hauback, "The crystal structure of γ-AlD_3," Journal of Alloys and Compounds, vol. 441, no. 1-2, pp. 364–367, 2007. ·
25. J. Graetz, J. J. Reilly, J. G. Kulleck, and R. C. Bowman, "Kinetics and thermodynamics of the aluminum hydride polymorphs," Journal of Alloys and Compounds, vol. 446-447, pp. 271–275, 2007.
26. S.-J. Hwang, R. C. Bowman, J. Graetz, J. J. Reilly, W. Langley, and C. M. Jensen, "NMR studies of the aluminum hydride phases and their stabilities," Journal of Alloys and Compounds, vol. 446, pp. 290–295, 2007.
27. M. Tkacz, T. Palasyuk, J. Graetz, and S. Saxena, "High-pressure Raman spectroscopy study of α and γpolymorphs of AlH_3," Journal of Raman Spectroscopy, vol. 39, no. 7, pp. 922–927, 2008.
28. J. Graetz, J. J. Reilly, V. A. Yartys et al., "Aluminum hydride as a hydrogen and energy storage material: past, present and future," Journal of Alloys and Compounds, vol. 509, no. 2, supplement, pp. S517–S528, 2011.
29. D. Blanchard, H. W. Brinks, B. C. Hauback, P. Norby, and J. Muller, "Isothermal decomposition of LiAlD4 with and without additives," Journal of Alloys and Compounds, vol. 404-406, pp. 743–747, 2005.
30. A. Andreasen, T. Vegge, and A. S. Pedersen, "Dehydrogenation kinetics of as-received and ball-milled LiAlH4," Journal of Solid State Chemistry, vol. 178, no. 12, pp. 3672–3678, 2005. ·
31. A. Andreasen, "Effect of Ti-doping on the dehydrogenation kinetic parameters of lithium aluminum hydride," Journal of Alloys and Compounds, vol. 419, no. 1-2, pp. 40–44, 2006.
32. K. M. Mackay, Hydrogen Compounds of the Metallic Elements, Spon, London, UK, 1966.
33. T. N. Dymova, D. P. Aleksandrov, V. N. Konoplev, T. A. Silina, and A. S. Sizareva, "Spontaneous and thermal-decomposition of lithium tetrahydroaluminate $LiAlH_4$—the promoting effect of mechanochemical action on the process," Koordinatsionnaya Khimiya, vol. 20, no. 4, pp. 279–285, 1994.
34. T. N. Dymova, D. P. Aleksandrov, V. N. Konoplev, and T. A. Silina, "Novel view of the nature of the chemical-composition and phase-composition modifications in lithium hydridoaluminates $LiAlH_4$and $LiAlH_6$ on heating," Koordinatsionnaya Khimiya, vol. 21, no. 3, pp. 175–182, 1995.
35. M. Fichtner and O. Fuhr, "Synthesis and structures of magnesium alanate and two solvent adducts,"Journal of Alloys and Compounds, vol. 345, no. 1-2, pp. 286–296, 2002.
36. O. M. Løvvik, S. M. Opalka, H. W. Brinks, and B. C. Hauback, "Crystal structure and thermodynamic stability of the lithium alanates $LiAlH_4$ and Li_3AlH_6," Physical Review B, vol. 69, no. 13, Article ID 134117, 9 pages, 2004.
37. M. Fichtner, O. Fuhr, and O. Kircher, "Magnesium alanate—a material for reversible hydrogen storage?" Journal of Alloys and Compounds, vol. 356, pp. 418–422, 2003.
38. M. Mamatha, B. Bogdanović, M. Felderhoff et al.,

"Mechanochemical preparation and investigation of properties of magnesium, calcium and lithium-magnesium alanates," Journal of Alloys and Compounds, vol. 407, no. 1-2, pp. 78–86, 2006.

39. E. Wiberg and R. U. Lacal, Zeitschrift für Naturforschung, vol. 6, p. 392, 1951.

40. D. M. Liu, Z. X. Qian, T. Z. Si, and Q. A. Zhang, "Synthesis, crystal structure and thermal decomposition of $LiCa(AlH_4)_3$," Journal of Alloys and Compounds, vol. 520, pp. 202–206, 2012.

41. S. I. Orimo, Y. Nakamori, J. R. Eliseo, A. Züttel, and C. M. Jensen, "Complex hydrides for hydrogen storage," Chemical Reviews, vol. 107, no. 10, pp. 4111–4132, 2007.

42. M. Mamatha, C. Weidenthaler, A. Pommerin, M. Felderhoff, and F. Schüth, "Comparative studies of the decomposition of alanates followed by in situ XRD and DSC methods," Journal of Alloys and Compounds, vol. 416, no. 1-2, pp. 303–314, 2006.

43. R. A. Varin and R. Parviz, "The effects of the micrometric and nanometric iron (Fe) additives on the mechanical and thermal dehydrogenation of lithium alanate ($LiAlH_4$), its self-discharge at low temperatures and rehydrogenation," International Journal of Hydrogen Energy, vol. 37, pp. 9088–9102, 2012.

44. M. J. Van Setten, G. A. De Wijs, V. A. Popa, and G. Brocks, "Ab initio study of Mg $(AlH_4)_4$," Physical Review B, vol. 72, no. 7, Article ID 073107, 2005.

45. G. C. Sinke, L. C. Walker, F. L. Oetting, and D. R. Stull, "Thermodynamic properties of aluminum hydride," The Journal of Chemical Physics, vol. 47, no. 8, pp. 2759–2761, 1967. ·

46. L. Schlapbach, "Hydrogen as a fuel and its storage for mobility and transport," MRS Bulletin, vol. 27, no. 9, pp. 675–676, 2002.

47. R. S. Irani, "Hydrogen storage: high-pressure gas containment," MRS Bulletin, vol. 27, no. 9, pp. 680–682, 2002.

48. J. Wolf, "Liquid-hydrogen technology for vehicles," MRS Bulletin, vol. 27, no. 9, pp. 684–687, 2002.

49. R. C. Bowman Jr. and B. Fultz, "Metallic hydrides I: hydrogen storage and other gas-phase applications," MRS Bulletin, vol. 27, no. 9, pp. 688–693, 2002.

50. J. M. Joubert, M. Latroche, and A. Percheron-Guégan, "Metallic hydrides II: materials for electrochemical storage," MRS Bulletin, vol. 27, no. 9, pp. 694–698, 2002. ·

51. E. Akiba and M. Okada, "Metallic hydrides III: body-centered-cubic solid-solution alloys," MRS Bulletin, vol. 27, no. 9, pp. 699–703, 2002.

52. A. Züttel and S. I. Orimo, "Hydrogen in nanostructured, carbon-related, and metallic materials," MRS Bulletin, vol. 27, no. 9, pp. 705–711, 2002.

53. B. Bogdanović and G. Sandrock, "Catalyzed complex metal hydrides," MRS Bulletin, vol. 27, no. 9, pp. 712–716, 2002.

54. O. Stecher and E. Wiberg, "A non-volatile, polymer aluminium hydrogen $(AlH_3)_x$ and several volatile compounds of monomeric AlH_3," Berichte der Deutschen Chemischen Gesellschaft, vol. 75, pp. 2003–2012, 1942.

55. A. E. Finholt, A. C. Bond, and H. I. Schlesinger, "Lithium aluminum hydride, aluminum hydride and lithium gallium hydride, and some of their applications in organic and inorganic chemistry," Journal of the American Chemical Society, vol. 69, no. 5, pp. 1199–1203, 1947.

56. G. Chizinsky, G. G. Evans, T. R. P. Gibb, and M. J. Rice, "Non-solvated aluminum hydride," Journal of the American Chemical Society, vol. 77, no. 11, pp. 3164–3165, 1955.

57. F. M. Brower, N. E. Matzek, P. F. Reigler et al., "Preparation and properties of aluminum hydride,"Journal of the American Chemical Society, vol. 98, no. 9, pp. 2450–2453, 1976. ·

58. H. Nöth and E. Wiberg, Forschr Und Forsch, vol. 8, no. 3, p. 344, 1966.

59. B. M. Bulychev, A. G. Burlakova, and P. A. Storozhenko, "Complex compounds of aluminum hydride ethoxide with mixed aluminum and boron hydrides with lithium and magnesium: compositions, physicochemical properties, and synthesis of unsolvated aluminum hydride," Zhurnal Neorganicheskoj Khimii, vol. 43, no. 5, pp. 829–836, 1998.

60. H. C. Brown and N. M. Yoon, "Selective reductions. X. Reaction of aluminum hydride with selected organic compounds containing representative functional groups. Comparison of the reducing characteristics of lithium aluminum hydride and its derivatives," Journal of the American Chemical Society, vol. 88, no. 7, pp. 1464–1472, 1966.

61. C. W. Heitsch, C. E. Nordman, and R. W. Parry, "The crystal structure and dipole moment in solution of the compound $AlH_3{\cdot}2n(CH_3)_3$," Inorganic Chemistry, vol. 2, no. 3, pp. 508–512, 1963.

62. F. Dallacker, K. W. Glombitza, and M. Lipp, "Derivate des methylendioxybenzols, III. Darstellung des 45-methylendioxy-phthalaldehyds," Justus Liebigs Annalen der Chemie, vol. 643, no. 1–3, pp. 67–82, 1961.

63. L. W. Trevoy and W. G. Brown, "Mechanism of lithium aluminum hydride reactions," Journal of the American Chemical Society, vol. 71, no. 5, pp. 1675–1678, 1949.

64. E. C. Ashby, J. R. Sanders, P. Claudy, and R. Schwartz, "Diethyl ether soluble aluminum hydride,"Journal of the American Chemical Society, vol. 95, no. 19, pp. 6485–6486, 1973.

65. A. E. Finholt, G. D. Barbaras, G. K. Barbaras, G. Urry, T. Wartik, and H. I. Schlesinger, "The preparation of sodium and calcium aluminium hydrides," Journal of Inorganic and Nuclear Chemistry, vol. 1, no. 4-5, pp. 317–325, 1955.

66. W. Schwab and K. Wintersberger, "Uber darstellung und eigenschaften von calcium-aluminiumhydrid $Ca(AlH_4)_4$," Zeitschrift für Naturforschung, vol. 8, no. 11, pp. 690–691, 1953.

67. B. M. Bulychev, A. V. Golubeva, P. A. Storozhenko, and K. N. Semenenko, "Synthesis of aluminum hydride

compounds from sodium hydride and aluminum trichloride in diethyl ether," Zhurnal Neorganicheskoj Khimii, vol. 43, no. 8, pp. 1242–1245, 1998.

68. E. C. Ashby, "A direct route to complex metal hydrides," Chemistry & Industry, vol. 5, pp. 208–209, 1962.

69. E. C. Ashby, G. J. Brendel, and H. E. Redman, "Direct synthesis of complex metal hydrides," Inorganic Chemistry, vol. 2, no. 3, pp. 499–504, 1963.

70. B. Bogdanović, R. A. Brand, A. Marjanović, M. Schwickardi, and J. Tölle, "Metal-doped sodium aluminium hydrides as potential new hydrogen storage materials," Journal of Alloys and Compounds, vol. 302, no. 1-2, pp. 36–58, 2000.

71. A. E. Finholt, A. C. Bond, and H. I. Schlesinger, "Lithium aluminum hydride, aluminum hydride and lithium gallium hydride, and some of their applications in organic and inorganic chemistry," Journal of the American Chemical Society, vol. 69, no. 5, pp. 1199–1203, 1947.

72. M. Fichtner, C. Frommen, and O. Fuhr, "Synthesis and properties of calcium alanate and two solvent adducts," Inorganic Chemistry, vol. 44, no. 10, pp. 3479–3484, 2005.

73. E. Wiberg and R. Bauer, "Neues zur kenntnis des magnesium-aluminium-wasserstoffs $Mg(AlH_4)_2$,"Zeitschrift für Naturforschung, vol. 7, no. 2, pp. 131–132, 1952.

74. M. Fichtner, C. Frommen, and O. Fuhr, "Synthesis and properties of calcium alanate and two solvent adducts," Inorganic Chemistry, vol. 44, no. 10, pp. 3479–3484, 2005.

75. T. N. Dymova, V. N. Konoplev, A. S. Sizareva, and D. P. Aleksandrov, "Magnesium tetrahydroaluminate: solid-phase formation with mechanochemical activation of a mixture of aluminum and magnesium hydrides," Koordinatsionnaya Khimiya, vol. 25, no. 5, pp. 312–315, 1999.

76. H. Kabbour, C. C. Ahn, S. J. Hwang, R. C. Bowman Jr., and J. Graetz, "Direct synthesis and NMR characterization of calcium alanate," Journal of Alloys and Compounds, vol. 446, pp. 264–266, 2007.

77. G. W. Schaeffer, J. S. Roscoe, and A. C. Stewart, "The reduction of iron (III) chloride with lithium aluminohydride and lithium borohydride: iron (II) borohydride," Journal of the American Chemical Society, vol. 78, no. 4, pp. 729–733, 1956.

78. W. Grochala and P. P. Edwards, "Thermal decomposition of the non-interstitial hydrides for the storage and production of hydrogen," Chemical Reviews, vol. 104, no. 3, pp. 1283–1315, 2004.

79. S. Orimo, Y. Nakamori, T. Kato, C. Brown, and C. M. Jensen, "Intrinsic and mechanically modified thermal stabilities of α-, β- and γ-aluminum trihydrides AlH_3," Applied Physics A, vol. 83, no. 1, pp. 5–8, 2006.

80. S. Sartori, S. M. Opalka, O. M. Løvvik, M. N. Guzik, X. Tang, and B. C. Hauback, "Experimental studies of α-AlD3 and α′-AlD_3 versus first-principles modelling of the alane isomorphs," Journal of Materials Chemistry, vol. 18, no. 20, pp. 2361–2370, 2008. ·

81. K. Tomiyasu, T. Sato, K. Horigane, S. Orimo, and K. Yamada, "Hydrogen release from Li alanates originates in molecular lattice instability emerging at ~100 K," Applied Physics Letters, vol. 100, Article ID 193901, 3 pages, 2012.

82. P. Claudy, B. Bonnetot, and J. M. Letoffe, "Preparation and physicochemical properties of magnesium alanate," Journal of Thermal Analysis, vol. 15, no. 1, pp. 119–128, 1979.

83. N. N. Mal›tseva, A. I. Golovanova, T. N. Dymova, and D. P. Aleksandrov, "Solid-phase formation of calcium hydridoaluminates $Ca(AlH_4)_2$ and $CaHAlH_4$ upon mechanochemical activation or heating of mixtures of calcium hydride with aluminum chloride," Russian Journal of Inorganic Chemistry, vol. 46, no. 12, pp. 1793–1797, 2001.

84. T. N. Dymova, D. P. Aleksandrov, V. N. Konoplev, T. A. Silina, and N. T. Kuznetsov, "Solid-phase reactions of aluminum $(AlH_3)_x$ crystalline hydride—formation of alkali-metal tetrahydroaluminates and hexahydroaluminates," Koordinatsionnaya Khimiya, vol. 19, no. 7, pp. 529–534, 1993.

85. B. Bogdanović, R. A. Brand, A. Marjanović, M. Schwickardi, and J. Tölle, "Metal-doped sodium aluminium hydrides as potential new hydrogen storage materials," Journal of Alloys and Compounds, vol. 302, no. 1-2, pp. 36–58, 2000.

86. G. S. Sandrock, L. Suda, and L. Schlapbach, "Applications," in Hydrogen in Intermetallic Compounds II, L. Schlapbach, Ed., pp. 197–258, Springer, Berlin, Germany, 1992.

87. W. de Graff, [Ph.D. thesis], University of Amsterdam, Amsterdam, The Netherlands, 1960.

88. B. Baranowski, "Thermodynamics of metal-hydrogen systems at high-pressures," The Journal of Physical Chemistry, vol. 76, no. 8, pp. 714–724, 1972.

89. J. Graetz, S. Chaudhuri, Y. Lee, T. Vogt, J. T. Muckerman, and J. J. Reilly, "Pressure-induced structural and electronic changes in α-AlH_3," Physical Review B, vol. 74, no. 21, Article ID 214114, 7 pages, 2006.

90. M. Tkacz, S. Filipek, and B. Baranowski, "High-pressure synthesis of aluminum-hydride from the elements," Polish Journal of Chemistry, vol. 57, no. 4–6, pp. 651–653, 1983.

91. B. Baranowski and M. Z. Tkacz, "The equilibrium between solid aluminum-hydride and gaseous-hydrogen," The Journal of Physical Chemistry, vol. 135, pp. 27–38, 1983.

92. S. K. Konovalov and B. M. Bulychev, "The P,T-state diagram and solid phase synthesis of aluminium hydride," Inorganic Chemistry, vol. 34, no. 1, pp. 172–175, 1995.

93. H. Saitoh, A. MacHida, Y. Katayama, and K. Aoki, "Formation and decomposition of AlH_3 in the aluminum-hydrogen system," Applied Physics Letters, vol. 93, no. 15, Article ID 151918, 3 pages, 2008.

94. J. Graetz, "New approaches to hydrogen storage," Chemical Society Reviews, vol. 38, no. 1, pp. 73–82, 2009.

95. G. Sandrock, K. Gross, and G. Thomas, "Effect of Ti-catalyst content on the reversible hydrogen storage properties of the sodium alanates," Journal of Alloys and Compounds, vol. 339, no. 1-2, pp. 299–308, 2002.

96. M. McCarty Jr., J. N. Maycock, and V. R. Pai Verneker, "Thermal decomposition of $LiAlH_4$," Journal of Physical Chemistry, vol. 72, no. 12, pp. 4009–4014, 1968.

97. G. Sandrock, K. Gross, G. Thomas, C. Jensen, D. Meeker, and S. Takara, "Engineering considerations in the use of catalyzed sodium alanates for hydrogen storage," Journal of Alloys and Compounds, vol. 330, pp. 696–701, 2002.

98. S.-S. Liu, L.-X. Sun, Y. Zhang et al., "Effect of ball milling time on the hydrogen storage properties of TiF_3-doped $LiAlH_4$," International Journal of Hydrogen Energy, vol. 34, no. 19, pp. 8079–8085, 2009.

99. Y. Kojima, Y. Kawai, M. Matsumoto, and T. Haga, "Hydrogen release of catalyzed lithium aluminum hydride by a mechanochemical reaction," Journal of Alloys and Compounds, vol. 462, no. 1-2, pp. 275–278, 2008.

100. J. Chen, N. Kuriyama, Q. Xu, H. T. Takeshita, and T. Sakai, "Reversible hydrogen storage via titanium-catalyzed $LiAlH_4$ and Li_3AlH_6," Journal of Physical Chemistry B, vol. 105, no. 45, pp. 11214–11220, 2001.

101. D. Blanchard, H. W. Brinks, B. C. Hauback, P. Norby, and J. Muller, "Isothermal decomposition of $LiAlD_4$ with and without additives," Journal of Alloys and Compounds, vol. 404–406, pp. 743–747, 2005.

102. J. R. A. Fernandez, F. Aguey-Zinsou, M. Elsaesser et al., "Mechanical and thermal decomposition of$LiAlH_4$ with metal halides," International Journal of Hydrogen Energy, vol. 32, no. 8, pp. 1033–1040, 2007.

103. Z. Xueping and L. Shenglin, "Study on hydrogen storage properties of $LiAlH_4$," Journal of Alloys and Compounds, vol. 481, no. 1-2, pp. 761–763, 2009.

104. M. Ismail, Y. Zhao, X. B. Yu, and S. X. Dou, "Effects of NbF5 addition on the hydrogen storage properties of $LiAlH_4$," International Journal of Hydrogen Energy, vol. 35, no. 6, pp. 2361–2367, 2010.

105. T. Sun, C. K. Huang, H. Wang, L. X. Sun, and M. Zhu, "The effect of doping $NiCl_2$ on the dehydrogenation properties of $LiAlH_4$," International Journal of Hydrogen Energy, vol. 33, no. 21, pp. 6216–6221, 2008.

106. R. A. Varin, L. Zbroniec, T. Czujko, and Z. S. Wronski, "The effects of nanonickel additive on the decomposition of complex metal hydride $LiAlH_4$ (lithium alanate)," International Journal of Hydrogen Energy, vol. 36, no. 1, pp. 1167–1176, 2011.

107. L. Hima Kumar, B. Viswanathan, and S. Srinivasa Murthy, "Dehydriding behaviour of $LiAlH_4$—the catalytic role of carbon nanofibres," International Journal of Hydrogen Energy, vol. 33, no. 1, pp. 366–373, 2008.

108. P. J. Herley and O. Christofferson, "Decomposition of α-aluminum hydride powder. 3. Simultaneous photolytic-thermal decomposition," Journal of Physical Chemistry, vol. 85, no. 13, pp. 1887–1892, 1981.

109. Rafi-Ud-Din, L. Zhang, L. Ping, and Q. Xuanhui, "Catalytic effects of nano-sized TiC additions on the hydrogen storage properties of $LiAlH_4$," Journal of Alloys and Compounds, vol. 508, no. 1, pp. 119–128, 2010.

110. H. Clasen, Germany Patent, 1141, 623, 1962.

111. H. K. Birnbaum, C. Buckley, F. Zeides et al., "Hydrogen in aluminum," Journal of Alloys and Compounds, vol. 253-254, pp. 260–264, 1997.

112. R. Zidan, B. L. Garcia-Diaz, C. S. Fewox, A. C. Stowe, J. R. Gray, and A. G. Harter, "Aluminium hydride: a reversible material for hydrogen storage," Chemical Communications, no. 25, pp. 3717–3719, 2009.

113. M. J. Martinez-Rodriguez, B. L. Garcia-Diaz, J. A. Teprovich, D. A. Knight, and R. Zidan, "Advances in the electrochemical regeneration of aluminum hydride," Applied Physics A, vol. 106, no. 3, pp. 545–550, 2012.

114. E. C. Ashby, "The direct synthesis of amine alanes," Journal of the American Chemical Society, vol. 86, no. 9, pp. 1882–1883, 1964.

115. J. Graetz, S. Chaudhuri, J. Wegrzyn et al., "Direct and reversible synthesis of AlH_3-triethylenediamine from Al and H_2," Journal of Physical Chemistry C, vol. 111, no. 51, pp. 19148–19152, 2007.

116. D. Lacina, J. Wegrzyn, J. Reilly, Y. Celebi, and J. Graetz, "Regeneration of aluminium hydride using dimethylethylamine," Energy and Environmental Science, vol. 3, no. 8, pp. 1099–1105, 2010.

117. D. Lacina, J. Reilly, J. Johnson, J. Wegrzyn, and J. Graetz, "The reversible synthesis of bis(quinuclidine) alane," Journal of Alloys and Compounds, vol. 509, no. 2, pp. S654–S657, 2011.

118. D. Lacina, J. Reilly, Y. Celebi, J. Wegrzyn, J. Johnson, and J. Graetz, "Regeneration of aluminum hydride using trimethylamine," Journal of Physical Chemistry C, vol. 115, no. 9, pp. 3789–3793, 2011.

119. H. Clasen, "Alanat-synthese aus den elementen und ihre bedeutung," Angewandte Chemie, vol. 73, no. 10, pp. 322–331, 1961.

120. J. Wang, A. D. Ebner, and J. A. Ritter, "Physiochemical pathway for cyclic dehydrogenation and rehydrogenation of $LiAlH_4$," Journal of the American Chemical Society, vol. 128, no. 17, pp. 5949–5954, 2006.

121. J. Graetz, J. Wegrzyn, and J. J. Reilly, "Regeneration of lithium aluminum hydride," Journal of the American Chemical Society, vol. 130, no. 52, pp. 17790–17794, 2008.

122. X. Liu, G. S. McGrady, H. W. Langmi, and C. M. Jensen, "Facile cycling of Ti-doped $LiAlH_4$ for high performance hydrogen storage," Journal of the American Chemical Society, vol. 131, no. 14, pp. 5032–5033, 2009.

123. X. F. Liu, H. W. Langmi, S. D. Beattie, F. F. Azenwi, G. S. McGrady, and C. M. Jensen, "Ti-Doped $LiAlH_4$ for hydrogen storage: synthesis, catalyst loading and cycling performance," Journal of the American Chemical Society, vol. 133, no. 39, pp. 15593–15597, 2011.

124. T. Q. Hua and R. K. Ahluwalia, "Alane hydrogen storage for automotive fuel cells—off-board regeneration processes and efficiencies," International Journal of Hydrogen Energy, vol. 36, no. 23, pp. 15259–15265, 2011.

125. R. K. Ahluwalia, T. Q. Hua, and J. K. Peng, "On-board and Off-board performance of hydrogen storage options for light-duty vehicles," International Journal of Hydrogen Energy, vol. 37, no. 3, pp. 2891–2910, 2012.

126. L. Klebanoff, "5-year review of Metal Hydride Center of Excellence," Presentation at DOE H_2 Program Annual Merritt Review, Washington, DC, USA, 2010.

CHAPTER 4 MATERIAL DEMANDS FOR STORAGE TECHNOLOGIES IN A HYDROGEN ECONOMY

M. Kunowsky, J. P. Marco-Lózar, and A. Linares-Solano

MOTIVATION

Today's energy sector is accompanied by a number of environmental inconveniences. In order to overcome those problems, future energy concepts have to be put into practice. In particular, renewable energies are needed, because (i) fossil energy promotes global warming and environmental contamination, (ii) the supply of nonrenewable energy sources is finite, and (iii) nuclear energy presents a serious danger due to its radioactive waste products. Among renewable energies, technologies for hydrogen storage will be an important piece of the jigsaw.

The world's current energy supplies are mainly based on fossil energy resources. These resources have their origin in organic (and therefore carbon-containing) compounds, which have been converted throughout millions of years. By burning them today, these resources are reintroduced into the natural carbon cycle and increase the CO_2 content of the atmosphere. CO_2 gas increases the world's greenhouse effect, leading to global warming [1]. In Figure 1 it can be seen that, since the industrialization in the 19th century, the atmospheric CO_2 content is continuously rising. Also the evolution of the global temperature is shown in Figure 1. For the same time period, an overall increasing tendency can be observed, which suggests a relationship between both trends that could be something more than casual.

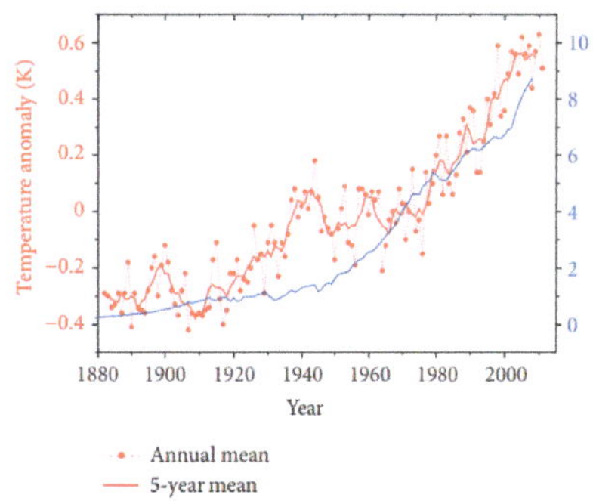

Figure 1: Global temperature (red) and CO_2 emissions (blue) over the past decades [14–16].

Another problem with natural and fossil energy resources is their limited availability. In the case of oil, this fact can be exemplified by the so-called peak oil phenomenon. The term "peak oil" was introduced by Shell-company scientist Hubbert more than 50 years ago, who correctly predicted that the oil production of the USA would have a maximum in the 1970s [2, 3]. Nowadays, this theory is applied to predict the worldwide production of oil. According to the peak oil theory, oil production reaches a maximum at a specific date and then drops by the same rate as its increase, leading to a symmetrical curve. While the main concept of peak oil finds wide acceptance, opinions diverge when it comes to questions of

Citation: M. Kunowsky, J. P. Marco-Lózar, and A. Linares-Solano, "Material Demands for Storage Technologies in a Hydrogen Economy," Journal of Renewable Energy, vol. 2013, Article ID 878329, 16 pages, 2013. doi:10.1155/2013/878329.

details. Particularly the present and future impacts of peak oil on science, technology, society, politics, business, and so forth are being lively discussed throughout the fields [4–13]. While some claim that peak oil already occurred, others speculate that this date is still far-off [6]. Indeed, the phenomenon turned out to come true for a number of single countries, and, apparently, it is also correct if only "conventional" oil fields are taken into account. This refers to oil fields of a considerable size which are easily accessible. In Figure 2, the world oil consumption is shown over the years, together with the discovery of conventional oil fields. It can be seen that the consumption increases continuously. The discovery of large oil reservoirs, on the other hand, increased rapidly in the beginning, having its peak in the 1960s. From then on, the curve decreases as rapidly as it had increased, leading to a bell-shaped curve. In the 1990s, the consumption overreached the discovery. Thus, the oil that is consumed today has been found already decades ago or emanates from less accessible, and therefore more costly, oil sources. Summing up, natural resources are finite and will run out sooner or later.

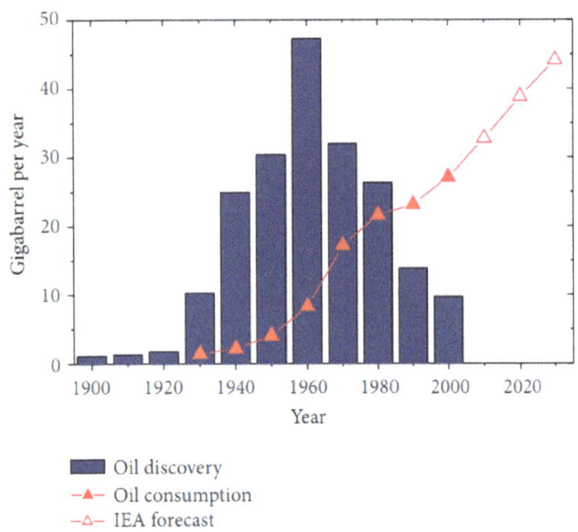

Figure 2: Discovery of conventional oil reserves (blue) and oil consumption (red) [5].

Based on the aforementioned facts, it can be concluded that alternatives need to be found for the future energy market and that, in the long run, only renewable energy sources can provide sustainability. However, the usage of renewable energy sources leads to a number of new problems. On one hand, there is the temporal misfit between energy generation and consumption. For example, electricity can be produced by solar energy at daytime, but not during the night when it is needed for illumination. On the other hand, primary renewable energy sources are not efficient enough to be used in mobile applications, where space is limited and weight has to be low. Both of these problems can be solved by means of the so-called "hydrogen economy".

INTRODUCTION

The Hydrogen Economy

Fundamental research on hydrogen was firstly made in the middle of the 18th century by Henry Cavendish, who discovered the element in 1766, and by Antoine Lavoisier, who gave it its name in 1783 [17]. Despite the attention which is paid to it today, the idea of using hydrogen as an energy carrier is not new at all. In fact, syngas and coal gas found widespread use for heating, illumination, and engines since around 1800, and for more than 150 years, until being replaced by natural gas [17, 18]. These gases are synthesized by steam gasification of different carbon sources and contain about 50% hydrogen mixed with other gases [17]. In fact, syngas is still being utilized today, in locations where natural gas is unavailable or costly [18]. Already in 1874, the novelist and science-fiction pioneer Jules Verne imagined a self-sustaining energy system based on electrolyzed hydrogen in his novel "The Mysterious Island".

However, the term "hydrogen economy" was not introduced until 1972 by Bockris [19]. The concept of the hydrogen economy gained great public attention as a consequence of the oil crises in the 1970s and moreover after the nuclear disasters of Chernobyl in 1986 and Fukushima in 2011. The basic idea behind the hydrogen economy is that hydrogen is produced by renewable energy sources, transported over large distances, stored in large amounts, and finally used to generate electricity, heat, and propulsion for the consumer [19]. Hydrogen could be produced centrally as well as locally and from numerous sources like wind, water, wave, solar, or biomass. Furthermore, its distribution could coexist with a conventional electricity network in order to provide households, industry, and vehicles with energy. For the distribution of hydrogen, a pipeline network, similar to today's natural gas system, could be used, if materials are found to be compatible [18].

Hydrogen Properties

Hydrogen is the lightest of all elements and the most abundant one in the universe [18, 20]. On earth, however, it only ranks the 15th element, and its concentration in the atmosphere is very low (0.5 ppm) [20,21]. Yet it forms the largest number of chemical compounds, water being the most abundant of them [20, p. 351 f.].

In its gaseous state, hydrogen forms the diatomic molecule H_2 (dihydrogen). Some basic properties of hydrogen are listed in Table 1. It can be seen that it is a solid below 13.85 K, and its liquid form exists only in a small range of temperatures up to 20.25 K. Above its critical point (T_C = 32.97 K; P_C = 1.276 MPa) hydrogen is a supercritical fluid. Hydrogen gas is transparent, odorless, and nontoxic.

Table 1: Hydrogen properties.

Atomic weight	1.0079	g mol^{-1}
Van der Waals radius	120	pm
Covalent radius	31	pm
ΔH ($H_2 \rightarrow 2H$)	436	kJ mol^{-1}
Density at 273 K and 0.1 MPa	0.0899	g l^{-1}
Liquid density (20.25 K)	70.8	g l^{-1}
Boiling point	20.25	K
Critical temperature	32.97	K
Critical pressure	1.276	MPa
Self-ignition temperature	747	K
Flammability limit in air	4.1–74.2	Vol.%
Explosive limits in air	18.3–59	Vol.%
Diffusion coefficient in air	0.634	cm^2 s^{-1}

Hydrogen burns with invisible flame and over a wide range of concentrations in air. These points call critics into action and provoke concerns about hydrogen safety. The Hindenburg Zeppelin disaster from 1937 is a popular example which is often cited by critics against hydrogen technology. However, the subject is controversially discussed, and some studies of the accident have revealed that, instead of its hydrogen charge, a highly flammable skin material of the airship may have caused the fire [22, 23]. Hydrogen is around 14 times lighter than air and has a high diffusion coefficient. In the case of an accidental release, this causes a strong buoyancy and a fast volatilization into the atmosphere. Gasoline, in contrast, does not disperse rapidly and builds nonbuoyant, toxic fumes. In Table 2, properties of different fuels are compared. It can be seen that the ignition energies at the lower flammability limit are in general very low. Thus, already electrostatic sparks may have sufficient energy for igniting fuel/air mixtures under these conditions [18]. Regarding the flammable and explosive ranges, the lower limits are generally considered more important in the case of an accident [18]. Despite the fact that hydrogen has a broad range of flammability, its lower flammable limit is significantly higher than for a number of other fuels. This includes widely used combustibles, like, for example, gasoline, butane, propane, and ethanol. Its autoignition temperature lies in a range similar to other fuels. According to a report on behalf of Ford Motor Company, the safety of a hydrogen FC vehicle would be potentially better than that of a gasoline or propane vehicle [18, 24]. For a large number of solid materials, hydrogen embrittlement is a problematic issue. Hydrogen can easily penetrate these materials via several mechanisms. Inside the host material, the hydrogen impedes the plastic flow of the material and decreases its ductility [25]. For high pressure applications and storage tank designs it is particularly important to take into account hydrogen embrittlement, in order to avoid the failure of system and tank components.

Table 2: Comparison of fuel safety properties in air [18, 24, 25, 27].

Fuel	Autoignition temperature K	Flammable limits Vol.%	Explosive limits Vol.%	Buoyant velocity $m\,s^{-1}$	Ignition energy[a] mJ
Hydrogen	673–858	4.1–74.2	18.3–59.0	1.2–9.0	10
Methane	923	5.3–15.0	6.3–13.5	0.8–6.0	20
Ethane	788	3.0–12.5	—	—[b]	—
Propane	723	2.3–9.5	3.1–7.0	—[b]	—
Butane	678	1.9–8.5	—	—[b]	—
Gasoline	553–729	1.4–7.6	1.1–3.3	—[b]	—
Ethanol	696	3.3–19.0	—	—[b]	—
Methanol	743	6.0–36.5	—	—[b]	—

[a]Ignition energy at the lower flammability limit. For stoichiometric mixtures, the value is significantly lower (e.g., 0.02 mJ for hydrogen) [18].
[b]Nonbuoyant, due to higher density than air ($1.2\,kg\,m^{-3}$) under standard conditions.

Due to its high conversion efficiency and its zero emission (referring to emissions of toxic pollution or greenhouse gases like, e.g., CO_2), hydrogen is regarded as an ideal secondary energy carrier. It can be produced using renewable energy and water, and, by means of storing it, the hydrogen can be retransformed to energy and water in another time and space. During those conversions no contamination is produced. For energy conversion, well-established technologies like internal combustion engines (ICEs) may be operated with hydrogen fuel if they are slightly modified [20]. By combustion in air, the emission is primarily water vapor. Only traces of nitrogen oxides are formed, which can be controlled to very low levels [18]. In ICEs, the conversion efficiency is thermodynamically limited by the Carnot efficiency to a relatively low value (below 30%) [25]. In terms of both, emission and efficiency, fuel cells (FCs) outperform classical combustion engines. Thus, they are able to convert the chemical energy of hydrogen to electricity with an efficiency of up to 60% [18, 22]. Thereby, water vapor is the sole emission [18]. Today, one of the main drawbacks of FCs is that they are expensive, amongst others because of their dependency on precious catalyst materials.

Aside from high costs of FCs, there are two other main obstacles which impede the introduction of hydrogen as an important energy vector today: hydrogen production and its storage. Thus, inexpensive methods for hydrogen mass production need to be found, and efficient, economic, and safe hydrogen storage technologies need to be developed in order to compete with less expensive fossil fuels and for introducing a hydrogen economy [18].

HYDROGEN STORAGE

As an energy carrier, hydrogen has to compete against other fuels. For the use in electric vehicles, hydrogen could outperform electric batteries which are relatively bulky and heavy [18]. According to General Motors, running an electric vehicle with hydrogen and fuel cells instead of batteries has a number of advantages [26]. For example, costs can be reduced, and considerably higher energy densities can be reached [26].

In Figures 3 and 4, the heat of combustion of hydrogen is compared with other gaseous, liquid, and solid fuels. On a gravimetric basis (Figure 3), the value of hydrogen clearly exceeds those of other alternatives, reaching more than twice of their energy content. This characteristic predestines hydrogen, for example, as a rocket fuel [18, 21]. For application in a FC car a mass of only around 4 kg of hydrogen is sufficient in order to reach a range of 400 km [22]. A contrary tendency can be observed for the heat of combustion on a volumetric basis (see Figure 4). This is why, under standard conditions, gaseous fuels are in a clear disadvantage. In particular hydrogen reaches a significantly lower volumetric value, due to its low density. Thus, if the 4 kg of hydrogen, which is needed for a FC car, would be carried under normal conditions (at room temperature and under atmospheric pressure), then the tank would occupy a volume of approximately $45\,m^3$; by way of illustration this would correspond to a balloon of around 4.4 m diameter.

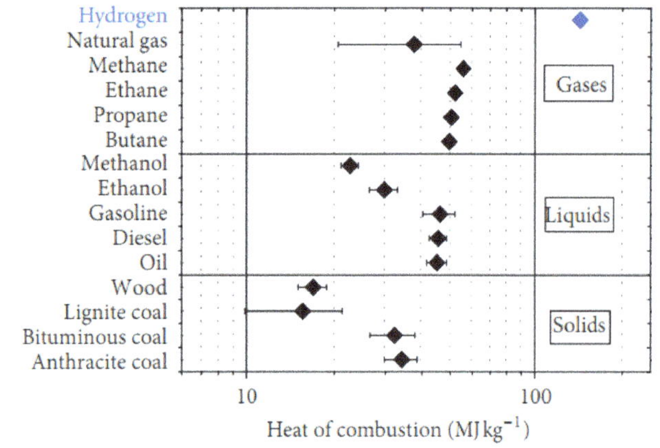

Figure 3: Heat of combustion on a gravimetric basis, shown for various gaseous, liquid, and solid fuels under standard conditions. The ranges of values were obtained by utilizing a multitude of sources [18, 28–38].

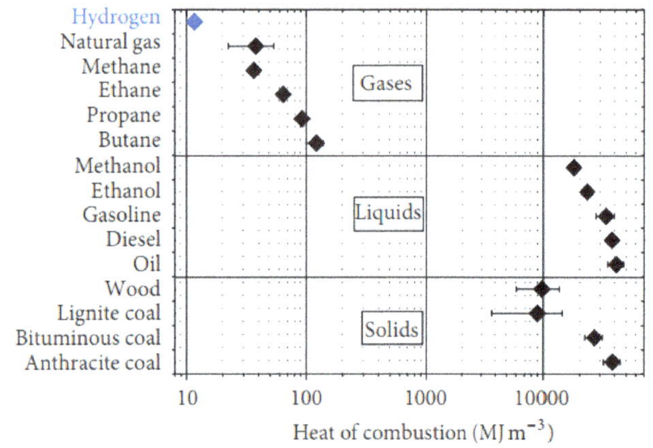

Figure 4: Heat of combustion on a volumetric basis, shown for various gaseous, liquid, and solid fuels under standard conditions. The ranges of values were obtained by utilizing a multitude of sources [18, 28–38].

The aforementioned reveals that, for hydrogen storage application, the storage volume needs to be reduced. Putting it in another way, this means that, for a given storage volume, the distances between hydrogen molecules need to be decreased. However, such an increase of the volumetric storage density has to come along with maintaining the gravimetric storage density on a high level. An important aspect is that, even though systems for hydrogen onboard storage in vehicles may be bulkier and heavier than for other fuels, these constraints can be acceptable, taking into account the highly efficient conversion of hydrogen and the fact that a relatively small amount of fuel is needed to travel a given distance [18].

Acceptable compromises need to be found regarding hydrogen storage systems which guarantee a successful introduction of hydrogen technology. In order to provide a framework for the development of hydrogen storage technology, governmental institutions and other funding bodies established a number of targets and milestones. These targets set limits to, for example, the refilling time, the cycle lifetime, the costs, and the densities of the storage system [22, 39, 40]. Usually, in the research of materials for hydrogen storage, special attention is drawn to the gravimetric and volumetric densities. In 2004, the DOE introduced a number of milestones, which also included targets concerning the storage densities [39]. Ever since their introduction, these guidelines actuated as an important stimulation for research in this field. However, despite the worldwide effort of scientists and engineers, none of the current technologies meets all of these demands. In 2009, the DOE revised their system density targets which are now based on light-duty vehicles and reduced them to 4.5 wt.% and 5.4 MJ kg^{-1} (target for 2010), 5.5 wt.% and 6.48 MJ kg^{-1} (target for 2015), and 7.5 wt.% and 9 MJ kg^{-1} (ultimate target), respectively [41]. Taking into account the lower heating value of hydrogen (33.33 kWh kg^{-1}), as well as the fact that 1 kWh = $3.6 \cdot 10^6$ Ws = 3.6 MJ, the volumetric storage densities corresponded to 28 g_{H_2} l^{-1}, 40 g_{H_2} l^{-1}, and 70 g_{H_2} l^{-1}, respectively (grams of H_2 per liter of storage volume) [41]. Some of the currently valid DOE key targets are listed in Table 3. Concerning the EU, the key research issue for volumetric storage density is 1.1 kWh l^{-1}, corresponding to 33 g_{H_2} l^{-1} [42].

Table 3: DOE hydrogen storage targets for light-duty vehicles [41].

Storage parameter	Units	2010	2015	Ultimate target
Gravimetric system density	kWh kg^{-1}	1.5	1.8	2.5
	MJ kg^{-1}	5.4	6.48	9
	wt.%	4.5	5.5	7.5
Volumetric system density	kWh l^{-1}	0.9	1.3	2.3
	MJ m^{-3}	3.24	4.68	8.28
	g_{H_2} l^{-1}	28	40	70
Fuel cost at pump	\$ GGE^{-1a}	3–7	2–6	2-3
System filling time for 5 kg H_2	min	4.2	3.3	2.5
	kg_{H_2} min^{-1}	1.2	1.5	2
Maximum loss of useable H_2	g (h kg_{H_2})$^{-1}$	0.1	0.05	0.05

[a] 1 gasoline gallon equivalent (GGE) ≈ 1 kg hydrogen [43].

Hydrogen can be stored, either in an empty tank (e.g., in the case of high pressure hydrogen storage or liquefied hydrogen storage) or in a tank containing a solid (solid-state H_2 storage). In the latter case, the storage density is often referred exclusively to the solid used, in order to facilitate the performance comparison of different materials.

Thus, it is extremely important to distinguish between storage densities on a material basis and those which are based on the whole storage system. It has to be pointed out that the technological requirements and targets, which were mentioned previously and which are listed in Table 3, refer to the whole storage system. This means that, apart from the fuel itself (and the storage material in the case of hydrogen storage in solids), also the tank shell, regulation, and system devices, as well as other auxiliaries, need to be taken into account. In the case of high pressure storage, suitable materials and sufficient dimensions need to be chosen for the tank shell, in order to withstand the high pressures. This reduces the effective storage density of the system [26]. Also in the case of liquid hydrogen storage, the storage density is reduced significantly, due to necessary adaptations like isolation, secondary devices, and so forth [26].

Overview of Hydrogen Storage Techniques

In the previous section, it was shown that the main problem for hydrogen storage is its extremely low density under normal conditions. In order to reduce the occupied volume by hydrogen gas, a number of different technologies are being investigated.

Compression

Compressing hydrogen up to high pressures is a relatively simple and cost-efficient technology [25]. This is why most of the current hydrogen car prototypes are based on storage tanks with high pressure vessels [26]. For the pressure vessels, adequate materials and dimensions are needed in order to withstand these high pressures, to accomplish safety issues, and to avoid hydrogen diffusion. In practice, this results in an important weight gain and, consequently, in a reduction of the system storage capacity. For pressurized vessels, spherical or cylindrical tank structures are necessary, in order to reduce the tank shell dimensions. According to the United Nations Economic Commission for Europe (UNECE), pressure vessels for hydrogen storage can be classified into four types [44]: Type 1: all metal cylinder, Type 2: partly fiber-wrapped, load-bearing metal liner, Type 3: fully fiber-wrapped, non-load-bearing metal liner, Type 4: fully fiber-wrapped, non-load-bearing nonmetal liner.

From type to type, the weight of the vessel is successively reduced by substituting metals with other lighter materials. The two basic components of a state-of-the-art type 4 pressure vessel are (a) a liner which prevents hydrogen permeation and (b) a composite structure which accounts for the mechanical stability of the structure. In Figure 5, a schematic diagram of a type 4 pressure vessel and its components is shown.

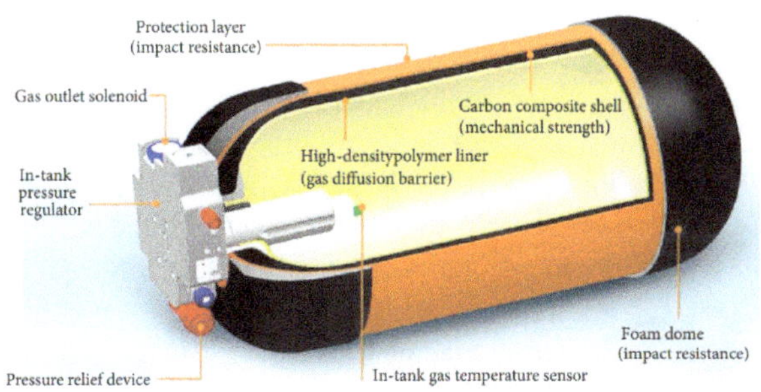

Figure 5: Schematic diagram of a type 4 pressure vessel and its components for gaseous hydrogen storage (Quantum Technologies, Inc., Irvine). Reprinted from [26], with permission from Elsevier.

An important concern regarding high pressure vessels is safety issues, especially in the case of mobile applications. Thus, pressure vessel has to be extensively tested regarding overpressure and accidental impacts. Another drawback of gaseous storage is the energy expenditure which is needed for the compression process. Thus, around 12% of the stored hydrogen energy is consumed for reaching 35 MPa and 15% for 70 MPa, respectively [26].

In order to determine gas densities at given pressure-temperature conditions, adequate models are needed. In general, the ideal gas equation (1) becomes invalid at elevated pressures:

$$p \cdot V = n \cdot R \cdot T \tag{1}$$

p: absolute gas pressure (Pa)

V: volume occupied by gas (m^3)

n: number of moles of gas (mol)

R: universal gas constant $(8.3145\,\mathrm{J(K\ mol)^{-1}})$

T: absolute temperature (K).

A better approximation for describing the thermodynamic behavior of real gases can be made by the van der Waals equation (see (2)). This approach is a modification of the ideal gas law which introduces two correction factors a and b:

$$\left(p + a \cdot \left(\frac{n}{V}\right)^2\right) \cdot (V - n \cdot b) = n \cdot R \cdot T \tag{2}$$

a: van der Waals correction for finite gas molecule size

b: van der Waals correction for volume of one mole of gas.

In Table 4, the van der Waals constants for hydrogen are given. The term $n \cdot b$ implements the volume occupied by the gas molecules, which has to be subtracted from the total volume available V. On the other hand, the term $a \cdot (n/V)^2$ represents the attraction due to van der Waals forces. It is added to the pressure component p, in order to account for the increasing attraction between the gas molecules, for increasing number of molecules n, or reduction of the available volume V.

Table 4: Van der Waals constants for hydrogen.

Description	Symbol	Value	Unit	Reference
VdW correction	a	0.2484	$(l^2\mathrm{bar})\ \mathrm{mol}^{-1}$	[27]
VdW correction	b	0.02651	$\mathrm{l\,mol}^{-1}$	[27]

Despite its descriptive character, the van der Waals equation turns out to be relatively imprecise if it is applied to hydrogen gas [45]. Hence, more precise equations of state (EOS) have to be used in order to estimate the density of hydrogen under given thermodynamic conditions [45]. Modifications of the van der Waals EOS have been proposed by Redlich-Kwong [46], Soave [47], and Peng-Robinson [48]. A different approach is done by empirical (virial) equations of state, based on the Benedict-Webb-Rubin (BWR) equation [49]. A comparison with the Peng-Robinson approach reveals that modified BWR (MBWR) equations give more precise values for hydrogen [50, 51]. Some common examples for MBWR equations are the Lee-Kesler equation [52] and the 32-term MBWR equation introduced by Younglove [53]. The latter is the basis of the NIST database of thermophysical properties of fluids which has been used throughout this study [36]. More recently, another approach was introduced by Leachman et al. [54] which claims to have an uncertainty in density of only 0.04% in the region between 250 and 450 K and for pressures up to 300 MPa [54].

Liquefaction

Another possibility is to store hydrogen in liquid form. However, hydrogen only condensates at very low temperature (20.4 K). The liquefaction process is based on the Joule-Thomson effect, which explains the temperature change of a pressurized real gas when it is throttled [25]. The work of liquefaction of an idealized process takes into account an isothermal compression followed by reversible isentropic expansion [25]. Furthermore, also the nuclear spin of hydrogen becomes important for the liquefaction process. The H_2molecule occurs in two forms: orthohydrogen (parallel spin) and parahydrogen (antiparallel spin). Under normal conditions, hydrogen gas is a mixture of 75% ortho- and 25% parahydrogen ("normal hydrogen"). With decreasing temperature, the amount of parahydrogen increases, reaching 99.8% at its liquid state [25]. The transformation of the ortho- to the parastate is accompanied by a heat release which prejudices the cooling process [25].

Taking into account the transformation from the ortho- to the para-state, the ideal work of liquefaction of hydrogen is 11.88 MJ kg^{-1} [25, 55]. This is about 15 times more than for nitrogen liquefaction [25, 55]. Industrial liquefaction plants are based on the Linde cycle or the Claude cycle which use prestage cooling with liquid nitrogen and subsequent recirculation of precooled hydrogen gas in a counterflow heat exchanger [25]. Their efficiency is significantly lower than the ideal process, consuming approximately 30% of the chemical energy stored inside the liquefied hydrogen which is an important disadvantage of this technology [26, 39, 56].

In Figure 6, a schematic diagram of a system for liquid hydrogen storage, designed for application in vehicles, is shown. Its main purpose is to thermally isolate the cryogenic hydrogen from ambient conditions. The tank consists of a double-walled metallic vessel whose interspace is evacuated and contains a multilayered isolation [57, 58]. Also in the case of liquid hydrogen storage, spherical or cylindrical tank structures are beneficial. Here, these configurations help to optimize the surface/volume ratio and to reduce heat transfer between the interior and exterior of the tank. In spite of the very complex and costly insulation, heat flow to the inside of the tank is unavoidable. This causes the constant evaporation of some of the liquid hydrogen, which leads to a pressure rise inside the tank. When about 1 MPa is reached, the hydrogen has to be released in order to avoid that the tank is damaged [26]. If not captured, this so-called hydrogen boil-off is lost to the environment. Hence, it cannot be used for propulsion anymore. Also, additional hydrogen is lost due to heat transfer in the course of the fueling process. Furthermore, it is claimed that large amounts of escaping hydrogen may contribute to global warming [59].

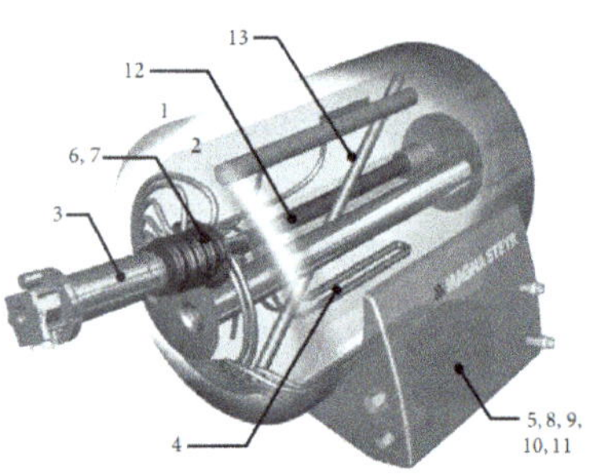

(1) Outer tank
(2) Inner tank
(3) Coupling (Johnston-Cox)
(4) Heater
(5) Heat exchanger
(6) Cryogenic filling valve
(7) Cryogenic return valve
(8) Pressure regulation valve
(9) Shut-off valve
(10) Boil-off valve
(11) Safety relief valve
(12) Support post
(13) Liquid level sensor

Figure 6: Schematic diagram of a liquid hydrogen tank and its components (Magna Steyr, BMW). Reprinted from [57], with permission from International Association of Hydrogen Energy.

Among the car manufacturers, opinions regarding liquid hydrogen storage are contradictory. Thus, BMW equipped a number of hydrogen prototypes with such systems [58]. According to General Motors Company, on the other hand, liquid hydrogen storage is not favorable over compressed gas storage [26]. Despite its disadvantages, liquid hydrogen is used as a fuel for spacecraft rocket propulsion. For this application, gravimetric and volumetric energy contents are the most critical parameters, and, because of the fast fuel consumption, isolation can be reduced to a minimum [60].

Chemisorption in Metal Hydrides

A third possibility is to store hydrogen through dissociative chemisorption, followed by the formation of a new compound. Among them, metallic hydrides are being widely studied. During chemisorption in metal hydrides, the hydrogen molecule is first dissociated on the surface, and then its atoms diffuse into the metal host [61]. Inside the bulk, chemical bonds are formed, resulting in the hydride phase. With respect to the bonding mechanism between the hydrogen and the host material, metal hydrides can be categorized, at a first glance, as saline hydrides (ionic bonding), covalent hydrides (covalent bonding), and interstitial hydrides (metallic bonding). Usually, the hydrogen compounds are formed and decomposed under determined pressure-temperature conditions. Thus, formation takes place under elevated hydrogen pressure whereby heat is generated due to the exothermic reaction (3). For releasing the hydrogen, the pressure is reduced and the material is heated. For reversible hydrides, these cycles can be repeated several times.

$$MH_n\,(s) + \text{heat} \rightleftharpoons M(s) + \frac{n}{2}H_n\,(g) \qquad (3)$$

Materials which have been investigated for storing chemically bonded hydrogen include numerous compounds. Although often more sophisticated alloys are investigated, some basic metal hydride systems are usually classified as AB_5 (e.g., $LaNi_5$), AB (e.g., FeTi), A_2B (e.g., Mg_2Ni), AB_2 (e.g., ZrV_2), and so forth [62–65]. Other candidates include simple hydrides which contain only one element different from hydrogen, for example, MgH_2 [61, 66] or AlH_3 [67]; complex metal hydrides such as alanates ($NaAlH_4$, $LiAlH_4$, $MgAlH_4$, etc.) [68–72] or borohydrides ($NaBH_4$, $LiBH_4$, etc.) [72], and alloyed combinations of them; amides and imides (metal-NH systems) [73–76]; reactive hydride composites [77–79]; as well as other chemical storage media like methanol (CH_4O), ammonia (NH_3), or B-N compounds in combination with on- or off-board reforming [72, 80–83]. Some of the classic metal hydride materials can be reversibly operated close to room temperature. However, these materials usually have very low gravimetric storage capacities of less than 3 wt.% [84]. On the contrary, light metal hydrides achieve promising gravimetric capacities (e.g., 18 wt.% in the case of $LiBH_4$); yet, they lack reversibility and/or need to be operated at high temperatures (e.g., around 573 K for MgH_2) [84]. In addition, some of the materials, like, for example, $NaAlH_4$, decompose by multistep reactions (see (4)) [69, 71, 72]:

$$6NaAlH_4 \xrightarrow[3.7\,wt.\%]{T_1} 2Na_3AlH_6 + 4Al + 6H_2$$

$$2Na_3AlH_6 + 4Al + 6H_2 \xrightarrow[1.9\,wt.\%]{T_2} 6NaH + 6Al + 9H_2$$

$$6NaH + 6Al + 9H_2 \xrightarrow[1.9\,wt.\%]{T_3} 6Na + 6Al + 12H_2 \quad (4)$$

The fact that the intermediate compounds decompose at different temperatures $T_3 > T_2 > T_1$ further complicates the heat management of the storage system. Currently, research is focused on improving kinetics and thermodynamics of the investigated materials and on finding new compounds and catalysts in order to obtain high capacities at lower temperatures [85].

Due to the Westlake Criterion, the H-H distance in stable hydrides can reach values of 2.1 Å [86, 87], and even lower distances have been reported [88, 89]. Thus, on a materials basis, very high volumetric storage capacities can be achieved with metal hydrides, when their crystal densities are taken into account. For example, $115\,g\,l^{-1}$ are reached in the case of $LaNi_5H_6$ and $150\,g\,l^{-1}$ for Mg_2FeH_6 or $Al(BH_4)_3$ [84]. In Figure7, the volumetric hydrogen densities of different technologies are compared. It can be seen that metal hydrides reach higher values than gaseous and even liquid hydrogen. This means that, in a determined volume, more hydrogen can be stored in form of a metal hydride than in form of compressed or liquefied hydrogen.

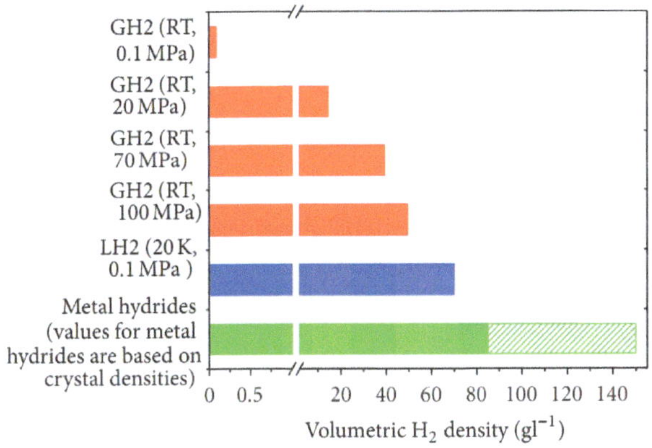

Figure 7: Comparison of volumetric fuel densities for different hydrogen storage technologies: compressed hydrogen (GH_2) at room temperature for 0.1, 20, 70, and 100 MPa [36]; liquid hydrogen (LH_2); and metal hydrides [22].

For hydrogen storage application, however, the hydride is usually available in powder form, in order to minimise diffusion paths and to increase reaction kinetics. Therefore, the value for the crystal density of the hydride should not be confused with the volumetric storage density, which is expected to be considerably lower. Furthermore, the material density changes in the course of the metal-hydride reaction. By continuously being loaded and discharged with hydrogen, the material matrix periodically expands and contracts for every hydrogen cycle, leading to a sintering of the material. Those density changes represent a major challenge for the practical implementation of this technique. In the case of AB_5-type metal hydrides this expansion can exceed 12% [90, 91]. Another problem with metal hydrides is their cycling stability due to segregation effects of their components or catalysts [92–99].

On a gravimetric basis, only very low values are reached with room temperature hydrides. Light metal hydrides perform much better. However, these materials involve high binding enthalpies which lead to a considerable amount of heat that needs to be transferred during hydrogen charging and discharging [26, 100,101]. As chemisorption is based on an exothermal reaction, a considerable amount of heat is generated upon hydrogenation. This heat has to be removed by an active cooling device. On the other hand, for utilizing the hydrogen, and in order to release it from the material, the tank needs to be heated up. Thus, a number of secondary devices are needed, which again occupy precious space and lead to low storage capacities, when the complete tank design is taken into account. For a typical heat of formation of $25\,MJ\,kg_{H_2}^{-1}$, a heat exchanger of several hundreds of kW would be necessary [26]. This significantly reduces the energy efficiency of the whole process, for example, in the case of MgH_2 by 30% [96].

Hydrogen Storage by Physisorption

Physisorption (or adsorption) fills the gap between the techniques described before. The H_2 molecule (adsorbate) is thereby bound by van der Waals forces on a porous material (the adsorbent) which has a high internal surface area. Physisorption is a non-activated process and the hydrogen-material-interactions are relatively weak (typically between 2 and $8\,kJ\,mol^{-1}$) [45, 102–105]. As a consequence, the sorption heat is much lower than for chemisorption, and no active cooling device is needed, nor does the material need to be heated up in order to release the major part of the stored hydrogen. On the other hand, the interaction is strong enough to enhance the amount of hydrogen stored in a given volume or to use cryogenic temperatures which are closer to ambient conditions (that means higher) than in the case of liquid hydrogen.

These relations among the different storage technologies are visualized in the thermophysical phase diagram in Figure 8 [106]. In this diagram, the volumetric density of parahydrogen is plotted over the temperature for different pressures. A wide range of temperatures from zero to 400 K is covered, and the temperatures which are most interesting for studying hydrogen adsorption (liquid nitrogen and room temperature) are indicated. A tank device for cryogenic adsorption could be operated within the range of these two temperatures [107]. Also the critical point and the boiling point of hydrogen, which is important for liquid hydrogen storage, are marked. The DOE target values for the volumetric storage density (see Table 3) are visualized by horizontal lines. It has to be reminded that these values refer to the whole storage system and are included only for orientation. It can be seen that, in order to achieve the 2015 target by compression, pressures superior to 72 MPa and 13.9 MPa would be needed at RT and 77 K, respectively. The ultimate DOE target for the volumetric density would require pressures of more than 187.7 MPa at RT or 44.6 MPa at 77 K. The challenge for hydrogen storage by physisorption is the reduction of these high pressures.

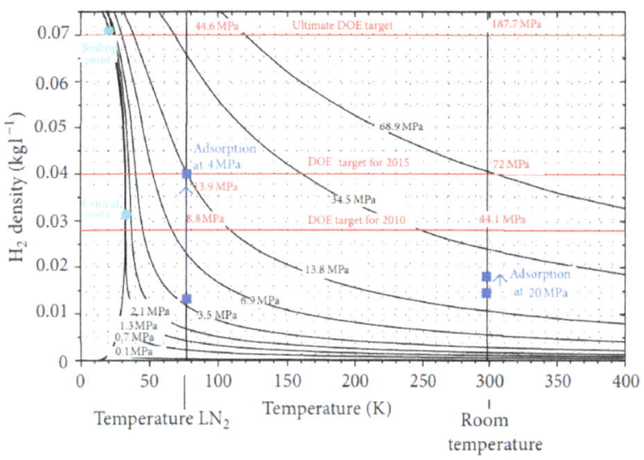

Figure 8: Density diagram for hydrogen, including the DOE targets for volumetric storage density. Thermo-physical data was obtained from NIST [106].

Basic Concepts for Hydrogen Storage by Physisorption

In general, two different strategies are considered for utilizing adsorbents in hydrogen storage tanks. In the first case, the tank is maintained at low temperatures ("cryoadsorption") [108]. At cryogenic temperatures, the thermal vibration of the adsorbate molecules is relatively low in comparison with the van der Waals forces, thus increasing the adsorption potential [84]. By using a temperature which is higher than in the case of liquid hydrogen storage, costs and isolation efforts can be significantly reduced. Usually, liquid nitrogen is used for cooling the tank down to 77 K. At this temperature, the adsorption potential is sufficiently high, so that the tank can be operated at relatively moderate pressures. Thus, many adsorbents reach their maximum excess adsorption of hydrogen at pressures of around 4 MPa. In Figure 8, a schematic example for adsorption at 77 K is included which exemplifies that, for reaching the 2015 DOE target, the operating pressure could be drastically reduced from 13.9 MPa (for a tank using compressed hydrogen) to 4 MPa.

From an application point of view, it would be more interesting to operate the tank at room temperature. In this second scenario, costs for cooling could be avoided and no isolation would be needed, thus reducing weight and volume of the storage device. However, at increased temperature, the influence of the van der Waals forces decreases due to an increased thermal vibration of the adsorbate molecules. Due to the low adsorption potential under these conditions, high pressures are needed to reach sufficient storage capacities. However, also at room temperature it is possible to reduce the operating pressure for the storage of a determined amount of hydrogen by using an adsorbent, in comparison with an empty vessel [109, 110] (see Figure 8).

Physisorption Fundamentals

Three different kinds of van der Waals forces can contribute to the interaction between adsorbent and adsorbate [111]:

- permanent dipole-permanent dipole forces (Keesom interaction),
- permanent dipole-induced dipole forces (Debye interaction),
- instantaneous induced dipole-induced dipole forces (London dispersion forces).

In the case of hydrogen adsorption on carbon surfaces, the first two contributions are usually not taken into account [103]. Mathematically, the van der Waals forces can be approximated by a Lennard-Jones potential (L-J potential) [112, 113]. Thereby, repulsion and attraction between adsorbent and adsorbate are represented as a function of their distance (see (5)):

$$L(r) = 4 \cdot \epsilon \left(\left(\frac{\sigma}{r} \right)^{12} - \left(\frac{\sigma}{r} \right)^{6} \right) \qquad (5)$$

(r): L-J potential between adsorbent and adsorbate

r: distance between adsorbent and adsorbate

ϵ: minimum value of $L(r)$

σ: finite distance at which $L(r) = 0$.

In Figure 9, the Lennard-Jones potential between carbon and hydrogen is shown. The values for ϵ and σ, corresponding to the interaction between carbon and hydrogen, are listed in Table 5 and were calculated using the formulas and data from [114]. It can be seen that the potential has a minimum at a distance of around 0.35 nm.

Table 5: Lenard-Jones constants for the carbon-hydrogen interaction.

Description	Symbol	Value	Unit	Reference
Lenard-Jones constant	ϵ	32.06	K	[114]
Lenard-Jones constant	σ	0.318	nm	[114]

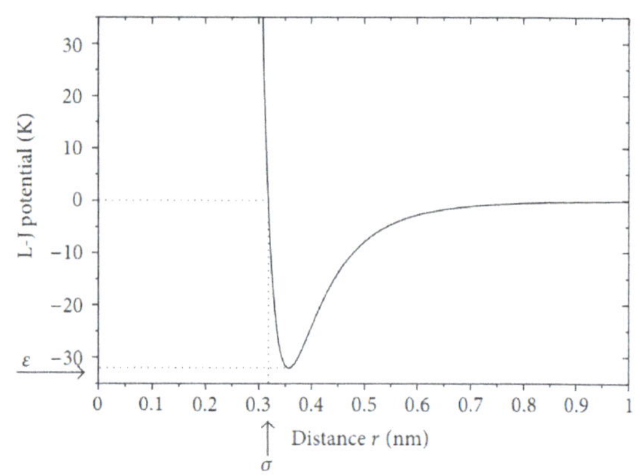

Figure 9: Lennard-Jones potential of the carbon-hydrogen interaction. Values for ϵ and σ were taken from [114].

Material Demands

Designing an optimized adsorbent for hydrogen storage is a challenging task. In general, there are two basic requirements which need to be taken into account [109, 115]. On the one hand, a large number of adsorption sites per weight and volume of adsorbate should be provided. This has to be done by developing a highly porous material with large surface areas and pore volumes. On the other hand, the heat of adsorption of the adsorbent should be increased. Thus, at room temperature, the optimum adsorption enthalpy of an adsorbent should be between 10 and 50 kJ mol^{-1} [102, 115], and theoretical calculations reveal a value of 15.1 kJ mol^{-1} [104].

In order to increase the adsorption enthalpy of adsorbents, three possibilities are generally considered [102,109, 115]: hydrogen spillover, unsaturated adsorption centers, and adsorption in very small pores. For hydrogen spillover, transition metals like, for example, platinum or palladium are introduced into the porous host material. Thereby, it is intended that these metal sites act as catalysts, in order to break the H–H bond, so that the adsorbed species are hydrogen atoms rather than the molecule itself. While theoretical studies claim the feasibility of this method, experiments have not yet been crowned with success [109, 115]. In addition, the formation of chemical bonds to the host material has to be taken into account so that high temperatures could be necessary for releasing the hydrogen [102].

Also in the case of unsaturated adsorption centers, the chemical composition of the material host is changed by introducing other species like, for example, metals. However, their interaction with the hydrogen molecule is weaker and limited to increase the adsorption potential. This is accomplished by donation of charge between the molecules and the functionalized interaction sites [102]. This approach has been proven to be partially successful. Thus, heats of adsorption could be increased to around 10 kJ mol^{-1} and even up to 22 kJ mol^{-1} [115]. However, these high values may be caused by mechanisms different from the adsorption process. The main disadvantage of those first two methods is that additional (heavy) species are introduced into the porous material. This decreases the specific surface area and pore volumes, prejudicing the gravimetric storage capacity of the material.

The third method, decreasing the pore sizes of the material, manages without the introduction of additional species. Here, the concept is to build a very narrow porosity in which the hydrogen molecule can simultaneously interact with multiple pore walls. Due to their proximity, the adsorption potentials of opposed pore walls overlap, resulting in an amplified attraction.

At 77 K, a correlation of hydrogen adsorption with the BET surface area has been reported [22, 116]. However, such correlation was found to be more precise with the total volume of micropores [45, 117, 118]. This indicates that, under cryogenic temperature conditions, adsorbents with a high amount of micropores smaller than 2 nm are beneficial [115–118].

At room temperature, the optimum pore size was found to be around 0.7 nm [102, 119]. This coincides with an overlapping of two carbon-hydrogen potentials, similar to the one shown in Figure 9. A pore of this size is sufficient for accommodating two layers of hydrogen molecules [119]. The amount of pores which are narrower than 0.7 nm can be obtained by calculating the DR micropore volume from CO_2 adsorption data, V_{DR} (CO_2) [120]. For hydrogen adsorption amounts at room temperature, a better correlation is achieved with this value than with the total micropore volume or the BET surface area [45, 116–118, 121–123].

Porous Materials for Hydrogen Storage

Numerous kinds of materials have been investigated as adsorbents for hydrogen. These include carbon-based materials (activated carbons, carbon nanotubes, nanofibers, fullerenes, carbons from templates, etc.), zeolites, silica, alumina, metal organic frameworks (MOFs), covalent organic frameworks (COFs), porous polymers, aerogels, and boron nitride materials. In the following, a brief overview of the different classes of materials as well as their performance for hydrogen storage is given.

Zeolites are aluminosilicates which can accommodate a wide variety of cations [112]. These materials can be found naturally or can be produced synthetically [112]. Due to their crystalline structure, zeolites have a very defined porous structure. However, they are composed of relatively heavy elements which limit their porosity, to maximum specific surface areas of less than 750 $m^2 g^{-1}$ [109]. They are used in numerous industrial applications, for example as detergents, catalysts, desiccants, molecular sieves, or for water purification as well as gas separation and purification through pressure swing adsorption (PSA) [124–126]. Zeolites were among the first porous materials which have been investigated for hydrogen storage purposes [127–131]. However, due to their low porosity, hydrogen adsorption amounts are limited. Thus, less than 1.5 wt.% are adsorbed at 77 K and 0.1 MPa, and, for higher pressures, maximum adsorption amounts are lower than 2 wt.% [131, 132]. Theoretical calculations reveal that, even under ideal conditions, adsorption amounts are limited to 2.86 wt.% [131].

Some natural or synthetic oxides like silica and alumina can also reveal high porosity which often has a high contribution of mesopores. Due to this characteristic, these materials find a lot of applications in catalysis, separations, sensors, drug delivery, optical devices, and molecular sieves [133]. Silica gel, for example, reaches a high surface area of more than 1000 $m^2 g^{-1}$ [132, 134], and some of them are commercially available, like, for instance, MCM-41 (with pore diameter limited to approximately 80 Å) [134–136], and SBA-15 (with pore sizes ranging from 20 to 300 Å) [137, 138]. However, their porous structure results inappropriate for hydrogen storage, and less than 0.6 wt.% were measured at 77 K and 0.1 MPa [132].

By means of sol-gel reaction, aerogels with large specific surface areas of more than 3000 $m^2 g^{-1}$ can be prepared [139, 140]. Hydrogen adsorption was reported to reach over 5 wt.% at 77 K and 3.5 MPa [141, 142]. However, their low material density of typically around 0.1 g cm^{-3} leads to low volumetric storage densities. Together with their inappropriate pore size distribution, these materials can be regarded unsuitable as hydrogen storage medium [139, 140].

More recently, also noncarbon nanotubes have been proposed for hydrogen storage [143]. Boron nitride (BN) nanotubes, for instance, were claimed to have advantages over carbon nanotubes (CNTs) and to adsorb up to 2.6 wt.% at RT and 10 MPa [144, 145]. However, other studies reveal that BN nanotubes only show insignificant improvements over CNTs at RT [146]. Other boron-carbon-nitride (B-C-N) materials reach specific surface areas of more than 1500 $m^2 g^{-1}$ and adsorb 1 wt.% hydrogen at 77 K and 0.1 MPa [147].

Another class of materials which recently were proposed for hydrogen storage application are porous polymers [148–150]. Thus, (hyper)crosslinked polymers can reach BET surface areas of up to 2000 $m^2 g^{-1}$ and adsorb 1.5 wt.% at 77 K and 0.12 MPa and 3.8 wt.% at 77 K and 4.5 MPa [115, 150–152]. Polymers of intrinsic microporosity (PIM) are another approach to introduce porosity into polymers. These materials reach BET surface areas of more than 1000 $m^2 g^{-1}$ and adsorb 1.7 wt.% at 77 K and 0.1 MPa [115, 148, 149].

Coordination polymers like metal-organic frameworks (MOFs) and covalent organic frameworks (COVs) are a different kind of porous polymer. Similar to zeolites, these materials are of highly crystalline nature. However, different from zeolites, the structure of these frameworks is composed of lighter elements, giving rise to higher specific porosity. Thus, MOF structures are built by metal ion clusters which are connected by organic linking groups. In the case of COFs, aromatic rings are the primary structural element [115]. The building blocks are linked by strong covalent bonds, resulting in a rigid structure which contains exclusively very light elements like H, B, C, N, or O. Coordination polymers can reach extremely high BET surface areas of more than 5000 $m^2 g^{-1}$ [109, 153]. For MOF-177, which has a BET surface area of 4600 $m^2 g^{-1}$, hydrogen uptake of 7.5 g(H_2)/g was reported at 77 K and 4 MPa [115]. The crystallinity of coordination polymers allows for an exact control of their porosity. A common strategy to achieve high surface areas in organic framework materials is to increase the size of the organic linkers [115]. However, this increases also the pore sizes of the structure which may be a problem for hydrogen adsorption at room temperature where small pores are advantageous. In addition, the low density of coordination polymers leads to relatively low volumetric storage density. Furthermore, MOFs are relatively sensitive to conditions of high temperature, mechanical pressure, and humidity which can lead to degradation of the material [115, 154].

Activated carbon (AC) is an interesting and well-known adsorbent with a highly adaptable porosity. Activated carbon materials are a form of carbon which does not occur naturally and which has to be synthesized technically. The technical process is called "activation" and can be performed utilizing different methods, as well as a large number of carbon-containing precursors. The porosity in ACs can cover a wide range of pore sizes and can be controlled by adjusting the activation parameters. Furthermore, the characteristics of the adsorbent can be adjusted by manipulating its surface chemistry. Activated carbon materials can be used in numerous applications, for example, in liquid and gas phase treatments (e.g., removal of volatile organic compounds (VOC) and SO_2 or for CO_2 and CH_4 separation and storage), energy storage (as electrical double layer capacitors), as adsorbents in cryocoolers, and so forth [155].

Together with zeolites, carbon materials were among the first adsorbents studied for hydrogen storage [156,157]. Thereby, research was focused on activated carbons (ACs) with high specific surface areas. Since the early days, a number of carbonaceous nanomaterials have been discovered, for example, fullerenes, multiwalled nanotubes (MWNTs), and single-walled nanotubes (SWNTs). In early studies, it was claimed that these fascinating materials would have a high potential for storing hydrogen, initializing a downright research "hype." To date, however, many of the early promises have not been fulfilled. What is more is that investigation of "classical" carbon adsorbents has suffered from this alteration of research focus. Currently, it is concluded that adsorption on nanomaterials is simply due to physisorption. Due to their low porosity in comparison with other carbon adsorbents like activated carbons, the latter perform better in terms of hydrogen storage [45]. Models for estimating the maximum hydrogen capacity of activated carbons often utilize graphene sheets. Such model has a surface area of 2630 $m^2 g^{-1}$ and a maximum hydrogen capacity of around 6 wt.% [102, 104]. If the graphene planes are separated by slip pores of idealized size, up to 9.2 wt.% are reachable [104]. Nevertheless, real activated carbons are highly disordered, giving rise to other adsorption sites, for example, at graphene edges, and reveal a distribution of different pore sizes.

A lot of data can be found for commercial activated carbon materials [116, 117]. Some of the most studied materials are Maxsorb as well as AX-21, a KOH activated carbon with a BET surface area of around 2500 $m^2 g^{-1}$ and a V_{DR} (CO_2) of 0.68 $cm^3 g^{-1}$ [117]. For this material, diverse maximum adsorption amounts between 3 and 6 wt.% were reported for a temperature of 77 K [56, 117, 123, 158]. At room temperature, AX-21 adsorbs 1 wt.% at 5 MPa and 1.6 wt.% at 70 MPa [119, 123].

However, fewer studies have been carried out regarding synthesis and analysis of activated carbon materials that are specially tailored for hydrogen storage. Measurements of metal-carbide-derived activated carbons revealed 3 wt.% at 77 K and 0.1 MPa [115, 159, 160]. On zeolite-templated carbon 6.9 wt.% were measured at 77 K and 2 MPa (2.6 wt.% at 77 K and 0.1 MPa) [115, 161]. For KOH activated anthracite, surface areas of more than 3000 $m^2 g^{-1}$ are obtained [123, 162]. At room temperature, maximum adsorption of almost 1.2 wt.% was measured for pressures of 20 MPa and 2.7 wt.% for 50 MPa [117, 119]. At 77 K, 5.3 wt.% were measured for 4 MPa [117] and around 6 wt.% for 8 MPa [123].

CONCLUSIONS

Strong efforts have been made in order to reach the targets for hydrogen storage in mobile applications which were established by bodies like the DOE. Yet, none of the investigated technologies currently fulfills all of the requirements. Today, the lion's share of hydrogen vehicles is based on pressure vessels with compressed hydrogen. However, this technology may not be appropriate for reaching the ambitious ultimate targets, and security concerns exist over the high pressures involved. The use of liquid hydrogen is handicapped by the energy intensive liquefaction process as well as hydrogen boil-off.

The storage of hydrogen inside materials could be advantageous in a number of aspects. Metal hydrides and other compounds with chemisorbed hydrogen can reach high capacities. However, one has to bear in mind that storage capacities of the materials alone have to be significantly higher according to the established targets which are system-based. In addition, materials with chemically bonded hydrogen often present other difficulties like heat management problems, low kinetics, and/or irreversibility. A promising alternative could be hydrogen storage by physisorption in porous materials. The challenges for designing suitable adsorbents are (1) increasing their porosity, (2) tuning of pore sizes, (3) optimization of adsorption potentials, and (4) enhancement of the bulk material density in order to reach high volumetric capacities.

ACKNOWLEDGMENTS

The authors would like to thank the Generalitat Valenciana and FEDER (Project PROMETEO/2009/047) for financial support.

REFERENCES

1. Core Writing Team, R. K. Pachauri, and A. Reisinger, "Climate change 2007: synthesis report," Tech. Rep., Intergovernmental Panel on Climate Change (IPPC), 2008.
2. M. K. Hubbert, "Energy from fossil fuels," Science, vol. 109, no. 2823, pp. 103–109, 1949.
3. M. K. Hubbert, "Nuclear energy and the fossil fuels," Tech. Rep. 95, Shell Development Company Publication, Exploration and Production Research Division, Shell Oil Company, 1956.
4. K. Aleklett and C. J. Campbell, "The peak and decline of world oil and gas production," Minerals and Energy—Raw Materials Report, vol. 18, no. 1, pp. 5–20, 2003.
5. ASPO, "Testimony on peak oil," Kjell Aleklett, 2005, http://www.peakoil.net/.
6. U. Bardi, "Peak oil: the four stages of a new idea," Energy, vol. 34, no. 3, pp. 323–326, 2009.
7. R. J. Brecha, "Emission scenarios in the face of fossil-fuel peaking," Energy Policy, vol. 36, no. 9, pp. 3492–3504, 2008.
8. A. J. Cavallo, "Predicting the peak in world oil production," Natural Resources Research, vol. 11, no. 3, pp. 187–195, 2002.
9. F. Curtis, "Climate change, peak oil, and globalization: contradictions of natural capital," Review of Radical Political Economics, vol. 39, no. 3, pp. 385–390, 2007.
10. R. L. Hirsch, R. Bezdek, and R. Wendling, "Peaking of world oil production and its mitigation," AIChE Journal, vol. 52, no. 1, pp. 2–8, 2006.
11. J. Kamalick, "Opinions divided over peak oil," ICIS Chemical Business, vol. 1, no. 41, pp. 18–19, 2006.
12. R. G. Nelson, "'Peak oil' is only a matter of time," Pipeline and Gas Journal, vol. 234, no. 2, pp. 45–46, 2007.
13. A. Witze, "Energy: that›s oil, folks...," Nature, vol. 445, no. 7123, pp. 14–17, 2007.
14. T. A. Boden, G. Marland, and R. J. Andres, "Global, regional, and national CO_2 emissions," Tech. Rep., Carbon Dioxide Information Analysis Center, Oak Ridge National Laboratory, U.S. Department of Energy, Oak Ridge, Tenn, USA, 2008.
15. J. Hansen, R. Ruedy, M. Sato, and K. Lo, "Global surface temperature change," Reviews of Geophysics, vol. 48, no. 4, Article ID RG4004, 2010.
16. NASA (National Aeronautics and Space Administration), "GISS (Goddard Institute for Space Studies) database," 2012, http://data.giss.nasa.gov/gistemp/graphs_v3/.
17. R. Weber, Wasserstoff—Wie Aus Ideen Chancen Werden, IZEAktuell, Informationszentrale der Elektrizitätswirtschaft (IZE), Frankfurt, Germany, 2nd edition, 1991.
18. J. Zumerchik, Ed., Macmillan Encyclopedia of Energy, Macmillan, New York, NY, USA, 2001.
19. J. O. M. Bockris, "The origin of ideas on a hydrogen economy and its solution to the decay of the environment," International Journal of Hydrogen Energy, vol. 27, no. 7-8, pp. 731–740, 2002.
20. P. Patnaik, Handbook of Inorganic Chemicals, McGraw-Hill, New York, NY, USA, 2002.
21. R. B. King, Ed., Encyclopedia of Inorganic Chemistry, John Wiley & Sons, Chichester, UK, 1994.
22. L. Schlapbach and A. Züttel, "Hydrogen-storage materials for mobile applications," Nature, vol. 414, no. 6861, pp. 353–358, 2001.
23. A. Bain and D. W. van Vorst, "Hindenburg tragedy revisited: the fatal flaw found," International Journal of Hydrogen Energy, vol. 24, no. 5, pp. 399–403, 1999.
24. C. E. Thomas (Ford Motor Company), "Direct-hydrogen-fueled proton-exchange-membrane fuel cell system for transportation applications—hydrogen vehicle safety report," Tech. Rep., U.S. Department of Energy, Washington, DC, USA, 1997, Contract no. DE-AC02-94CE50389, http://www.directedtechnologies.com/publications/storage/H2VehicleSafetyReport97-05.pdf.
25. E. Tzimas, C. Filiou, S. D. Peteves, and J. B. Vryret, "Hydrogen storage: state-of-the-art and future perspective," Technical Report EUR 20995 EN, European Commision, Joint Research Centre (JRC), 2003, http://publications.jrc.ec.europa.eu/repository/bitstream/111111111/6013/1/EUR%2020995%20EN.pdf.
26. R. von Helmolt and U. Eberle, "Fuel cell vehicles: status 2007," Journal of Power Sources, vol. 165, no. 2, pp. 833–843, 2007.
27. J. A. Dean, Ed., Lange›s Handbook of Chemistry, McGraw-Hill, New York, NY, USA, 15th edition, 1999.
28. K. H. Grote and J. Feldhusen, Eds., DUBBEL—Taschenbuch für den Maschinenbau, Springer, Berlin, Germany, 22nd edition, 2007.
29. H. D. Baehr, Thermodynamik: Grundlagen und technische Anwendungen, Springer, Berlin, Germany, 11th edition, 2002.
30. "Energy statistics manual," Tech. Rep., International Energy Agency (IEA), Paris, France, 2005.
31. M. J. O›Neil, P. E. Heckelman, C. B. Koch, and K. J. Roman, Eds., The Merck Index—An Encyclopedia of Chemicals, Drugs, and Biologicals, Merck Research Laboratories, Merck and Company, Whitehouse Station, NJ, USA, 14th edition, 2006.
32. D. R. Lide, Ed., CRC Handbook of Chemistry and Physics—A Ready—Reference Book of Chemical and Physical Data, CRC Press, Boca Raton, Fla, USA, 2004.
33. E. Meyer, Chemistry of Hazardous Materials, Pearson, Upper Saddle River, NJ, USA, 4th edition, 2004.
34. L. Bretherick, Bretherick›s Handbook of Reactive Chemical Hazards, Butterworth and Company, London, UK, 4th edition, 1990.
35. R. A. Bailey, H. M. Clark, J. P. Ferris, and S. Krause, Chemistry of the Environment, Academic Press, San Diego, Calif, USA, 2nd edition, 2002.

36. NIST National Institute of Standards and Technology, "Thermophysical properties of fluid systems," 2012.
37. IFA Institut für Arbeitsschutz der Deutschen Gesetzlichen Unfallversicherungg, "GESTIS-database on hazardous substances," 2012, http://www.dguv.de/ifa/en/gestis/stoffdb/index.jsp.
38. T. Fischer and H. J. Dorn, Physikalische Formeln und Daten, Ernst Klett Verlag GmbH, Stuttgart, Germany, 1st edition, 2002.
39. DOE, "Basic research needs for the hydrogen economy," Tech. Rep., U.S. Department of Energy, Office of Basic Energy Sciences US DOE, Washington, DC, USA, 2004, http://www.sc.doe.gov/bes/hydrogen.pdf.
40. S. Satyapal, J. Petrovic, C. Read, G. Thomas, and G. Ordaz, "The U.S. department of energy›s national hydrogen storage project: progress towards meeting hydrogen-powered vehicle requirements,"Catalysis Today, vol. 120, no. 3-4, pp. 246–256, 2007.
41. DOE. U.S. Department of Energy, "Targets for onboard hydrogen storage systems for light-duty vehicles," 2012, http://www1.eere.energy.gov/hydrogenandfuelcells/storage/pdfs/targets_onboard_hydro_storage.pdf.
42. S. Barrett, "The European hydrogen and fuel cell strategic research agenda and deployment strategy,"Fuel Cells Bulletin, vol. 2005, no. 5, pp. 12–19, 2005.
43. U.S. National Work Group, "Meeting for the Development of Commercial Hydrogen Measurement Standards, Fuel Specifications Subcommittee, Appendix E, The Starting Point: A Discussion Paper Describing a Proposed Method of Sale and Quality Specification for Hydrogen Vehicle Fuel," 2008, http://ts.nist.gov/WeightsAndMeasures/upload/H2-Laws-and-Reg-Paper-USNWG-JUN2008.pdf.
44. United Nations Economic Commission for Europe (UNECE), "UN ECE WP. 29 GRPE Working Doc. 2004/3—Proposal for a new draft Regulation: uniform provisions concerning the approval of: 1. Specific components of motor vehicles using compressed gaseous hydrogen, 2.Vehicles with regard to the installation of specific components for the use of compressed gaseous hydrogen," 2004, http://www.unece.org/trans/doc/2004/wp29grpe/TRANS-WP29-GRPE-2004-03e.doc.
45. A. Linares-Solano, M. Jordá-Beneyto, M. Kunowsky, D. Lozano-Castelló, F. Suarez-García, and D. Cazorla-Amorós, "Hydrogen storage in carbon materiales," in Carbon Materials—Theory and Practice, A. P. Terzyk, P. A. Gauden, and P. Kowalczyk, Eds., Research Signpost, Kerala, India, 2008.
46. O. Redlich and J. N. S. Kwong, "On the thermodynamics of solutions. V: an equation of state. Fugacities of gaseous solutions," Chemical Reviews, vol. 44, no. 1, pp. 233–244, 1949.
47. G. Soave, "Equilibrium constants from a modified Redlich-Kwong equation of state," Chemical Engineering Science, vol. 27, no. 6, pp. 1197–1203, 1972.
48. D. Y. Peng and D. B. Robinson, "A new two-constant equation of state," Industrial and Engineering Chemistry Fundamentals, vol. 15, no. 1, pp. 59–64, 1976.
49. M. Benedict, G. B. Webb, and L. C. Rubin, "An empirical equation for thermodynamic properties of light hydrocarbons and their mixtures. I. Methane, ethane, propane and n-butane," The Journal of Chemical Physics, vol. 8, no. 4, pp. 334–345, 1940.
50. L. Zhou and Y. Zhou, "Determination of compressibility factor and fugacity coefficient of hydrogen in studies of adsorptive storage," International Journal of Hydrogen Energy, vol. 26, no. 6, pp. 597–601, 2001.
51. C. Zhang, X. Lu, and A. Gu, "How to accurately determine the uptake of hydrogen in carbonaceous materials," International Journal of Hydrogen Energy, vol. 29, no. 12, pp. 1271–1276, 2004.
52. B. I. Lee and M. G. Kesler, "Generalized thermodynamic correlation based on three-parameter corresponding states," AIChE Journal, vol. 21, no. 3, pp. 510–527, 1975.
53. B. A. Younglove, "Thermophysical properties of fluids. I. Argon, ethylene, parahydrogen, nitrogen, nitrogen trifluoride, and oxygen," Journal of Physical and Chemical Reference Data, vol. 11, supplement 1, pp. 1–349, 1982.
54. J. W. Leachman, R. T. Jacobsen, S. G. Penoncello, and E. W. Lemmon, "Fundamental equations of state for parahydrogen, normal hydrogen, and orthohydrogen," Journal of Physical and Chemical Reference Data, vol. 38, no. 3, pp. 721–748, 2009.
55. M. T. Syed, S. A. Sherif, T. N. Veziroglu, and J. W. Sheffield, "An economic analysis of three hydrogen liquefaction systems," International Journal of Hydrogen Energy, vol. 23, no. 7, pp. 565–576, 1998.
56. E. Poirier, R. Chahine, P. Bénard et al., "Storage of hydrogen on single-walled carbon nanotubes and other carbon structures," Applied Physics A, vol. 78, no. 7, pp. 961–967, 2004. ·
57. D. Mori and K. Hirose, "Recent challenges of hydrogen storage technologies for fuel cell vehicles,"International Journal of Hydrogen Energy, vol. 34, no. 10, pp. 4569–4574, 2009.
58. T. Wallner, H. Lohse-Busch, S. Gurski et al., "Fuel economy and emissions evaluation of BMW hydrogen 7 mono-fuel demonstration vehicles," International Journal of Hydrogen Energy, vol. 33, no. 24, pp. 7607–7618, 2008.
59. T. K. Tromp, R. L. Shia, M. Allen, J. M. Eiler, and Y. L. Yung, "Potential environmental impact of a hydrogen economy on the stratosphere," Science, vol. 300, no. 5626, pp. 1740–1742, 2003.
60. N. Eigen, MT Aerospace AG, D-86153 Augsburg, Germany, private communication, 2008.
61. W. Oelerich, T. Klassen, and R. Bormann, "Metal oxides as catalysts for improved hydrogen sorption in nanocrystalline Mg-based materials," Journal of Alloys and Compounds, vol. 315, no. 1-2, pp. 237–242, 2001.
62. D. G. Ivey and D. O. Northwood, "Storing energy in metal hydrides: a review of the physical metallurgy," Journal of Materials Science, vol. 18, no. 2, pp. 321–347, 1983.

63. G. Sandrock, "Panoramic overview of hydrogen storage alloys from a gas reaction point of view,"Journal of Alloys and Compounds, vol. 293, pp. 877–888, 1999.

64. R. A. Varin, T. Czujko, and Z. S. Wronski, Eds., Nanomaterials for Solid State Hydrogen Storage, Springer Science, Business Media, New York, NY, USA, 2009.

65. K. K. Pant and R. B. Gupta, "Hydrogen fuel: production, transport, and storage," in Hydrogen Storage in Metal Hydrides, pp. 381–408, CRC Press, Taylor & Francis Group, Boca Raton, Fla, USA, 2009.

66. L. Belkbir, E. Joly, and N. Gerard, "Comparative study of the formation-decomposition mechanisms and kinetics in $LaNi_5$ and magnesium reversible hydrides," International Journal of Hydrogen Energy, vol. 6, no. 3, pp. 285–294, 1981.

67. G. Sandrock, J. Reilly, J. Graetz, W. M. Zhou, J. Johnson, and J. Wegrzyn, "Accelerated thermal decomposition of AlH_3 for hydrogen-fueled vehicles," Applied Physics A, vol. 80, no. 4, pp. 687–690, 2005.

68. B. Bogdanović and M. Schwickardi, "Ti-doped alkali metal aluminium hydrides as potential novel reversible hydrogen storage materials," Journal of Alloys and Compounds, vol. 253-254, pp. 1–9, 1997.

69. K. J. Gross, G. J. Thomas, and C. M. Jensen, "Catalyzed alanates for hydrogen storage," Journal of Alloys and Compounds, vol. 330–332, pp. 683–690, 2002.

70. M. Fichtner, O. Fuhr, and O. Kircher, "Magnesium alanate—a material for reversible hydrogen storage?" Journal of Alloys and Compounds, vol. 356-357, pp. 418–422, 2003.

71. N. Eigen, M. Kunowsky, T. Klassen, and R. Bormann, "Synthesis of $NaAlH_4$-based hydrogen storage material using milling under low pressure hydrogen atmosphere," Journal of Alloys and Compounds, vol. 430, no. 1-2, pp. 350–355, 2007.

72. S. I. Orimo, Y. Nakamori, J. R. Eliseo, A. Züttel, and C. M. Jensen, "Complex hydrides for hydrogen storage," Chemical Reviews, vol. 107, no. 10, pp. 4111–4132, 2007.

73. P. Chen, Z. Xiong, J. Luo, J. Lin, and K. L. Tan, "Interaction of hydrogen with metal nitrides and imides," Nature, vol. 420, no. 6913, pp. 302–304, 2002. ·

74. Z. Xiong, G. Wu, J. Hu, and P. Chen, "Ternary imides for hydrogen storage," Advanced Materials, vol. 16, no. 17, pp. 1522–1525, 2004.

75. P. Chen, Z. Xiong, G. Wu, Y. Liu, J. Hu, and W. Luo, "Metal-N-H systems for the hydrogen storage,"Scripta Materialia, vol. 56, no. 10, pp. 817–822, 2007. ·

76. D. H. Gregory, "Lithium nitrides, imides and amides as lightweight, reversible hydrogen stores,"Journal of Materials Chemistry, vol. 18, no. 20, pp. 2321–2330, 2008.

77. J. J. Vajo, S. L. Skeith, and F. Mertens, "Reversible storage of hydrogen in destabilized $LiBH_4$," Journal of Physical Chemistry B, vol. 109, no. 9, pp. 3719–3722, 2005. ·

78. G. Barkhordarian, T. Klassen, M. Dornheim, and R. Bormann, "Unexpected kinetic effect of MgB_2 in reactive hydride composites containing complex borohydrides," Journal of Alloys and Compounds, vol. 440, no. 1-2, pp. L18–L21, 2007.

79. U. Bösenberg, J. W. Kim, D. Gosslar et al., "Role of additives in $LiBH_4$-MgH_2 reactive hydride composites for sorption kinetics," Acta Materialia, vol. 58, no. 9, pp. 3381–3389, 2010.

80. L. F. Brown, "A comparative study of fuels for on-board hydrogen production for fuel-cell-powered automobiles," International Journal of Hydrogen Energy, vol. 26, no. 4, pp. 381–397, 2001.

81. J. M. Ogden, M. M. Steinbugler, and T. G. Kreutz, "Comparison of hydrogen, methanol and gasoline as fuels for fuel cell vehicles: implications for vehicle design and infrastructure development," Journal of Power Sources, vol. 79, no. 2, pp. 143–168, 1999.

82. C. W. Hamilton, R. T. Baker, A. Staubitz, and I. Manners, "B-N compounds for chemical hydrogen storage," Chemical Society Reviews, vol. 38, no. 1, pp. 279–293, 2009.

83. J. Graetz, "New approaches to hydrogen storage," Chemical Society Reviews, vol. 38, no. 1, pp. 73–82, 2009.

84. L. Zhou, "Progress and problems in hydrogen storage methods," Renewable and Sustainable Energy Reviews, vol. 9, no. 4, pp. 395–408, 2005.

85. J. Yang, A. Sudik, C. Wolverton, and D. J. Siegel, "High capacity hydrogen storage materials: attributes for automotive applications and techniques for materials discovery," Chemical Society Reviews, vol. 39, no. 2, pp. 656–675, 2010.

86. D. G. Westlake, "Site occupancies and stoichiometries in hydrides of intermetallic compounds: geometric considerations," Journal of The Less-Common Metals, vol. 90, no. 2, pp. 251–273, 1983.

87. K. J. Gross, A. Züttel, and L. Schlapbach, "On the possibility of metal hydride formation—part II: geometric considerations," Journal of Alloys and Compounds, vol. 274, no. 1-2, pp. 239–247, 1998.

88. V. A. Yartys, R. V. Denys, B. Hauback et al., "Short hydrogen-hydrogen separations in novel intermetallic hydrides, $RE_3Ni_3In_3D_4$ (RE=La, Ce and Nd)," Journal of Alloys and Compounds, vol. 330-332, pp. 132–140, 2002.

89. V. A. Yartys, A. B. Riabov, R. V. Denys, M. Sato, and R. G. Delaplane, "Novel intermetallic hydrides,"Journal of Alloys and Compounds, vol. 408–412, pp. 273–279, 2006, Proceedings of Rare Earths›04 in Nara, Japan.

90. V. Iosub, M. Latroche, J. M. Joubert, and A. Percheron-Guégan, "Optimisation of $MmNi_{5-x}Sn_x$ (Mm=La, Ce, Nd and Pr, $0.27<x<0.5$) compositions as hydrogen storage materials," International Journal of Hydrogen Energy, vol. 31, no. 1, pp. 101–108, 2006.

91. M. Botzung, S. Chaudourne, O. Gillia et al., "Simulation and experimental validation of a hydrogen storage tank

with metal hydrides," International Journal of Hydrogen Energy, vol. 33, no. 1, pp. 98–104, 2008.

92. L. Schlapbach, A. Seiler, and F. Stucki, "Surface segregation in FeTi and its catalytic effect on the hydrogenation II: AES and XPS studies," Materials Research Bulletin, vol. 13, no. 10, pp. 1031–1037, 1978.

93. Z. Dehouche, T. Klassen, W. Oelerich, J. Goyette, T. K. Bose, and R. Schulz, "Cycling and thermal stability of nanostructured MgH_2-Cr_2O_3 composite for hydrogen storage," Journal of Alloys and Compounds, vol. 347, no. 1-2, pp. 319–323, 2002.

94. G. J. Thomas, K. J. Gross, N. Y. C. Yang, and C. Jensen, "Microstructural characterization of catalyzed $NaAlH_4$," Journal of Alloys and Compounds, vol. 330–332, pp. 702–707, 2002.

95. H. Pan, Y. Zhu, M. Gao et al., "A study on the cycling stability of the Ti-V-based hydrogen storage electrode alloys," Journal of Alloys and Compounds, vol. 364, no. 1-2, pp. 271–279, 2004.

96. V. Bérubé, G. Radtke, M. Dresselhaus, and G. Chen, "Size effects on the hydrogen storage properties of nanostructured metal hydrides: a review," International Journal of Energy Research, vol. 31, no. 6-7, pp. 637–663, 2007.

97. S. V. Alapati, J. K. Johnson, and D. S. Sholl, "Stability analysis of doped materials for reversible hydrogen storage in destabilized metal hydrides," Physical Review B, vol. 76, no. 10, Article ID 104108, 8 pages, 2007.

98. Y. Liu, H. Pan, M. Gao, R. Li, and Q. Wang, "Intrinsic/extrinsic degradation of Ti-V-based hydrogen storage electrode alloys upon cycling," Journal of Physical Chemistry C, vol. 112, no. 42, pp. 16682–16690, 2008.

99. Y. Suttisawat, P. Rangsunvigit, B. Kitiyanan, and S. Kulprathipanja, "Effect of co-dopants on hydrogen desorption/absorption of $HfCl_4$ and TiO_2-doped $NaAlH_4$," International Journal of Hydrogen Energy, vol. 33, no. 21, pp. 6195–6200, 2008.

100. P. Pfeifer, C. Wall, O. Jensen, H. Hahn, and M. Fichtner, "Thermal coupling of a high temperature PEM fuel cell with a complex hydride tank," International Journal of Hydrogen Energy, vol. 34, no. 8, pp. 3457–3466, 2009.

101. C. Veerraju and M. R. Gopal, "Heat and mass transfer studies on elliptical metal hydride tubes and tube banks," International Journal of Hydrogen Energy, vol. 34, no. 10, pp. 4340–4350, 2009.

102. R. Ströbel, J. Garche, P. T. Moseley, L. Jörissen, and G. Wolf, "Hydrogen storage by carbon materials,"Journal of Power Sources, vol. 159, no. 2, pp. 781–801, 2006.

103. A. W. C. van den Berg and C. O. Areán, "Materials for hydrogen storage: current research trends and perspectives," Chemical Communications, no. 6, pp. 668–681, 2008.

104. S. K. Bhatia and A. L. Myers, "Optimum conditions for adsorptive storage," Langmuir, vol. 22, no. 4, pp. 1688–1700, 2006.

105. U. Eberle, M. Felderhoff, and F. Schüth, "Chemical and physical solutions for hydrogen storage,"Angewandte Chemie—International Edition, vol. 48, no. 36, pp. 6608–6630, 2009.

106. R. D. McCarty, J. Hord, and H. M. Roder, Selected Properties of Hydrogen (Engineering Design Data), vol. 168 of NBS Monograph, Center for Chemical Engineering, National Engineering Laboratory, National Bureau of Standards, National Bureau of Standards, Boulder, Colo, USA, 1981.

107. S. M. Aceves, F. Espinosa-Loza, E. Ledesma-Orozco et al., "High-density automotive hydrogen storage with cryogenic capable pressure vessels," International Journal of Hydrogen Energy, vol. 35, no. 3, pp. 1219–1226, 2010.

108. R. K. Ahluwalia and J. K. Peng, "Automotive hydrogen storage system using cryo-adsorption on activated carbon," International Journal of Hydrogen Energy, vol. 34, no. 13, pp. 5476–5487, 2009.

109. R. E. Morris and P. S. Wheatley, "Gas storage in nanoporous materials," Angewandte Chemie—International Edition, vol. 47, no. 27, pp. 4966–4981, 2008.

110. L. Zhou, Y. Zhou, and Y. Sun, "Enhanced storage of hydrogen at the temperature of liquid nitrogen,"International Journal of Hydrogen Energy, vol. 29, no. 3, pp. 319–322, 2004.

111. A. W. Adamson and A. P. Gast, Physical Chemistry of Surfaces, John Wiley & Sons, New York, NY, USA, 6th edition, 1997.

112. D. D. Do, Adsorption Analysis: Equilibria and Kinetics, vol. 2 of Series on Chemical Engineering, Imperial College Press, London, UK, 1998.

113. J. E. Lennard-Jones, "Processes of adsorption and diffusion on solid surfaces," Transactions of the Faraday Society, vol. 28, pp. 333–359, 1932.

114. F. D. Lamari, B. Weinberger, M. Kunowsky, and D. Levesque, "Material design using molecular modeling for hydrogen storage," AIChE Journal, vol. 55, no. 2, pp. 538–547, 2009.

115. J. Germain, J. M. J. Fréchet, and F. Svec, "Nanoporous polymers for hydrogen storage," Small, vol. 5, no. 10, pp. 1098–1111, 2009.

116. K. M. Thomas, "Hydrogen adsorption and storage on porous materials," Catalysis Today, vol. 120, no. 3-4, pp. 389–398, 2007.

117. M. Jordá-Beneyto, F. Suárez-García, D. Lozano-Castelló, D. Cazorla-Amorós, and A. Linares-Solano, "Hydrogen storage on chemically activated carbons and carbon nanomaterials at high pressures,"Carbon, vol. 45, no. 2, pp. 293–303, 2007.

118. M. Kunowsky, J. P. Marco-Lozar, D. Cazorla-Amorós, and A. Linares-Solano, "Scale-up activation of carbon fibres for hydrogen storage," International Journal of Hydrogen Energy, vol. 35, no. 6, pp. 2393–2402, 2010.

119. M. A. de la Casa-Lillo, F. Lamari-Darkrim, D. Cazorla-Amorós, and A. Linares-Solano, "Hydrogen storage in activated carbons and activated carbon fibers," Journal of Physical Chemistry B, vol. 106, no. 42, pp. 10930–10934, 2002.

120. D. Cazorla-Amorós, J. Alcaniz-Monge, and A. Linares-Solano, "Characterization of activated carbon fibers by CO_2 adsorption," Langmuir, vol. 12, no. 11, pp. 2820–2824, 1996.

121. N. Texier-Mandoki, J. Dentzer, T. Piquero, S. Saadallah, P. David, and C. Vix-Guterl, "Hydrogen storage in activated carbon materials: role of the nanoporous texture," Carbon, vol. 42, no. 12-13, pp. 2744–2747, 2004.

122. B. Panella, M. Hirscher, and S. Roth, "Hydrogen adsorption in different carbon nanostructures,"Carbon, vol. 43, no. 10, pp. 2209–2214, 2005.

123. V. Fierro, A. Szczurek, C. Zlotea et al., "Experimental evidence of an upper limit for hydrogen storage at 77 K on activated carbons," Carbon, vol. 48, no. 7, pp. 1902–1911, 2010.

124. F. Rouquerol, J. Rouquerol, and K. Sing, Adsorption by Powders and Porous Solids: Principles, Methodology and Applications, Academic Press, London, UK, 1999.

125. M. W. Ackley, S. U. Rege, and H. Saxena, "Application of natural zeolites in the purification and separation of gases," Microporous and Mesoporous Materials, vol. 61, no. 1–3, pp. 25–42, 2003.

126. Y. Tao, H. Kanoh, L. Abrams, and K. Kaneko, "Mesopore-modified zeolites: preparation, characterization, and applications," Chemical Reviews, vol. 106, no. 3, pp. 896–910, 2006.

127. D. Fraenkel and J. Shabtai, "Encapsulation of hydrogen in molecular sieve zeolites," Journal of the American Chemical Society, vol. 99, no. 21, pp. 7074–7076, 1977.

128. J. Weitkamp, M. Fritz, and S. Ernst, "Zeolites as media for hydrogen storage," International Journal of Hydrogen Energy, vol. 20, no. 12, pp. 967–970, 1995.

129. H. W. Langmi, A. Walton, M. M. Al-Mamouri et al., "Hydrogen adsorption in zeolites A, X, Y and RHO," Journal of Alloys and Compounds, vol. 356-357, pp. 710–715, 2003.

130. H. W. Langmi, D. Book, A. Walton et al., "Hydrogen storage in ion-exchanged zeolites," Journal of Alloys and Compounds, vol. 404–406, pp. 637–642, 2005.

131. J. G. Vitillo, G. Ricchiardi, G. Spoto, and A. Zecchina, "Theoretical maximal storage of hydrogen in zeolitic frameworks," Physical Chemistry Chemical Physics, vol. 7, no. 23, pp. 3948–3954, 2005.

132. M. G. Nijkamp, J. E. M. J. Raaymakers, A. J. van Dillen, and K. P. de Jong, "Hydrogen storage using physisorption-materials demands," Applied Physics A, vol. 72, no. 5, pp. 619–623, 2001.

133. A. Katiyar, S. Yadav, P. G. Smirniotis, and N. G. Pinto, "Synthesis of ordered large pore SBA-15 spherical particles for adsorption of biomolecules," Journal of Chromatography A, vol. 1122, no. 1-2, pp. 13–20, 2006.

134. T. Yanagisawa, T. Shimizu, K. Kuroda, and C. Kato, "The preparation of alkyltrimethylammonium-kanemite complexes and their conversion to microporous materials," Bulletin of the Chemical Society of Japan, vol. 63, no. 4, pp. 988–992, 1990.

135. J. S. Beck, J. C. Vartuli, W. J. Roth et al., "A new family of mesoporous molecular sieves prepared with liquid crystal templates," Journal of the American Chemical Society, vol. 114, no. 27, pp. 10834–10843, 1992.

136. C. T. Kresge, M. E. Leonowicz, W. J. Roth, J. C. Vartuli, and J. S. Beck, "Ordered mesoporous molecular sieves synthesized by a liquid-crystal template mechanism," Nature, vol. 359, no. 6397, pp. 710–712, 1992.

137. D. Zhao, J. Feng, Q. Huo et al., "Triblock copolymer syntheses of mesoporous silica with periodic 50 to 300 angstrom pores," Science, vol. 279, no. 5350, pp. 548–552, 1998.

138. D. Zhao, Q. Huo, J. Feng, B. F. Chmelka, and G. D. Stucky, "Nonionic triblock and star diblock copolymer and oligomeric sufactant syntheses of highly ordered, hydrothermally stable, mesoporous silica structures," Journal of the American Chemical Society, vol. 120, no. 24, pp. 6024–6036, 1998.

139. R. W. Pekala, "Organic aerogels from the polycondensation of resorcinol with formaldehyde," Journal of Materials Science, vol. 24, no. 9, pp. 3221–3227, 1989.

140. M. S. Dresselhaus, "Future directions in carbon science," Annual Review of Materials Science, vol. 27, no. 1, pp. 1–34, 1997.

141. H. Kabbour, T. F. Baumann, J. H. Satcher, A. Saulnier, and C. C. Ahn, "Toward new candidates for hydrogen storage: high-surface-area carbon aerogels," Chemistry of Materials, vol. 18, no. 26, pp. 6085–6087, 2006.

142. H. Y. Tian, C. E. Buckley, S. B. Wang, and M. F. Zhou, "Enhanced hydrogen storage capacity in carbon aerogels treated with KOH," Carbon, vol. 47, no. 8, pp. 2128–2130, 2009.

143. C. N. R. Rao, B. C. Satishkumar, A. Govindaraj, and M. Nath, "Nanotubes," ChemPhysChem, vol. 2, no. 2, pp. 78–105, 2001.

144. R. Ma, Y. Bando, H. Zhu, T. Sato, C. Xu, and D. Wu, "Hydrogen uptake in boron nitride nanotubes at room temperature," Journal of the American Chemical Society, vol. 124, no. 26, pp. 7672–7673, 2002.

145. D. Golberg, Y. Bando, C. Tang, and C. Zni, "Boron nitride nanotubes," Advanced Materials, vol. 19, no. 18, pp. 2413–2432, 2007.

146. L. Xiu-Ying, W. Chao-Yang, T. J. Yong-Jian, S. Wei-Guo, W. Wei-Dong, and X. Jia-Jing, "Theoretical studies on hydrogen adsorption of single-walled boron-nitride and carbon nanotubes using grand canonical Monte Carlo method," Physica B, vol. 404, no. 14-15, pp. 1892–1896, 2009.

147. D. Portehault, C. Giordano, C. Gervais et al., "High-surface-area nanoporous boron carbon nitrides for hydrogen storage," Advanced Functional Materials, vol. 20, no. 11, pp. 1827–1833, 2010.

148. P. M. Budd, B. S. Ghanem, S. Makhseed, N. B. McKeown, K. J. Msayib, and C. E. Tattershall, "Polymers of intrinsic microporosity (PIMs): robust, solution-processable, organic nanoporous materials,"Chemical Communications, vol. 10, no. 2, pp. 230–231, 2004.

149. N. B. McKeown, B. Gahnem, K. J. Msayib et al.,

"Towards polymer-based hydrogen storage materials: engineering ultramicroporous cavities within polymers of intrinsic microporosity," Angewandte Chemie—International Edition, vol. 45, no. 11, pp. 1804–1807, 2006.

150. J. Germain, J. Hradil, J. M. J. Fréchet, and F. Svec, "High surface area nanoporous polymers for reversible hydrogen storage," Chemistry of Materials, vol. 18, no. 18, pp. 4430–4435, 2006.

151. J. Germain, F. Svec, and J. M. J. Fréenet, "Preparation of size-selective nanoporous polymer networks of aromatic rings: potential adsorbents for hydrogen storage," Chemistry of Materials, vol. 20, no. 22, pp. 7069–7076, 2008.

152. J. Y. Lee, C. D. Wood, D. Bradshaw, M. J. Rosseinsky, and A. I. Cooper, "Hydrogen adsorption in microporous hypercrosslinked polymers," Chemical Communications, no. 25, pp. 2670–2672, 2006.

153. H. Furukawa, N. Ko, Y. B. Go et al., "Ultrahigh porosity in metal-organic frameworks," Science, vol. 329, no. 5990, pp. 424–428, 2010.

154. J. Juan-Juan, J. P. Marco-Lozar, F. Suárez-García, D. Cazorla-Amorós, and A. Linares-Solano, "A comparison of hydrogen storage in activated carbons and a metal-organic framework (MOF-5),"Carbon, vol. 48, no. 10, pp. 2906–2909, 2010.

155. A. Linares-Solano, D. Lozano-Castelló, M. A. Lillo-Ródenas, and D. Cazorla-Amorós, "Carbon activation by alkaline hydroxides preparation and reactions, porosity and performance," Chemistry and Physics of Carbon, vol. 30, pp. 1–62, 2008.

156. J. S. Noh, R. K. Agarwal, and J. A. Schwarz, "Hydrogen storage systems using activated carbon,"International Journal of Hydrogen Energy, vol. 12, no. 10, pp. 693–700, 1987. ·

157. R. K. Agarwal and J. A. Schwarz, "Analysis of high pressure adsorption of gases on activated carbon by potential theory," Carbon, vol. 26, no. 6, pp. 873–887, 1988.

158. M. Hirscher and B. Panella, "Nanostructures with high surface area for hydrogen storage," Journal of Alloys and Compounds, vol. 404–406, pp. 399–401, 2005. ·

159. Y. Gogotsi, A. Nikitin, H. Ye et al., "Nanoporous carbide-derived carbon with tunable pore size,"Nature Materials, vol. 2, no. 9, pp. 591–594, 2003.

160. Y. Gogotsi, C. Portet, S. Osswald et al., "Importance of pore size in high-pressure hydrogen storage by porous carbons," International Journal of Hydrogen Energy, vol. 34, no. 15, pp. 6314–6319, 2009.

161. Z. Yang, Y. Xia, and R. Mokaya, "Enhanced hydrogen storage capacity of high surface area zeolite-like carbon materials," Journal of the American Chemical Society, vol. 129, no. 6, pp. 1673–1679, 2007.

162. D. Lozano-Castelló, M. A. Lillo-Ródenas, D. Cazorla-Amorós, and A. Linares-Solano, "Preparation of activated carbons from Spanish anthracite—I. Activation by KOH," Carbon, vol. 39, no. 5, pp. 741–749, 2001

CHAPTER 5

DEVELOPMENT OF HYDROGEN STORAGE TANK SYSTEMS BASED ON COMPLEX METAL HYDRIDES

Morten B. Ley, Mariem Meggouh, Romain Moury, Kateryna Peinecke and Michael Felderhoff

Keywords: Hydrogen storage, complex hydrides, modeling, thermolysis, tank design

INTRODUCTION

Over the last 15 years complex aluminum and boron hydrides have been investigated as possible hydrogen storage materials [1,2,3]. Although the composition of these materials seems to be similar, the chemical behavior is entirely different. Several complex aluminum hydrides ($NaAlH_4$, $KAlH_4$, Na_3AlH_6 *etc.*) can be decomposed at elevated temperatures and rehydrogenation is possible under technically relevant conditions using catalysts [4]. However, the decomposition temperature for complex boron hydrides ($LiBH_4$, $NaBH_4$) is often much higher and reversibility cannot be observed under conditions used for complex aluminum hydrides [5]. Consequently, complex aluminum hydrides can be used for technical applications where rehydrogenation of the hydrogen storage material is an important prerequisite [6]. In contrast, complex boron hydrides are favored for disposal cartridge systems, releasing hydrogen in a hydrolysis reaction at ambient temperatures [7]. These different properties result in completely different engineering as well as technical requirements. This makes the development of suitable tank systems based on aluminum or boron hydride compounds challenging.

This review describes recent research in the development of tank systems based on complex hydrides for thermolysis and hydrolysis. Thermolysis requires a heat input and care must be taken in the design of the storage tank in order to effectively distribute the heat. Hydrolysis, on the other hand, requires not only effective mixing of the complex hydride and water but also separation of the produced hydrogen gas and the slurry consisting of the decomposition product and water. While thermolysis tank systems are developed in demonstration projects, hydrolysis tank systems have already found real life applications. The review is divided into two parts, the first focuses on thermolysis, while the second concerns hydrolysis. Initially, reactor, kinetic and system models will be introduced in each part. Hereafter, specific examples with the most common complex aluminum or boron hydrides are given together with their applications.

THERMOLYSIS

Tank Developments for Thermal Hydrogen Evolution

Development of a hydrogen storage tank based on complex hydrides involves different levels of research. The first level provides a screening of materials at laboratory scale. Once relevant candidates have been selected, a methodology to evaluate their performances must be established. The evaluation is based on two steps. The first one is modeling of the full-scale system and the second one is manufacturing of a prototype to experimentally evaluate the performance of the material and validate the results of the simulation. The

Citation: Morten B. Ley, Mariem Meggouh, Romain Moury, Kateryna Peinecke and Michael Felderhoff, Development of Hydrogen Storage Tank Systems Based on Complex Metal Hydrides, doi:10.3390/ma8095280.

Department of Energy (DoE) in the United States of America has given several milestones for the research on the optimal complex hydride/tank design pair (Table 1). All these technical specifications are material dependent, thus the first screening starts with a selection among available complex hydrides. Nevertheless, improvement of the scale-up system through several strategies is still possible. Among all the complex hydrides claimed for hydrogen storage, just a few of them have been chosen to develop a full-scale tank system. Since Bogdanović and Schwickardi discovered that sodium alanate can reversibly store hydrogen using a Ti-based catalyst [4], its hydrogen storage properties have been widely studied. The absorption/desorption mechanism has been found to occur through a two-step process (Equation (1)). Another system based on a mixture of magnesium nitride and lithium hydride ($Mg(NH_2)_2$-2LiH), is also reviewed. This system can reversibly store hydrogen following the mechanism described in Equation (2). The last materials studied for tank development are ammonia borane (AB, NH_3BH_3) and alane (AlH_3). They are not reversible by direct hydrogenation but instead need a chemical treatment to be recovered. Nevertheless, they have attracted a lot of attention owing to their high gravimetric hydrogen capacity (19.5 wt % H_2 for AB and 10 wt % for AlH_3). The decomposition mechanism is described in Equations (3) and (4) for AB and AlH_3respectively.

$$NaAlH_4 \overset{1}{\rightleftharpoons} 1/3\, Na_3AlH_6 + 2/3\, Al + H_2 \overset{2}{\rightleftharpoons} NaH + Al + 3/2\, H_2 \tag{1}$$

$$2\, Mg(NH_2)_2 + 4\, LiH \overset{1}{\rightleftharpoons} LiH + LiNH_2 + Li_2Mg_2(NH)_3 + 3\, H_2 \overset{2}{\rightleftharpoons} 2\, Li_2Mg(NH)_2 + 4\, H_2 \tag{2}$$

$$NH_3BH_3 \rightarrow [NH_2BH_2]_n + H_2 \rightarrow [NHBH]_n + 2H_2 \rightarrow BN + 3H_2 \tag{3}$$

$$AlH_3 \rightarrow Al + 3/2\, H_2 \tag{4}$$

In this part, we focus on the optimization of the tank system for thermolysis by reviewing the different strategies based on the aforementioned complex hydrides selected for modeling and manufacture. Firstly, an overview is given of modeling aspects developed in order to simulate tank systems. Secondly, a focus is put on the manufacturing of scaled-up tank systems to experimentally evaluate the effects of optimization parameters and the viability of the system according to the DoE requirements.

Table 1. Department of Energy (DoE) requirements for hydrogen storage system for mobile application [8]. PEM: proton exchange membrane.

System Storage Parameter	Original 2010 Target	Revised 2010 Target	2017 Target	Ultimate Target
Gravimetric capacity ($kg\cdot H_2\cdot kg^{-1}$ system)	6%	4.5%	5.5%	7.5%
Volumetric capacity ($g\cdot H_2\cdot L^{-1}$ system)	45	28	40	70
Operational cycle life	1000	1000	1500	1500
Filling time (min for 5 kg)	3	4.2	3.3	2.5
Min full flow rate ($g\cdot H_2\cdot s^{-1}/kW$)	0.02	0.02	0.02	0.02
Min delivery pressure at 85 °C PEM fuel cell (atm)	8	5	4	3
Fuel purity	99.99%	99.97%	99.97%	99.97%

Modeling of Thermolysis Tank Systems

Lab-scale evaluation of materials for hydrogen storage differs significantly from tank development in the kg scale. In order to avoid unnecessary work in the development and construction of metal hydrides tank systems, modeling is often used as a starting process for tank evolution. Simulations are therefore performed to optimize tank design and operating parameters, but also to allow evaluation of the complete performances. For the modeling of a tank system based on complex hydrides, three sub-processes that occur during hydrogen sorption require implementation in the calculations: (i) hydrogen transport, (ii) chemical reaction and (iii) heat transfer [9]. Hydrogen transport and heat transfer are modeled through mass, momentum and energy balance equations and all simulation are performed using the same assumptions and equations. Chemical reactions are modeled using empirical kinetic models; therefore the next section focuses on the different models found in the literature. After the final model is proposed, parametric studies can be performed to gain insight into the influence of different parameters on the system performance.

The first step in building a kinetics model is determination of the dehydrogenation mechanism of the considered material. Once the mechanism is known, the rate for each step is defined by a general model parameterized using three variables: the temperature T, the conversion α, and the pressure p (Equation (5)) [10]:

$$d\alpha/dt = (Ae^{-Ea/RT}) \times f(p, p_{eq}) \times g(\alpha) \tag{5}$$

where E_a is the activation energy, A is a pre-exponential factor, and R is the universal gas constant. The function $g(\alpha)$ and $f(p, p_{eq})$ (p_{eq} is the equilibrium pressure) must be determined experimentally at isobaric and isothermal conditions. The choice of the function $g(\alpha)$ is determined by fitting experimental data with a chosen kinetics model. Typically, two models are used: (i) decelerating model (reaction order model) (Equation (6)) and (ii) sigmoidal model described by Johnson-Mehl-Avrami (JMA) equations, (Equation (7)).

$$f(\alpha) = (1-\alpha)^n;\ n = \text{reaction order} \tag{6}$$

$$f(\alpha) = n(1-\alpha)(-\ln(1-\alpha))^{(n-1)/n};\ n = \text{constant} \tag{7}$$

Once a proper model is chosen, an assumption on the function $f(p, p_{eq})$ has to be made, in order to determine both the activation energy E_a and pre-exponential factor A. Usually $f(p, p_{eq})$ is described by the logarithm $\pm\ln(p/p_{eq})$ or approximated by its Taylor development. In the literature numerous pressure dependent models have been proposed. Table 2 summarizes all the kinetic models proposed for the materials used in tank modeling.

Table 2. Summary of the main kinetics models with pre-exponential factor A and activation energy E_a.

Material	Reference	Step	$g(\alpha)$	$f(p, p_{eq})$	A (s^{-1})	E_a ($kJ \cdot mol^{-1}$)
$NaAlH_4$	[11]	1_{des}	1st order	$\ln(p_{eq}/p)$	1.9×10^{11}	85.6
		2_{des}	1st order	$\ln(p_{eq}/p)$	2.9×10^{10}	88.3
		1_{abs}	2nd order	$\ln(p/p_{eq})$	6.2×10^{8}	61.6
		2_{abs}	1st order	$\ln(p/p_{eq})$	1.0×10^{8}	56.2
	[12]	1_{abs}	JMA $n = 1.4$	$\ln(p/p_{eq})(p/p_0)^{2.7}$	3.6×10^{-1}	54.7
			1st order	$\ln(p/p_{eq})(p/p_0)^{1.5}$	1.9×10^{3}	62.9
		2_{abs}	JMA $n = 1.4$	$(p - p_{eq})/p_{eq}$	2.0×10^{11}	117.5
			1st order	$((p - p_{eq})/p)^{1.5}$	1.2×10^{8}	86.4
	[13]	1_{des}	2nd order	$(p_{eq} - p)/p_{eq}$	4.0×10^{5}	110
		2_{des}	1st order	$(p_{eq} - p)/p_{eq}$	6.0×10^{12}	110
		1_{abs}	2nd order	$(p - p_{eq})/p_{eq}$	1.0×10^{8}	80
		2_{abs}	1st order	$(p - p_{eq})/p_{eq}$	1.5×10^{5}	70
	[14]	1_{abs}	JMA $n = 1.33$	$(p - p_{eq})/p_{eq}$	1.5×10^{9}	91.5
		2_{abs}	JMA $n = 1.33$	$(p - p_{eq})/p_{eq}$	2.3×10^{8}	91.7
	[15]	1_{des}	0 order	$((p_{eq} - p)/p_{eq})^2 + 1.04((p_{eq} - p)/p_{eq})$	5.4×10^{10}	105.8
		2_{des}	1st order	$((p_{eq} - p)/p_{eq})^2 + 0.46((p_{eq} - p)/p_{eq})$	3.4×10^{8}	91.5
$Li_2Mg(NH)_2$	[16]	1_{des}	JMA $n = 1.5$	$\ln(p_{eq}/p)$	2.3×10^{12}	131.8
		2_{des}	0 order	$1 - (0.001515/0.33wt_{max})(p - 1.1)$	3.0×10^{15}	161.4
		1_{abs}	0 order	$(p - p_{eq})/p_{eq}$	2.7×10^{17}	164.8
		2_{abs}	1st order	$(p_{eq} - p)/p_{eq}$	4.7×10^{14}	147.8

des = desorption, abs = absorption; definition of the steps see Equations (1) and (2); wt_{max} is the maximum weight fraction.

Simulation of Thermolysis Tank Systems

Simulations on Sodium Alanate

Na Ranong *et al.*, investigated the effects of the outer diameter in a multi tubular tank on refueling time, the heat transfer units and the total weight of the system [17]. The kinetics model from Franzen was used in the calculation [12]. Two 8 kg tank units with a reactor outer diameter of 16 mm and 60 mm were compared. Although the 16 mm reactor showed better results than the 60 mm reactor for both refueling time (290 s (16 mm) *vs.* 540 s (60 mm)) and heat transfer (610 $W \cdot m^{-1} \cdot K^{-1}$ (16 mm) *vs.* 319 $W \cdot m^{-1} \cdot K^{-1}$ (60 mm)), the 60 mm reactor was chosen for manufacture due to less number of tubes needed within the tank leading to a reduced overall weight (68 kg (16 mm) *vs.*38 kg (60 mm)).

Hardy and Anton developed a hierarchical methodology for hydrogen loading process to refine operating parameters [18,19]. The methodology consists of four sub-models: (i) A kinetics scoping model [20]; (ii) A geometry scoping model was developed to refine length scale in a given tank geometry (the geometry is available in [21] and the calculated geometry parameters in [18]); (iii) a heat transfer removal scoping model developed to calculate coolant flowrate and temperature, pressure drop, convection heat coefficients and temperature increases along the cooling channel (the results are described in [19]); and (iv) a 2- and 3-dimensional finite element model built with previous optimized parameters (from (i) to (iii)) as input parameters. The model was used to assess the detailed performances for a $NaAlH_4$ system, but could be also adapted for other systems, if a proper kinetic model would be provided.

Pfeifer *et al.*, investigated the thermal coupling between a high temperature proton exchange membrane (HT-PEM) fuel cell with a 2 kg sodium alanate tank using simulation studies in order to identify the parameters influencing efficiency and start-up during hydrogen discharge [22]. The Luo and Gross kinetic model was used in the simulation [11]. The authors highlighted that the decomposition kinetic was the limiting parameter on the efficiency of the cell. An optimum operating temperature of 185 °C with a fuel cell power of 1 kW was found. The minimum operating temperature was evaluated at 120 °C with a cumulative output of 0.8 kWh.

Bhouri *et al.*, performed optimizations of two different tank configurations: a shell, tube and fin hydride bed system (1) and a multi-tubular reactor equipped with fins (2) [23,24]. The hierarchical methodology developed by Hardy and Anton [18,19] was used with a UTRC™ kinetic model [20]. The purpose for configuration (1) was to optimize the fin thickness and number of heat exchanger on both loading and discharging processes. It was demonstrated that the fin thickness had no influence during the loading process. However, during discharge the rate and amount of hydrogen release were improved by using thicker fins. After all, this decreased the gravimetric and volumetric capacities [23]. The effects of the number, thickness and tip clearance of the fins on hydrogen uptake were evaluated for configuration (2). The authors concluded that an increase in the number of fins and decrease in the tip clearance had a beneficial effect on the loading rate at the expense of hydrogen capacity. The optimized configuration (six fins, 0.1 cm thickness and 0 cm of tips clearance) led to an improvement of the loading rate by 41%. On the contrary, fin thickness had no impact on the loading rate. Thus, optimization of the system was based on a compromise between loading rate and fins contribution to the overall weight [24].

Effects of operating parameters during refueling in a cylindrical bed with longitudinal fins and cooling heat exchanger were reviewed by Raju and Kumar [25]. The geometry of the tank was adopted from Hardy and Anton [18,19]. A sensitivity analysis was performed to estimate the effect of the initial bed temperature, supply pressure, coolant flow rate, coolant temperature and thermal conductivity. The kinetic model was adopted from Luo and Gross [11]. An operating pressure of 15 MPa or above was chosen to keep the temperature in the optimal range 420–450 K and a coolant flow rate of 25 lpm/tube at 380 K was found to be optimal. The best configuration required an increase of 8 $W \cdot m^{-1} \cdot K^{-1}$ in the thermal conductivity by compaction with an addition of a conductivity enhancer (aluminum or graphite) [25].

Simulation was performed for a multi-tubular tank using the kinetic model from Lozano *et al.*[14]. Effect of the inner diameter, compaction and addition of expanded natural graphite (ENG) were evaluated according to hydrogen capacity during the loading (here 4.5 kg H_2 in 10 min was used as a constraint). The optimized configuration, for a 35 mm inner diameter tank with compacted powder and no ENG addition, was determined at 10 MPa and 130 °C [26]. The effect of graphite addition and the design of the heat exchanger were also studied by Johnson *et al.* [27]. The simulation was performed using a first principle kinetic model that relied on experimentally determined parameters (full explanation of the model can be found in [20]).

In 2014, a comparative resistance analysis was developed [9], using a kinetic model from Lozano*et al.* [14,15]. The authors estimated, based on the driving force for each sub-process (hydrogen transport, intrinsic kinetics and heat transfer), the resistance during hydrogen sorption. The authors highlighted that hydrogen transport has a negligible effect (independent of the size of the reactor), the intrinsic kinetics play a decisive role in a small cell (2 mm), and heat transfer is the main resistance during absorption in a scale-up tank. Consequently, improvement of heat transfer is necessary in order to obtain a tank system viable for application.

Simulations on Other Complex Hydrides

Ammonia borane (AB, NH_3BH_3) has a very high practical gravimetric density of about 16 wt % and both solid and liquid AB have been tested for application purposes [28,29]. Unfortunately, AB requires an off-board refueling system, since chemical treatment is a necessity for regeneration. The off-board regeneration system would be the limiting step in the development of AB storage tanks. A model was developed to estimate the AB efficiency in Auger and slurry reactors [28], together with a set of guidelines for the fluid-phase hydrogen storage properties [29].

The complex hydride (CH) $2LiNH_2$-$1.1MgH_2$-$0.1LiBH_4$-3 wt % $ZrCoH_3$ and the metal hydride (MH) $LaNi_{4.3}Al_{0.4}Mn_{0.3}$ were used in tandem for scale-up tank simulation studies [30,31,32]. At first, a new reactor concept was designed by combining advantages of both CH (high hydrogen content) and MH (high reaction rate) [30,31]. The authors used a cylindrical tank heated through the outer wall. The MH bed was positioned in the center of the tank separated from the CH bed by a membrane. In the absorption step, the heat of absorption of the MH bed (low temperature) was transferred to the CH bed (higher temperature). The kinetics for the simulation was based on the model by Bürger *et al.* [16]. The viability of the concept was demonstrated by comparing the temperature front in a pure CH bed and in a MH/CH combined bed. At 7 MPa and 130 °C, the absorption in the pure CH bed started from the annulus to the core. However, the heat removal (from the outer wall) was difficult when the temperature increased in the core. In the combined MH/CH bed, the MH part started to absorb at 7 MPa and room temperature. When the inner temperature exceeded 200 °C, the heat was transferred from the core to the annulus, which in turn facilitated heat removal from the outside of the tank [30]. In regard to the desorption step, the MH stabilized the pressure in the reactor around the equilibrium pressure of the CH, whereby a pressure drop was avoided during desorption [31]. In conclusion, the MH bed had a beneficial effect on the system by introducing a better temperature management during absorption and a better pressure management during desorption. In a later work, the tank concept was used in simulation studies to optimize operating parameters. The lower loading time was determined for the ratio between the thicknesses of the CH/MH beds; the optimal CH thickness was 12.5 mm. A sensitivity study demonstrated that the limiting parameter on the loading rate was the thermal conductivity of the complex hydride [32].

Developed Complex Metal Hydride Based Tank Systems

Na Ranong *et al.* [17] have given a comprehensive overview of design solutions for the different components of metal hydride storage tanks. A higher density of the powdered material can improve the thermal conductivity and volumetric capacity limitations [33,34]. Lozano *et al.*, found that compaction of the material is the most influential factor for the optimization of the hydrogen capacity of the storage system [26]. Some researches and developers have been concentrating their efforts on finding a suitable material and designing a system for onboard vehicular applications [27,35]. Others have focused on increasing the efficiency of the combined heat-power units for the stationary hydrogen storing systems [36,37,38], or on improving different properties and kinetics of the complete system [35,39,40]. Here, a brief overview is presented on hydrogen storage systems that have been published, based on sodium alanate, metal amides, ammonia borane and alane. The research and development efforts are aimed at improving the gravimetric/volumetric capacity, efficiency of the heat management and lowering the overall cost of the systems.

Sodium Alanate Based Tank Systems

Bellosta von Colbe *et al.*, demonstrated the functionality of an 8 kg tank during absorption and desorption with a peak technical absorption time below 10 min, thereby validating the design and simulation work carried out previously [15,41]. The ease of use, speed of charging, and reproducibility of results of the 8 kg sodium alanate tank were highly promising [36]. This has paved the way for the application of complex hydrides based hydrogen storage tanks wherever a source of available waste heat is compatible with the energy needs of the tank (especially regarding the temperature level). This is the case especially in large scale and stationary applications [36].

Utz *et al.*, studied the behavior of a powdered bed containing $NaAlH_4$ doped with 4 mol % $CeCl_3$in a lab-scale hydrogen storage tank with flow-through mode. The results showed a significant influence of the cooling by excess hydrogen on the flow-directional temperature profiles. The initial thermal conductivity of the bed increased by a factor of 1.3 compared to values reported in literature (0.67 $W \cdot m^{-1} \cdot K^{-1}$). This caused significantly lower temperature peaks in the center of the reaction bed. The value for the permeability decreased by 50% and led to increase in the pressure drop. No cycling degradation after 36 cycles was observed and a storage capacity of approx. 3.9 wt % H_2 was reached with this material [39].

Modularity was a key concept in the design of a General Motors R&D/Sandia National Laboratories hydrogen storage system (Figure 1). The system was designed to be refueled in approximately 10 min and to deliver hydrogen at up to 2.0 $g \cdot s^{-1}$. A heat amount of 60 MJ had to be removed during refueling by means of a circulating heat transfer fluid. However, for hydrogen delivery, heat must be supplied to the hydrogen storage system *in-situ* and this was achieved by means of a catalytic heater [27].

Urbanczyk *et al.*, designed a hydrogen storage tank based on 2.7 kg of $NaAlH_4$ doped with 4 mol % $TiCl_3$ that was thermally coupled with a high temperature proton exchange membrane (HT-PEM) fuel cell (Figure 2). The waste heat of the fuel cell was used to heat up the storage tank during dehydrogenation, which was consequently fed into the fuel cell. The desorbed amount of hydrogen gas was enough for 3 h of fuel cell operation and the system produced 940 Wh of cumulative energy [37].

Na_3AlH_6 doped with 4 mol % $TiCl_3$ was used as a storage material in an Al-alloy tank system developed at IUTA (Institut für Energie- und Umwelttechnik, Germany) and Max-Planck-Institut für Kohlenforschung, in order to decrease the overall weight, (Figure 3). The hexahydride was operated at a lower pressure compared to $NaAlH_4$. The heat transfer was realized through an oil flow in a bayonet heat exchanger, manufactured by extrusion molding from an aluminum alloy. The 0.21 kg hydrogen storage tank released and absorbed 3.6 g (1.7 wt %) of hydrogen at approximately 450 K. A test with 45 cycles (hydrogenation and dehydrogenation) was carried out without any failure of the tank or its components. Operation of the tank under real conditions indicated the possibility for applications with stationary HT-PEM fuel cell systems [38].

The same partners at IUTA and Max-Planck-Institut für Kohlenforschung used Na_3AlH_6 doped with 4 mol % $TiCl_3$, 8 mol % Al and 8 mol % activated carbon as a hydrogen storage material for a scaled up 1.9 kg Al-alloy storage tank with corrugated heat exchangers. Up to 31 hydrogenation and dehydrogenation cycles were performed without degradation. The aim of the demonstration project was to produce a lightweight system that could be connected to a HT-PEM fuel cell. The system could improve efficiency of a combined heat-power unit for household applications [42].

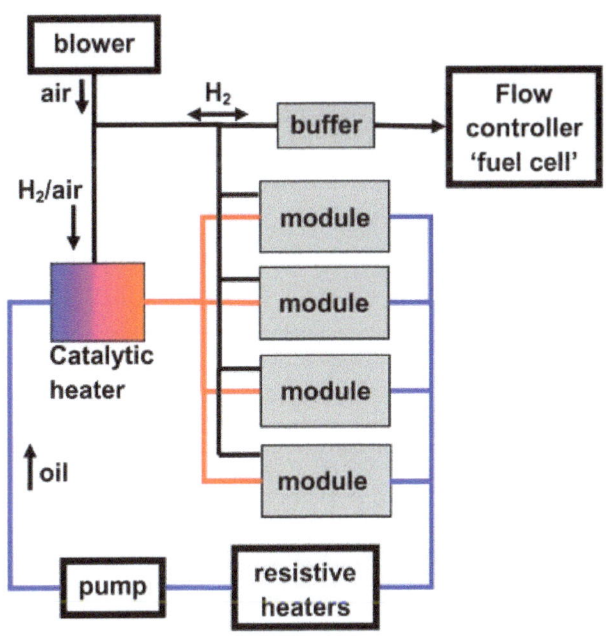

Figure 1. Process flow diagram of General Motors R&D/Sandia National Laboratories hydrogen storage system. Reprinted with permission from reference [27]. Copyright 2012 Elsevier.

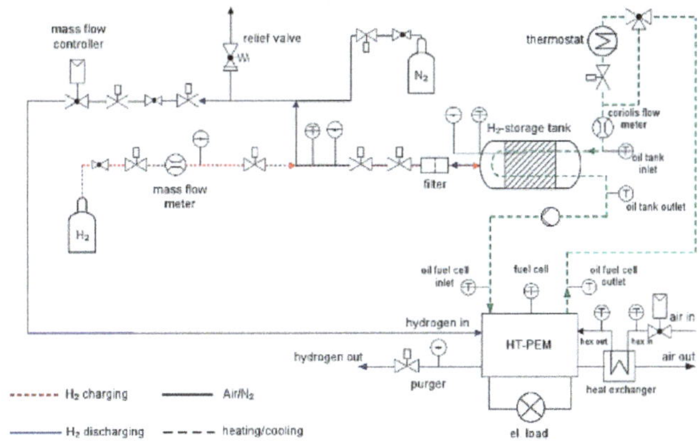

Figure 2. Sketch of the test rig where the storage tank is coupled with high temperature proton exchange membrane (HT-PEM) fuel cell. Reprinted with permission from reference [37]. Copyright 2011 Wiley.

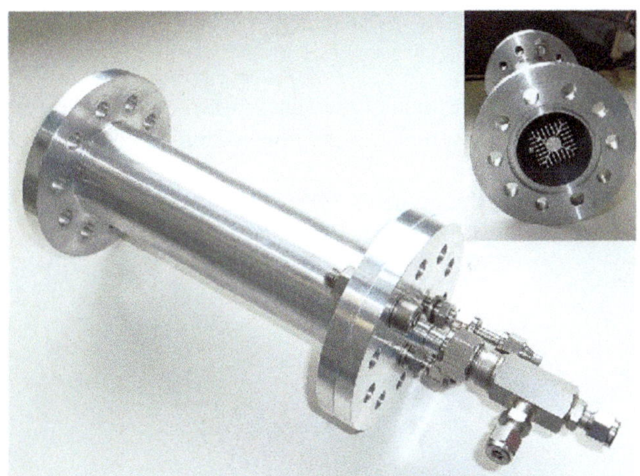

Figure 3. Al-Alloy hydrogen storage tank developed by Institut für Energie- und Umwelttechnik/Max-Planck-Institut für Kohlenforschung (IUTA/MPI). Reprinted with permission from reference [38]. Copyright 2014 Elsevier.

A lightweight tank for hydrogen storage containing 4.4 kg of $NaAlH_4$ doped with $TiCl_3{\cdot}1/3AlCl_3$and expanded natural graphite (ENG) was designed, built and tested [36]. The tank improved the gravimetric capacity by 83% and the volumetric capacity by 49% compared to a previous tank [36]. Additionally, the heat evolution and temperature spikes during hydrogen absorption were studied and since $NaAlH_4$ has a high specific heat, the material itself acted as a heat sink, aiding in the heat management of the system. The authors also performed the first-ever radiography with fast neutrons on the operational complex-hydride based test tank [43].

Metal Amide Based Tank Systems

Yan *et al.*, demonstrated that changes in the compaction pressure and the graphite content of $Mg(NH_2)_2$-2LiH-0.07KOH pellets in a cylindrical lab-scale tank affected the hydrogen storage properties. The addition of up to 9 wt % ENG significantly enhanced the heat transfer in the hydride bed, which resulted in the rise of the central temperature. This improved the hydrogen desorption property of the tank and also increased the hydrogen desorption time at a constant flow rate (Table 3). The volumetric storage density of the composite reached

47 g·L^{-1} after compression at 365 MPa [40]. The hydrogen permeability decreased with increasing compression pressure, while the entire desorption process only changed slightly. The results indicated that the heat transfer behavior of the hydride bed dominated the hydrogen desorption of the tank [44].

Bürger *et al.*, designed a 600 g lab scale combination reactor where two materials, $2LiNH_2$-$1.1MgH_2$-$0.1LiBH_4$-3 wt % $ZrCoH_3$ and $LaNi_{4.3}Al_{0.4}Mn_{0.3}$, are separated by a gas permeable layer. As it was mentioned in the modeling part, the reason for the combined reactor concept was that the fast desorption reaction of the metal hydride stabilizes the pressure of the reactor at the equilibrium pressure of the complex metal hydride. The model was developed and validated by experimental data. This reactor has been developed in order to be coupled with a HT-PEM fuel cell running at technically relevant conditions, *i.e.*, allowing for 2 h operation of 1 kW_e fuel cell [31].

Ammonia Borane and Alane Based Tank Systems

Brooks *et al.*, investigated the slurry-based ammonia borane (AB) and alane system for hydrogen storage in light-duty vehicles and developed a lab-scale slurry based reactor. AB was selected because of its promising practical hydrogen storage capacity of approximately 16 wt %, while alane has a practical hydrogen storage capacity of 10 wt % [45]. Once these chemicals have reacted to produce hydrogen, the byproduct must be removed from the system and off-board regenerated before it can be reused. This differs from the previous system described that can be regenerated directly onboard. Although slurries can be difficult to handle, they can be transported on and off the vehicle using a flow-through process [35]. There are several drawbacks in handling these materials. The high densities of both slurries result in need for mixing the feed and product prior to movement. Alane was found to have some advantages over AB, which generates fuel cell impurities (*i.e.*, ammonia, borazine, diborane) that require removal before reaching the fuel cell [46].

A summary of the developed hydrogen storage systems based on thermolysis is shown in Table 3.

Table 3. Summary of the developed hydrogen storage systems.

Storage Material	Weight of Storage Material (kg)	Design	Capacity (wt. % H_2)	T, P Conditions Charging/Discharging	Kinetic/Cycles	Purpose
$NaAlH_4$ doped with 2 mol % ($TiCl_3$-$0.3AlCl_3$), 5 mol % carbon	8	Tubular reactor with porous sintered metal tube	3.7	charging: 125 °C, 10 MPa; discharging: 160–175 °C, 0.02–1 MPa	1–10 activation cycles after 10 min 80% capacity achieved	Large scale and stationary applications [36].
$NaAlH_4$ doped with 2 mol % $CeCl_3$	0.087	Hydride bed reactor with flow-thru mode	3.9	charging: 130 °C, 10 MPa; discharging: 180 °C, 0.13 MPa	36 experiments; decrease in permeability; increased thermal conductivity $\lambda_{eff} = 0.67\ W \cdot m^{-1} \cdot K^{-1}$	To investigate an operational principle, changes in heat transfer, permeability and reaction kinetics [39].
$NaAlH_4$, Al, 10 wt % ENG	4 × 21.5	Modular system of 12 tubular vessels	3.2 in 10 min	charging: 120–150 °C, 5.52–6.89 MPa, oil temperature: 120–140 °C	40 absorption/desorption cycles	To be refueled in 10 min and to deliver H_2 up to 2.0 $g \cdot s^{-1}$ [27].
$NaAlH_4$ doped with 4 mol % $TiCl_3$	2.7	Stainless steel tank with double wounded helical coil heat exchanger	2.24	charging: 135 °C, 10 MPa; discharging: 120–180 °C, 0.1 MPa	7 two or more hour desorption cycles coupled with FC, that supplied 165–240 W power	To couple with HT-PEM and use waste heat from HT-PEM for desorbing H_2 from the tank [37].
Na_3AlH_6 doped with 4 mol % $TiCl_3$	0.213	Al-alloy tank with bayonet heat exchanger	1.7	charging: 150–170 °C, 2.5 MPa; discharging: 177–180 °C, 0.65 MPa	10 absorption/desorption cycles	To develop and test lightweight Al-alloy storage tank [38].
Na_3AlH_6 doped with 4 mol % $TiCl_3$, 8 mol % Al and 8 mol % activated carbon	1.9	Al-alloy tank with corrugated heat exchanger	2.1	charging: 160 °C, 2.5 MPa; discharging: 180 °C, 1.6 MPa	31 absorption/desorption cycles	To develop the lightweight Al-tank that is produced by extrusion molding [42].
$NaAlH_4$ doped with 2 mol % ($TiCl_3$-$0.3AlCl_3$) and 5 mol % ENG	4.4	Ti-alloy tube-and-shell tank system	4	charging: 124 °C, 10 MPa; discharging: 120–170 °C, 9 MPa (constant flow)	33 cycles, 120 min of absorption with restricted H_2 flow of 245 $L_n \cdot min^{-1}$; 200 min desorption 3.7 $L \cdot min^{-1}$	To improve gravimetric and volumetric capacity [43].
$Mg(NH_2)_2$-$2LiH$-$0.07KOH$ with 9 wt % ENG	0.098	Cylindrical lab-scale hydrogen storage tank with porous sintered metal tube as H_2 supply	N/A	charging: 220 °C, 8 MPa; discharging: 220 °C at constant H_2 flow rate of 0.6 $L \cdot min^{-1}$	Desorption duration of 79.5 min at 0.6 $L \cdot min^{-1}$ H_2	To investigate influence of graphite content and compaction pressure on desorption properties [44].
$LiNH_2$-MgH_2-$LiBH_4$ 3 wt % $ZrCoH_3$ (in annulus) $LaNi_{4.3}Al_{0.4}Mn_{0.3}$ (in core)	0.6	Tubular reactor, two materials separated by a gas permeable layer	N/A	charging: 165–170 °C up to 0.17 MPa; discharging at constant and periodic H_2 mass flow	10% of H_2 desorbed in about 30 min; majority of H_2 desorbed in about 1 h	To validate a model and study effects of the reactor concept on desorption performance [31].

HYDROLYSIS

Hydrogen release can also occur apart from being subjected to heat via hydrolysis by reaction with water. A considerable amount of work has been invested in experimental testing of hydrolysis systems for complex metal hydrides [47]. Sodium borohydride, $NaBH_4$, is one of the most tested materials for hydrolysis, (Equation (8)) [48,49,50]. $NaBH_4$ was first discovered in 1940 by Schlesinger and Brown, but the work was classified and not published until 1953 [51]. $NaBH_4$ has a hydrogen content of 10.8 wt % and is the only complex hydride that so far has a viable field of applications.

$$NaBH_4 + 2H_2O \rightarrow NaBO_2 + 4H_2 \quad (8)$$

Ammonia borane (NH_3BH_3, AB) has received considerable interest as a hydrogen storage material owing to its high hydrogen content (19.5 wt %) and the ability to release hydrogen at low temperatures. When in contact with water, ammonia borane releases hydrogen according to Equation (9), which is accelerated in acidic conditions and by the use of metal catalysts [52,53,54,55,56]. Some work has been reported for hydrolysis of other metal hydrides or complex metal hydrides, e.g., LiH, MgH_2, CaH_2, $LiBH_4$, KBH_4 and $LiAlH_4$ [57,58,59,60,61,62,63].

$$BH_3NH_3 + 2H_2O \rightarrow NH_4BO_2 + 3H_2 \quad (9)$$

Tank Developments for Hydrolysis Based Reactors

Experimental hydrolysis systems are based on two different reactor designs, a batch or flow reactor. In a batch reactor, the complex boron hydride and the catalyst are mixed together in water and the reactor is refueled by exchanging the entire solution including remaining boron hydride, byproduct and the catalyst [64,65,66,67]. In a flow reactor, the boron hydride solution is passed over or through a solid catalyst bed followed by separation of the produced hydrogen and the byproduct solution [48,68,69]. In both reactor systems, the concentration of the boron hydride solution as well as handling of the waste solution poses problems, e.g., for $NaBH_4$ and $NaBO_2$ since the solubility of $NaBH_4$ in alkaline water solutions is much higher than for $NaBO_2$ [50,70]. The hydrolysis reaction can also occur by a reaction between steam and a solid boron hydride [57,71,72]. The use of steam can possibly reduce the amount of water otherwise used in the boron hydride solutions, thus improving the total energy storage of the system [57].

Flow Reactors

In 2000, a first example of a flow catalyzed hydrolysis reaction was published [48]. The hydrolysis reaction was carried out over resin beads coated with Ru metal. In another study, a conversion close to 100% was achieved with a small reactor filled with an alumina support covered with Ru metal. However, the catalyst was present in excess, which may have enabled the high conversion [73]. High conversion of 95% over 700 h was reported for an auto-thermal fixed-bed reactor using a Ru-based catalyst from Millennium Cell Inc. (Eatontown, NJ, USA) [74]. The reactor used the heat evolved from the exothermic reaction between $NaBH_4$ and water to sustain the hydrolysis reaction [69]. This system was further developed by integrating a heat exchanger to heat up the fuel feed prior to contact with the catalyst bed. The integrated reactor equipped with a heat exchanger gave a more uniform heat distribution over a range of fuel flow rates, which increased the catalyst effectiveness [68].

Recently, a new concept reactor with a magnetic containment of permanent trade magnets to recollect the catalyst made out of a magnetic support coated with Ru particles was described [75]. The reactor should allow better control of the hydrogen production kinetics. The reactor had an efficiency up to 90% and the catalyst showed an increased stability and activity for hydrogen generation [76]. The magnetic reactor was also tested in combination with a fuel cell [77].

The flow reactors discussed so far have a tubular design. Recently, a new π-shaped reactor was proposed [78]. A gas channel was situated above the cavity for the catalytic hydrolysis reaction. The design of the reactor decreased the impact of the hydrogen on the surface of the catalyst, improving the conversion efficiency and allowing for a stable hydrogen generation rate. The hydrogen gas was collected in the gas channel and was quickly moved away from the catalyst. This stopped the reverse flow of the $NaBH_4$ solution and a continuous flow to the catalyst was achieved. The π-reactor improved the conversion with approximately 10% compared to a conventional tubular reactor [78].

Micro-reactors have also been proposed for production of hydrogen from hydrolysis of chemical hydrides for portable customer electronics [79,80,81,82]. The micro reactor described by Kim *et al.*, was made from three layers of photosensitive glass [79]. The catalyst support made of nickel foam was positioned between the glass plates. After assembly of the micro reactor, a Co-P-B catalyst was deposited on the catalyst support by reduction of a solution of 0.4 M $CoCl_2$ and 0.4 M NaH_2PO_2 by $NaBH_4$. The fuel solution was a mixture between 15 wt % $NaBH_4$ and 5 wt % NaOH and the fuel feeding rate was 30 $\mu L\cdot min^{-1}$. The micro reactor had an average conversion efficiency of 93%. The micro reactor was used in combination with a micro fuel cell. The system had a maximum power output of 157 mW at a current of 0.5 A. There was no difference in performance when hydrogen supplied from the micro reactor was exchanged with pure hydrogen, which showed the micro reactor generated enough hydrogen for the fuel cell [79,80].

Zhu *et al.* introduced a micro reactor system for generating hydrogen from a NH_3BH_3 solution. The micro reactor employed self-circulation by a micro pumping mechanism, which was made possible by the hydrogen bubbles produced in the hydrolysis reaction [82]. This meant hydrogen was generated without parasitic power consumption. In the system, the dissolved NH_3BH_3 was feed into the micro reactor from a rechargeable fuel reservoir. The system was made from 375 μm thick silicon wafers using a microfabrication process and the channels were drawn by a reactive ion etching using SF_6 and finally etched with a KOH solution. Platinum black was electroplated into the channels to serve as catalyst for the hydrolysis of NH_3BH_3. A dilute solution of 1 or 5 wt % NH_3BH_3 was used for hydrogen production and conversions up to ~52% and 57%, respectively, were recorded. A small PEM fuel cell was also used in combination with the micro reactor and allowed the generation of power for micro-scale energy devices [82].

Batch Reactors

Ferreira *et al.*, tested two different batch reactors with a difference in volume and in the shape of the bottom of the reactor. The first reactor had a flat bottom and $V = 0.646$ L, while second reactor had a conical bottom geometry and $V = 0.229$ L [64]. Their tests were carried out with the same amount of catalyst in each reactor and the results indicated that a higher pressure in the smallest reactor ($p_{small} = 1.14$ MPa, $p_{big} = 0.4$ MPa) promoted the hydrolysis reaction. The conical shape of the smallest reactor was thought to increase the hydrogen production rate (rH2, small ≈ 0.00077 $MPa\cdot s^{-1}$, rH2, big ≈ 0.00036 $MPa\cdot s^{-1}$) [64]. In a later publication, Ferreira *et al.*, tested three different reactors, a flat bottom reactor with $V = 0.646$ L, a flat bottom reactor with $V = 0.229$ L and a conical shape reactor with $V = 0.229$ L. The conical shaped reactor was shown to be superior and increased both the hydrogen yield and generation time while decreasing the induction time. The better performance of the conical shaped reactor may be attributed to an enhanced contact between the $NaBH_4$ and the catalyst [70].

Reactor Modeling of Hydrolysis Tank Systems

Only limited reports have been published of reactor modeling for hydrolysis purposes. The physical processes inside a hydrolysis reactor are complex and multi-phase flow, heat and mass transfer need to be evaluated. Additionally, the reaction kinetics has to be considered, making the theoretical prediction/description of a reactor a complex task.

Zhang *et al.*, designed, built and tested a 1 kW_e $NaBH_4$ hydrogen generation system (Figure 4) [83,84]. Experimental data were first collected and evaluated for a stainless steel packed-bed reactor with a commercial catalyst, 3% 2 mm Ru on carbon extrude and using solutions of $NaBH_4$ with different concentrations. The 15 % concentrated $NaBH_4$ solution gave a 95% chemical conversion with a maximum gravimetric capacity of 3.1 wt % H_2. However, above a $NaBH_4$ concentration of 15% the discharge product ($NaBO_2$) would crystallize upon cooling to room temperature making it difficult to be removed from the discharge tank. The ideal $NaBH_4$ concentration was suggested to be between 10% and 15%. Thermal runaway was encountered but could be avoided by continuously cooling the system. The tests also showed that different flow rates had an impact on the temperature profile inside the reactor [83].

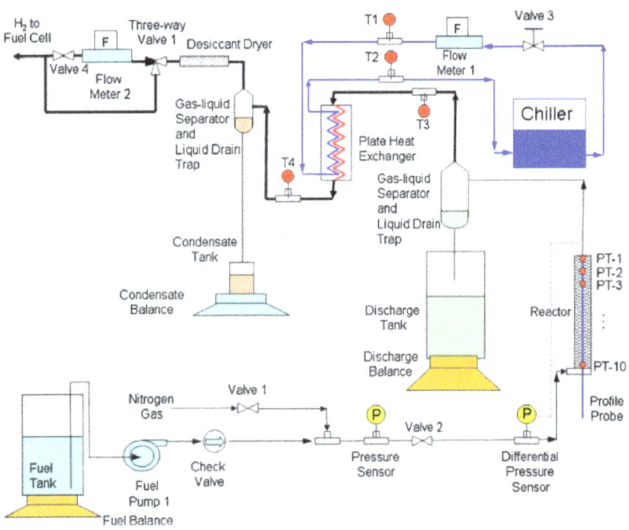

Figure 4. Schematic of the 1 kW_e $NaBH_4$ hydrogen generation system. Reprinted with permission from reference [83]. Copyright 2007 Elsevier.

The results from Zhang *et al.*, were further evaluated and used in a one-dimensional numerical reactor model [84]. The model assumed homogeneous catalysis and adiabatic operational conditions. The data used for the geometrical evaluation of the model was the temperature inside the reactor at different positions. The model produced outputs in the form of temperature, chemical conversion, relative humidity and molar flows of hydrogen and water vapor, which all matched the experimental data well at different $NaBH_4$ concentrations. Additionally, the model included two sub-models, which allowed the evaluation of non-isothermal water evaporation processes and the pressure drop of the two-phase flow through the porous catalyst. The model was one-dimensional and therefore the geometrical parameters of the reactor were not evaluated in detail and only in the flow direction. The system was initially built for vehicle applications, but it could have potential for other small portable applications [84].

In 2012, Sousa *et al.*, published theoretical work on a 3-dimensional batch reactor model [85]. Experimental data were collected from a batch reactor at different temperatures with a very dilute alkaline solution of $NaBH_4$ in order to avoid problems with solubility of the hydrolysis reaction product. The data were fitted using three different kinetic models in order to evaluate the kinetics of the hydrolysis reaction. The catalyst used in the reactor was a Ru catalyst supported on Ni-foam. At low temperatures or high $NaBH_4$ concentrations the kinetics followed a zero-order model, while at high temperatures or low $NaBH_4$ concentrations, the reaction depended on the $NaBH_4$ concentration and the kinetics were best described by a first order model. The experimental data were further used in a numerical 3-dimensional non-isothermal model for a pilot scale reactor. This model described the transport phenomena and also included the kinetic model for the hydrolysis reaction. Mechanical stirring in the batch reactor was shown to have an impact on the mass transfer, since $NaBH_4$ was more easily allowed to reach the catalytic sites. Furthermore, the model showed that $NaBH_4$ in the bottom of the reactor was used less efficiently, because the produced hydrogen decreased the stirring in the bottom. The model assumed a single-phase flow during the hydrolysis reaction, which neglected the flow of hydrogen from the catalyst after production. The gas bubbles of hydrogen were responsible for the movement of the fluid and without the incorporated artificial mixing the performance predicted by the model would be extremely low and not compatible with the experimental data obtained with the reactor. In conclusion, the 3-dimensional model was able to simulate the reactor processes and validate the kinetic model [85].

In order to evaluate a two-phase flow inside a batch reactor and better describe the impact of hydrogen formation, a new two-dimensional reactor model was recently described [86]. Data obtained from a previous described experimental setup was reused for the new model [85]. The catalyst was again a Ru catalyst supported on Ni-foam. The previous model only focused on the area, where the reaction took place and the hydrogen storage region was not considered, whereas in the new model this is taken into account [85]. The new model used a two-dimensional plane through three reaction tubes, which enabled the analysis of three regions namely the reactive solution, thermal fluid and the metal walls. Therefore, the model allowed for better description of the impact of hydrogen production (bubbles) from the hydrolysis reaction. The computational methods indicated that even though a porous support (Ni-foam) was used for the catalyst, the hydrolysis reaction occurred mainly on the surface of the catalyst foam. Additionally, the results showed that only 30% of the catalyst was effectively used. Different positions of the catalyst foam as well as a plastic support for the Ni-foam inside the reactor were also studied in order to obtain information about the optimal position for reaching the quickest conversion. The optimal position was in the central region of the reactor chamber. Interestingly, the plastic support seemed to have a positive effect on the mixing during hydrogen production. The best result was obtained for a plastic support with multiple small holes. The authors concluded that a two-phase flow model approach gave a better representation of the real system compared to a single-phase flow model [86].

As mentioned earlier the hydrolysis reaction can also be carried out using steam and solid state $NaBH_4$ with an added catalyst. In one report, the reactor for this type of reaction consisted of brass mesh to contain the solid $NaBH_4$ inside a cobber reactor. The steam was then passed into the reactor, where the hydrolysis reaction occurred [57]. A two-part dissolution-reaction model was proposed for a steam hydrolysis system [72]. The model accounted for deliquescence, solid dissolution and hydrolysis in order to release hydrogen and the model could give reasonable estimates for the kinetic constant and the mass transfer coefficients. However, the model could as such not be used for reactor design.

System Modeling for Hydrolysis Setups

The kinetic model described by Sousa *et al.*, was further used in a model for a PEM fuel cell power system using a $NaBH_4$ batch reactor as hydrogen generator [85,87]. However, the actual model of the reactor was described with molar balance equations (zero dimension model) for the number of moles of $NaBH_4$, H_2O and $NaBO_2$. This was possible because the only variable in the reactor was the hydrogen pressure. The Langmuir-Hinshelwood model was selected for describing the kinetics of the hydrogen generation rate over a range of temperatures and in time [85]. The results showed the possibility of using a $NaBH_4$ hydrolysis reaction as the hydrogen source for the PEM fuel cell system. Indeed, this study had a bigger perspective, as a complete $NaBH_4$-PEM fuel cell power system was simulated. The magnetically supported reactor described by Pozio *et al.*, was also simulated with a simple mathematically model. This was used to evaluate the process of catalytic hydrolysis and in a later publication the complete hydrogen generator was simulated [75,76].

Applications Using Sodium Borohydride

$NaBH_4$ has been given a no-go recommendation as a solid-state hydrogen storage material for automotive applications by the US DoE due to its high decomposition temperature. However, since $NaBH_4$ releases hydrogen by hydrolysis at room temperature and the solution is both stable and non-toxic, $NaBH_4$ has a large potential for defense and civil applications. For application purposes, hydrogen is usually catalytically generated from alkaline $NaBH_4$ solutions. An advantage of using $NaBH_4$ is that no external heat supply is required due to the exothermic hydrolysis reaction. A disadvantage is that the $NaBO_2$ byproduct (although environmentally friendly), needs to be collected in a separate tank in order to be recycled off board into $NaBH_4$. For more information on the role of $NaBH_4$ in the field of hydrogen fueled applications, the reader is referred to more in-depth articles and reviews [7,88,89,90,91].

Small-Unmanned Aerial Vehicles

Small-unmanned aerial vehicles (UAVs) have gained much interest in the fields of defense and security in order to minimize human loses. They have been used for surveillance missions owing to the low noise and heat emissions, minimizing detection by enemies. Instead of internal combustion engines or secondary batteries, which have low thermal efficiencies and face short flight endurance (60–90 min), fuel cell systems are regarded as a more suitable alternative power source. Fuel cell systems show high thermal efficiencies, have an electrochemical reaction rather than a combustion reaction, and silent mode of operation [92].

As fuel cells operate on hydrogen, compressed hydrogen [93], liquefied hydrogen [94,95,96,97], and chemical hydrides [94] have all been used for UAV applications. For example, AeroVironment Inc. (Monrovia, CA, USA) was the first to use liquefied hydrogen for a high-altitude long-endurance (HALE) aircraft vehicle (Global Observer), which operated for over an hour. However, the manufacturers claimed 24 h flight time with a full tank [94]. In the case of compressed hydrogen, Bradley *et al.*, reported 43 min of cruising flight, based on the measured capacity of the on-board hydrogen tank (192 L), which could be increased to 52 min, if the hydrogen utilization was increased to 99% [93]. Since chemical hydrides have high energy densities, the combination with a proton exchange membrane (PEM) fuel cell stack can significantly improve the flight duration of UAVs compared to liquefied or compressed hydrogen [98]. In this section various small UAVs, which use $NaBH_4$ as their hydrogen source, are discussed as to their design, manufacture, and flight tests.

The US Air Force Research Laboratory in conjunction with Small Business Innovation Research partners AeroVironment, Protonex Technology Corporation (Southborough, MA, USA) and Millennium Cell Inc. successfully demonstrated a nearly 5 h flight with Puma, an UAV, operating on a lithium ion battery and a PEM fuel cell system using $NaBH_4$ as a hydrogen source [94]. The flight duration was increased to 9 h through further technological developments, demonstrating the improvement in endurance using chemical hydrides instead of compressed hydrogen [99]. The lithium ion battery provided the power for take-off and maneuvers whereas the fuel cell recharged the battery and provided a continuous power supply for the plane during cruise flight. Weighing only 5.7 kg and with an aerial observation at line-of-sight up to 10 km, the Puma was ideal for military applications [99].

Kim *et al.*, also reported the use of a PEM fuel cell system operating with $NaBH_4$ for UAVs, which consisted of a fuel cell stack, hydrogen generator and a hybrid power management system (Figure 5) [98]. Catalytic hydrolysis resulted in the formation of hydrogen gas and borate by feeding the 15 wt % $NaBH_4$ alkaline solution through the catalytic reactor. The generated hydrogen was purified by a dehumidifier and used to power up the fuel cell stack and Li battery, whereas $NaBO_2$ was separated in a gas-liquid separator where it remained for collection [98].

To ensure a stable performance of the fuel cell system, cooling of the stacks becomes necessary. Kim *et al.*, recommended air-cooling systems opposed to water-cooling systems to reduce complexity and overall bulkiness of the cooling system. Air flowed through the intakes, cooled the stacks and then hot air left through the exits located near the propeller. Additionally, both ground and flight tests were performed, showing a flight endurance of 2.5 h with 360 g of a 15 wt % $NaBH_4$ solution [98]. A flight endurance of 2 h was achieved using 300 g of a 15 wt % $NaBH_4$aqueous solution, although, the authors claimed a flight endurance of more than 3 h would be achievable based on the energy density of the fuel cell system. However, various problems were encountered during the flight tests, which occurred in the pump, reactor and filters [92]. The design was later improved by replacing the two tank system for a volume-exchange fuel tank where the fuel ($NaBH_4$ solution) and spent fuel ($NaBO_2$) exchange volume within one tank hence minimizing the volumetric density (Figure 6) [100].

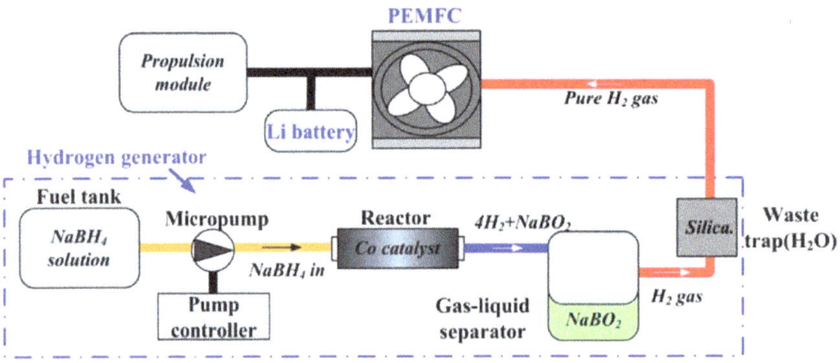

Figure 5. Operating principle of fuel cell system equipped in the unmanned aerial vehicle (UAV) platform. Reprinted with permission from reference [98]. Copyright 2011 Elsevier.

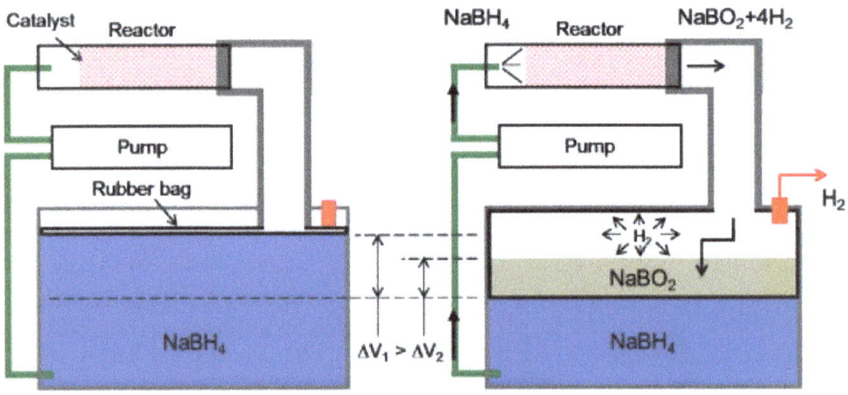

Figure 6. Schematic of a volume-exchange fuel tank. Reprinted with permission from reference [100]. Copyright 2014 Elsevier.

In order to obtain a high gravimetric energy density, the fuel cell stack and hydrogen generator should be ideally reduced to a minimum. However, the weight of the fuel cell stack typically takes up between 43% and 49.6% of the total weight in the fuel cell of the UAV [92,98,100]. Developing new materials to reduce the weight of the graphite bipolar plates incorporated in the fuel cell stacks would increase the flight endurance significantly [92].

Small Portable Applications

Fuel cells as an alternative for lithium-ion batteries for laptops and mobile phone applications have attracted considerable research interest. This is mainly due to the low capacity of the current lithium-ion batteries, resulting in short operating times. Fuel cell technologies are regarded as the most suitable for small portable applications. Although direct methanol fuel cells are typically used, issues regarding fuel crossover and low system volumetric densities remain to be resolved [101]. The direct hydrogen fuel cell systems that use a metal hydride based hydrogen storage tank have gained much attention owing to the high power and volumetric energy densities that could be obtained [102,103]. Unfortunately, their size requirements were too large to surpass current Li-ion secondary batteries and therefore miniaturization of the fuel cells is important to achieve commercial implementation [104].

$NaBH_4$-based refuel devices for laptops and mobile phones have also been reported [105]. For example, Prosini *et al.*, designed a refill device to generate hydrogen from HCl/H_2O/$NaBH_4$solution for a fuel cell powered mobile phone (Figure 7). With a 0.8 V fuel cell, a 1 W hydrogen mass flow was released for over 2.5 h. The refill case was made out of plastic and therefore the refill costs were based on the price of $NaBH_4$ (1.5 g for 9–10 days). The authors claimed an estimated energy density of roughly 720 Wh·kg^{-1} instead of 170 Wh·kg^{-1} typically for commercial lithium-ion batteries, making the device competitive with existing batteries [105]. As far as we are aware, no other publications on sodium borohydride based refuel devices for small cellular or laptop applications have been published.

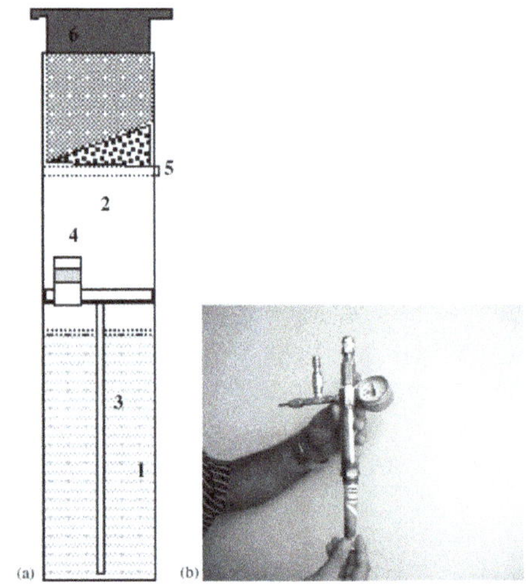

Figure 7. **(a)** A schematic of the hydrogen producing device: 1, HCl chamber; 2, reaction chamber; 3, connection channel; 4, pressure regulation valve; 5, diaphragm; 6, activation button; and **(b)** insertion of the device in the reaction chamber. Reprinted with permission from reference [105]. Copyright 2006 Elsevier.

Submarine Applications

Submarines are valuable assets for naval applications owing to their invisibility when submerged under water [106,107]. Unfortunately, most of the submarines are fitted with diesel-electric propulsion resulting in limited submerged time due to the limited capacity of the batteries. In addition, when charging the batteries only a few meters below the surface (snorkeling) the submarine is susceptible to detection. Therefore, the development of air independent propulsion in order to prolong the underwater performance becomes crucial [107,108]. Among the various investigated systems, low temperature PEM fuel cells were shown to be the most promising. A schematic diagram for a conventional battery-based system and fuel cell/battery-based system is shown in Figure 8. $NaBH_4$ allows a higher gravimetric hydrogen storage capacity compared to compressed hydrogen and does not have the safety and reliability constrains of compressed gases [108,109]. Liquid oxygen can be used as the oxidant, although compressed air, which is often used for other purposes on-board, is also considered [108]. System modeling for different fuel cell/battery-based systems showed that the total submerged endurance was substantially enhanced using fuel cells [108]. Furthermore, with the $NaBH_4$/liquid oxygen system the submarine could remain submerged minimally for 6.5 days with a total distance of 3500 km compared to an only 20 h submerged period for the conventional system [107].

Apart for space applications, direct borohydride fuel cells (DBFCs) in combination with $H_2O_2$are also investigated as high-density power sources for autonomous underwater vehicles [110,111,112]. In liquid form, H_2O_2 is a thousand times denser than oxygen [113]. The US Office of Naval Research (Arlington, VA, USA) and the UK Defence Science and Technology Laboratory (Salisbury, UK) designed a fuel cell operating on the $NaBH_4/H_2O_2$ system [110]. Protonex is also currently developing an underwater fuel cell power system using the Millennium Cell Inc. hydrogen generation system [114].

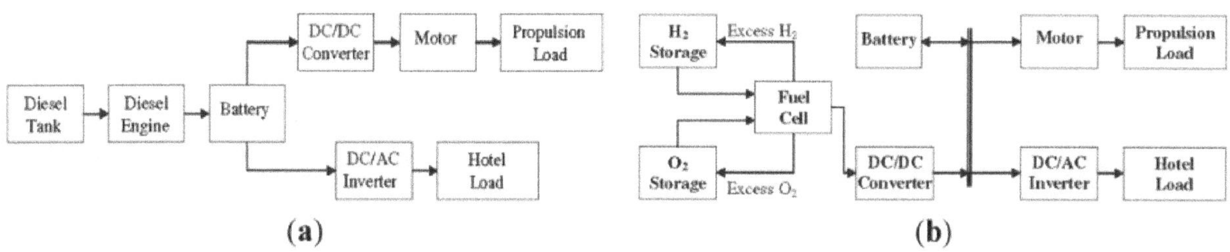

Figure 8. Schematic diagram for the (**a**) conventional battery based system and (**b**) fuel cell/battery-based system. Reprinted with permission from reference [107]. Copyright 2011 Elsevier.

Recycling of $NaBH_4$ Spent Fuel

$NaBH_4$ is non-reversible and recycling of the byproduct sodium metaborate ($NaBO_2$) becomes critical. A cheap and efficient route for recycling the spent fuel ($NaBO_2$ dissolved in alkaline aqueous solution) still has to be developed. For $NaBH_4$ decomposed in water solution, its regeneration starts with dry $NaBO_2$. Therefore, the dissolved $NaBO_2$ has to be dried, which is an energy intense process. $NaBH_4$ can be recovered from $NaBO_2$ through ball milling using MgH_2/Mg. Here, the small particle size and high surface area increase the yield of the reaction [53,112,115].

For the direct borohydride/peroxide fuel cell, Miley *et al.*, developed a regenerative fuel cell, which can transform the end products back into the reactants, using an aqueous $NaBH_4$ solution and H_2O_2 [113]. Prevention of the generation of hydrogen and oxygen gas during regeneration is crucial as the regeneration of $NaBH_4$ and H_2O_2 (Equation (10)), which occurs at 2.23 V, is in competition with the electrolysis reaction (Equation (11)) that occurs at 1.23 V.

$$NaBO_2 + 6\,H_2O \rightarrow NaBH_4 + 4\,H_2O_2 \tag{10}$$

$$2\,H_2O \rightarrow 2\,H_2 + O_2 \tag{11}$$

The authors circumvented this issue by using high over-potential materials for the electrodes and exploring the formation of perborate that hydrolyze to peroxide at 1.7 V (Equation (12)), greatly reducing the gas evolution.

$$NaBO_3 + H_2O \rightarrow NaBO_2 + H_2O_2 \tag{12}$$

Furthermore, the inhibition of the $NaBH_4$ crossover, a critical issue often faced for DBFCs, can be reduced by improving the membrane electrode assembly or the liquid diffusion layer [112]. In order to further develop $NaBH_4$ hydrolysis systems for portable applications, additional research is needed to reduce the cell volume, increase the gravimetric hydrogen storage capacity and reach efficient recycling of the spent fuel.

Applications Using Ammonia Borane

The development of a continuous hydrogen generator, using a continuous feeding of ammonia borane as both fuel and spent fuel, has been the focus of many researchers [28,116,117,118]. Unfortunately, high costs and raw material scarcity only allow ammonia borane to be used only for small-scale applications. Seo *et al.*, were the first to report on the design and testing of an ammonia borane hydrogen power-pack, which operated an UAV for 57 min using a 200 W PEM fuel cell. The system was operated by delivering spherical solid ammonia borane beads into a semi-batch type reactor filled with tetraethylene glycol dimethyl ether, a liquid promoter. Hydrogen purification equipment with a high filter capacity and efficient drainage system for spent fuel were also integrated for application purposes. The hydrogen generation system was the heaviest part of the hydrogen power pack taking up 64% of the total weight, whereas the fuel cell stack took up only 17.6% and the ammonia borane beads 4%. Based on the obtained results, the authors proposed an advanced reactor concept (Figure 9). They also suggested optimization for the discharging of liquid/solid spent fuels by using triethylene glycol dimethyl ether as the liquid promoter as it could enhance the fluidity of the spent fuels [116].

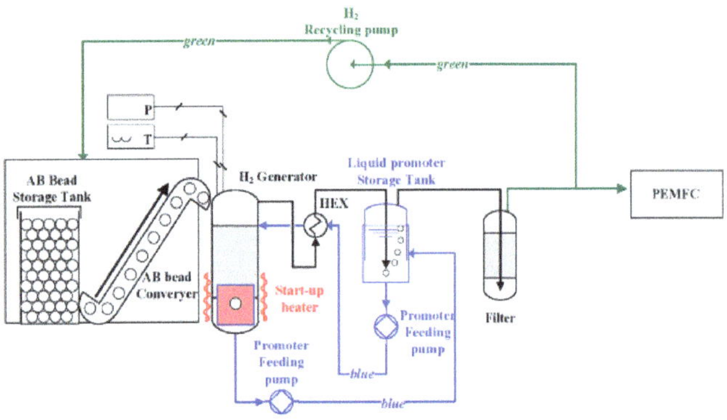

Figure 9. Schematic of the advanced fuel cell system powered by ammonia borane for prolonged operation. Reprinted with permission from reference [116]. Copyright 2014 Elsevier.

SUMMARY AND OUTLOOK

There are several limitations for metal hydride-based onboard storage systems that prevent their implementation: low gravimetric and volumetric capacities, insufficient kinetics within appropriate temperatures and pressure ranges, and high cost of the overall engineering system. Modeling provides a powerful tool for the development of strategies and improvement of full-scale tanks. Several studies on tank modeling for thermolysis have been proposed, mainly based on sodium alanate with few examples of others complex hydrides. The main purpose of these simulations was the optimization of some operating parameters (temperature, pressure, thermal conductivity, coolant flow rate and coolant temperature) as well as the tank design (length scale, geometry and fins content). Optimizations were then proposed on the basis of the simulation results. Several modeling tools have been built on the basis of the hierarchical methodology and resistance analysis in order to estimate performances of the system and the limiting factors. The simulations showed that a good thermal management was necessary for the absorption, whereas the pressure control was important for desorption. Still, differences exist between the simulations, which come from the choice of the kinetic model that is implemented as the governing equation for the simulations. Therefore, precautions have to be taken, since the kinetic model is built through experimental considerations. Heat transfer in the metal hydride tank can be improved using heat exchangers, multi-tubular tank geometries and heat transfer enhancers. However, this will also increase the weight of the system. Hence, optimizations are often a compromise between heat transfer and hydrogen content. The strategy based on CH/MH beds could provide a new step in tank design leading to better temperature and pressure management without decreasing the hydrogen capacity excessively.

Although ample research has been performed with complex metal hydride materials on small-scale laboratory batches, considerably fewer full-scale thermolysis tanks/systems have been developed and analyzed. Improving the ratio between the mass of the complex metal hydride bed to the mass of the tank wall, by screening lighter materials for the tank wall and developing hydrogen storage materials exhibiting both higher gravimetric and volumetric storage properties, should be a goal in order to obtain lightweight storage systems. There are several design principles that must be taken into account while developing a complex metal hydride storage tank, such as packing arrangement of the material, hydrogen supply, heat transfer to the heat transfer medium, effective heat conductivity of the metal hydride bed and volume expansion. Additionally, efficient removal of the reaction heat from the metal hydride tank during refueling and how to provide heat during hydrogen delivery to the fuel cell remain an unsolved problem. Thermolysis of complex metal hydrides has yet to demonstrate the fulfillment of the DoE requirements (Table 1). Especially concerning the gravimetric capacity of approximately 1.5 wt % and refueling time of 10 min.

In hydrolysis, recycling of the spent fuel remains the single biggest challenge. For $NaBH_4$, it requires the recycling of $NaBO_2$ back into $NaBH_4$, either starting from dry $NaBO_2$ or through the use of a regenerative fuel cell. This often means additional costs cannot be avoided. Hence, the use of $NaBH_4$ for portable applications becomes less attractive. As the byproduct from the hydrolysis reaction can decrease the catalyst efficiency, flow reactors holds the biggest potential for reaching commercial applications as these allow to collect the byproduct in a separate tank [98,100] The flow reactor with a π-shape design contained a gas channel for more efficient collection of the hydrogen gas. This design may also improve larger scale systems without sacrificing the overall volume and mass of the system. A batch reactor for hydrolysis is simple and does not allow for major design changes. Indeed, batch reactors are probably best suited for convenient tests of new catalysts and for small systems, where volume constraints do not allow for a second tank for collection of the byproduct. Implementation of a conical bottom in the batch reactor was shown to have a positive effect on the hydrolysis reaction. Tank modeling for hydrolysis systems has only received limited attention and was mainly performed on systems with a batch reactor. The proposed models concern mainly the transport phenomena and kinetics for the reaction in the liquid. Further research in modeling of flow reactors may help to improve systems meant for applications.

Although hydrolysis technology using complex metal hydrides (e.g., $NaBH_4$ and AB) have found its way into UAVs and submarine applications, the majority of these applications are currently still driven by fossil fuels. These application may benefit from the installation of filters to obtain desired the hydrogen purity, efficient drainage of spent fuel, and operation conditions (humidity, vibrations, temperature *etc.*). All these improvements may affect the fuel cell operation and consequently, the success of the application. Additionally, further research is needed to create a more straightforward hydrolysis system preferably with on-board recycling of the spent fuel and more robustness towards operation conditions. Indeed, the advantages such as silent mode of operation and thermal signature can be quickly overshadowed by disadvantages such as limited space in a UAVs and the weight of both the energy storage system and the auxiliary system (which can easily account for roughly one third of the total weight of the aircraft). Therefore, any weight reduction of the hydrolysis technology that can be achieved will translate directly into an increase in flight duration.

CONCLUSIONS

After 20 years of intensive research on complex metal hydrides, a widespread commercial penetration of these hydrogen storage materials in technical applications is not in sight for the next years. The research attention over the last years has been given to fundamental understanding of crystal structures, reaction kinetics and thermodynamics of these materials as well as development of materials with high hydrogen content. The development of tank systems and demonstration projects with complex metal hydrides has not always been part of the past research activities because of the challenges that come along with engineering development.

Several technical barriers still exist for successful implementation of solid-state hydrogen storage systems based on thermolysis for stationary and portable applications and remain a significant challenge for transportation applications. Currently available storage tank systems typically require large volume, high weight, and have slow hydrogen charging/discharging rates and high system cost. Long-term research goals for the complex hydride materials should be to improve the gravimetric and volumetric capacities, to obtain sufficient kinetics within appropriate temperatures and pressure ranges and to lower the cost of the overall systems.

It is obvious that commercial success is not only a question of high hydrogen content of hydrogen storage materials. The overall performance of the whole system, consisting of hydrogen storage material, as well as storage tank design and integration of all the components into a technical system, plays a much more important role than one particular property. Taking this into account, engineering developments for the optimization of solid-state hydrogen storage tank systems must be intensified to overcome technical barriers and to construct safe and consumer-friendly system based on renewable energy carriers.

ACKNOWLEDGMENTS

The basic funding of the Max-Planck-Society is gratefully acknowledged.

AUTHOR CONTRIBUTIONS

Morten B. Ley, Mariem Meggouh, Romain Moury, Kateryna Peinecke and Michael Felderhoff contributed equally in writing and correcting the paper.

REFERENCES

1. Schlapbach, L.; Züttel, A. Hydrogen-storage materials for mobile applications. *Nature* **2001**,*414*, 353–358.
2. Eberle, U.; Felderhoff, M.; Schüth, F. Chemical and physical solutions for hydrogen storage.*Angew. Chem. Int. Ed.* **2009**, *48*, 6608–6630.
3. Ley, M.B.; Jepsen, L.H.; Lee, Y.-S.; Cho, Y.W.; Bellosta von Colbe, J.M.; Dornheim, M.; Rokni, M.; Jensen, J.O.; Sloth, M.; Filinchuk, Y.; *et al.* Complex hydrides for hydrogen storage—New perspectives. *Mater. Today* **2014**, *17*, 122–128.
4. Bogdanović, B.; Schwickardi, M. Ti-doped alkali metal aluminium hydrides as potential novel reversible hydrogen storage materials. *J. Alloys Compd.* **1997**, *253–254*, 1–9.
5. Rude, L.H.; Nielsen, T.K.; Ravnsbæk, D.B.; Bösenberg, U.; Ley, M.B.; Richter, B.; Arnbjerg, L.M.; Dornheim, M.; Filinchuk, Y.; Besenbacher, F.; *et al.* Tailoring properties of borohydrides for hydrogen storage: A review. *Phys. Status Solidi A* **2011**, *208*, 1754–1773.
6. Mosher, D.A.; Arsenault, S.; Tang, X.; Anton, D.L. Design, fabrication and testing of $NaAlH_4$based hydrogen storage systems. *J. Alloys Compd.* **2007**, *446–447*, 707–712.
7. Demirci, U.B. The hydrogen cycle with the hydrolysis of sodium borohydride: A statistical approach for highlighting the scientific/technical issues to prioritize in the field. *Int. J. Hydrogen Energy* **2015**, *40*, 2673–2691.
8. Technical system targets: Onboard hydrogen storage for light-duty fuel cell vehicles 2012. Available online: http://www1.eere.energy.gov/vehiclesandfuels/pdfs/program/hstt_roadmap_june2013.pdf (accessed on 25 August 2015).
9. Lozano, G.A.; Bellosta von Colbe, J.M.; Klassen, T.; Dornheim, M. Transport phenomena*versus* intrinsic kinetics: Hydrogen sorption limiting sub-process in metal hydride beds. *Int. J. Hydrogen Energy* **2014**, *39*, 18952–18957.
10. Vyazovkin, S.; Burnham, A.K.; Criado, J.M.; Pérez-Maqueda, L.A.; Popescu, C.; Sbirrazzuoli, N. Ictac kinetics committee recommendations for performing kinetic computations on thermal analysis data. *Thermochim. Acta* **2011**, *520*, 1–19.
11. Luo, W.; Gross, K.J. A kinetics model of hydrogen absorption and desorption in Ti-doped $NaAlH_4$. *J. Alloys Compd.* **2004**, *385*, 224–231.
12. Franzen, J. *Modellierung und Simulation eines Wasserstoffspeichers auf der Basis von Natriumalanat*; Hamburg, Techn. Univ.: Hamburg, Germany, 2009.
13. Hardy, B.J. Geometry, heat removal and kinetic scoping models for hydrogen storage systems. Available online: http://www1.eere.energy.gov/hydrogenandfuelcells/pdfs/bruce_hardy_srnl-2007-0043_part1.pdf (accessed on 25 February 2015).
14. Lozano, G.A.; Ranong, C.N.; Bellosta von Colbe, J.M.; Bormann, R.; Fieg, G.; Hapke, J.; Dornheim, M. Empirical kinetic model of sodium alanate reacting system (I). Hydrogen absorption. *Int. J. Hydrogen Energy* **2010**, *35*, 6763–6772.
15. Lozano, G.A.; Ranong, C.N.; Bellosta von Colbe, J.M.; Bormann, R.; Fieg, G.; Hapke, J.; Dornheim, M. Empirical kinetic model of sodium alanate reacting system (II). Hydrogen desorption. *Int. J. Hydrogen Energy* **2010**, *35*, 7539–7546.
16. Bürger, I.; Hu, J.J.; Vitillo, J.G.; Kalantzopoulos, G.N.; Deledda, S.; Fichtner, M.; Baricco, M.; Linder, M. Material properties and empirical rate equations for

hydrogen sorption reactions in $2LiNH_2$–$1.1MgH_2$–$0.1LiBH_4$–3wt % $ZrCoH_3$. *Int. J. Hydrogen Energy* **2014**, *39*, 8283–8292.

17. Na Ranong, C.; Höhne, M.; Franzen, J.; Hapke, J.; Fieg, G.; Dornheim, M.; Eigen, N.; Bellosta von Colbe, J.M.; Metz, O. Concept, design and manufacture of a prototype hydrogen storage tank based on sodium alanate. *Chem. Eng. Technol.* **2009**, *32*, 1154–1163.
18. Hardy, B.J.; Anton, D.L. Hierarchical methodology for modeling hydrogen storage systems. Part i: Scoping models. *Int. J. Hydrogen Energy* **2009**, *34*, 2269–2277.
19. Hardy, B.J.; Anton, D.L. Hierarchical methodology for modeling hydrogen storage systems. Part ii: Detailed models. *Int. J. Hydrogen Energy* **2009**, *34*, 29923004.
20. Dedrick, D.E.; Kanouff, M.P.; Larson, R.S.; Johnson, T.A.; Jorgensen, S.W. Heat and mass transport in metal hydride based hydrogen storage systems. In Proceeding of the ASME Summer Heat Transfer Conference, San Francisco, CA, USA, 19–23 July 2009; pp. 249–260.
21. Mosher, D.A.; Tang, X.; Arsenault, S.; Laube, B.; Cao, M.; Brown, R. High density hydrogen storage system demonstration using NaAlH4 complex compound hydrides. Available online: http://www.hydrogen.energy.gov/pdfs/progress05/vi_a_2_anton.pdf (accessed on 22 February 2015).
22. Pfeifer, P.; Wall, C.; Jensen, O.; Hahn, H.; Fichtner, M. Thermal coupling of a high temperature pem fuel cell with a complex hydride tank. *Int. J. Hydrogen Energy* **2009**, *34*, 3457–3466.
23. Bhouri, M.; Goyette, J.; Hardy, B.J.; Anton, D.L. Sensitivity study of alanate hydride storage system. *Int. J. Hydrogen Energy* **2011**, *36*, 621–633.
24. Bhouri, M.; Goyette, J.; Hardy, B.J.; Anton, D.L. Numerical modeling and performance evaluation of multi-tubular sodium alanate hydride finned reactor. *Int. J. Hydrogen Energy***2012**, *37*, 1551–1567.
25. Raju, M.; Kumar, S. System simulation modeling and heat transfer in sodium alanate based hydrogen storage systems. *Int. J. Hydrogen Energy* **2011**, *36*, 1578–1591.
26. Lozano, G.A.; Ranong, C.N.; Bellosta von Colbe, J.M.; Bormann, R.; Hapke, J.; Fieg, G.; Klassen, T.; Dornheim, M. Optimization of hydrogen storage tubular tanks based on light weight hydrides. *Int. J. Hydrogen Energy* **2012**, *37*, 2825–2834.
27. Johnson, T.A.; Kanouff, M.P.; Dedrick, D.E.; Evans, G.H.; Jorgensen, S.W. Model-based design of an automotive-scale, metal hydride hydrogen storage system. *Int. J. Hydrogen Energy* **2012**, *37*, 2835–2849.
28. Devarakonda, M.; Brooks, K.; Rönnebro, E.; Rassat, S. Systems modeling, simulation and material operating requirements for chemical hydride based hydrogen storage. *Int. J. Hydrogen Energy* **2012**, *37*, 2779–2793.
29. Semelsberger, T.A.; Brooks, K.P. Chemical hydrogen storage material property guidelines for automotive applications. *J. Power Sources* **2015**, *279*, 593–609.
30. Bürger, I.; Komogowski, L.; Linder, M. Advanced reactor concept for complex hydrides: Hydrogen absorption from room temperature. *Int. J. Hydrogen Energy* **2014**, *39*, 7030–7041.
31. Bürger, I.; Luetto, C.; Linder, M. Advanced reactor concept for complex hydrides: Hydrogen desorption at fuel cell relevant boundary conditions. *Int. J. Hydrogen Energy* **2014**, *39*, 7346–7355.
32. Bhouri, M.; Bürger, I.; Linder, M. Optimization of hydrogen charging process parameters for an advanced complex hydride reactor concept. *Int. J. Hydrogen Energy* **2014**, *39*, 17726–17739.
33. Van Hassel, B.A.; Mosher, D.; Pasini, J.M.; Gorbounov, M.; Holowczak, J.; Tang, X.; Brown, R.; Laube, B.; Pryor, L. Engineering improvement of $NaAlH_4$ system. *Int. J. Hydrogen Energy***2012**, *37*, 2756–2766.
34. Sulic, M.; Cai, M.; Kumar, S. Cycling and engineering properties of highly compacted sodium alanate pellets. *Int. J. Hydrogen Energy* **2012**, *37*, 15187–15195.
35. Brooks, K.P.; Semelsberger, T.A.; Simmons, K.L.; van Hassel, B. Slurry-based chemical hydrogen storage systems for automotive fuel cell applications. *J. Power Sources* **2014**, *268*, 950–959.
36. Bellosta von Colbe, J.M.; Metz, O.; Lozano, G.A.; Pranzas, P.K.; Schmitz, H.W.; Beckmann, F.; Schreyer, A.; Klassen, T.; Dornheim, M. Behavior of scaled-up sodium alanate hydrogen storage tanks during sorption. *Int. J. Hydrogen Energy* **2012**, *37*, 2807–2811.
37. Urbanczyk, R.; Peil, S.; Bathen, D.; Heßke, C.; Burfeind, J.; Hauschild, K.; Felderhoff, M.; Schüth, F. HT-PEM fuel cell system with integrated complex metal hydride storage tank.*Fuel Cells* **2011**, *11*, 911–920.
38. Urbanczyk, R.; Peinecke, K.; Felderhoff, M.; Hauschild, K.; Kersten, W.; Peil, S.; Bathen, D. Aluminium alloy based hydrogen storage tank operated with sodium aluminium hexahydride Na_3AlH_6. *Int. J. Hydrogen Energy* **2014**, *39*, 17118–17128.
39. Utz, I.; Linder, M.; Schmidt, N.; Hu, J.J.; Fichtner, M.; Wörner, A. Experimental study of powder bed behavior of sodium alanate in a lab-scale H_2 storage tank with flow-through mode. *Int. J. Hydrogen Energy* **2012**, *37*, 7645–7653.
40. Yan, M.-Y.; Sun, F.; Liu, X.-P.; Ye, J.-H. Effects of compaction pressure and graphite content on hydrogen storage properties of $Mg(NH_2)_2$–2LiH hydride. *Int. J. Hydrogen Energy* **2014**, *39*, 19656–19661.
41. Eigen, N.; Keller, C.; Dornheim, M.; Klassen, T.; Bormann, R. Industrial production of light metal hydrides for hydrogen storage. *Scr. Mater.* **2007**, *56*, 847–851.
42. Urbanczyk, R.; Peinecke, K.; Meggouh, M.; Felderhoff, M.; Peil, S.; Bathen, D. Solid hydrogen storage realized by doped Na_3AlH_6 in kg scale charged and operated in an Al alloy tank. 2015. to be published.
43. Bellosta von Colbe, J.M.; Lozano, G.; Metz, O.; Bücherl, T.; Bormann, R.; Klassen, T.; Dornheim, M. Design, sorption behaviour and energy management in a sodium alanate-based lightweight hydrogen storage tank. *Int. J. Hydrogen Energy* **2015**, *40*, 2984–2988.
44. Yan, M.-Y.; Sun, F.; Liu, X.-P.; Ye, J.-H.; Wang, S.-M.; Jiang, L.-J. Effects of graphite content and

compaction pressure on hydrogen desorption properties of $Mg(NH_2)_2$–2LiH based tank.*J. Alloys Compd.* **2015**, *628*, 63–67.

45. Ismail, I.M.K.; Hawkins, T. Kinetics of thermal decomposition of aluminium hydride: I-non-isothermal decomposition under vacuum and in inert atmosphere (argon). *Thermochim. Acta***2005**, *439*, 32–43.
46. Nakagawa, T.; Burrell, A.K.; Del Sesto, R.E.; Janicke, M.T.; Nekimken, A.L.; Purdy, G.M.; Paik, B.; Zhong, R.-Q.; Semelsberger, T.A.; Davis, B.L. Physical, structural, and dehydrogenation properties of ammonia borane in ionic liquids. *RSC Adv.* **2014**, *4*, 21681–21687.
47. Sankir, M.; Serin, R.B.; Semiz, L.; Sankir, N.D. Unusual behavior of dynamic hydrogen generation from sodium borohydride. *Int. J. Hydrogen Energy* **2014**, *39*, 2608–2613.
48. Amendola, S.C.; Sharp-Goldman, S.L.; Janjua, M.S.; Kelly, M.T.; Petillo, P.J.; Binder, M. An ultrasafe hydrogen generator: Aqueous, alkaline borohydride solutions and Ru catalyst. *J. Power Sources* **2000**, *85*, 186–189.
49. Liu, B.H.; Li, Z.P. A review: Hydrogen generation from borohydride hydrolysis reaction. *J. Power Sources* **2009**, *187*, 527–534.
50. Demirci, U.B.; Akdim, O.; Andrieux, J.; Hannauer, J.; Chamoun, R.; Miele, P. Sodium borohydride hydrolysis as hydrogen generator: Issues, state of the art and applicability upstream from a fuel cell. *Fuel Cells* **2010**, *10*, 335–350.
51. Schlesinger, H.I.; Brown, H.C.; Hoekstra, H.R.; Rapp, L.R. Reactions of diborane with alkali metal hydrides and their addition compounds. New syntheses of borohydrides. Sodium and potassium borohydrides. *J. Am. Chem. Soc.* **1953**, *75*, 199–204.
52. Demirci, U.B.; Miele, P. Sodium borohydride *versus* ammonia borane, in hydrogen storage and direct fuel cell applications. *Energ. Environ. Sci.* **2009**, *2*, 627–637.
53. Weidenthaler, C.; Felderhoff, M. Complex hydrides. In *Handbook of Hydrogen Storage*; Hirscher, M., Ed.; Wiley-VCH Verlag GmbH & Co. KGaA: Weinheim, Germany, 2010.
54. Metin, Ö.; Mazumder, V.; Özkar, S.; Sun, S. Monodisperse nickel nanoparticles and their catalysis in hydrolytic dehydrogenation of ammonia borane. *J. Am. Chem. Soc.* **2010**, *132*, 1468–1469.
55. Sun, D.; Mazumder, V.; Metin, Ö.; Sun, S. Catalytic hydrolysis of ammonia borane via cobalt palladium nanoparticles. *ACS Nano* **2011**, *5*, 6458–6464.
56. Demirci, U.B.; Miele, P. Cobalt-based catalysts for the hydrolysis of $NaBH_4$ and NH_3BH_3.*Phys. Chem. Chem. Phys.* **2014**, *16*, 6872–6885.
57. Aiello, R.; Sharp, J.H.; Matthews, M.A. Production of hydrogen from chemical hydrides via hydrolysis with steam. *Int. J. Hydrogen Energy* **1999**, *24*, 1123–1130.
58. Chen, J.; Fu, H.; Xiong, Y.; Xu, J.; Zheng, J.; Li, X. $MgCl_2$ promoted hydrolysis of MgH_2nanoparticles for highly efficient H_2 generation. *Nano Energy* **2014**, *10*, 337–343.
59. Kojima, Y.; Kawai, Y.; Kimbara, M.; Nakanishi, H.; Matsumoto, S. Hydrogen generation by hydrolysis reaction of lithium borohydride. *Int. J. Hydrogen Energy* **2004**, *29*, 1213–1217.
60. Kojima, Y.; Suzuki, K.-I.; Kawai, Y. Hydrogen generation by hydrolysis reaction of magnesium hydride. *J. Mater. Sci.* **2004**, *39*, 2227–2229.
61. Kong, V.; Foulkes, F.; Kirk, D.; Hinatsu, J. Development of hydrogen storage for fuel cell generators I: Hydrogen generation using hydrolysis hydrides. *Int. J. Hydrogen Energy* **1999**,*24*, 665–675.
62. Kong, V.; Kirk, D.; Foulkes, F.; Hinatsu, J. Development of hydrogen storage for fuel cell generators II: Utilization of calcium hydride and lithium hydride. *Int. J. Hydrogen Energy***2003**, *28*, 205–214.
63. Laversenne, L.; Goutaudier, C.; Chiriac, R.; Sigala, C.; Bonnetot, B. Hydrogen storage in borohydrides comparison of hydrolysis conditions of $LiBH_4$, $NaBH_4$ and KBH_4. *J. Therm. Anal. Calorim.* **2008**, *94*, 785–790.
64. Ferreira, M.J.F.; Coelho, F.; Rangel, C.M.; Pinto, A.M.F.R. Batch sodium borohydride hydrolysis systems: Effect of sudden valve opening on hydrogen generation rate. *Int. J. Hydrogen Energy* **2012**, *37*, 1947–1953.
65. Ferreira, M.J.F.; Fernandes, V.R.; Gales, L.; Rangel, C.M.; Pinto, A.M.F.R. Effects of the addition of an organic polymer on the hydrolysis of sodium tetrahydroborate in batch reactors. *Int. J. Hydrogen Energy* **2010**, *35*, 11456–11469.
66. Pinto, A.; Falcao, D.; Silva, R.; Rangel, C. Hydrogen generation and storage from hydrolysis of sodium borohydride in batch reactors. *Int. J. Hydrogen Energy* **2006**, *31*, 1341–1347.
67. Pinto, A.M.F.R.; Ferreira, M.J.F.; Fernandes, V.R.; Rangel, C.M. Durability and reutilization capabilities of a Ni–Ru catalyst for the hydrolysis of sodium borohydride in batch reactors.*Catal. Today* **2011**, *170*, 40–49.
68. Zhang, Q.; Smith, G.; Wu, Y. Catalytic hydrolysis of sodium borohydride in an integrated reactor for hydrogen generation. *Int. J. Hydrogen Energy* **2007**, *32*, 4731–4735.
69. Zhang, Q.; Smith, G.; Wu, Y.; Mohring, R. Catalytic hydrolysis of sodium borohydride in an auto-thermal fixed-bed reactor. *Int. J. Hydrogen Energy* **2006**, *31*, 961–965.
70. Ferreira, M.J.F.; Rangel, C.M.; Pinto, A.M.F.R. Water handling challenge on hydrolysis of sodium borohydride in batch reactors. *Int. J. Hydrogen Energy* **2012**, *37*, 6985–6994.
71. Beaird, A.M.; Davis, T.A.; Matthews, M.A. Deliquescence in the hydrolysis of sodium borohydride by water vapor. *Ind. Eng. Chem. Res.* **2010**, *49*, 9596–9599.
72. Yu, L.; Matthews, M.A. A reactor model for hydrogen generation from sodium borohydride and water vapor. *Int. J. Hydrogen Energy* **2014**, *39*, 3830–3836.
73. Gervasio, D.; Tasic, S.; Zenhausern, F. Room temperature micro-hydrogen-generator. *J. Power Sources* **2005**, *149*, 15–21.

74. Amendola, S.C.; Binder, M.; Sharp-Goldman, S.L.; Kelly, M.T.; Petillo, P.J. System for hydrogen generation. *US Patent* **2003**.

75. Pozio, A.; Defrancesco, M.; Monteleone, G.; Oronzio, R.; Galli, S.; Dangelo, C.; Marrucci, M. Apparatus for the production of hydrogen from sodium borohydride in alkaline solution.*Int. J. Hydrogen Energy* **2008**, *33*, 51–56.

76. Oronzio, R.; Monteleone, G.; Pozio, A.; De Francesco, M.; Galli, S. New reactor design for catalytic sodium borohydride hydrolysis. *Int. J. Hydrogen Energy* **2009**, *34*, 4555–4560.

77. Galli, S.; De Francesco, M.; Monteleone, G.; Oronzio, R.; Pozio, A. Development of a compact hydrogen generator from sodium borohydride. *Int. J. Hydrogen Energy* **2010**, *35*, 7344–7349.

78. Li, H.-Y.; Chen, Y.-T.; Lu, M.-T.; Lai, Y.-H.; Yang, J.-T. Design and testing of a novel catalytic reactor to generate hydrogen. *Int. J. Hydrogen Energy* **2014**, *39*, 11945–11954.

79. Kim, T. Hydrogen generation from sodium borohydride using microreactor for micro fuel cells. *Int. J. Hydrogen Energy* **2011**, *36*, 1404–1410.

80. Kim, T. Fully-integrated micro PEM fuel cell system with $NaBH_4$ hydrogen generator. *Int. J. Hydrogen Energy* **2012**, *37*, 2440–2446.

81. Zhu, L.; Meng, D.D.; Kroodsma, N.; Yeom, J.; Shannon, M.A. An integrated microfluidic self-regulating and self-circulating hydrogen generator for fuel cells. In Proceedings of International Solid-State Sensors, Actuators and Microsystems Conference, Denver, CO, USA, 21–25 June 2009; pp. 652–655.

82. Zhu, L.; Kroodsma, N.; Yeom, J.; Haan, J.L.; Shannon, M.A.; Meng, D.D. An on-demand microfluidic hydrogen generator with self-regulated gas generation and self-circulated reactant exchange with a rechargeable reservoir. *Microfluid Nanofluidics* **2011**, *11*, 569–578.

83. Zhang, J.; Zheng, Y.; Gore, J.P.; Fisher, T.S. 1 kW_e sodium borohydride hydrogen generation system: Part I: Experimental study. *J. Power Sources* **2007**, *165*, 844–853.

84. Zhang, J.; Zheng, Y.; Gore, J.P.; Mudawar, I.; Fisher, T.S. 1 kW_e sodium borohydride hydrogen generation system: Part II: Reactor modeling. *J. Power Sources* **2007**, *170*, 150–159.

85. Sousa, T.; Fernandes, V.R.; Pinto, P.J.R.; Slavkov, Y.; Bosukov, L.; Rangel, C.M. A sodium borohydride hydrogen generation reactor for stationary applications: Experimental and reactor simulation studies. *Chem. Eng. Sci.* **2012**, *84*, 70–79.

86. Sousa, T.; Rangel, C.M. A dynamic two phase flow model for a pilot scale sodium borohydride hydrogen generation reactor. *Int. J. Hydrogen Energy* **2014**, *39*, 5291–5300.

87. Pinto, P.J.R.; Sousa, T.; Fernandes, V.R.; Pinto, A.M.F.R.; Rangel, C.M. Simulation of a stand-alone residential PEMFC power system with sodium borohydride as hydrogen source. *Int. J. Electr. Power Energy Syst.* **2013**, *49*, 57–65.

88. Lee, C.J.; Kim, T. Hydrogen supply system employing direct decomposition of solid-state $NaBH_4$. *Int. J. Hydrogen Energy* **2015**, *40*, 2274–2282.

89. Ma, J.; Choudhury, N.A.; Sahai, Y. A comprehensive review of direct borohydride fuel cells.*Renew. Sustain. Energy Rev.* **2010**, *14*, 183–199.

90. Merino-Jiménez, I.; de León Ponce, C.; Shah, A.A.; Walsh, F.C. Developments in direct borohydride fuel cells and remaining challenges. *J. Power Sources* **2012**, *219*, 339–357.

91. Jamard, R.; Salomon, J.; Martinent-Beaumont, A.; Coutanceau, C. Life time test in direct borohydride fuel cell system. *J. Power Sources* **2009**, *193*, 779–787.

92. Kim, T.; Kwon, S. Design and development of a fuel cell-powered small unmanned aircraft.*Int. J. Hydrogen Energy* **2012**, *37*, 615–622.

93. Bradley, T.H.; Moffitt, B.A.; Mavris, D.N.; Parekh, D.E. Development and experimental characterization of a fuel cell powered aircraft. *J. Power Sources* **2007**, *171*, 793–801.

94. McConnell, V.P. Military UAVs claiming the skies with fuel cell power. *Fuel Cells Bull.* **2007**,*2007*, 12–15.

95. Stroman, R.O.; Jackson, G.S. Modeling the performance of an ideal $NaBH_4$–H_2O_2 direct borohydride fuel cell. *J. Power Sources* **2014**, *247*, 756–769.

96. Stroman, R.O.; Jackson, G.S.; Garsany, Y.; Swider-Lyons, K. A calibrated hydrogen-peroxide direct-borohydride fuel cell model. *J. Power Sources* **2014**, *271*, 421–430.

97. Stroman, R.O.; Schuette, M.W.; Swider-Lyons, K.; Rodgers, J.A.; Edwards, D.J. Liquid hydrogen fuel system design and demonstration in a small long endurance air vehicle. *Int. J. Hydrogen Energy* **2014**, *39*, 11279–11290.

98. Kim, K.; Kim, T.; Lee, K.; Kwon, S. Fuel cell system with sodium borohydride as hydrogen source for unmanned aerial vehicles. *J. Power Sources* **2011**, *196*, 9069–9075.

99. AeroVironment. Aerovironment Puma small UAS achieves record flight of over nine hours using fuel cell battery hybrid system. Available online: https://www.avinc.com/public-safety/news-events/article/aerovironment_puma_small_uas_achieves_record_flight (accessed on 25 August 2015).

100. Kim, T. $NaBH_4$ (sodium borohydride) hydrogen generator with a volume-exchange fuel tank for small unmanned aerial vehicles powered by a PEM (proton exchange membrane) fuel cell. *Energy* **2014**, *69*, 721–727.

101. Liu, J.G.; Zhao, T.S.; Chen, R.; Wong, C.W. The effect of methanol concentration on the performance of a passive DMFC. *Electrochem. Commun.* **2005**, *7*, 288–294.

102. Akiyama, K.; Matsumoto, S.; Miyasaka, A.; Shodai, T. An air-breathing single cell small proton exchange membrane fuel cell system with AB_5-type metal hydride and an ultra-low voltage input boost converter. *J. Power Sources* **2009**, *186*, 37–44.

103. Han, H.S.; Cho, C.; Kim, S.Y.; Hyun, J.M. Performance evaluation of a polymer electrolyte membrane fuel cell system for powering portable freezer. *Appl. Energy* **2013**, *105*, 125–137.

104. Kim, S.H.; Miesse, C.M.; Lee, H.B.; Chang, I.W.; Hwang, Y.S.; Jang, J.H.; Cha, S.W. Ultra compact direct hydrogen fuel cell prototype using a metal hydride hydrogen storage tank for a mobile phone. *Appl. Energy* **2014**, *134*, 382–391.

105. Prosini, P.P.; Gislon, P. A hydrogen refill for cellular phone. *J. Power Sources* **2006**, *161*, 290–293.

106. McConnell, V.P. Now, voyager? The increasing marine use of fuel cells. *Fuel Cells Bull.* **2010**,*2010*, 12–17.

107. Ghosh, P.C.; Vasudeva, U. Analysis of 3000 t class submarines equipped with polymer electrolyte fuel cells. *Energy* **2011**, *36*, 3138–3147.

108. Psoma, A.; Sattler, G. Fuel cell systems for submarines: From the first idea to serial production. *J. Power Sources* **2002**, *106*, 381–383.

109. Lakeman, J.B.; Rose, A.; Pointon, K.D.; Browning, D.J.; Lovell, K.V.; Waring, S.C.; Horsfall, J.A. The direct borohydride fuel cell for uuv propulsion power. *J. Power Sources* **2006**, *162*, 765–772.

110. Swider-Lyons, K.E.; Carlin, R.T.; Rosenfeld, R.L.; Nowak, R.J. Technical issues and opportunities for fuel cell development for autonomous underwater vehicles. In Proceedings of the 2002 Workshop on Autonomous Underwater Vehicles, San Antonio, TX, USA, 20–21 June 2002; pp. 61–64.

111. Luo, N.; Miley, G.H.; Kim, K.-J.; Burton, R.; Huang, X. $NaBH_4/H_2O_2$ fuel cells for air independent power systems. *J. Power Sources* **2008**, *185*, 685–690.

112. Santos, D.M.F.; Sequeira, C.A.C. Sodium borohydride as a fuel for the future. *Renew. Sustain. Energy Rev.* **2011**, *15*, 3980–4001.

113. Miley, G.H.; Kim, K.J.; Luo, N.; Shrestha, P.J. Advanced $NaBH_4/H_2O_2$ fuel cell for space applications. *AIP Conf. Proc.* **2009**, *1103*, 157–163.

114. Protonex. Underwater applications. Available online: http://www.protonex.com/solutions/underwater-applications/ (accessed on 25 August 2015).

115. Kojima, Y.; Haga, T. Recycling process of sodium metaborate to sodium borohydride. *Int. J. Hydrogen Energy* **2003**, *28*, 989–993.

116. Seo, J.-E.; Kim, Y.; Kim, Y.; Kim, K.; Lee, J.H.; Lee, D.H.; Kim, Y.; Shin, S.J.; Kim, D.-M.; Kim, S.-Y.; *et al.* Portable ammonia-borane-based H_2 power-pack for unmanned aerial vehicles. *J. Power Sources* **2014**, *254*, 329–337.

117. Nakagawa, T.; Paik, B.; Davis, B.; Semelsberger, T.; Baker, T.; Sneddon, L. Annual merit review, May 17th Los Alamos National Laboratory la-ur 12-20427. Available online: http://www.hydrogen.energy.gov/pdfs/review12/st040_davis_2012_o.pdf (accessed on 21 August 2015).

118. Semelsberger, T.A.; Davis, B.L.; Rekken, B.D.; Paik, B.; Tafoya, J.I. Chemical hydrogen rate modeling, validation, and system demonstration. In Proceedings of DOE Fuel Cell Technologies Program Annual Merit Review, EERE: Hydrogen, Fuel Cells and Infrastructure Technologies Program, Washington, DC, USA, 13–17 May 2013.

CHAPTER 6

RECENT PROGRESS IN METAL BOROHYDRIDES FOR HYDROGEN STORAGE

Hai-Wen Li, Yigang Yan, Shin-ichi Orimo, Andreas Züttel and Craig M. Jensen

Keywords: Hydrogen, hydride, borohydride, hydrogen storage

INTRODUCTION

Development of advanced hydrogen storage materials for onboard hydrogen storage systems is regarded as a key prerequisite for widespread adoption of fuel cell vehicles. For commercial vehicle applications, hydrogen storage materials must possess all the following capabilities: high gravimetric hydrogen density, adequate hydrogenation-dehydrogenation temperature/rate, cycling stability, and low cost [1].

Metal borohydrides $M(BH_4)_n$ (n indicates the valence of M) have high gravimetric hydrogen densities and thus have attracted great interest for use in hydrogen storage [2]. The gravimetric hydrogen densities and physical properties of various $M(BH_4)_n$ [3] are shown in Figure 1 and Table 1, respectively.

1	2	3	4	5	6	7	8	9	10	11	12	13	14
Li 18.5	Be 20.8												
Na 10.7	Mg 14.9											[Al] 16.9	
K 7.5	Ca 11.6	Sc 13.5	Ti III 13.1 IV 15.0	V 12.7	Cr 9.9	[Mn] 9.5	[Fe] 9.4	[Co] 9.1	[Ni] 9.1	[Cu]* 5.1	Zn 4.4	[Ga] 10.6	Ge 12.2
Rb 4.0	Sr 6.9	Y 9.1	Zr 10.7	Nb 8.8						[Ag]* 3.3	[Cd] 2.9	[In] 7.6	[Sn] 9.1
Cs 2.7	Ba 4.8	Ln	Hf 6.8							[Au] 1.9	Hg 1.8	Tl 4.9	
		Ac											

Ln	La 6.6	Ce 6.6		Nd 6.4		Sm 6.2	Eu 6.2	Gd 6.0	Tb 5.9	Dy 5.8	Ho 5.8	Er 5.7	Tm 5.7	Yb 5.6	Lu 5.5
Ac		Th 5.5	Pa 5.6	U III 4.3 IV 5.4	Np 5.4	Pu 5.4									

Figure 1. Elements M formed $M(BH_4)_n$, their gravimetric hydorgne densities in the unit of mass%. Asterisks indicate compounds stabilizaed at room temperature by coordiantion with ligands. Brackets indicate compounds reported to be unstable at room temperature but may be isolated at low temperature [3]. Reproduced with permission from reference 3.

Table 1. Density of $M(BH_4)_n$ and hydrogen density in $M(BH_4)_n$[3]. Reproduced with permission from reference 3.

$M(BH_4)_n$	CAS No.	Density (g/mol)	Density (g/cm^3)	Hydrogen Density (mass%)	Hydrogen Density (kg/m^3)
$LiBH_4$	16949-15-8	21.78	0.66	18.5	122.1
$NaBH_4$	16940-66-2	37.83	1.07	10.7	114.5
KBH_4	13762-51-1	53.94	1.17	7.5	87.8
$RbBH_4$	20346-99-0	100.31	1.92	4.0	76.8
$CsBH_4$	-	147.75	2.42	2.7	65.3
$Be(BH_4)_2$	17440-85-6	38.70	0.702	20.8	146.0
$Mg(BH_4)_2$	16903-37-0	53.99	0.989	14.9	147.4
$Ca(BH_4)_2$	17068-95-0	69.76	(1.07)	11.6	(124.1)
$Mn(BH_4)_2$	-	84.62	(1.24)	9.5	(117.8)
$Al(BH_4)_3$	16962-07-5	71.51	0.79	16.9	133.5
$Zr(BH_4)_4$	12370-59-1	150.6	1.179	10.7	126.2
$Hf(BH_4)_4$	53608-70-1	237.6	1.65	6.8	112.2

Densities in brackets shows the value calculated from the crystal structure.

First-principles calculations of the electronic structures of $M(BH_4)_n$ show that they are nonmetallic and have relatively large energy gaps of 1.8–6.8 eV. A boron atom forms covalent bonds with four surrounding H atoms to form $[BH_4]^-$ anions, the charge of which is compensated by metal cations M^{n+}. The electronic structures indicate that charge transfer from the metal cations M^{n+} to the $[BH_4]^-$ anions is a key feature determining the thermodynamic stability of $M(BH_4)_n$[4,5,6,7,8].

Some $M(BH_4)_n$ have been used as "one-way" hydrogen sources that release the hydrogen on contact with water (via hydrolysis) [9,10,11]. Because the hydrolysis reaction is highly irreversible, such materials are certainly not candidates for reversible hydrogen storage. In this review, we focus on the recent progress in the dehydrogenation and rehydrogenation reactions of $M(BH_4)_n$ at controlled temperature and hydrogen

Citation: Hai-Wen Li, Yigang Yan, Shin-ichi Orimo, Andreas Züttel and Craig M. Jensen, Recent Progress in Metal Borohydrides for Hydrogen Storage, doi:10.3390/en4010185.

pressure. Some several excellent reviews on $M(BH_4)_n$ are also available [2,12,13,14,15,16,17,18].

FUNDAMENTALS OF HYDROGEN STORAGE PROPERTIES

The study of $M(BH_4)_n$ as a candidate hydrogen storage material started with $LiBH_4$ [19]. It releases 13.8 mass% of hydrogen by decomposition into LiH and B [19,20,21,22,23,24,25]. The dehydrogenation process at a low heating rate (0.5 K/min) exhibits three distinct desorption peaks, which indicates that it involves several intermediate steps [25]. This was recently explained by the formation of $Li_2B_{12}H_{12}$ as an intermediate compound, evidenced by Raman and solid-state CP/MAS NMR analyses, as well as theoretical prediction [26,27,28,29]. $Li_2B_{12}H_{12}$ has a cubic crystal structure with Pa3 symmetry [30]. A mechanistic explanation of the formation of $Li_2B_{12}H_{12}$ indicates that the diborane or higher borane species formed during the decomposition of $LiBH_4$ would react with the remaining $LiBH_4$ to afford $Li_2B_{12}H_{12}$ and possibly even $Li_2B_{10}H_{10}$ [31]. To understand the mechanism of dehydrogenation better, intensive investigations on the dynamics have been performed, including on the rotational and vibration motions of BH_4 tetrahedra; the atomic motions of H, Li, and B; H–D exchange in BH_4 units, etc. [32,33,34,35,36,37,38,39,40,41,42,43,44,45,46,47,48,49, 50]. The enthalpy ΔH and entropy ΔS of dehydrogenation were found to be 74 kJ mol^{-1} H_2 and 115 J K^{-1} mol^{-1} H_2, respectively, based on the pressure-concentration-temperature (PCT) isotherm measurement [51].

Dehydrogenation of $LiBH_4$ is reversible, because the end products lithium hydride (LiH) and boron absorb hydrogen at 873 K in a hydrogen pressure of 35 MPa for 12 h [23] or at 1000 K under a hydrogen pressure of 15 MPa for over 10 h [25] to form $LiBH_4$. The high rehydrogenation temperature of above 873 K might be due to the sluggish kinetics of rehydrogenation, which requires recombination of the dehydrogenated LiH and B, that is, breaking of the rigid boron lattice, and the subsequent diffusion of Li and B atoms toward each other would be mainly responsible for the sluggish kinetics. The sluggish kinetics can be enhanced by preparing a homogenous dispersion of Li and B atoms on an atomic level. For example, forming a binary LiB_x compound (e.g., Li_7B_6, LiB_3, and LiB, etc.) has been proved to significantly improve the reaction kinetics of $LiBH_4$ formation over that of the Li/B mixture [51,52,53]. In addition, the formation of $LiBH_4$ might be related to diborane B_2H_6 [53,54,55].

$Mg(BH_4)_2$ releases approximately 14.9 mass% of hydrogen when heated up to 870 K [56,57,58,59,60,61,62,63,64,65,66]. The dehydrogenation process is found to also proceed through multiple steps together with the formation of intermediate compounds [62,63,65], as summarized in Figure 2. Thus far, one of the intermediate compounds has been theoretically predicted and experimentally confirmed as being $MgB_{12}H_{12}$ [28,29,62,65,66,67,68,69,70] similar to that in the case of $LiBH_4$ [26]. The lowest energy configuration of $MgB_{12}H_{12}$ was predicted as being monoclinic in the space group C2/m (No. 12) [29]. Anhydrous $MgB_{12}H_{12}$ cannot be obtained from simple thermal decomposition of $Mg(H_2O)_6B_{12}H_{12}\cdot 6H_2O$ [71]. The apparent enthalpy and entropy for the conversion of $Mg(BH_4)_2$ to MgH_2 were estimated from PCT measurement to be 39 kJ mol^{-1} H_2 and 91 J K^{-1} mol^{-1} H_2 [56], respectively. In a later work, different values were reported: ΔH = 57 kJ mol^{-1} H_2 and ΔS = 128 J K^{-1} mol^{-1} H_2 [62]. These differences might originate from differences in the measurement conditions. For example, a shorter equivalent judgment time will lead to a higher plateau due to the sluggish kinetics.

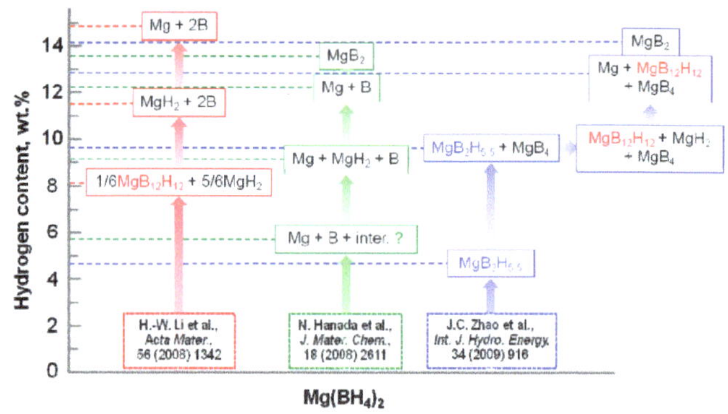

Figure 2. The dehydrogenation processes of $Mg(BH_4)_2$ based on the respective reports. Question mark indicates the possibility of the formation of an intermediate compound during the dehydrogenation.

Rehydrogenation of $Mg(BH_4)_2$ occurs at 543 K in a hydrogen pressure of 40 MPa for 48 h. Approximately 6.1 mass% of hydrogen is rehydrogenated through the formation of $MgB_{12}H_{12}$[62,66]. The formation of $MgB_{12}H_{12}$, rather than $Mg(BH_4)_2$, after rehydrogenation may be explained from the viewpoint of the crystal (cluster) structure. The formation of $Mg(BH_4)_2$ is more thermodynamically favorable than that of $MgB_{12}H_{12}$ from the dehydrogenated products (MgH_2 and boron) at 573 K. However, for the formation of $Mg(BH_4)_2$, the B–B bonds in the icosahedral cluster of B should be broken and the B atoms should spatially migrate to form $[BH_4]^-$. The barrier against breaking of the B–B bonds and migration of the B atoms is probably too high to overcome under the present rehydrogenation conditions [66]. However, it is almost the same boron icosahedral clusters in between B and $MgB_{12}H_{12}$ that facilitate the formation of $[B_{12}H_{12}]^{2-}$. In other words, the formation of $Mg(BH_4)_2$ requires higher temperature and pressure to overcome the aforementioned barriers. This was proven in a recent study: partial formation of $Mg(BH_4)_2$ was confirmed in rehydrogenation experiments carried out at 663 K in a hydrogen pressure of 90 MPa for 3 days [72]. In addition, $Mg(BH_4)_2$ can be prepared by hydrogenation of MgB_2 at 673 K under a hydrogen pressure of 90 MPa for a few days [73]. This hydrogenation reaction can be greatly promoted by using nanostructures with grain boundary defects introduced by mechanical milling [73,74,75]. Recently, $Mg(BH_4)_2$ was found to be formed by hydrogenation of $Mg(B_3H_8)_2$ at 523 K under a hydrogen pressure of 12 MPa [76].

$Ca(BH_4)_2$ was theoretically predicted to release 9.6 mass% of hydrogen according to the following reaction [77]:

$$Ca(BH_4)_2 = 2/3\ CaH_2 + 1/3\ CaB_6 + 10/3\ H_2 \tag{1}$$

Experimentally, approximately 9.0 mass% of hydrogen was released when $Ca(BH_4)_2$ was heated to 800 K [78,79,80]. This value shows good agreement with that predicted by Equation 1. The presence of two endothermic peaks corresponding to dehydrogenation suggests the formation of intermediate compounds, which is confirmed by the powder X-ray diffraction (XRD) analyses of the dehydrogenated products that were heated to approximately 663 K [78] and after the plateau in the PCT profile measured at 593 K [79]. The intermediate compound is suggested to be a CaB_2H_xcompound having a $HgCl_2$-type structure with Pnma symmetry according to the high-resolution synchrotron radiation powder XRD measurement [81]. By contrast, the proposed CaB_2H_x was predicted to be both structurally unstable and too high in energy to be a dehydrogenation intermediate based on the plane wave density functional theory calculations [82]. The formation of $CaB_{12}H_{12}$ as an intermediate compound is suggested by the experiment and first-principle calculations [29,69,70,83]. A pure $CaB_{12}H_{12}$ compound has been synthesized and the crystal structure has been determined to be monoclinic with C2/c symmetry [84]. Identification of the intermediate compound is expected to increase our understanding of the dehydrogenation process of $Ca(BH_4)_2$. The enthalpy and entropy change for the first step of dehydrogenation are estimated to be $\Delta H = 87$ kJ mol^{-1} H_2 and $\Delta S = 158$ J K^{-1} mol^{-1} H_2 [80], respectively.

Regarding the rehydrogenation properties of pure $Ca(BH_4)_2$, no tangible data are available. $Ca(BH_4)_2$ has been predicted theoretically to have limited viability as a reversible storage material for on-board storage due to the formation of a stable intermediate compound of $CaB_{12}H_{12}$ [83]. It was found that the ball-milled mixtures of $CaB_{12}H_{12}$ and CaH_2 do not produce $Ca(BH_4)_2$ during the rehydrogenation test at 670 K in a hydrogen pressure of 100 MPa [84]. With additives, approximately 57% of the $Ca(BH_4)_2$ is obtained by rehydrogenation at 623 K in a hydrogen pressure of 10 MPa [85,86,87].

$Y(BH_4)_3$ releases approximately 7.8 mass% of hydrogen when heated to 773 K [88,89,90,91,92,93]. This value is almost equal to the theoretical dehydrogenation content (7.5 mass%) according to the assumed reaction:

$$Y(BH_4)_3 = YH_2 + 3\,B + 5\,H_2 \tag{2}$$

The whole reaction process of $Y(BH_4)_3$ on heating to 773 K is summarized as follows: (1) phase transformation, (2) melting, (3) decomposition of $Y(BH_4)_3$ into an intermediate compound and YH_3, (4) decomposition of the intermediate compound, and (5) decomposition of YH_3 into YH_2[89]. Approximately 1.1–1.3 mass% of hydrogen was rehydrogenated at 573 K in a hydrogen pressure of 35 MPa for 24 h [89]. Similar to that of $Y(BH_4)_3$, a multistep dehydrogenation process were also confirmed in $Ce(BH_4)_3$ [94]. In addition, there are several other $M(BH_4)_n$ (e.g., M = Zn, Al, Ti, Mn and Zr) [95,96,97,98,99,100 ,101,102,103]. Most of these were found to be irreversible due to their low melting temperatures and the release of diborane with hydrogen.

IMPROVEMENT OF HYDROGEN STORAGE PROPERTIES

Tailoring Thermodynamics

In principle, the Gibbs free energy for a certain reaction determines the reaction temperature. Taking into account that the entropy change ΔS mainly comes from the gaseous hydrogen (i.e., a constant $S^0_{H2} = 130$ J · K^{-1} mol^{-1} H_2), enthalpy change ΔH becomes the specific indicator to evaluate thermodynamic stability. For metal borohydrides, ΔH is calculated from the difference in the heat of formation between the product and starting materials. Therefore, two main approaches to tailor the thermodynamic stabilities from both sides—the starting materials and products—are expected: (a) destabilization of $M(BH_4)_n$ and (b) stabilization of dehydrogenation products, as shown in Figure 3.

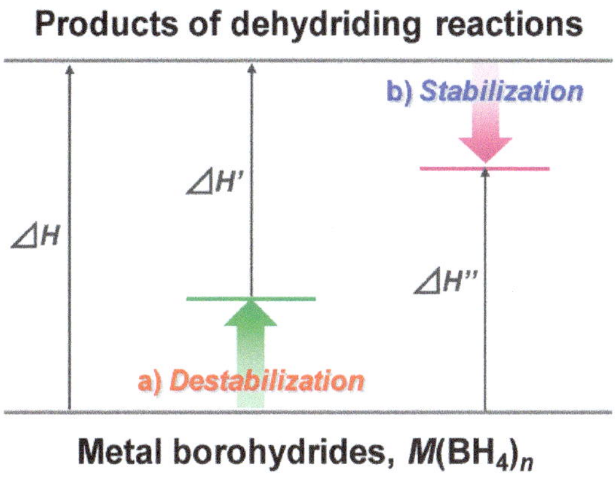

Figure 3. Schematic illustration of two main approaches to tailor the thermodynamic stabilities of metal borohydrides$M(BH_4)_n$: (a) destabilization of $M(BH_4)_n$ and (b) stabilization of dehydrogenation products.

One approach to reducing the enthalpy change ΔH_{deh} is the destabilization of $M(BH_4)_n$, as illustrated in Figure 3. The thermodynamic stabilities for the series of metal borohydrides $M(BH_4)_n$(M = Li, Na, K, Mg, Ca, Sc, Zr, Hf, Cu, Zn, and Al; n = 1–4) have been systematically investigated using first-principles calculations [104]. A good correlation between the predicted heat of formation ΔH_{form} of $M(BH_4)_n$ and the Pauling electronegativity χ_P of M is found, which suggests that the χ_P is a useful indicator to estimate the thermodynamic stability of $M(BH_4)_n$.

The enthalpy change ΔH_{deh} of the dehydrogenation of $M(BH_4)_n$ correlates not only with the stability of $M(BH_4)_n$, but also with the stability of the products. The dehydrogenation reactions were assumed as the following equation:

$$M(BH_4)_n = MH_m + n\,B + (4n - m)/2\,H_2 \tag{3}$$

In the case of no MH_m formation, direct decomposition into elements was assumed. Then, the ΔH_{deh} of the dehydrogenation was estimated using the reported values of ΔH_{prod} and predicted ΔH_{form}:

$$\Delta H_{deh} = \Delta H_{prod} - \Delta H_{form} \tag{4}$$

A good correlation between ΔH_{deh} and χ_P was also obtained according to the theoretical prediction. Furthermore, it is expected that $M(BH_4)_n$ with $\chi_P \geq 1.5$ are thermodynamically unstable [3]. The dehydrogenation properties were experimentally investigated using thermal desorption analysis during the heating process for milling samples [104,105,106]. A good correlation betweenT_d (defined as the temperature of the first peak) and χ_P was confirmed experimentally, and was similar to that predicted by the first-principles calculations. Therefore, both the theoretical prediction and the experimental results consistently indicate that T_d can be roughly estimated by considering χ_P as an indicator.

Inspired by this finding, an approach of producing multi cation borohydrides $MM'(BH_4)_n$, in which M and M' have different electronegativities has been proposed to precisely tailor the thermodynamic stability [107]. Several typical samples has been synthesized, including $LiZr(BH_4)_5$, $Li_2Zr(BH_4)_6$, $LiK(BH_4)_2$, $LiSc(BH_4)_4$, $KSc(BH_4)_4$, and so on [107,108,109,110,111,112,113,114,115,116,117]. Note that some samples were reported to form a large $[M'(BH_4)_n]^{m-}$ complex anion units such as $[Sc(BH_4)_4]^-$ [109,117], rather than the bimetallic borohydrides. Most of the $MM'(BH_4)_n$ exhibit moderate thermodynamic stabilities between $M(BH_4)_n$ and $M'(BH_4)_n$. The appropriate combination of cations might be an effective method for adjusting the thermodynamic stability of metal borohydrides, similar to the conventional "alloying" method for hydrogen storage alloys. The detailed dehydrogenation-rehydrogenation properties of the recently synthesized $MM'(BH_4)_n$ are given in Table 2.

Table 2. Hydrogen storage properties of bimetallic borohydrides, $MM'(BH_4)_n$.

Borohydrides	Hydrogen (mass%)			Conditions: temp (K) [pressure (MPa)]		Reference
	Ideal	Obs (First Dehyd)	Obs (Rehyd)	First Dehyd	Rehyd	
$ZrLi(BH_4)_5$	11.7			595–873		[107]
$ZrLi_2(BH_4)_6$	12.5			650–873		[107]
$LiK(BH_4)_2$	10.6			653–		[108]
$LiSc(BH_4)_4$	14.5	4.38–6.4 ^{m}		415–673	673 [7]	[109,111]
$LiZn_2(BH_4)_5$ *	9.5			238–473		[112]
$NaZn_2(BH_4)_5$ *	8.8			241–		[112]
$NaSc(BH_4)_4$	12.7	0.97 ^{m}		440–750		[113]
$Na_2Mn(BH_4)_4$	10.1	2.9		393		[114]
$KSc(BH_4)_4$	11.2	4.4 ^{p}		470–580		[117]
$LiBH_4/Mg(BH_4)_2$	16.0	12.5	2.5	513–773	673 [10]	[115]
$xLiBH_4 + (1\text{-}x)Ca(BH_4)_2$ $(0 < x < 1)$	9.6–18.5	10 (x = 0.4, 0.6)	4 (x = 0.4)	473–773	673 [9]	[116]

* represents that the release of B_2H_6 in the decomposition process; ^{m} indicates the value referred to the mixture of metal borohydride with side products, while ^{p} indicates the value referred to the pure metal borohydride.

The other approach to reducing enthalpy change ΔH_{deh} is the stabilization of dehydrogenated products by the combination of $M(BH_4)_n$ with elements, metal hydrides, complex hydrides, and so on. A typical example for this approach is the 2 $LiBH_4$ + MgX system [118]. Pure $LiBH_4$ releases approximately 13.8 mass% of hydrogen according to the following reaction:

$LiBH_4 = LiH + B + 3/2\ H_2$

$$LiBH_4 = LiH + B + 3/2\ H_2 \qquad (5)$$

The enthalpy and entropy for Equation 5 have been predicted to be 66.6 kJ mol^{-1} H_2 and 97.4 J K^{-1} mol^{-1} [119], respectively, which means T_d (in 0.1 MPa H_2) is approximately 683 K. This temperature is too high for practical applications, although the combination with Mg-based compounds clearly results in a large reduction of ΔH_{deh}. Dehydrogenation of the $LiBH_4/MgH_2$ system proceeds as follows:

$$2\ LiBH_4 + MgH_2 = 2\ LiH + MgB_2 + 4\ H_2 \qquad (6)$$

In this case, due to the formation of MgB_2, ΔH_{deh} was reduced to 45.8 kJ mol^{-1} H_2 and T_d (in 0.1 MPa H_2) decreased to approximately 441 K [120]. At the same time, a significant destabilization of $LiBH_4$ due to combination with the $LiNH_2$ system was reported [121], as indicated by the much higher dehydrogenation pressure of the combined materials than that of $LiBH_4$ alone. Due to these significant destabilization effects, a large number of combination materials have been developed. The recent progress on these combination systems [120,121,122,123,124,125,126,127,128,129,130,131,132,133,134,135,136,137,138,139,140,141,142,143,144,145,146,147,148,149,150,151,152,153,154,155,156,157,158,159,160,161,162,163,164,165,166,167,168,169,170,171,172,173,174,175,176,177,178,179,180,181,182,183,184,185] is summarized in Table 3.

This approach not only reduces the enthalpy of the dehydrogenation, which decreases T_d, but also kinetically enhances the rehydrogenation reaction. That is, in contrast to the formation of pure $LiBH_4$ (from right to left in Equation 5) discussed in Section 2, $LiBH_4$ and MgH_2 form simultaneously (from right to left in Equation 6) under fairly moderate conditions: 5 MPa hydrogen pressure in the temperature range 523–573 K [136]. The difference between these two reactions is attributed to the different crystal structures and boron bonding of pure boron and MgB_2, as shown in Figure 4. That is, in pure boron, the common building blocks are icosahedral units that consist of 12 boron atoms, and each atom is connected to five other atoms via covalent bonds. By contrast, in MgB_2, each boron atom is connected to a maximum of three other boron atoms. The recent activities of the above-mentioned metal boride systems [136,186,187,188,189,190,191,192] are summarized in Table 4.

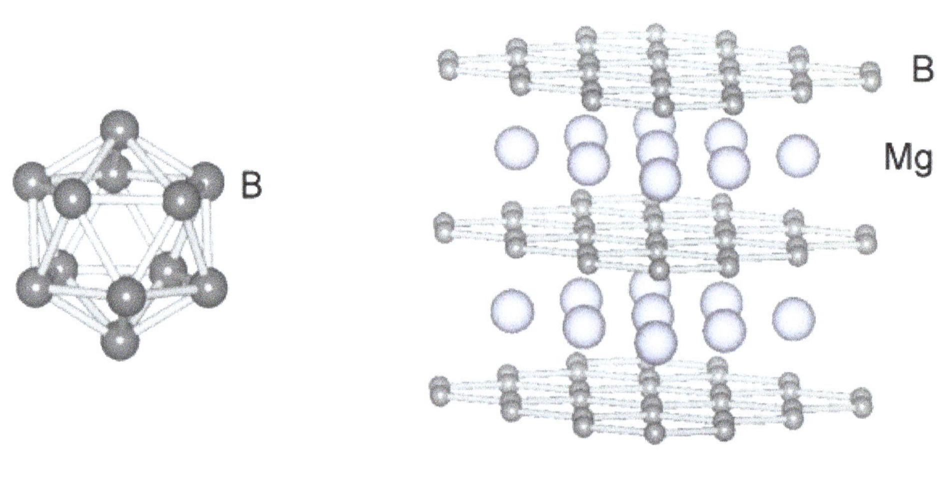

(a) (b)

Figure 4. Schematic illustrations of (a) B_{12} icosahedral unit in boron and (b) layer structure of MgB_2.

Table 3. Hydrogen storage properties of the combination systems of $M(BH_4)_n$ with metal hydrides, complex hydrides, metals, etc.

Reaction	Hydrogen (mass%)			Conditions: temp (K) [pressure (MPa)]		Theor/Exp ΔH_{dehyd} (kJ mol^{-1} H_2)	Reference
	Ideal	Obs (First Dehyd)	Obs (Rehyd)	First Dehyd	Rehyd		
$LiBH_4$ + 1/2 MgH_2 = LiH + 1/2 MgB_2 + 2 H_2	11.5	8.0–10.6	8.0–10	543–723	503–723 [10]	T 50.4–66.8/ E 40.5	[29,120,124, 134–136,142,145,156, 168,172,186]
$LiBH_4$ + 2 $LiNH_2$ = Li_3BN_2 + 4 H_2	11.9	7.8–12	0.1	622–673	573 [5]	E 23	[121,123,124, 127–130,132,149, 160,166]
$LiBH_4$ + 1/2 Al = LiH + 1/2 AlB_2 + 3/2 H_2	8.6	6.8–7.2	5–7.6	553–823	573–773 [10–15.5]	T 18.8–57.9	[134,135,137,163–165]
$LiBH_4$ + 1/2 $LiAlH_4$ = 3/2 LiH + 1/2 AlB_2 + 9/4 H_2	11.1	6–10	4.8–5.1	327–773	623–873 [4–7]		[157]
$LiBH_4$ + 1/2 Mg = LiH + 1/2 MgB_2 + 3/2 H_2	8.9	5.6		648–773		T 46.4	[134,135]
$LiBH_4$ + 1/6 CaH_2 = LiH + 1/6 CaB_6 + 10/6 H_2	11.7	5.1–11.1	9–11.1	423–773	673 [10]	T 45.4–66.5	[19,135,137,142,144, 168]
$LiBH_4$ + 1/2 ScH_2 = LiH + 1/2 ScB_2 + 2 H_2	8.9	4.5		553–723		T 34.1	[134,135,142,143]
$LiBH_4$ + 1/6 CeH_2 = LiH + 1/6 CeB_6 + 10/6 H_2	7.4	6.1–6.2	6.0	443–723	623 [10]	T 44.1	[144,168,177]
$LiBH_4$ + 1/4 YH_3 = LiH + ¼ YB_4 + 15/8 H_2	8.4	7.2		623			[168]
$LiBH_4$ + 1/4 MgH_2 + 1/4 Al = LiH + 1/4 $MgAlB_4$ + 7/4 H_2	10.0	9.4	6	533–673	673 [4]	E 57	[159]
$Ca(BH_4)_2$ + MgH_2 = CaH_2 + MgB_2 + 4 H_2	8.4	7.1		623–723		T 47	[135]
$Ca(BH_4)_2$ + MgH_2 = 2/3 CaH_2 + 1/3 CaB_6 + Mg + 13/3 H_2	9.1	8.1	5.5	593–773	623 [9]	T 45	[187]
$Mg(BH_4)_2$ + $LiNH_2$ = Li-Mg + BN-related + 5 H_2	13.1	11.4		433–873			[174]
$NaBH_4$ + 1/2 MgH_2 = Na + 1/2 MgB_2 + 5/2 H_2	9.9	9	6	330–723	723 [5]	T 62	[183,186]
$NaBH_4$ + 2 $NaNH_2$ = Na_3BN_2 + 4 H_2 *	7.0			500–773			[173]
x $LiBH_4$ + *y* $(LiNH_2)_2$ + *z* (MgH_2) = Li_3BN_2 + Mg_3N_2 + LiH + H_2					453 [15]		[140,141,155]
x:*y*:*z* = 2:1:1	13.0	8.5	2.9	413–743			
2:0.5:1	13.6	8.6	3.7	413–743			
2:1:2	11.8	6.6	3.1	428–743			
1:1:1	11.3	5.6	2.7	428–743			
3:1:1.5	13.4	9.1		413–743			

* represents that the release of NH_3 in the decomposition process: T and E represent the theoretic and experimental values, respectively.

Table 4. Hydrogen storage properties of metal borides and metal hydrides systems.

Reaction	Hydrogen (mass%)			Conditions: temp (K) [pressure (MPa)]		Reference
	Ideal	Obs (First Hyd)	Obs (Dehyd)	First Hyd	Dehyd	
LiH + 1/2 MgB_2 + 2 H_2 = $LiBH_4$ + 1/2 MgH_2	11.5	3.2–11	3.0–8.0	538–673 [9–35]	538–723 [0–0.6]	[120,145, 182,186,187]
LiF + 1/2 MgB_2 + H_2 = $LiBH_{4-y}F_y$ + 1/2 MgF_2 + $LiH_{1-x}F_x$		6.6	6.4	663 [6]	693 [0.5]	[189]
Li_7Sn_2 + 7/2 MgB_2 + 14 H_2 = 7 $LiBH_4$ + 7/4 Mg_2Sn + 1/4 Sn	5.9	2.5–3		573–673 [20–30]		[186]
CaH_2 + MgB_2 + $4H_2$ = $Ca(BH_4)_2$ + MgH_2	8.4	4.7–7		573–673 [20–35]		[186,187]
NaH + 1/2 MgB_2 + 2 H_2 = $NaBH_4$ + 1/2 MgH_2	7.9	6.2–6.7		573–673 [20–35]		[186]
1/3 CaB_6 + 2/3 CaH_2 + 10/3 H_2 = $Ca(BH_4)_2$	9.6	4.8		623 [10]	543–573	[186]
2/3 CaH_2 + 1/3 CaB_6 + 1/2 Mg + 23/6 H_2 = $Ca(BH_4)_2$ + 1/2 MgH_2	9.3		4.9–5.9	623–673 [9]		[192]
$MgNi_{2.5}B_2$ + 2 LiH + 4 MgH_2 + $4H_2$ = 2 $LiBH_4$ + 5/2 Mg_2NiH_4	2.5	1.0	1.0	623 [16]	613 [0.4]	[167]
MgB_2 + 4 H_2 = $Mg(BH_4)_2$	14.9		11.0	673 [95]		[73,74]

Promoting Kinetics

$M(BH_4)_n$ dehydrogenates via stepwise reactions accompanied by the formation of intermediate compounds. Although the rate-controlling step has not been determined, the complicated dehydrogenation feature makes the promotion of kinetics rather challenging. There are two approaches that have been extensively studied to reduce the reaction barrier: (a) appropriate additives with catalytic abilities and (b) the nanoconfinement approach.

Complex hydrides including alanates, amides, and borohydrides are commonly known to exhibit sluggish kinetics. Since Bogdanovic et al. found that the addition of Ti-based compounds significantly promoted the dehydrogenation-rehydrogenation reactions of $NaAlH_4$ [193], the development of complex hydrides for hydrogen storage has significantly increased. Stimulated by this finding, a large number of additives from oxides, halides, metals, and carbon-based materials to $M(BH_4)_n$ have been examined [19,20,58,87,101,130,144,150,161,168,170,171,191,194,195,196,197,198,199,200,201,202,203,204,205,206,207,208,209,210]; the corresponding dehydrogenation and rehydrogenation properties are summarized in Table 5. For instance, the most effective additive for $LiBH_4$ was found to be the mixture of 0.2 $MgCl_2$ + 0.1 $TiCl_3$, in which approximately 5 mass% of hydrogen was released from 333 K and 4.5 mass% of hydrogen was rehydrogenated at 873 K in 7 MPa H_2 [195]. The addition of $TiCl_3$ was also proved to be effective for $Mg(BH_4)_2$, the starting dehydrogenation temperature of which was reduced to 361 K [58]. Titanium isopropoxide was reported to be an effective additive for the combination system of 2 $LiBH_4$ + MgH_2 [136]. According to the X-ray absorption spectroscopy analysis, titanium isopropoxide was demonstrated to form a disordered TiO_2 anatase during ball milling with 2 $LiBH_4$ + MgH_2. After several dehydrogenation-rehydrogenation cycles, Ti species were demonstrated to change to Ti_2O_3 and TiB_2 [181]. Similar phenomena were reported for most additives, and the oxidation states of the initial additives vary with cycling experiments, though the catalytic mechanism is still uncertain.

The dehydrogenation-rehydrogenation reactions of $M(BH_4)_n$ are accompanied by the diffusion of constituent elements such as M, B, and H. The relatively high reaction temperature further prolongs the diffusion distance and enlarges the crystallite size of M and B, thus further degrading the hydrogen storage properties. Fabricating and maintaining the nanosized metal borohydrides is expected to be an effective way to overcome the aforementioned barriers. One known approach is to incorporate hydrogen storage materials into nanoporous materials. It was first applied in the NH_3BH_3 system [211] and good improvement was obtained: The dehydrogenation temperature was largely decreased and the amount of borazine gas (an impurity) was significantly reduced. Recently, this approach was introduced for $M(BH_4)_n$ materials, and some recent achievements [160,212,213,214,215,216,217,218,219,220,221,222] are summarized in Table 6.

By the incorporation into nanoporous carbon scaffolds with a 13-nm pore size, the dehydrogenation rates of $LiBH_4$ were found to be up to 50 times faster than those in the bulk materials measured at 573 K [212]. In addition, the capacity loss over three cycles was reduced from 72% for bulk $LiBH_4$ to ~40% for nanoconfined $LiBH_4$. Further, a recent report confirmed the synergetic effects of nanoconfinement and Ni addition on the dehydrogenation and rehydrogenation properties of $LiBH_4$. The nanoconfinement of the mixture of $LiBH_4$ and Ni addition in a nanoporous carbon scaffold shows a higher rehydrogenation rate and larger rehydrogenation amount than those of the samples without Ni addition at 593 K in a hydrogen pressure of 4 MPa [220]. This suggests that the combination of nanoconfinement and additives would be a valid way to improve the hydrogen storage properties of $M(BH_4)_n$.

Table 5. Hydrogen storage properties of $M(BH_4)_n$ with additives (oxides, halides, carbon, etc.).

$M(BH_4)_n$	Additives		Hydrogen (mass%)		Conditions: temp (K) [pressure (MPa)]		Toxic Byproduct	Reference
	Type	Amount	Obs (First Dehyd)	Obs (Rehyd)	First Dehyd	Rehyd		
$LiBH_4$	SiO_2	10–25 mass%	9–10 ^{m}		423–873			[19,20,204]
	TiO_2	25–80 mass%	4–9 ^{m}	3.5–8.3 ^{m}	373–873	873 [7–10]		[197,209]
	ZrO_2	25 mass%	8–9 ^{m}		448–873			[197]
	V_2O_3	25 mass%	8–9 ^{m}	8 ^{m}	448–873	873 [10]		[197]
	SnO_2	25 mass%	8–9 ^{m}		448–873			[197]
	Nb_2O_5	50–80 mass%	4–6 ^{m}		373–873			[209]
	Fe_2O_3	50–66.7 mass%	5.7–9 ^{m}		373–873			[209]
	V_2O_5	50–66.7 mass%	5.7–9 ^{m}		373–873			[209]
	$TiCl_3$	10–88 mass%	2.8–9.2 ^{m}	3.4 ^{m}	373–873	773 [7]	B_2H_6	[197,208]
	$CoCl_2$	5–100 mol%	10.5–18.3 ^{p}		503–873		B_2H_6	[213]
	TiH_2	10–50 mol%	6–15 ^{m}	2.5–4.5 ^{m}	573–873	773 [7]		[208]
	TiF_3	10–50 mol%	6.4–14 ^{m}	0.2–4.0 ^{m}	373–773	623–773 [7–10]	B_2H_6	[208,211]
	ZnF_2	10–50 mol%	3.7–7 ^{m}	1–4 ^{m}	393–773	773 [7]	B_2H_6	[208]
	mixture of $MgCl_2/TiCl_3$	30 mol%	5 ^{m}	4.5 ^{m}	333–873	873 [7]		[198]
	Mg	10–20 mol%	9 ^{m}		333–873			[198]
	Al	20 mol%	7.8 ^{m}	3.5 ^{m}	353–873	873 [10]		[198]
	Sc	33 mol%	2.9 ^{m}		673–773			[134]
	Ti	33 mol%	2.5 ^{m}		673–773			[134]
	V	33 mol%	4.4 ^{m}		673–773			[134]
	Cr	33 mol%	4.4 ^{m}		673–773			[134]
$LiBH_4$	MgH_2	80 mass%	8.8–9.2 ^{m}	8.5 ^{m}	627–853	673 [10]		[126,172]
	graphite	30 mass%	9.9 ^{p}	2.6 ^{p}	663–773	673 [10]		[212]
	activated carbon	30 mass%	11.2 ^{p}	4.6 ^{p}	623–773	673 [10]		[212]
	single-walled carbon nanotubes	30 mass%	11.4 -12.3 ^{p}	3.7 ^{p}	553–773	673 [10]		[205,212]
	single-walled carbon nanotubes	9.1–50 mass%	5–11.8 ^{p}	3.7–6.1 ^{p}	723 (iso)	673 [10]		[207]
	mesoporous carbon	50 mass%	7 ^{m}	6 ^{m}	605–873	623 [3]		[200]
	mixture of TiF_3/SiO_2	50 mass%	8.3 ^{m}	4 ^{m}	343–823	773 [4.5]		[204]
	Pt/C	10–50 mass%	9.2–15.7 ^{m}	6.1 ^{m}	553–973	873 [3]		[206]
$LiBH_4$ + 1/2MgH_2	Ti-iso	5–10 mol%	6.5–8.4 ^{m}	6.0 ^{m}	673 (iso)	623 [5]		[170,136]
	Zr-iso	10 mol%	5.5 ^{m}		673 (iso)			[150]
	$ZrCl_4$	10 mol%	7.5 ^{m}		673 (iso)			[150]
	SiO_2	5 mol%	9.3 ^{m}		673 (iso)			[136]
	VCl_3	5 mol%	9.1 ^{m}		673 (iso)			[136]
	graphite	10 mass%	9.5 ^{p}		723 (iso)			[168]
	carbon nanofibers	10 mass%	10.0 ^{p}		723 (iso)			[168]
	activated carbon	10 mass%	10.0 ^{p}		723 (iso)			[168]
	single-walled carbon nanotubes	10 mass%	10.0 ^{p}	6.7 ^{p}	~573–773	673 [7.5]		[168]
	muti-walled carbon nanotubes	10 mass%	10.0 ^{p}		723 (iso)			[168]
	TiF_3	5 mol%	9.7 ^{p}		573 (iso)			[72]
$Li_3BN_2H_8$	Pt/Vulcan carbon	1–10 mass%	9–13 ^{m}	1.4 ^{m}	388–673	423 [8.4]	NH_3	[130]
	Pd	5–10 mass%	11.8–13 ^{m}		473–673			[130]
	$PdCl_2$	8.3 mass%	10.4 ^{m}		473–673			[130]
	$CoCl_2$	5 mass%	8–10 ^{m}		388–493			[149]
$Li_4BN_3H_{10}$	$NiCl_2$	11 mass%	7.6 ^{p}		433–673		NH_3	[161]
$Ca(BH_4)_2$ + MgH_2	Ti-iso	1 mass%	7.1 ^{m}		523–723			[190]
$Ca(BH_4)_2$	NbF_5	2 mol%	8.3 ^{m}	4.6–5.0 ^{m}	~593–823	623 [9]		[87]
	$NbCl_5$	2 mol%	4.1–5.0 ^{m}	3.1–4.5 ^{m}	~593–823	623 [9]		[87]
	TiF_3	2 mol%	4.1–5.0 ^{m}	2.5–4.2 ^{m}	~593–823	623 [9]		[87]
	$TiCl_3$	2 mol%	4.1–5.0 ^{m}	3.5–4.4 ^{m}	~593–823	623 [9]		[87]
$Mg(BH_4)_2$	$TiCl_3$	25 mass%	13.7 ^{p}		361–800			[58]
	TiO_2	25 mass%	13.7 ^{p}		483–800			[58]

^{m} indicates the value referred to the mixture of metal borohydride with side products, while ^{p} indicates the value referred to the pure metal borohydride.

Table 6. Hydrogen storage properties of $M(BH_4)_n$ confined by nanoporous materials.

Reaction	Type of Nano Scaffold (size (nm))	Loading Ratio (mass%)	Hydrogen (mass%)		Conditions: temp (K) [pressure (MPa)]		Reference
			Obs (First Dehyd)	Obs (Rehyd)	First Dehyd	Rehyd	
$LiBH_4 = LiH + B + 3/2\ H_2$	Nanoporous Carbon (13–25)	25–50	4.6–6.4 ^{m}		503–873	673 [10]	[212]
	Activated carbon (1.75–3.2)	28.4	11.2 ^{p}	6.6 ^{p}	493–773	573 [5]	[213]
	Mesoporous carbon (4)	33	3.4 ^{m}		473–773		[214]
	Nanoporous Carbon (2)		8.8 ^{p}		493–673		[219]
$Li_3BN_2H_8 = Li_3BN_2 + 4\ H_2$ *	Nanoporous carbon scaffolds (16 ± 3)	33	11.1 ^{p}	3.8 ^{p}	523–673	573 [5]	[160]
	activated carbon AX-21 (2)	33	10.7 ^{p}	4.0 ^{p}	438–673	573 [5]	[160]
$Mg(BH_4)_2 = MgB_2 + 4\ H_2$	Activated carbon (<2)	44 ± 3	6.0 ^{m}		443–773		[216]
$LiBH_4$ + 3.75 mass%Ni	Nanoporous carbon (2–3)	25	14 ^{p}	10 ^{p}	473–673	593 [4]	[220]
$LiBH_4 + 1/2\ MgH_2 = LiH + 1/2\ MgB_2 + 2\ H_2$	Nanoporous carbon aerogel (~21)		4.7 ^{m}	4 ^{m}	533–743	643–663 [5–10]	[221]

* represents that the release of NH_3 in the decomposition process; ^{m} indicates the value referred to the mixture of metal borohydride with side products, while ^{p} indicates the value referred to the pure metal borohydride.

CONCLUSIONS

Metal borohydrides $M(BH_4)_n$ with an extremely high hydrogen density, have been regarded as potential candidates for on-board hydrogen storage. The high reaction temperature and sluggish kinetics, however, decrease their potential for practical applications. In order to overcome these barriers, a number of efforts have been devoted to improving the hydrogen storage properties in both thermodynamic and kinetic areas. The thermodynamic stability of $M(BH_4)_n$ can be predicted and tailored by considering the electronegativity of M. The multistep reaction pathway together with the formation of an intermediate compound such as $M(B_{12}H_{12})_{n/2}$, might account for the high reaction temperature and slow kinetics. The suppression of the formation of $M(B_{12}H_{12})_{n/2}$ can be realized by the combination of $M(BH_4)_n$ with other hydrides or metals. This approach changes the reaction pathway, and therefore, improves the hydrogen storage properties in both thermodynamic and kinetic areas. Regarding the improvement of reaction kinetics, both additives and nanoconfinement have been proved to be vital.

None of the current materials can fulfill the requirements of on-board hydrogen storage for fuel cell vehicles; therefore, continued efforts are required to develop novel materials. According to theoretical predications, many $M(BH_4)_n$ have the potential to release hydrogen at a moderate temperature, and the sluggish kinetics are considered to be mainly responsible for the high reaction temperature. Effective strategies to achieve fast reaction kinetics will be an important research direction to achieve practical success with metal borohydrides.

ACKNOWLEDGMENTS

Financial support received from NEDO (New Energy and Industrial Technology Development Organization, in collaboration with Toyota Central R&D Labs., Inc.); GCOE (Global COE program, Tohoku University); DOE (Department of Energy) and European Commission is greatly appreciated.

REFERENCES

1. Schlapbach, L.; Züttel, A. Hydrogen-storage materials for mobile applications. Nature **2001**,414, 353–358.
2. Orimo, S.; Nakamori, Y.; Eliseo, J.; Züttel, A.; Jensen, C.M. Complex Hydrides for Hydrogen Storage. Chem. Rev. **2007**, 107, 4111–4132.
3. Nakamori, Y.; Orimo, S. Borohydrides as hydrogen storage materials. In Solid-State Hydrogen Storage; Walker, G., Ed.; Woodhead Publishing Limited: Cambridge, UK, 2008; pp. 420–449.
4. Miwa, K.; Ohba, N.; Towata, S.; Nakamori, Y.; Orimo, S. First-principles Study on Lithium Borohydride $LiBH_4$. Phys. Rev. B **2004**, 69, 245120.
5. Ge, Q. Structure and Energetics of $LiBH_4$ and Its surfaces: A First-Principles Study. J. Phys. Chem. A **2004**, 108, 8682–8690.
6. Vajeeston, P.; Ravindran, P.; Kjekshus, A.; Fjellvåg. Structrual Stability of Alkali Boron Tetrahydrides ABH_4 (A = Li, Na, K, Rb, Cs) From First Principle Calculation. J. Alloys Compd.**2005**, 387, 97–104.
7. Łodziana, Z.; Vegge, T. Structural Stability of Complex Hydrides: $LiBH_4$ Revisited. Phys. Rev. Lett. **2004**, 93, 14550.
8. Orgaz, E.; Membrillo, A.; Castañeda, R.; Aburto, A. Electronic Structure of Ternary Hydrides Based on Light Elements. J. Alloys Compd. **2005**, 404–406, 176–180.
9. Schlesinger, H.I.; Brown, H.C.; Finholt, A.E.; Gilbreath, J.R.; Hoekstra, H.R.; Hyde, E.K. Sodium Borohydride, Its Hydrolysis and its Use as a Reducing Agent and in the Generation of Hydrogen. J. Am. Chem. Soc. **1953**, 75, 215–219.
10. Liu, B.H.; Li, Z.P.; Suda, S. Nickel- and cobalt-based catalysts for hydrogen generation by hydrolysis of borohydride. J. Alloys Compd. **2006**, 415, 288–293.
11. Wang, P.; Kang, X.D. Hydrogen-rich boron-containing materials for hydrogen storage.Dalton Trans. **2008**, 40, 5400–5413.
12. Grochala, W.; Edwards, P.P. Thermal Decomposition of the Non-Interstitial Hydrides for the Storage and Production of Hydrogen. Chem. Rev. **2004**, 104, 1283–1315.
13. Sakintuna, B.; Lamari-Darkrimb, F.; Hirscherc, M. Metal Hydride Materials for Solid Hydrogen Storage: A review. Int. J. Hydrogen Energy **2007**, 32, 1121–1140.
14. Chen, P.; Zhu, M. Recent progress in hydrogen storage. Mater. Today. **2008**, 11, 36–43.
15. Rönnebro, E. Development of group II borohydrides as hydrogen storage materials. Curr. Opin. Solid State Mater. Sci. **2010**.
16. George, L.; Saxena, S.K. Structural Stability of Metal Hydrides, Alanates and Borohydrides of Alkali and Alkali-Earth Elements: A Review. Int. J. Hydrogen Energy **2010**, 35, 5454–5470.
17. Jain, I.P.; Jain, P.; Jain, A. Novel hydrogen storage materials: A review of lightweight complex hydrides. J. Alloys Compd. **2010**, 503, 303–339.
18. Hagemann, H.; Černý, R. Synthetic approaches to inorganic borohydrides. Dalton Trans.**2010**, 39, 6006–6012.
19. Züttel, A.; Wenger, P.; Rentsch, S.; Sudan, P. $LiBH_4$ A New Hydrogen Storage Material. J. Power Sources **2003**, 118, 1–7.
20. Züttel, A.; Rentsch, S.; Fischer, P.; Wenger, P.; Sudan, P.; Mauron, Ph.; Emmenegger, Ch. Hydrogen Storage Properties of $LiBH_4$. J. Alloys Compd. **2003**, 356-357, 515–520.
21. Nakamori, Y.; Orimo, S. Destabilization of Li-based Complex Hydrides. J. Alloys Compd.**2004**, 370, 271–275.
22. Orimo, S.; Nakamor, Y.; Züttel, A. Material Properties of MBH_4 (M = Li, Na, and K). Mater.Sci. Eng. B **2004**, 108, 51–53.
23. Orimo, S.; Nakamori, Y.; Kitahara, G.; Miwa, K.; Ohba, N.; Towata, S.; Züttel, A. Dehydriding and Rehydriding Reactions of $LiBH_4$. J. Alloys Compd. **2005**, 404-406, 427–430.

24. Nakamori, Y.; Orimo, S.; Tsutaoka, T. Dehydriding Reaction of Metal Hydrides and Alkali Borohydrides Enhanced by Microwave Irradiation. Appl. Phys. Lett. **2006**, 88, 112104.

25. Züttel, A.; Borgschulte, A.; Orimo, S. Tetrahydroborates as New Hydrogen Storage Materials. Scripta Mater. **2007**, 56, 823–828.

26. Orimo, S.; Nakamori, Y.; Ohba, N.; Miwa, K.; Aoki, M.; Miwa, S.; Züttel, A. Experimental Studies on Intermediate Compound of $LiBH_4$. Appl. Phys. Lett. **2006**, 89, 021920.

27. Ohba, N.; Miwa, K.; Aoki, M.; Noritake, T.; Towata, S.; Nakamori, Y.; Orimo, S.; Züttel, A. First-Principles Study on the Stability of Intermediate Compounds of $LiBH_4$. Phys. Rev. B**2006**, 74, 075110.

28. Hwang, S.-J.; Bowman, R.C., Jr.; Reiter, J.W.; Rijssenbeek, J.; Soloverchik, G.L.; Zhao, J.-C.; Kabbour, H.; Ahn, C.C. NMR Confirmation for Formation of $[B_{12}H_{12}]^{2-}$ Complexes during Hydrogen Desorption from Metal Borohydrides. J. Phys. Chem. C **2008**, 112, 3164–3169.

29. Ozolin, V.; Majzoub, E.H.; Wolverton, C. First-Principles Predition of Thermodynamically Reversible Hydrogen Storage Reactions in the Li-Mg-Ca-B-H. System. J. Am. Chem. Soc. **2009**,131, 230–237.

30. Her, J.-H.; Yousufuddin, M.; Zhou, W.; Jalisatgi, S.S.; Kulleck, J.G.; Zan, J.A.; Hwang, S.-J.; Bowman, R.C., Jr.; Udovic, T.J. Crystal Structure of $Li_2B_{12}H_{12}$: A Possible Intermediate Phase in the Decomposition of $LiBH_4$. Inorg. Chem. **2008**, 47, 9757–9759.

31. Friedrichs, O.; Remhof, A.; Hwang, S.-J.; Züttel, A. Role of $Li_2B_{12}H_{12}$ for the Formation and Decomposition of $LiBH_4$. Chem. Mater. **2009**, 22, 3265–3268.

32. Soulié, J-Ph.; Renaudin, G.; Černy, R.; Yvon, K. Lithium Boro-hydride $LiBH_4$. I. Crystal Structure. J. Alloys Compd. **2002**, 346, 200–205.

33. Renaudin, G.; Gomes, S.; Hagemann, H.; Keller, L.; Yvon, K. Structural and Spectroscopic Studies on the Alkali Borohydrides MBH_4 (M = Na, K, Rb, Cs). J. Alloys Compd. **2004**, 375, 98–106.

34. Frankcombe, T.J.; Kroes, G.-J. Quasiharmonic Approximation Applied to $LiBH_4$ and Its Decomposition Products. Phys. Rev. B **2006**, 73, 174302.

35. Hartman, M.R.; Rush, J.J.; Udovic, T.J.; Bowman, R.C., Jr.; Hwang, S.-J. Structure and Vibrational Dynamics of Isotopically Labeled Lithium Borohydride Using Neutron Diffraction and Spectroscopy. J. Solid State Chem. **2007**, 180, 1298–1305.

36. Łodziana, Z.; Züttel, A.; Zielinski, P. Titanium and Native Defects in $LiBH_4$ and $NaAlH_4$. J. Phys: Condens. Matter. **2008**, 20, 465210.

37. Borgschulte, A.; Züttel, A.; Hug, P.; Racu, A.-M.; Schoenes, J. Hydrogen-Deuterium Exchange in Bulk $LiBH_4$. J. Phys. Chem. A **2008**, 112, 4749–4753.

38. Buchter, F.; Łodziana, Z.; Mauron, Ph.; Remhof, A.; Friedrichs, Q.; Borgschulte, A.; Züttel, A.; Sheptyakov, D.; Strässle, Th.; Ramirez-Cuesta, A.J. Dynamical Properties and Temperature Induced Molecular Disordering of $LiBH_4$ and $LiBD_4$. Phys. Rev. B **2008**, 78, 094302.

39. Skripov, A.V.; Soloninin, A.V.; Filinchuk, Y.; Chernyshov, D. Nuclear Magnetic Resonance Study of Rotational Motion and The Phase transition in $LiBH_4$. J. Phys. Chem. C **2008**, 112, 18701–18705.

40. Corey, R.L.; Shane, D.T.; Bowman, R.C., Jr.; Conradi, M.S. Atomic Motions in $LiBH_4$ by NMR. J. Phys. Chem. C **2008**, 112, 18706–18710.

41. Gremaud, R.; Łodziana, Z.; Hug, P.; Willenberg, B.; Racu, A.-M.; Schoenes, J.; Ramirez-Cuesta, A.J.; Clark, S.J.; Refson, K.; Züttel, A.; Borgschulte, A. Evidence for Hydrogen Transport in Deuterated $LiBH_4$ from Low-Temperature Raman-Scattering Measurement and First-Principles Calculations. Phys. Rev. B **2008**, 80, 100301.

42. Filinchuk, Y.; Chernyshov, D.; Nevidomskyy, A.; Dmitriev, V. High-Pressure Polymorphism as a Step towards Destabilization of $LiBH_4$. Angew. Chem. Int. Ed. **2008**, 47, 529–532.

43. Shane, D.T.; Bowman, R.C., Jr.; Conradi, M.S. Exchange of Hydrogen Atoms between BH_4 in $LiBH_4$. J. Phys. Chem. C **2009**, 113, 5039–5042.

44. Ramzan, M.; Ahuja, R. Ab initio Molecular Dynamics Study of the Hydrogen-Deuterium Exchange in Bulk Lithiumborohydride ($LiBH_4$). Appl. Phys. Lett. **2009**, 94, 141903.

45. Hagemann, H.; Filinchuk, Y.; Chernyshov, D.; van Beek, W. Lattice Anharmonicity and Structural Evolution of $LiBH_4$: an Insight from Raman and X-ray Diffraction Experiments.Phase Transitions **2009**, 82, 344–355.

46. Andresen, E.R.; Gremaud, R.; Borgschulte, A.; Ramirez-Cuesta, A.J.; Züttel, A.; Hamm, P. Vibrational Dynamics of $LiBH_4$ by Infrared Pump-Probe and 2D Spectroscopy. J. Phys. Chem. A **2009**, 113, 12838–12846.

47. Hao, S.; Sholl, D.S. The Role of Interstitial H_2 in Hydrogen Diffusion in Light Metal Borohydrides. Phys. Chem. Chem. Phys. **2009**, 11, 11106–11109.

48. Galvez-Ruiz, J.C.; Sanchez, M. Structural Analysis of Alkali Metal Tetrahydroborates: The Role of Metal and Coordination from in the $[BH_4]^-$ Anion Structure. J. Mol. Struc.: Theochem.**2009**, 908, 114–116.

49. Remhof, A.; Gremaud, R.; Buchter, F.; Lodziana, Z.; Embs, J.P.; Ramirez-Cuesta, A.J.; Borgschulte, A.; Züttel, A. Hydrogen Dynamics in Lightweight Tetrahydroborates. Z. Phys. Chem. **2010**, 224, 263–278.

50. Remhof, A.; Lodziana, Z.; Martelli, P.; Friedrichs, O.; Züttel, A.; Skripov, A.V.; Embs, J.P.; Strässle, T. Rotational Motion of BH_4 Units in MBH_4 (M = Li, Na, K) from Quasielastic Neutron Scattering and Density Functional Calculations. Phys. Rev. B **2010**, 81, 214304.

51. Mauron, P.; Buchter, F.; Friedrichs, O.; Remhof, A.; Bielmann, M.; Zwichy, C.N.; Züttel, A. Stability and Reversibility of $LiBH_4$. J. Phys. Chem. B **2008**, 112, 906–910.

52. Friedrichs, O.; Buchter, F.; Borgschulte, A.; Remhof, A.; Zwicky, C.N.; Mauron, P.H.; Bielmann, M.; Züttel, A. Direct Synthesis of $Li[BH_4]$ and $Li[BD_4]$ from the Elements. Acta Mater. **2007**, 56, 949–954.

53. Remhof, A.; Friedrichs, O.; Buchter, F.; Mauron, P.H.; Züttel, A.; Wallacher, D. Solid-State Synthesis of $LiBD_4$ Observed by In Situ Neutron Diffraction. Phys. Chem. Chem. Phys. **2008**,10, 5859–5862.

54. Friedrichs, O.; Borgschulte, A.; Kato, S.; Buchter, F.; Gremaud, R.; Remhof, A.; Züttel, A. Low-Temperature Synthesis of $LiBH_4$ by Gas-Solid Reaction. Chem. Eur. J. **2009**, 15, 5531–5534.

55. Friedrichs, O.; Remhof, A.; Borgschulte, A.; Buchter, F.; Orimo, S.; Züttel, A. Breaking the passivation—the road to a solvent fee borohydride synthesis. Phys. Chem. Chem. Phys. **2010**,12, 10919–10922.

56. Matsunaga, T.; Buchter, F.; Mauron, P.; Bielman, M.; Nakamori, Y.; Orimo, S.; Ohba, N.; Miwa, K.; Towata, S.; Züttel, A. Hydrogen Storage Properties of $Mg(BH_4)_2$. J. Alloys Compd.**2008**, 459, 583–588.

57. Chłopek, K.; Frommen, C.; Léon, A.; Zabara, O.; Fichtner, M. Synthesis and Properties of Magnesium Tetrahydroborate, $Mg(BH_4)_2$. J. Mater. Chem. **2007**, 17, 3496–3503.

58. Li, H.-W.; Kikuchi, K.; Nakamori, Y.; Miwa, K.; Towata, S.; Orimo, S. Effects of Ball Milling and Additives on Dehydriding Behaviors of Well-crystallized $Mg(BH_4)_2$. Scripta Mater. **2007**,57, 679–682.

59. Riktor, M.D.; Sørby, M.H.; Chłopek, K.; Fichtner, M.; Buchter, F.; Züttel, A.; Hauback, B.C. In Situ Synchrotron Diffraction Studies of Phase Transitions and Thermal Decomposition of $Mg(BH_4)_2$ and $Ca(BH_4)_2$. J. Mater. Chem. **2007**, 17, 4939–4942.

60. Matsunaga, T.; Buchter, F.; Miwa, K.; Towata, S.; Orimo, S.; Züttel, A. Magnesium borohydride: A New Hydrogen Storage Material. Renew. Energy **2008**, 33, 193–196.

61. Varin, R.A.; Chiu, Ch.; Wronski, Z.S. Mechano-Chemical Activation Synthesis (MCAS) of Disordered $Mg(BH_4)_2$ Using $NaBH_4$. J. Alloys Compd. **2008**, 462, 201–208.

62. Li, H.-W.; Kikuchi, K.; Nakamori, Y.; Ohba, N.; Miwa, K.; Towata, S.; Orimo, S. Dehydriding and Rehydriding Processes of Well-Crystallized $Mg(BH_4)_2$ Accompanying with Formation of Intermediate Compounds. Acta Mater. **2008**, 56, 1342–1347.

63. Hanada, N.; Ch1opek, K.; Frommen, C.; Lohstroh, W.; Fichtner, M. Thermal Decomposition of $Mg(BH_4)_2$ Under He flow and H_2 Pressure. J. Mater. Chem. **2008**, 18, 2611–2614.

64. Yan, Y.; Li, H.-W.; Nakamori, Y.; Ohba, N.; Miwa, K.; Towata, S.; Orimo, S. Differential Scanning Calorimetry Measurements of Magnesium Borohydride $Mg(BH_4)_2$. Mater. Trans.**2008**, 49, 2751–2752.

65. Soloveichika, G.L.; Gao, Y.; Rijssenbeek, J.; Andrus, M.; Kniajanski, S.; Bowman, R.C., Jr.; Hwangc, S.-J.; Zhao, J.-C. Magnesium Borohydride as a Hydrogen Storage Material: Properties and Dehydrogenation Pathway of Unsolvated $Mg(BH_4)_2$. Int. J. Hydrogen Energy**2009**, 34, 916–928.

66. Li, H.-W.; Miwa, K.; Ohba, N.; Fujita, T.; Sato, T.; Yan, Y.; Towata, S.; Chen, M.W.; Orimo, S. Formation of Intermediate Compound with $B_{12}H_{12}$ Cluster: Experimental and Theoretical Studies on Magnesium Borohydride $Mg(BH_4)_2$. Nanotechnology **2009**, 20, 204013.

67. Ozolins, V.; Majzoub, E.H.; Wolverton, C. First-Principles Prediction of a Ground State Crystal Structure of Magnesium Borohydride. Phys. Rev. Lett. **2008**, 100, 135501.

68. van Setten, M.J.; Lohstroh, W.; Fichtner, M. A New Phase in the Decomposition of $Mg(BH_4)_2$: First-Principles Simulated Annealing. J. Mater. Chem. **2009**, 19, 7081–7087.

69. Kulkarni, A.D.; Wang, L.L.; Johnson, D.D.; Sholl, D.S.; Johnson, J.K. First-principles Characterization of Amorphous Phases of $MB_{12}H_{12}$, M = Mg, Ca. J. Phys. Chem. C **2010**, 114, 14601–14605.

70. Li, S.; Willis, M.; Jena, P. Reaction Intermediates during the Dehydrogenation of Metal Borohydrides: A Cluster Perspective. J. Phys. Chem. C **2010**, 114, 16849–16854.

71. Chen, X.; Lingam, H.K.; Huang, Z.; Yisgedu, T.; Zhao, J.-C.; Shore, S.G. Thermal Decomposition Behavior of Hydrated Magnesium Dodecahydrododecaborates. J. Phys. Chem. Lett. **2010**, 1, 201–204.

72. Newhouse, R.J.; Stavila, V.; Hwang, S.-J.; Klebanoff, L.E.; Zhang, J.Z. Reversibility and Improved Hydrogen Release of Magnesium Borohydride. J. Phys. Chem. C **2010**, 114, 5224–5232.

73. Severa, G.; Rönnebro, E.; Jensen, C.M. Direct Hydrogenation of Magnesium Boride to Magnesium Borohydride: Demonstration of >11 Weight Percent Reversible Hydrogen Storage. Chem. Commun. **2010**, 46, 421–423.

74. Li, H.-W.; Matsunaga, T.; Yan, Y.; Maekawa, H.; Ishikiriyama, M.; Orimo, S. Nanostructure-induced hydrogenation of layered compound MgB_2. J. Alloys Compd. **2010**, 505, 654–656.

75. Pistidda, C.; Garroni, S.; Dolci, F.; Bardají, E.G.; Khandelwal, A.; Nolis, P.; Dornheim, M.; Gosalawit, R.; Jensen, T.; Cerenius, Y.; Suriñach, S.; Baró, M.D.; Lohstroh, W.; Fichtner, M. Synthesis of amorphous $Mg(BH_4)_2$ from MgB_2 and H_2 at room temperature. J. Alloys Compd.**2010**, 508, 212–215.

76. Chong, M.; Karkamkar, A.; Autrey, T.; Orimo, S.; Jalisatgi, S.; Jensen, C.M. Reversible dehydrogenation of magnesium borohydride to magnesium triborane in the solid state under moderate conditions. Chem. Commun. **2011**, 47, 1330–1332.

77. Miwa, K.; Aoki, M.; Noritake, T.; Ohba, N.; Nakamori, Y.; Towata, S.; Züttel, A.; Orimo, S. Thermodynamical Stability of Calcium Borohydride $Ca(BH_4)_2$. Phys. Rev. B **2006**, 74, 155122.

78. Kim, J.-H.; Jin, S.-A.; Shim, J.-H.; Cho, Y.W. Thermal Decomposition Behavior of Calcium Borohydride $Ca(BH_4)_2$. J. Alloys Compd. **2008**, 461, L20–L22.

79. Aoki, M.; Miwa, K.; Noritake, T.; Ohba, N.; Matsumoto, M.; Li, H.-W.; Nakamori, Y.; Towata, S.; Orimo, S. Structural and Dehydriding Properties of $Ca(BH_4)_2$. Appl. Phys. A **2008**, 92, 601–605.

80. Mao, J.; Guo, Z.; Poh, C.K.; Ranjbar, A.; Guo, Y.; Yu, X.; Liu, H. Study on the dehydrogenation kinetics and thermodynamics of $Ca(BH_4)_2$. J. Alloys Compd. **2010**, 500, 200–205.

81. Riktor, M.D.; Sørby, M.H.; Chopek, K.; Fichtner, M.; Haubac, B.C. The Identification of a Hitherto Unknown Intermediate Phase CaB_2H_X from Decomposition of $Ca(BH_4)_2$. J. Mater. Chem. **2009**, 19, 2754–2759.

82. Frankcombe, T.J. Calcium Borohydride for Hydrogen Storage: A Computational Study of $Ca(BH_4)_2$ Crystal Structures and the CaB_2H_x Intermediate. J. Phys. Chem. C **2010**, 114, 9503–9509.

83. Wang, L.-L.; Graham, D.D.; Robertson, I.M.; Johnson, D.D. On the Reversibility of Hydrogen-Storage Reactions in $Ca(BH_4)_2$: Characterization via Experiment and Theory. J. Phys. Chem. C **2009**, 113, 20088–20096.

84. Stavila, V.; Her, J.-H.; Zhou, W.; Hwang, S.-J.; Kim, C.; Ottley, L.A.M.; Udovic, T.J. Probing the structure, stability and hydrogen storage properties of calcium dodecahydro-closo-dodecaborate. J. Solid State Chem. **2010**, 183, 1133–1140.

85. Rönnebro, E.; Majzoub, E.H. Calcium Borohydride for Hydrogen Storage: Catalysis and Reversibility. J. Phys. Chem. B **2007**, 111, 12045–12047.

86. Kim, J.-H.; Jin, S.-A.; Shim, J.-H.; Cho, Y.W. Reversible Hydrogen Storage in Calcium Borohydride $Ca(BH_4)_2$. Scripta Mater. **2008**, 58, 481–483.

87. Kim, J.-H.; Shim, J.-H.; Cho, Y.W. On the Reversibility of Hydrogen Storage in Ti- and Nb-catalyzed $Ca(BH_4)_2$. J. Power Sources. **2008**, 181, 140–143.

88. Sato, T.; Miwa, K.; Nakamori, Y.; Ohoyama, K.; Li, H.-W.; Noritake, T.; Aoki, M.; Towata, S.; Orimo, S. Experimental and computational studies on solvent-free rare-earth metal borohydrides $R(BH_4)_3$ (R = Y, Dy, and Gd). Phys. Rev. B **2008**, 77, 104114.

89. Yan, Y.; Li, H.-W.; Sato, T.; Umeda, N.; Miwa, K.; Towata, S.; Orimo, S. Dehydriding and rehydriding properties of yttrium borohydride $Y(BH_4)_3$ prepared by liquid-phase synthesis.Int. J. Hydrogen Energy **2009**, 34, 5732–5736.

90. Frommen, C.; Aliouane, N.; Deledda, S.; Fonnelop, J.E.; Grove, H.; Lieutenant, K.; Llamas-Jansa, I.; Sartori, S.; Sorby, M.H.; Hauback, B.C. Crystal structure, polymorphism, and thermal properties of yttrium borohydride $Y(BH_4)_3$. J. Alloys Compd. **2010**, 496, 710–716.

91. Ravnsbæk, D.B.; Filinchuk, Y.; Cerny, Radovan; Ley, M.B.; Haase, D.; Jakobsen, H.J.; Skibsted, J.; Jensen, T.R. Thermal Polymorphism and Decomposition of $Y(BH_4)_3$. Inorg. Chem. **2010**, 49, 3801–3809.

92. Lee, Y.-S.; Shim, J.-H.; Cho, Y.W. Polymorphism and Thermodynamics of $Y(BH_4)_3$ from First Principles. J. Phys. Chem. C **2010**, 114, 12833–12837.

93. Jaroń, T.; Grochala, W. $Y(BH_4)_3$—an Old–New Ternary Hydrogen Store aka Learning from a Multitude of Failures. Dalton Trans. **2010**, 39, 160–166.

94. Gennari, F.C.; Esquivel, M.R. Synthesis and Dehydriding Process of Crystalline $Ce(BH_4)_3$. J. Alloys Compd. **2009**, 485, L47–L51.

95. Jeon, E.; Cho, Y.W. Mechanochmical Synthesis and Thermal Decomposition of Zinc Borohydride. J. Alloys Compd. **2006**, 422, 273–275.

96. Srinivasan, S.; Escobar, D.; Jurczyk, M.; Goswami, Y.; Stefanakos, E. Nanocatalyst Doping of $Zn(BH_4)_2$. J. Alloys. Compd. **2008**, 462, 294–302.

97. Titov, L.V.; Eremin, E.R. Complex of Aluminum Borohydride with Calcium Borohydride.Russ. Chem. B. **1975**, 24, 1095–1096.

98. Fang, Z.Z.; Ma, L.P.; Kang, X.D.; Wang, P.J.; Wang, P.; Cheng, H.M. In situ Formation and Rapid Decomposition of $Ti(BH_4)_3$ by Mechanical Milling $LiBH_4$ with TiF_3. Appl. Phys. Lett.**2009**, 94, 044104.

99. Gennari, F.C.; Albanesi, L.F.; Rios, I.J. Synthesis and Thermal Stability of $Zr(BH_4)_4$ and $Zr(BD_4)_4$ Produced by Mechanochemical Processing. Inorg. Chim. Acta. **2009**, 362, 3731–3737.

100. Łodziana, Z. Multivalent metal tetrahydroborides of Al, Sc, Y, Ti, and Zr. Phy. Rev. B **2010**,81, 144108.

101. Choudhury, P.; Srinivasan, S.S; Bhethanabotla, V.R.; Goswami, Y.; McGrath, K.; Stefanakos, E.K. Nano-Ni Doped Li-Mn-B-H System as a New Hydrogen Storage Candidate. Int. J. Hydrogen Energy. **2009**, 34, 6325–6334.

102. Varin, R.A.; Zbroniec, L. The Effects of Ball Milling Nanometric Nickel on the Hydrogen Desorption from Lithium Borohydride and Manganese Chloride (3 $LiBH_4$ + $MnCl_2$) Mixture.Int. J. Hydrogen Energy. **2010**, 35, 3588–3597.

103. Černý, R.; Penin, N.; Hagemann, H.; Filinchuk, Y. The First Crystallographic and Spectroscopic Characterization of a 3d-Metal Borohydride: $Mn(BH_4)_2$. J. Phys. Chem. C **2009**,113, 9003–9007.

104. Nakamori, Y.; Miwa, K.; Ninoyiya, A.; Li, H.; Ohba, N.; Towata, S.; Züttel, A.; Orimo, S. Correlation between Thermodynamical Stabilities of Metal Borohydrides and Cation Electronegativities: First-Principles Calculations and Experiments. Phys. Rev. B **2006**, 74, 045126.

105. Nakamori, Y.; Li, H.-W.; Mastuo, M.; Miwa, K.; Towata, S.; Orimo, S. Development of Metal Borohydrides for Hydrogen Storage. J. Phys. Chem. Sol. **2008**, 69, 2292–2296.

106. Sugiyama, J.; Ikedo, Y.; Noritake, T.; Ofer, O.; Goko, T.; Månsson, M.; Miwa, K.; Ansaldo, E.J.; Brewer, J.H.; Chow, K.H.; Towata, S. Microscopic Indicator for Thermodynamic Stability of Hydrogen Storage Materials Provided by Positive Muon-Spin Rotation. Phys. Rev. B **2010**, 81, 092103.

107. Li, H.-W.; Orimo, S.; Nakamori, Y.; Miwa, K.; Ohba, N.; Towata, S.; Züttel, A. Materials Designing of Metal Borohydrides: Viewpoints from Thermodynamically Stabilities. J. Alloys Compd. **2007**, 446-447, 315–318.

108. Nickels, E.A.; Jones, M.O.; David, W.I.F.; Johnson, S.R.; Lowton, R.L.; Sommariva, M.; Edwards, P.P. Tuning the Decomposition Temperature in Complex Hydrides:

Synthesis of a Mixed Alkali Metal Borohydride. Angew. Chem. Int. Ed. **2008**, 47, 2817–2819.

109. Hagemann, H.; Longhini, M.; Kaminski, J.W.; Wesolowski, T.A.; Černy, R.; Penin, N.; Sørby, M.H.; Hauback, B.C.; Severa, G.; Jensen, C.M. $LiSc(BH_4)_4$: A Novel Salt of Li^+ and Discrete $Sc(BH_4)_4^-$ Complex Anions. J. Phys. Chem. A **2008**, 112, 7551–7555.

110. Seballos, J.; Zhang, J.Z.; Rönnebro, E.; Herberg, J.L.; Majzoub, E.H. Metastabilility and Crystal Structure of the Bialkali Complex Metal Borohydride $NaK(BH_4)_2$. J. Alloys Compd.**2009**, 476, 446–450.

111. Kim, C.; Hwang, S.-J.; Bowman, R.C., Jr.; Reiter, J.W.; Zan, J.A.; Kulleck, J.G.; Kabbour, H.; Majzoub, E.H.; Ozolins, V. $LiSc(BH_4)_4$ as a Hydrogen Storage Material: Multinuclear High-Resolution Solid-State NMR and First-Principles Density Functional Theory Studies. J. Phys. Chem. C **2009**, 113, 9956–9968.

112. Ravnsbæk, D.; Filinchuk, Y.; Cerenius, Y.; Jakobsen, H.J.; Besenbacher, F.; Skibsted, J.; Jensen, T.R. A Series of Mixed-Metal Borohydrides. Angew. Chem. Int. End. **2009**, 48, 6659–6663.

113. Černý, R.; Severa, G.; Ravnsbæk, D.B.; Filinchuk, Y.; D'Anna, V.; Hagemann, H.; Haase, D.; Jensen, C.M.; Jensen, T.R. $NaSc(BH_4)_4$: A Novel Scandium-Based Borohydride. J. Phys. Chem. C **2010**, 114, 1357–1364.

114. Severa, G.; Hagemann, H.; Longhini, M.; Kaminski, J.W.; Wesolowski, T.A.; Jensen, C.M. Thermal Desorption, Vibrational Spectroscopic, and DFT Computational Studies of the Complex Manganese Borohydrides $Mn(BH_4)_2$ and $[Mn(BH_4)_4]^{2-}$. J. Phys. Chem. C **2010**, 114, 15516–15521.

115. Fang, Z.Z.; Kang, X.D.; Wang, P.; Li, H.-W.; Orimo, S. Unexpected Dehydrogenation Behavior of $LiBH_4$/$Mg(BH_4)_2$ Mixture Associated with the In Situ Formation of Dual-Cation Borohydride. J. Alloys Compd. **2010**, 491, L1–L4.

116. Lee, J.Y.; Ravnsbæk, D.; Lee, Y.-S.; Kim, Y.; Cerenius, Y.; Shim, J.-H.; Jensen, T.R.; Hur, N.H.; Cho, Y.W. Decomposition Reactions and Reversibility of the $LiBH_4$-$Ca(BH_4)_2$ Composite. J. Phys. Chem. C **2009**, 113, 15080–15086.

117. Černý, R.; Ravnsbæk, D.B.; Severa, G.; Filinchuk, Y.; D'Anna, V.; Hagemann, H.; Haase, D.; Skibsted, J.; Jensen, C.M.; Jensen, T.R. Structure and Characterization of $KSc(BH_4)_4$. J. Phys. Chem. C **2010**, 114, 19540–19549.

118. Vajo, J.J.; Olson, G.L. Hydrogen Storage in Destabilized Chemical Systems. Scripta Mater.**2007**, 56, 829–834.

119. NIST Webbook. Available online: http://webbook.nist.gov/chemistry (accessed on 12 January 2011).

120. Vajo, J.J.; Skeith, S.L.; Mertens, F. Reversible Storage of Hydrogen in Destabilized $LiBH_4$. J. Phys. Chem. B **2005**, 109, 3719–3722.

121. Aoki, M.; Miwa, K.; Noritake, T.; Kitahara, G.; Nakamori, Y.; Orimo, S.; Towata, S. Destabilization of $LiBH_4$ by Mixing with $LiNH_2$. Appl. Phys. A **2005**, 80, 1409–1412.

122. Johnson, S.R.; Anderson, P.A.; Edwards, P.P.; Gameson, I.; Prendergast, J.W.; Al-Mamouri, M.; Book, D.; Harris, I.R.; Speight, J.D.; Walton, A. Chemical Activation of MgH_2, a New route to Superior Hydrogen Storage Materials. Chem. Commun. **2005**, 22, 2823–2825.

123. Pinkerton, F.E.; Meisner, G.P.; Meyer, M.S.; Balogh, M.P.; Kundrat, M.D. Hydrogen Desorption Exceeding Ten Weight Percent from the New Quaternary Hydride $Li_3BN_2H_8$. J. Phys. Chem. B **2005**, 109, 6–8.

124. Alapati, S.V.; Johnson, J.K.; Sholl, D.S. Identification of Destabilized Metal Hydrides for Hydrogen Storage Using First Principles Calculations. J. Phys. Chem. B **2006**, 110, 8769–8776.

125. Cho, Y.W.; Shim, J.-H.; Lee, B.-J. Thermal Destabilization of Binary and Complex Metal Hydrides by Chemical Reaction: A Thermodynamic Analysis. Comput. Coupling Phase Diagr. Thermochem. **2006**, 30, 65–59.

126. Yu, X.B.; Grant, D.M.; Walker, G.S. A New Dehydrogenation Mechanism for Reversible Multicomponent Borohydride Systems—the Role of Li-Mg Alloys. Chem. Commun. **2006**, 37, 3906–3908.

127. Nakamori, Y.; Ninomiya, A.; Kitahara, G.; Aoki, M.; Noritake, T.; Miwa, K.; Kojima, Y.; Orimo, S. Dehydriding Reactions of Mixed Complex Hydrides. J. Power Sources. **2006**, 155, 447–455.

128. Meisner, G.P.; Scullin, M.L.; Balogh, M.P.; Pinkerton, F.E.; Meyer, M.S. Hydrogen Release from Mixtures of Lithium Borohydride and Lithium Amide: A Phase Diagram Study. J. Phys. Chem. B **2006**, 110, 4189–4192.

129. Noritake, T.; Aoki, M.; Towata, S.; Ninimiya, A.; Nakamori, Y.; Orimo, S. Crystal Structure Analysis of Novel Complex Hydrides Formed by Combination of $LiBH_4$ an $LiNH_2$. Appl. Phys. A **2006**, 83, 277–279.

130. Pinkerton, F.E.; Meyer, M.S.; Meisner, G.P.; Balogh, M.P. Improved Hydrogen Release from $LiB_{0.33}N_{0.67}H_{2.67}$ with Novel Metal Additives. J. Phys. Chem. B **2006**, 110, 7967–7974.

131. Chater, P.A.; David, W.I.F.; Johnson, S.R.; Edwards, P.P.; Anderson, P.A. Synthesis and Crystal Structure of $Li_4BH_4(NH_2)_3$. Chem. Commun. **2006**, 23, 2439–2441.

132. Noritake, T.; Aoki, M.; Towata, S.; Ninomiya, A.; Nakamori, Y.; Orimo, S. Crystal Structure Analysis of Novel Complex Hydrides Formed by the Combination of $LiBH_4$ and $LiNH_2$.Appl. Phys. A **2006**, 83, 277–279.

133. Alapati, S.V.; Johnson, J.K.; Sholl, D.S. Stability Analysis of Doped Materials for Reversible Hydrogen Storage in Destabilized Metal Hydrides. Phys. Rev. B **2007**, 76, 104108.

134. Yang, J.; Sudik, A.; Wolverton, C. Destabilizing $LiBH_4$ with a Metal (M = Mg, Al, Ti, V, Cr, or Sc) or Metal Hydride (MH_2 = MgH_2, TiH_2, CaH_2). J. Phys. Chem. **2007**, 111, 19134–19140.

135. Siegel, D.J.; Wolverton, C.; Ozoliņš, V. Thermodynamics Guidelines for the Prediction of Hydrogen Storage Reactions and Their Application to Destabilized Hydride Mixtures. Phys. Rev. B **2007**, 76, 134102.

136. Bösenberg, U.; Doppiu, S.; Mosegaard, L.; Barkhordarian, G.; Eigen, N.; Borgschulte, A.; Jensen, T.R.; Cerenius, Y.; Gutfleisch, O.; Klassen, T.; Dornheim, M.; Bormann,

R. Hydrogen Sorption Properties of MgH_2–$LiBH_4$ Composites. Acta Mater. **2007**, 55, 3951–3958.

137. Kang, X.-D.; Wang, P.; Ma, L.-P.; Cheng, H.-M. Reversible Hydrogen Storage in $LiBH_4$Destabilized by Milling with Al. Appl. Phys. A **2007**, 89, 963–966.

138. Pinkerton, F.E.; Meyer, M.S. Reversible Hydrogen Storage in the Lithium Borohydride—Calcium Hydride Coupled System. J. Alloys Compd. **2008**, 464, L1–L4.

139. Wolverton, C.; Siegel, D.J.; Akbarzadeh, A.R.; Ozoliņš, V. Discovery of Novel Hydrogen Storage Materials: An Atomci Scale Computional Approach. J. Phys.: Condens. Matter **2008**,20, 064228.

140. Sudik, A.; Yang, J.; Halliday, D.; Wolverton, C. Hydrogen Storage Properties in $(LiNH_2)_2$-$LiBH_4$-$(MgH_2)_x$ Mixture (x = 0.0–1.0). J. Phys. Chem. C **2008**, 112, 4384–4390.

141. Yang, J.; Sudik, A.; Siegel, D.J.; Halliday, D.; Drews, A.; Carter, R.O., III; Wolverton, C.; Lewis, G.J.; Sachtler, J.W.A.; Low, J.J.; Faheem, S.A.; Lesch, D.A.; Ozoliņš, V. A Self-Catalyzing Hydrogen-Storage Material. Angew. Chem. Int. Ed. **2008**, 47, 882–887.

142. Alapati, S.V.; Johnson, J.K.; Sholl, D.S. Large-Scale Screening of Metal Hydride Mixtures for High-Capacity Hydrogen Storage from First-Principle Calculations. J. Phys. Chem. C **2008**,112, 5258–5262.

143. Purewal, J.; Hwang, S.-J.; Bowman, R.C., Jr.; Rönnebro, E.; Fultz, B.; Ahn, C. Hydrogen Sorption Behavior of the ScH_2-$LiBH_4$ System: Experimental Assesment of Chemical Destabilization Effects. J. Phys. Chem. C **2008**, 112, 8481–8485.

144. Jin, S.-A.; Lee, Y.-S.; Shim, J.-H.; Cho, Y.W. Reversible Hydrogen Storage in $LiBH_4$-MgH_2 (M = Ce, Ca). J. Phys. Chem. C **2008**, 112, 9520–9524.

145. Wan, X.; Markmaitree, T.; Osborn, W.; Shaw, L.L. Nanoengineering-Enabled Solid-State Hydrogen Uptake and Release in the $LiBH_4$ Plus MgH_2 System. J. Phys. Chem. C **2008**, 112, 18232–18243.

146. Langmi, H.W.; McGrady, G.S. Ternary Nitrides for Hydrogen Storage: Li-B-N, Li-Al-N and Li-Ga-N Systems. J. Alloys Compd. **2008**, 466, 287–292.

147. Jin, S.-A.; Shim, J.-H.; Cho, Y.W.; Yi, K.-W.; Zabara, O.; Fichtner, M. Reversible Hydrogen Storage in $LiBH_4$-Al-LiH Composite Powder. Scripta Mater. **2008**, 58, 963–965.

148. Fan, M.-Q.; Sun, L.-X.; Zhang, Y.; Xu, F.; Zhang, J.; Chu, H.-L. The Catalytic Effect of Additive Nb_2O_5 on the Reversible Hydrogen Storage Performances of $LiBH_4$-MgH_2Composite. Int. J. Hydrogen Energy. **2008**, 33, 74–80.

149. Tang, W.S.; Wu, G.T.; Liu, T.; Wee, A.T.S.; Yong, C.K.; Xiong, Z.; Hor, T.S.; Chen, P. Cobalt-Catalyzed Hydrogen Desorption from the $LiNH_2$-$LiBH_4$ System. Dalton Trans. **2008**, 18, 2395–2399.

150. Bösenberg, U.; Vainio, U.; Pranzas, P.K.; Bellosta von Colbe, J.M.; Goerigk, G.; Welter, E.; Dornheim, M.; Schreyer, A.; Bormann, R. On the Chemical State and Distribution of Zr- and V-Based Additives in Reactive Hydride Composites. Nanotechnology **2009**, 20, 204003.

151. Wang, P.J.; Ma, L.P.; Fang, Z.Z.; Kang, X.D.; Wang, P. Improved Hydrogen Storage Properties of Li-Mg-B-H System by Milling with Titanium Trifluoride. Energy Environ. Sci.**2009**, 2, 120–123.

152. Du, A.J.; Smith, S.C.; Yao, X.D.; Sun, C.H.; Li, L.; Lu, G.Q. First Principle Study of Hydrogen of MgB_2: An Important Step toward Reversible Hydrogen Storage in the Coupled $LiBH_4$/MgH_2 System. J. Nanosci. Nanotechnol. **2009**, 9, 4388–4391.

153. Yang, J.; Hirano, S. Improving the Hydrogen Reaction Kinetics of Complex Hydrides. Adv. Mater. **2009**, 21, 1–6.

154. Graetz, J.; Chaudhuri, S.; Salguero, T.T.; Vajo, J.J.; Meyer, M.S.; Pinkerton, F.E. Local bonding and atomic environments in Ni-catalyzed complex hydrides. Nanotechnology **2009**, 20, 204007.

155. Sudik, A.; Yang, J.; Siegel, D.J.; Wolverton, C.; Carter, R.O., III; Drews, A.R. Impact of Stoichiometry on the Hydrogen Storage Properties of $LiNH_2$-$LiBH_4$-MgH_2 Ternary Composites. J. Phys. Chem. C **2009**, 113, 2004–2013.

156. Walker, G.S.; Grant, D.M.; Price, T.C.; Yu, X.B.; Legrand, V. High Capacity Multicomponent Hydrogen Storage Materials: Investigation of the Effect of Stoichimetry and Decomposition Conditions on the Cycling Behaviour of $LiBH_4$–MgH_2. J. Power Sources **2009**, 194, 1128–1134.

157. Mao, J.F.; Guo, Z.P.; Liu, H.K.; Yu, X.B. Reversible Hydrogen Storage in Titanium-Catalyzed $LiAlH_4$-$LiBH_4$. J. Alloys. Compd. **2009**, 487, 434–438.

158. Puszkiel, J.A.; Gennari, F.C. Reversible Hydrogen Storage in Metal-Doped Mg-$LiBH_4$Composites. Scripta Mater. **2009**, 60, 667–670.

159. Zhang, Y.; Tian, Q.; Chu, H.L.; Zhang, J.; Sun, L.; Sun, J.; Wen, Z. Hydogen De/Resorption Properties of the $LiBH_4$-MgH_2-Al System. J. Phys. Chem. C **2009**, 113, 21964–21969.

160. Wu, H.; Zhou, MI.; Wang, K.; Udovic, T.J.; Rush, J.J.; Yildirim, T.; Bendersky, L.A.; Gross, A.F.; van Atta, S.L.; Vajo, J.J.; Pinkerton, F.E.; Meyer, M.S. Size Effects on the Hydrogen Storage Properties of Nanoscaffolded $Li_3BN_2H_8$. Nanotechnology **2009**, 20, 204002.

161. Pinkerton, F.E.; Meyer, M.S. Hydrogen Desorption Behavior of Nickel-Chloride-Catalyzed Stoichiometric $Li_4BN_3H_{10}$. J. Phys. Chem. C **2009**, 113, 11172–11176.

162. Kim, J.W.; Friedrichs, O.; Ahn, J.-P.; Kim, D.H.; Kim, S.C.; Remhof, A.; Chung, H.-S.; Lee, J.; Shim, J.-H.; Cho, Y.W.; Züttel, A.; Oh, K.H. Microstructural Change of 2$LiBH_4$/Al with Hydrogen Sorption cycling: Separation of Al and B. Scripta Mater. **2009**, 60, 1089–1092.

163. Friedrichs, O.; Kim, J.W.; Remhof, A.; Buchter, F.; Borgschulte, A.; Wallacher, D.; Cho, Y.W.; Fichtner, M.; Oh, K.H.; Züttel, A. The Effect of Al on the Hydrogen Sorption Mechanism of $LiBH_4$. Phys. Chem. Chem. Phys. **2009**, 11, 1515–1520.

164. Blanchard, D.; Shi, Q.; Boothroyd, C.B.; Vegge, T. Reversibility of Al/Ti Modified $LiBH_4$. J. Phys. Chem. C **2009**, 113, 14059–14066.

165. Zhang, Y.; Tian, Q.; Zhang, J.; Liu, S.-S.; Sun, L.-X. The

Dehydrogenation Reactions and Kinetics of $2LiBH_4$-Al Composite. J. Phys. Chem. C **2009**, 113, 18424–18430.

166. Singer, J.P.; Meyer, M.S.; Speer, R.M., Jr.; Fischer, J.E.; Pinkerton, F.E. Determination o the Phase Behavior of $(LiNH_2)_c(LiBH_4)_{1-c}$ Quaternary Hydride through in Situ X-ray Diffraction. J. Phys. Chem. C **2009**, 113, 18927–18934.

167. Li, W.; Vajo, J.J.; Cumberland, R.W.; Liu, P.; Hwang, S.-J.; Kim, C.; Bowman, R.C., Jr. Hydrogenation of Magnesium Nickel Boride for Reversible Hydrogen Storage. Phys. Chem. Lett. **2010**, 1, 69–72.

168. Wang, P.-J.; Fang, Z.-Z.; Ma, L.-P.; Kang, X.-D.; Wang, P. Effect of Carbon Addition on Hydrogen Storage Behaviors of Li-Mg-B-H System. Int. J. Hydrogen Energy **2010**, 35, 3072–3075.

169. Shim, J.-H.; Lim, J.-H.; Rather, S.-U.; Lee, Y.-S.; Reed, D.; Kim, Y.; Book, D.; Cho, Y.W. Effect of Hydrogen Back Pressure on Dehydrogenation Behavior of $LiBH_4$-Based Reactive Hydride Composites. Phys. Chem. Lett. **2010**, 1, 59–63.

170. Deprez, E.; Muñoz-Márquez, M.A.; Roldán, M.A.; Prestipino, C.; Javier Palomares, F.; Bonatto Minella, C.; Bösenberg, U.; Dornheim, M.; Bormann, R.; Fernández, A. Oxidation State and Local Structure of Ti-Based Additives in the Reactive Hydride Composite 2 $LiBH_4$+ MgH_2. J. Phys. Chem. **2010**, 114, 3309–3317.

171. Bösenberg, U.; Kim, J.W.; Gosslar, D.; Eigen, N.; Jensen, T.G.; Bellosta von Colbe, J.M.; Zhou, Y.; Dahms, M.; Kim, D.H.; Günther, R.; Cho, Y.W.; Oh, K.H.; Klassen, T.; Bormann, R.; Dornheim, M. Role of Additives in $LiBH_4$-MgH_2 Reactive Hydride Composites for Sorption Kinetics. Acta Mater. **2010**, 58, 3381–3389.

172. Price, T.C.; Grant, D.M.; Legrand, V.; Walker, G.S. Enhanced Kinetics for the $LiBH_4$:MgH_2Multi-Component Hydrogen Storage System–The Effects of Stoichiometry and Decomposition Environment on Cycling Behaviour. Int. J. Hydrogen Energy **2010**, 35, 4154–4161.

173. Somer, M.; Acar, S.; Koz, C.; Kokal, I.; Höhn, P.; Cardoso-Gil, R.; Aydemir, U.; Akselrud, L. α- and β-$Na_2[BH_4][NH_2]$: Two Modifications of a Complex Hydride in the System $NaNH_2$-$NaBH_4$; Syntheses, Crystal Structures, Thermal Analyses, Mass and Vibrational Spectra. J. Alloys Compd. **2010**, 491, 98–105.

174. Yu, X.B.; Guo, Y.H.; Sun, D.L.; Yang, Z.X.; Ranjbar, A.; Guo, Z.P.; Liu, H.K.; Dou, S.X. A Combined Hydrogen Storage System of $Mg(BH_4)_2$-$LiNH_2$ with Favorable Dehydrogenation.J. Phys. Chem. C **2010**, 114, 4733–4737.

175. Mao, J.; Guo, Z.; Leng, H.; Wu, Z.; Guo, Y.; Yu, X.; Liu, H. Reversible Hydrogen Storage in Destabilized $LiAlH_4$–MgH_2–$LiBH_4$ Ternary-Hydride System Doped with TiF_3. J. Phys. Chem. C **2010**, 114, 11643–11649.

176. Ravnsbæk, D.B.; Jensen, T.R. Tuning hydrogen storage properties and reactivity: Investigation of the $LiBH_4$-$NaAlH_4$ sytem. J. Phys Chem. Sol. **2010**, 71, 1144–1149.

177. Mauron, P.; Bielmann, M.; Remhof, A.; Züttel, A.; Shim, J.-K.; Cho, Y.W. Stability of the $LiBH_4$/CeH_2 Composite System Determined by Dynamic pcT Measurements. J. Phys. Chem. C **2010**, 114, 16801–16805.

178. Bösenberg, U.; Ravnsbæk, D.B.; Hagemann, H.; D'Anna, V.; Minella, C.B.; Pistidda, C.; van Beek, W.; Jensen, T.R.; Bormann, R.; Dornheim, M. Pressure and Temperature Influence on the Desorption Pathway of the $LiBH_4$-MgH_2 Composite System. J. Phys. Chem. C **2010**, 114, 15212–15217.

179. Zeng, L.; Miyaoka, H.; Ichikawa, T.; Kojima, Y. Superior Hydrogen Exchange Effect in the MgH_2-$LiBH_4$ System. J. Phys. Chem. C **2010**, 114, 13132–13135.

180. Crosby, K.; Shaw, L.L. Dehydriding and re-hydriding properties of high-energy ball milled $LiBH_4$ + MgH_2 mixture. Int. J. Hydrogen Energy **2010**, 35, 7519–7529.

181. Choudhury, P.; Bhethanabotla, V.R.; Stefanakos, E. First principles study to identify the reversible reaction step of a multinary hydrogen storage "Li-Mg-B-N-H" system. Int. J. Hydrogen Energy **2010**, 35, 9002–9011.

182. Shaw, L.L.; Wan, X.; Hu, J.Z.; Kwak, J.H.; Yang, Z. Solid-State Hydriding Mechanism in the $LiBH_4$-MgH_2 System. J. Phys. Chem. C **2010**, 114, 8089–8098.

183. Garroni, S.; Milanese, C.; Girella, A.; Marini, A.; Mulas, G.; Menéndez, E.; Pistidda, C.; Dornheim, M.; Suriñach, S.; Baró, M.D. Sorption Properties of $NaBH_4$/MH_2 (M = Mg, Ti) Powder System. Int. J. Hydrogen. Energy **2010**, 35, 5434–5441.

184. Gennari, F.C.; Puszkiel, J.A. Enhanced Hydrogen Sorption Kinetics of $Mg_{50}Ni$-$LiBH_4$. J. Power Sources. **2010**, 195, 3266–3274.

185. Hattrick-Simpers, J.R.; Maslar, J.E.; Niemann, M.U.; Chium, C.; Srinivasan, S.S.; Stefanakos, E.K.; Bendersky, L.A. Raman Spectroscopic Observation of Dehydrogenation in Ball-Milled $LiNH_2$-$LiBH_4$-MgH_2 Nanoparticles. Int. J. Hydrogen Energy. **2010**, 35, 6323–6331.

186. Barkhordarian, G.; Klassen, T.; Dornheim, M.; Bormann, R. Unexpected Kinetic Effect of MgB_2 in Reactive Composites Containing Complex borohydrides. J. Alloys Compd. **2007**, 440, L18–L21.

187. Barkhordarian, G.; Jensen, T.R.; Doppiu, S.; Bo1senberg, U.; Borgschulte, A.; Gremaud, R.; Cerenius, Y.; Dornheim, M.; Klassen, T.; Bormann, R. Formation of $Ca(BH_4)_2$ from Hydrogenation of CaH_2 + MgB_2 Composite. J. Phys. Chem. C **2008**, 112, 2743–2749.

188. Hu, J.Z.; Kwak, J.H.; Y, Z.; Wan, X.; Shaw, L.L. Direct Observation of Ion Exchange in Mechanically Activated LiH + MgB_2 System Using Ultrahigh Field Nuclear Magnetic Resonance Spectroscopy. Appl. Phys. Lett. **2009**, 94, 141905.

189. Gosalawit-Utke, R.; Bellosta von Colbe, J.M.; Dornheim, M.; Jensen, T.R.; Cerenius, Y.; Minella, C.B.; Peschke, M.; Bormann, R. LiF–MgB_2 System for Reversible Hydrogen Storage.J. Phys. Chem. C **2010**, 114, 10291–10296.

190. Gonzalez-Silveira, M.; Gremaud, R.; Schreuders, H.; van Setten, M.J.; Batyrev, E.; Rougier, A.; Dupont, L.; Bardají, E.G.; Lohstroh, W.; Dam, B. In-Situ Deposition of Alkali and Alkaline Earth Hydride Thin

Films To Investigate the Formation of Reactive Hydride Composites. J. Phys. Chem. C **2010**, 114, 13895–13901.

191. Rongeat, C.; D'Anna, V.; Hagemann, H.; Borgschulte, A.; Züttel, A.; Schultz, L.; Gutfleisch, O. Effect of Additives on the Synthesis and Reversibility of $Ca(BH_4)_2$. J. Alloys Compd. **2010**,493, 281–287.

192. Kim, Y.; Reed, D.; Lee, Y.-S.; Shim, J.-H.; Han, H.N.; Book, D.; Cho, Y.W. Hydrogenation Reaction of CaH_2-CaB_6-Mg Mixture. J. Alloys Compd. **2010**, 492, 597–600.

193. Bogdanović, B.; Schwickardi, M. Ti-doped alkali metal aluminium hydrides as potential novel reversible hydrogen storage materials. J. Alloy. Compd. **1997**, 253-254, 1–9.

194. Au, M.; Jurgensen, A. Modified Lithium Borohydrides for Reversible Hydrogen Storage. J. Phys. Chem. B **2006**, 110, 7062–7067.

195. Au, M.; Jurgensen, A.; Zeigler, K. Modified Lithium Borohydrides for Reversible Hydrogen Storage (2). J. Phys. Chem. B **2006**, 110, 26482–26487.

196. Yu, X.B.; Wu, Z.; Chen, Q.R.; Li, Z.L.; Weng, B.C.; Huang, T.S. Improved Hydrogen Storage Properties of $LiBH_4$ Destabilized by Carbon. Appl. Phys. Lett. **2007**, 90, 034106.

197. Zhang, Y.; Zhang, W.-S.; Wang, A.-Q.; Sun, L.-X.; Fan, M.-Q.; Chu, H.-L.; Sun, J-C.; Zhang, T. $LiBH_4$ Nanoparticles Supported by Disordered Mesoporous Carbon: Hydrogen Storage Performances and Destabilization Mechanisms. Int. J. Hydrogen Energy. **2007**, 32, 3976–3980.

198. Varin, R.A.; Chiu, C.; Wronski, Z.S.; Calka, A. The Effects of Oxidized and Oxide-Free Boron on the Mg-B-H Nanohydrides Transformation in the Nearly Nanosized Powders. Solid State Phenom. **2007**, 128, 47–52.

199. Mosegaard, L.; Møller, B.; Jørgensen, J.-E.; Filinchuk, Y.; Cerenius, Y.; Hanson, J.; Dimasi, E.; Besenbacher, F.; Jensen, T.R. Reactivity of $LiBH_4$: In Situ Synchrotron Radian Powder X-ray Diffraction Study. J. Phys. Chem. C **2008**, 112, 1299–1303.

200. Yin, L.; Wang, P.; Fang, Z.Z.; Cheng, H. Thermodynamically Tuning $LiBH_4$ by Fluorine Anion Doping for Hydrogen Storage: A Density functional Study. Chem. Phys. Lett. **2008**,450, 318–321.

201. Zhang, Y.; Zhang, W.-S.; Fan, M.-Q.; Liu, S.-S.; Chu, H.-L.; Zhang, Y.-H.; Gao, X-Y.; Sun, L.-X. Enhanced Hydrogen Storage Performance of $LiBH_4$-SiO_2-TiF_3 Composite. J. Phys. Chem. C**2008**, 112, 4005–4110.

202. Fang, Z.Z.; Kang, X.D.; Dai, H.B.; Zhang, M.J.; Wang, P.; Cheng, H.M. Reversible Dehydrogenation of $LiBH_4$ Catalyzed by As-Prepared Sing-Walled Carbon Nanotubes.Scripta Mater. **2008**, 58, 922–925.

203. Xu, J.; Yu, X.B.; Zou, Z.Q.; Li, Z.L.; Wu, Z.; Akins, L.D.; Yang, H. Enhanced Dehydrogenation of $LiBH_4$ Catalyzed by Carbon-Supported Pt Nanoparticles. Chem. Commun. **2008**, 44, 5740–5742.

204. Fang, Z.-Z.; Kang, X.-D.; Wang, P.; Cheng, H.-M. Improved Reversible Dehydrogenation of Lithium Borohydride by Milling with As-Prepared Single-Walled Carbon Nanotubes. J. Phys. Chem. C **2008**, 112, 17023–17029.

205. Au, M.; Jurgensen, A.R.; Spencer, W.A.; Anto, D.A.; Pinkerton, F.E.; Hwang, S.-J.; Kim, C.; Bowman, R.C., Jr. Stability and Reversibility of Lithium Borohydrides Doped by Metal Halides and Hydrides. J. Phys. Chem. C **2008**, 112, 18661–18671.

206. Yu, X.B.; Grant, D.M.; Walker, G.S. Dehydrogenation of $LiBH_4$ Destabilized with Various Oxides. J. Phys. Chem. C **2009**, 113, 17945–17949.

207. Arnbjerg, L.M.; Ravnsbæk, D.B.; Filinchuk, Y.; Vang, R.T.; Cerenius, Y.; Besenbacher, F.; Jørgensen, J.-E.; Jakobsen, H.J.; Jensen, T.R. Structure and Dynamics for $LiBH_4$-LiCl Solid Solutions. Chem. Mater. **2009**, 21, 5772–5782.

208. Guo, Y.H.; Yu, X.B.; Gao, L.; Xia, G.L.; Guo, Z.P.; Liu, H.K. Significantly Improved Dehydrogenation of $LiBH_4$ Destabilized by TiF_3. Energy Environ. Sci. **2010**, 3, 465–470.

209. Fang, Z.Z.; Kang, X.D.; Wang, P. Improved Hydrogen Storage Properties of $LiBH_4$ by Mechanical Milling with Various Carbon Additives. J. Int. Hydrogen Energy **2010**, 35, 8247–8252.

210. Zhang, B.J.; Liu, B.H. Hydrogen Desorption from $LiBH_4$ destabilized by chlorides of transition metal Fe, Co, and Ni. Int. J. Hydrogen Energy **2010**, 35, 7288–7294.

211. Gutowska, A.; Li, L.; Shin, Y.; Wang, C.M.; Li, X.S.; Linehan, J.C.; Smith, R.S.; Kay, B.D.; Schmid, B.; Shaw, W.; Gutowski, M.; Autrey, T. Nanoscaffold Mediates Hydrogen Release and the Reactivity of Ammonia Borane. Angew. Chem. Int. Ed. **2005**, 44, 3578.

212. Gross, A.F.; Vajo, J.J.; van Atta, S. L.; Olson, G.L. Enhanced Hydrogen Storage Kinetics of $LiBH_4$ in Nanoporous Carbon Scaffolds. J. Phys. Chem. C **2008**, 112, 5651–5657.

213. Fang, Z.Z.; Wang, P.; Rufford, T.E.; Kang, X.D.; Lu, G.Q.; Cheng, H.M. Kinetic- and Thermodynamic-Based Improvements of Lithium Borohydride Incorporated Into Activated Carbon. Acta Mater. **2008**, 56, 6257–6263.

214. Cahen, S.; Eymery, J.-B.; Janot, R.; Tarascon, J.-M. Improvement of the $LiBH_4$ Hydrogen Desorption by Inclusion into Mesoporous Carbons. J. Power Sources **2009**, 189, 902–908.

215. Ingleson, M.J.; Barrio, J.P.; Bacsa, J.; Steiner, A.; Darling, G.R.; Jones, J.T.A.; Khimyak, Y.Z.; Rosseinsky, M.J. Magnesium Borohydride Confined in a Metal–Organic Framework: A Preorganized System for Facile Arene Hydroboration. Angew. Chem. Int. Ed. **2009**, 48, 2012–2016.

216. Fichtner, M.; Zhao-Karger, Z.; Hu, J.; Roth, A.; Weidler, P. The Kinetic Properties of $Mg(BH_4)_2$ Infiltrated in Activated Carbon. Nanotechnology **2009**, 20, 204029.

217. Ngene, P.; Adelhelm, P.; Beale, A.M.; de Jong, K.P.; de Jongh, P.E. $LiBH_4$/SBA-15 Nanocomposites Prepared by Melt infiltration under Hydrogen pressure: Synthesis and Hydrogen Sorption Properties. J. Phys. Chem. C **2010**, 114, 6163–6168.

218. Shane, D.T.; Corey, R.L.; McIntosh, C.; Rayhel, L.H.; Bowman, R.C., Jr.; Vajo, J.J.; Gross, A.F.; Conradi, M.S. $LiBH_4$ in Carbon Aerogel Nanoscaffolds: An NMR Study of Atomic Motions.J. Phys. Chem. C **2010**, 114, 4008–4014.

219. Liu, X.; Peaslee, D.; Jost, C.Z.; Majzoub, E.H. Controlling the Decomposition Pathway of $LiBH_4$ via Confinement in Highly Ordered Nanoporous Carbon. J. Phys. Chem. C **2010**, 114, 14036–14041.

220. Ngene, P.; van Zwienen, M.R.; de Jongh, P.E. Reversibility of the hydrogen desorption from $LiBH_4$: A synergetic effect of nanoconfinement and Ni addition. Chem. Comm. **2010**, 46, 8201–5203.

221. Nielsen, T.K.; Bösenber, U.; Gosalawit, R.; Dornheim, M.; Cerenius, Y.; Besenbacher, F.; Jensen, T.R. A Reversible Nanoconfined Chemical Reaction. ACS NANO **2010**, 4, 3903–3908.

222. Sartori, S.; Knudsen, K.D.; Zhao-karger, Z.; Bardaji, E.G.; Muller, J.; Fichtner, M.; Hauback, B.C.; Sartori, S.; Knudsen, K.D.; Zhao-karger, Z.; Bardaji, E.G.; Muller, J.; Fichtner, M.; Hauback, B.C. Nanoconfined Magnesium Borohydride for Hydrogen Storage Application Investigation by SANS and SAXS. J. Phys. Chem. C **2010**, 114, 18785–18789.

CHAPTER 7

THE CURRENT STATUS OF HYDROGEN STORAGE ALLOY DEVELOPMENT FOR ELECTROCHEMICAL APPLICATIONS

Kwo-hsiung Young and Jean Nei

Keywords: Metal hydride, hydrogen storage alloy, NiMH battery, alkaline fuel cell, electrochemical reaction

INTRODUCTION

Hydrogen storage alloys are important for a few electrochemical applications, especially in the energy storage area. The basic of electrochemical use of the hydrogen storage alloy can be described as follows: when hydrogen enters the lattice of most transition metals, interstitial metal hydride (MH) is formed. The electrons accompanying the hydrogen atoms form a metal-hydrogen band right below the Fermi level, which indicates that the interstitial MH is metallic in nature. While protons in the interstitial MH hop between neighboring occupation sites by quantum mechanical tunneling, the electrons remain within a short distance (3–10 angstroms) of the protons to maintain local charge neutrality. Under the influence of an electric field, electrons and protons will move in opposite directions. In an electrochemical environment, a voltage is applied to cause electrons to flow, and the charges are balanced out by moving conductive ions through a highly alkaline aqueous electrolyte with good ionic conductivity. During charge, a negative voltage (with respect to the counter electrode) is applied to the metal/metal hydride electrode current collector, and electrons enter the metal through the current collector to neutralize the protons from the splitting of water that occurs at the metal/electrolyte interface (Figure 1a). This electrochemical charging process is characterized by the half reaction:

$$M + H_2O + e^- \rightarrow MH + OH^- \quad (1)$$

During discharge, protons in the MH leave the surface and re-combine with OH^- in the alkaline electrolyte to form H_2O, and charge neutrality pushes the electrons out of the MH through the current collector, performing electrical work in the attached circuitry (Figure 1b). The electrochemical discharge process is given by the half reaction:

$$MH + OH^- \rightarrow M + H_2O + e^- \quad (2)$$

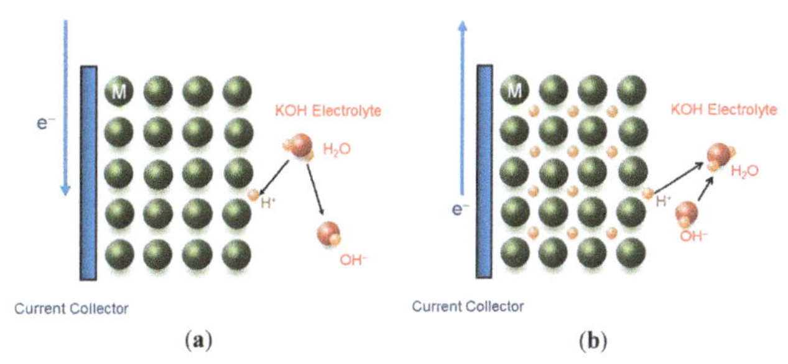

Figure 1. Schematics showing the electrochemical reactions between water and metal hydride during charge (**a**) and discharge (**b**). Due to the alkaline nature of the electrolyte, protons cannot desorb or absorb from the surface of metal without the incorporation of water and OH^-.

The standard potential of this redox half reaction depends on the chosen MH and is usually as low as possible to maximize the amount of stored energy without exceeding the hydrogen evolution potential (−0.83 V *versus* standard hydrogen electrode). Zn is an exception. With a complete 3d

Citation: Kwo-hsiung Young and Jean Nei, The Current Status of Hydrogen Storage Alloy Development for Electrochemical Applications, doi:10.3390/ma6104574.

shell, Zn is a natural prohibitor for hydrogen evolution and thus a more negative voltage is possible, which increases the operation voltage of Ni-Zn battery.

The most important electrochemical application for MH is the negative electrode material for nickel metal hydride (NiMH) batteries. Together with a counter electrode from the $Ni(OH)_2$/NiOOH system, which has been used in NiCd and NiFe batteries as early as 1901 by Thomas Edison, the NiMH battery was first demonstrated by researchers in Battelle in 1967 with a mixed TiNi + Ti_2Ni alloy as the negative electrode [1]. Commercialization of the NiMH battery was independently realized by Ovonic Battery Company, Sanyo, and Matsushita in 1989 with AB_2 and AB_5 MH alloys. NiMH battery development started from small cylindrical cells (0.7 to 5 Ah) for portable electronic devices and progressed to 100 Ah prismatic cells for electric vehicle applications. The first commercially available electric vehicle in the modern era was the EV1 produced by General Motors in 1999. It was powered by a 26.4 kWh NiMH battery pack. Since then, NiMH batteries have powered more than 5 million hybrid electric vehicles made by Toyota, Honda, Ford, and other automakers, demonstrating the robustness and longevity of the NiMH battery. Recently, the NiMH battery has ventured into the stationary application market with advantages in long service life, a wide temperature range, low costs averaged over the service life, abuse immunity, and environmental friendliness. Several reviews on the topic of MH used in NiMH batteries are available [2,3,4,5,6,7,8,9,10]. In this report, we present the recent progress since the last review made in 2010 [10].

Besides NiMH batteries, MH (most commonly the misch metal-based AB_5 MH alloy) can also be used in other electrochemical applications such as lithium-ion based batteries and metal-air batteries. Metal hydride electrodes have a potential window of 0.1 to 0.5 V *versus* Li+/Li and the lowest polarization among conversion electrodes. These MH electrodes have shown the capability for greater capacity and can be used as anode electrodes in lithium-ion battery [10,11,12]. An air-MH battery that utilizes a misch metal-based AB_5 alloy in conjunction with a perovskite oxide-based cathode has been demonstrated by several research groups [13,14,15]. New types of V-flow/NiMH [16,17] and lead-acid/NiMH hybrid batteries [18] have been developed at the University of Hong Kong. Pd-treated $LaNi_{4.7}Al_{0.3}$ has been used in a Ni-hydrogen battery [19]. Another application of $LaNi_5$ is the use as a cathode in a photo-electrochemical cell for water decomposition [20].

HYDROGEN STORAGE ALLOYS FOR NIMH BATTERY NEGATIVE ELECTRODES

The use of MH in NiMH batteries started with research conducted by Schmitt and Beccu at Battelle Memorial Institute (TiNi-based) [1,21] and William and Buschow at Philips Research Laboratories ($LaNi_5$-based) [2,22]. Almost 50 years of research on this subject have been conducted. In 2001, nine requirements were established for a suitable MH alloy for NiMH batteries: high capacity, good electrochemical catalysis, easy formation, excellent corrosion resistance, suitable hydrogen equilibrium, good kinetics and efficiency, long cycle life, small pressure-concentration-temperature (PCT) hysteresis, and low cost [6]. Today, the main implementation of NiMH batteries has shifted from consumer portable devices to hybrid electric vehicle and stationary applications. Accordingly, new criteria for suitable MH alloy design requirements, such as low self-discharge, good kinetics at low temperature, fast proton diffusion in the bulk, low pulverization rate during service life, and endurance of high temperature storage, are required for these applications. The amount of metallic inclusions in the MH surface oxide after activation and the high-rate dischargeability (HRD) are characteristics that are essential in meeting many of these new requirements. While the former can be quantified by magnetic susceptibility, the latter is studied by analyzing the surface reaction current density and bulk diffusion constant. Typical values from several alloy systems are compared and listed in Table 1 for reference. Recent progress of MH alloys for NiMH battery applications is reviewed in the following sub-sections categorized by alloy system.

Table 1. Properties comparison of several metal hydride (MH) alloys. Saturated magnetic susceptibility (M_S) is proportional to the total amount of metallic nickel in the surface after activation. Applied magnetic field corresponding to half of the saturated magnetic susceptibility ($H_{1/2}$) is inversely proportional to the average number of Ni atoms in a cluster. Surface exchange current (I_0) and diffusion constant (D) are qualitative measurements of the catalytic nature of the surface reaction and the proton transportation in the bulk of the alloy, respectively.

Alloy system	Composition	M_S ($memu \cdot g^{-1}$)	$H_{1/2}$ (kOe)	I_0 ($mA \cdot g^{-1}$)	D ($\times 10^{-11}$ $cm^2 \cdot s^{-1}$)	Reference
AB_2	$Ti_{12}Zr_{21.5}Ni_{36.2}V_{9.5}Cr_{4.5}Mn_{13.6}Sn_{0.3}Co_2Al_{0.4}$	33	0.162	32.1	9.7	[23]
AB_5	$La_{10.5}Ce_{4.3}Pr_{0.5}Nd_{1.4}Ni_{60.0}Co_{12.7}Mn_{5.9}Al_{4.7}$	434	0.173	43.2	25.5	[23]
A_2B_7	$La_{16.3}Mg_{7.0}Ni_{65.1}Co_{11.6}$	369	0.125	41.0	30.8	[23]
A_2B_7	$Nd_{18.8}Mg_{2.5}Ni_{65.1}Al_{13.6}$	132	0.171	22.7	11.4	[23]
A_2B_7	$La_{3.8}Pr_{7.7}Nd_{7.7}Mg_{4.0}Ni_{72.1}Al_{4.7}$	314	0.128	51.5	31.9	This work
A_2B_7	$Nd_{18.4}Zr_{0.2}Mg_{3.6}Ni_{74.1}Co_{0.1}Al_{3.5}$	679	0.102	52.5	64	[24]
Zr-A_2B_7	Zr_2Ni_7	213	0.281	22.3	41	This work, [25]
Zr-AB_5	$ZrNi_{4.5}$	2286	0.400	20.1	60.6	This work, [26]

Rare Earth-Based AB_5 Alloys

Vucht *et al.* first reported the hydrogen storage capability of rare earth-based AB_5 intermetallic alloy in 1970 [27]. Since then, this alloy system has become the most widely used intermetallic alloy in all metal hydride applications. After over 40 years of research, many compositions, structures, processes, and electrode fabrication modifications have been performed. Recent efforts have focused on (1) cost reduction by introducing Fe and Cu into the alloy formula to reduce/

eliminate expensive Co and (2) improvement in HRD and low-temperature performance. The effects of substituting Ni by Cu, Fe, and Mo on charge-transfer resistance at −40 °C are summarized in Figure 2. The unit of Ω·g for charge-transfer resistance is used instead of the traditionally used Ω to rid the contribution from amount of electrode material; therefore, the values can be compared fairly among various MH alloys. The effect of Fe on charge-transfer resistance performance is interesting and can be explained by the evolutions of two factors: surface reaction area and catalytic ability. While surface catalytic ability is increased monotonically as the level of Fe-addition increases, surface reaction area is increased by 1% of Fe but decreased by further addition [28]. Consequently, charge-transfer resistance decreases with 1% Fe due to the improvements in both surface reaction area and catalytic ability, increases with a little higher Fe-content as the surface area diminishes, and finally decreases (but not to the level achieved with 1% Fe) with high amount of Fe when the surface catalytic ability takes on a more dominating role. Other research works on AB_5 alloys are summarized in Table 2.

Table 2. Summary of recent research on misch metal-based AB_5 MH alloys. S: substitution, P: process, A: additives.

Method	Alloy formula/process/additives	Secondary phase (s)	Range of *x*, *etc.*	Capacity	HRD	Cycle life	Charge retention	Low temperature	Reference
S	$La_{10.5}Ce_{4.3}Pr_{0.5}Nd_{1.4}Ni_{64.3-x}Co_{5.0}Mn_{4.6}Al_{6.0}Zr_{0.2}Fe_x$	–	0 to 1.5	down	up	down	down	up	[28]
S	$La_{10.5}Ce_{4.3}Pr_{0.5}Nd_{1.4}Ni_{64.3}Co_{5.0-x}Mn_{4.6}Al_{6.0}Cu_x$	(Al, Mn)Ni	0 to 5.4	down	up	down	up	up	[29]
S	$La_{10.5}Ce_{4.3}Pr_{0.5}Nd_{1.4}Ni_{64.3-x}Co_{5.0}Mn_{4.6}Al_{6.0}Zr_{0.2}Mo_x$	Mo	0 to 4	down	down	same	same	up	[30]
S	$La_{10.5}Ce_{4.3}Pr_{0.5}Nd_{1.3}Ni_{67.7-x-y}Mn_xAl_y$	–	Mn (0–0.6), Al (0–3.4)	–	–	up	up	–	[31]
S	$NdNi_{5-x-y-z}Co_xAl_yMn_z$	–	Co (0–0.5), Al (0–0.5), Mn (0–0.8)	up	down	up	down	–	[32]
S	$La_{0.7}Ce_{0.3}Ni_{3.75}Mn_{0.35}Al_{0.15}Cu_{0.75-x}Fe_x$	–	0 to 0.2	down	down	up	–	–	[33]
S	$La_{0.7}Ce_{0.3}Ni_{3.85}Mn_{0.3}Cu_{0.4}Fe_{0.15-x}(Fe_{0.43}B_{0.57})_x$	$La_3Ni_{12}B_2$	0 to 0.15	down	up	down	–	–	[34]
S	$LaNi_{3.55}Co_{0.2-x}Mn_{0.35}Al_{0.15}Cu_{0.75}(Fe_{0.43}B_{0.57})_x$	$La_3Ni_{12}B_2$	0 to 0.1	down	up	down	–	–	[35]
S	$LaNi_{3.55}Co_{0.2-x}Mn_{0.35}Al_{0.15}Cu_{0.75}(V_{0.81}Fe_{0.19})_x$	Ni-rich, La-rich	0 to 0.05	up	up	down	–	–	[36]
S	$La_{0.7}Ce_{0.3}Ni_{3.75-x}Mn_{0.35}Al_{0.15}Cu_{0.75}(Fe_{0.43}B_{0.57})_x$	–	0 to 0.15	up	up	up	–	–	[37]
S	$La_{0.7}Ce_{0.3}Ni_{3.83-x}Mn_{0.35}Al_{0.15}Cu_{0.75}(Fe_{0.43}B_{0.57})_x$	$La_3Ni_{12}B_2$	0 to 0.15	down	up	down	–	–	[38]
S	$La_{0.7}Ce_{0.3}Ni_{3.75-x}Mn_{0.35}Al_{0.15}Cu_{0.75}(V_{0.81}Fe_{0.19})_x$	–	0 to 0.05	down	up	up	–	–	[39]
S	$La_{0.7}Ce_{0.3}Ni_{4.2}Mn_{0.9-x}Cu_{0.37}(V_{0.81}Fe_{0.19})_x$	(V, Mn, Ni)	0 to 0.1	same	up	down	–	–	[40]
S	$La_{0.7}Ce_{0.3}Ni_{4.2}Mn_{0.9-x}Cu_{0.37}(Fe_{0.43}B_{0.57})_x$	$La_3Ni_{12}B_2$	0 to 0.1	down	up	down	–	–	[41]
S	$La_{0.7}Ce_{0.3}Ni_{3.75}Mn_{0.35}Al_{0.15}Cu_{0.75-x}(Fe_{0.43}B_{0.57})_x$	$La_3Ni_{12}B_2$	0 to 0.1	down	up	up	–	–	[42]
S	$La_{0.7}Ce_{0.3}Ni_{3.75}Mn_{0.35}Al_{0.15}Cu_{0.75-x}(V_{0.81}Fe_{0.19})_x$	–	0 to 0.1	same	up	up	–	–	[43]
S	$La_{0.7}Ce_{0.3}(Ni_{3.65}Mn_{0.35}Al_{0.15}Cu_{0.75}(Fe_{0.43}B_{0.57})_{0.10})_x$	$La_3Ni_{12}B_2Ce_2Ni_7$	0.9 to 1.0	up	up	up	–	–	[44]
S	$MlNi_{3.55}Co_{0.75-x}Mn_{0.4}Al_{0.3}(Cu_{0.75}P_{0.25})_x$	P-rich, Mn-rich	0 to 0.5	down	up	up then down	–	–	[45,46]
S	$LaNi_{5-x}In_x$	–	0.1 to 0.5	down	–	up	–	–	[47]
S	$LaNi_{4.3}(Co,Al)_{0.7-x}In_x$	–	0 to 0.1	up	up	–	–	–	[48]
S	$LaNi_{4.1-x}Co_{0.6}Mn_{0.3}Al_x$	–	0 to 0.45	down	down	up	up	–	[49]
S	$La_{0.78}Ce_{0.22}Ni_{3.73}Mn_{0.30}Al_{0.17}Fe_xCo_{0.8-x}$	–	0 to 0.8	down	down	up	up	–	[50]
S	$LaNi_{4.4-x}Co_{0.3}Mn_{0.3}Al_x$	–	0 to 0.2	up	Up then down	up	up	–	[51]
S	$MmNi_{3.79-x}Mn_{0.35}Co_{0.60}Al_{0.25}B_x$	$CeCo_4B$	0 to 0.2	down	up	–	–	–	[52]
S	$La_{0.35}Ce_{0.65}Ni_{3.54}Mn_{0.35}Co_{0.80-x}Al_{0.32}Mo_x$	–	0 to 0.25	up	up	up	–	–	[53]
S	$Mm_{0.8-x}Ti_xLa_{0.2}Ni_{3.7}Mn_{0.3}Co_{0.3}Al_{0.38}Mo_{0.02}$	–	0 to 0.05	up	–	up	up	–	[54]
S	$La_{0.65-x}Ce_{0.25-x}Pr_{0.03}Nd_{0.07}Y_{2x}Ni_{3.65}Mn_{0.3}Co_{0.75}Al_{0.3}$	–	0 to 0.04	down	down	up	–	–	[55]
S	$La_{1-x}Y_xNi_{3.55}Mn_{0.4}Co_{0.75}Al_{0.3}$	–	0 to 0.1	up	–	up	–	–	[56]
S	Eliminates Co, Mn	–	–	up	–	–	up	–	[57]
P	Pre-treatment if 12 M NaOH + 0.05 M $NaBH_4$	–	–	–	up	up	–	up	[58]
P	Melt-spin	$LaNi_5$, La_2Ni_3	–	up	–	down	–	–	[59]
P	Gas Atomization	–	–	down	same	up	up	same	[29]
P	Annealing temperature increase	–	–	up	down	up	up	–	[60]
A	Ni-PTFE plating	–	–	same	–	potentially up	–	–	[61]
A	Carbon nanosphere	–	–	up	up	down	–	–	[62]
A	Graphite	–	–	down	up		–	–	[63]
A	Co nano and Y_2O_3	–	–			up	–	–	[64]
A	Co_3O_4	–	–	up	up		–	–	[65]
A	Co_3O_4	–	–	up	up	up	–	–	[66]
A	$Ni(OH)_2$	–	–	up	up	down	–	–	[67]

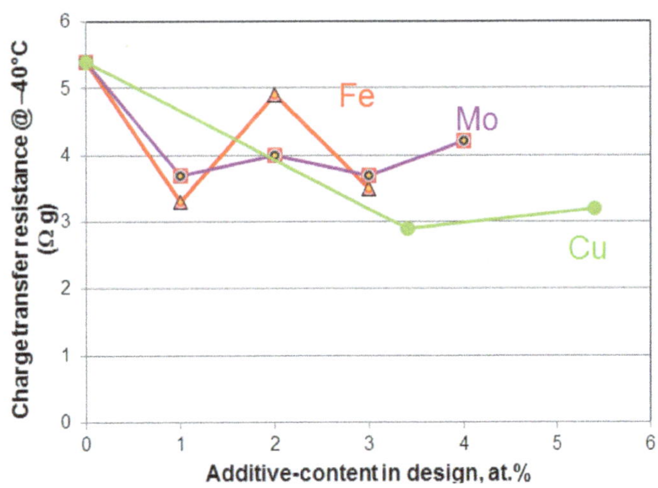

Figure 2. Plot of charge-transfer resistance measured at −40 °C by AC impedance as a function of Fe-, Mo-, or Cu-substitution in AB_5 MH alloy [28,29,30]. All three additives at the lowest substitution level contribute positively in lowering the resistance.

Laves Phase-Based AB_2 Alloys

AB_2 alloys for NiMH battery applications are composed of main phases belonging to a family of materials known as Laves phases: hexagonal C14 phase and face-center-cubic C15 phase. C36 phase may also be present but is difficult to distinguish from C14 in XRD analysis. The main controlling factor for the C14/C15 ratio is the average electron density (*e*/*a*). Nei *et al.* reported that chemical potential can be used to fine-tune the C14/C15 threshold when Ti, Zr, and Hf, which have the same number of outer-shell electrons, are used together [68]. The common minor phases are Zr_7Ni_{10}, Zr_9Ni_{11}, ZrNi, and TiNi. The microstructures [69,70,71,72] and the contributions [73,74,75,76,77] of these phases were studied extensively in recent years.

While many studies on AB_2 alloys were reported, most of them focused on the influences of A- or B-site substitutions on electrochemical properties. These results are summarized in Table 3. The effect of partial substitution of Ni by various modifiers on −40 °C charge transfer resistance is summarized in Figure 3.

As shown in Figure 3, both La and Si are very effective in improving the low-temperature performance. Additionally, the influence of stoichiometry (B/A ratio) on electrochemical properties was reported [78,79]. PCT hysteresis was correlated to the pulverization rate in AB_2 alloys [80,81,82]. Gas atomization and hydrogen annealing were introduced to produce AB_2 metal powder directly [83,84]. V-free AB_2 alloys were also developed to reduce raw material costs [85,86].

Table 3. Summary of recent research on Laves phase-based AB_2 MH alloys.

Base alloy	Substitution	Major effects	Reference
C14-domintaed	Al	Al improves bulk diffusion and surface reactivity. Al and Co together improves all electrochemical performances	[87,88]
C14-domintaed	B	B improves HRD and low-temperature performance but decreases charge retention, capacity, and cycle life	[85]
C14-domintaed	C	C increases HRD and charge retention but decreases low-temperature, capacity and cycle life	[85]
C14-domintaed	Co	Co provides easy activation, improves/decreases capacity, better cycle life and charge retention, but impedes HRD	[87,89,90]
C14-domintaed	Cr	Cr improves charge retention but impedes HRD	[89]
C14-domintaed	Mo	Mo improves HRD, low-temperature performance, charge retention, and cycle life	[91]
C14-domintaed	Cu	Cu increases capacity, facilitates activation, but decreases HRD.	[92]
C14-domintaed	Fe	Fe facilitates activation, increases total electrochemical capacity and effective surface reaction area, decreases HRD and bulk diffusion, and deteriorates low-temperature performance	[87,93]
C14-domintaed	Gd	Gd improves low-temperature performance, but decreases charge retention, HRD, capacity, and cycle life	[85]
C14-domintaed	La	La improves capacity, HRD, and low-temperature performance with a trade-off of inferior cycle stability	[94]
C14-domintaed	Mg	Mg improves charge retention, deteriorates capacity, low-temperature performance, and cycle life	[85]
C14-domintaed	Mn	Mn increases capacity, facilitates activation, but decreases cycle life	[89,95]
C14-domintaed	Ni	Ni improves cycle life and HRD but reduces capacity	[89]
C14-domintaed	Pt	Pt improves capacity and HRD	[96]
C14-domintaed	Si	1 at % of Si is beneficial to HRD and low-temperature performance	[97]
C14-domintaed	Sn	Sn improves charge retention but deteriorates HRD and cycle life	[87,98]
C14-domintaed	Ti	Ti increases HRD and facilitates activation	[99]
C14-domintaed	V	V increases capacity but decreases HRD and charge retention	[100]
Both C14- and C15-dominated	Y	Y improves activation, HRD, and low-temperature performance by increasing reaction surface area	[101,102]
C14-domintaed	Zr	Zr increases capacity	[99]

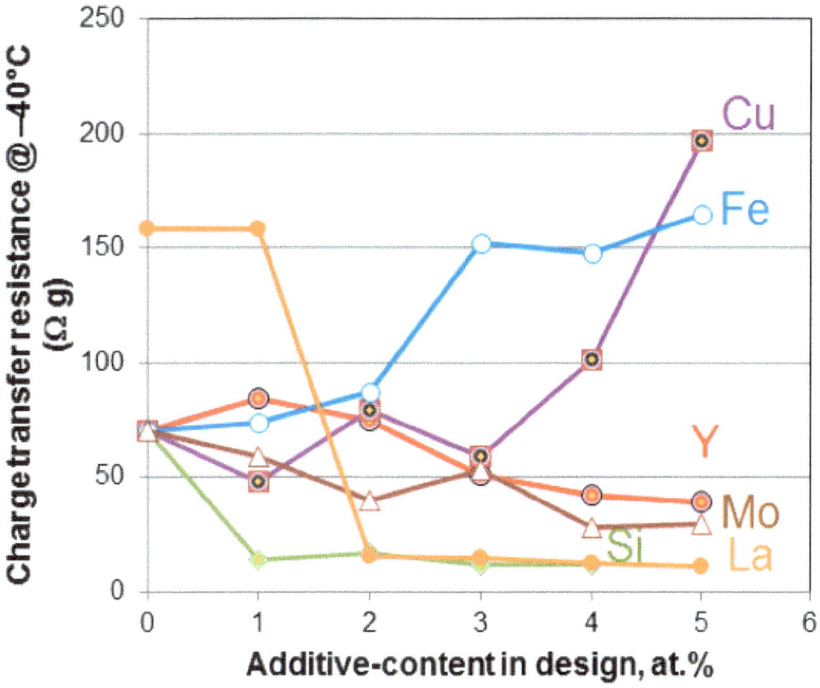

Figure 3. Plot of charge-transfer resistance measured at −40 °C by AC impedance as function of Cu-, Fe-, Y-, Mo-, La-, or Si-substitution in AB_2 MH alloy [91,92,93,94,97,102]. La- and Si-modified alloys demonstrate the lowest resistance.

Superlattice A_2B_7-Type (A_2B_4-AB_5-Hybrid-Type, such as AB_3, A_2B_7, A_5B_{19}, and AB_4) Alloys

Mg-containing superlattice alloys have been used extensively in Japanese-made NiMH batteries for retail market since the introduction by Sanyo (eneloop) [103,104,105]. Batteries with superlattice A_2B_7-type alloys as negative electrode exhibit much lower self-discharge compared to ones with traditionally used AB_5 alloys [105]. Therefore, NiMH battery's storing and ready-to-use-out-of-the-pack capabilities can be much improved by the use of superlattice alloy, which enable NiMH battery to grow much more competitive compared to primary battery. Furthermore, superlattice alloys have been reported to have superior capacity, cycle life, and high-rate performances compared to the conventional alloys and are becoming a dominating force in both consumer and hybrid electric vehicle markets. The multi-phase MH system, with an overall B/A ratio between 3 and 4, is composed of a number of AB_5 slabs placed between two A_2B_4 slabs (Figure 4). Mg is needed to lower the average metal-hydrogen bond strength in order to obtain the appropriate heat of hydride formation suitable for NiMH battery applications (≈ −39 kJ·mol^{-1} H_2 at room temperature and 1 atm), and its replacement amount for rare earth is about 30% and 15% for La and Nd, respectively. Mg mainly replaces rare earth elements on the A_2B_4 slab. The distribution of Mg (which has a high vapor pressure during melting) in the alloy is particularly important because Mg-lean regions have a tendency to form AB_5 phase. Lin *et al.* presented a brief review of different types of A_2B_7 alloys [106]. The superlattice MH alloys can be classified into three groups: La-only, La-Pr-Nd, and Nd-only. The addition of Ce promotes AB_5 phase and therefore is rarely used. The La-only group has the highest capacity but the shortest cycle life due to the easy oxidation of La. The Nd-only group has the best charge retention and cycle life performance but the lowest capacity. Properties of the La-Pr-Nd group fall between the properties for La- and Nd-only alloys. Annealing is the most convenient method to manipulate phase distribution, and its effect was reported on La-only [107,108,109,110,111,112,113], Nd-only [24], La-Gd [114], La-Nd [115], La-Pr-Nd [116,117], and La-Ce-Pr-Nd [118] alloys. While Nd-only and La-Pr-Nd superlattice alloys are used in low self-discharge and high capacity NiMH batteries, respectively, La-only superlattice alloy is not used in commercial products due to its low cycle life from pulverization [119]. However, La-only alloy is studied extensively in the research society [120] due to easy sample preparation. The recent progress in the NiMH battery applications of superlattice alloys is summarized in Table 4.

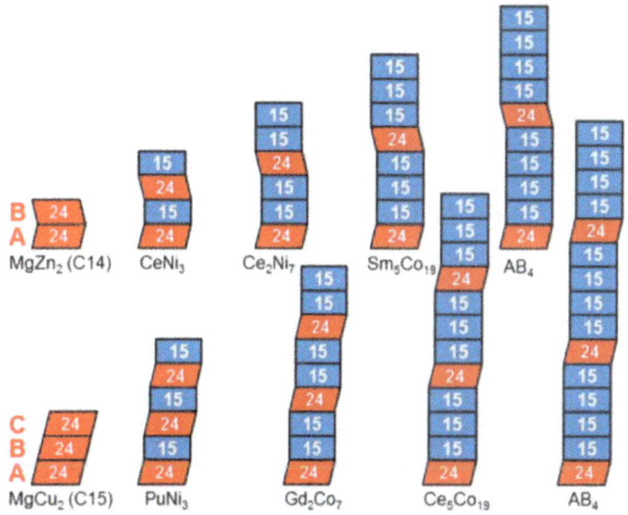

Figure 4. Schematics of stacking sequences of superlattice alloy systems. The stacking sequence is constructed with one to four AB_5 (blue 15) slabs in between slabs of A_2B_4 (red 24). Two structures are available for each stacking sequence depending on the direction of the A_2B_4 slab shifts. The tilted stacking of A_2B_4 in a C14 structure first shifts (1/3, 1/3) and then shifts back (−1/3, −1/3) while C15 structure shifts (1/3, 1/3) consecutively on the *a-b* plane.

Table 4. Summary of recent progress in electrochemical property improvement in superlattice MH alloys.

Substitution/Process	Alloy formula	Range of x	Capacity	HRD	Cycle life	Charge retention	Comment	Reference
Ce	$(La_{0.7}Mg_{0.3})_{1-x}Ce_xNi_{2.8}Co_{0.5}$	0 to 0.1	down	up	up	–	–	[121]
Dy	$(La_{1-x}Dy_x)_{0.8}Mg_{0.2}Ni_{3.4}Al_{0.1}$	0 to 0.2	up	–	same	down	–	[122]
Gd	$(La_{2-x}Gd_xMg)(NiCoAlZn)_{3.5}$	0 to 1	up	down	up	–	–	[123]
Nd	$La_{0.8-x}Nd_xMg_{0.2}Ni_{3.35}Al_{0.1}Si_{0.05}$	0 to 0.2	up	up	up	–	–	[124]
Nd	$(La_{1-x}Nd_x)_2Mg(Ni_{0.8}Co_{0.15}Mn_{0.05})_9$	0 to 0.3	down	–	up	–	–	[125]
Pr	$La_{0.75-x}Pr_xMg_{0.25}Ni_{3.2}Co_{0.2}Al_{0.1}$	0 to 0.4	–	–	up	–	–	[126]
Pr	$La_{0.8-x}Pr_xMg_{0.2}Ni_{3.15}Co_{0.2}Al_{0.1}Si_{0.05}$	0 to 0.3	up	up	up	–	–	[127]
Pr	$La_{0.75-x}Pr_xMg_{0.25}Ni_{3.2}Co_{0.2}Al_{0.1}$	0 to 0.2	down	–	up	–	–	[128]
Sc	$(La_{2-x}Gd_xMg)(NiCoAlZn)_{3.5}$	0 to 1	up	up	same	–	–	[123]
Sm	$La_{0.8-x}Sm_xMg_{0.2}Ni_{3.15}Co_{0.2}Al_{0.1}Si_{0.05}$	0 to 0.1	up	up	up	–	–	[129]
Ti	$(La_{0.67}Mg_{0.33})_{1-x}Ti_xNi_{2.75}Co_{0.25}$	0 to 0.05	down	up	up	–	–	[130]
Ti	$(La_{1-x}Ti_x)_2MgNi_{8.25}Co_{0.75}$	0 to 0.1	down	up	up	–	–	[131]
Zr	$La_{0.75-x}Pr_xMg_{0.25}Ni_{3.2}Co_{0.2}Al_{0.1}$	0 to 0.2	up	–	up	–	–	[128]
Zr	$La_{0.75-x}Zr_xMg_{0.25}Ni_{3.2}Co_{0.2}Al_{0.1}$	0 to 0.2	–	down	–	–	–	[132]
Mg	$La_{1.7+x}Mg_{1.3-x}(NiCoMn)_{9.3}$	0 to 0.4	up	down	down	–	Improves activation	[133]
Mg	$La_{0.85}Pr_{0.15}Mg_x(Ni_{0.7}Co_{0.2}Mn_{0.1})_9$	0.5 to 1.0	up	up	–	–	–	[134]
Mg	$La_{0.8-x}Gd_{0.2}Mg_xNi_{3.1}Co_{0.3}Al_{0.1}$	0.1 to 0.15	up	–	up	–	–	[135]
Mg	$La_{0.8-x}Gd_{0.2}Mg_xNi_{3.1}Co_{0.3}Al_{0.1}$	0 to 0.15	up	–	up	–	–	[136]
Mg	$La_{0.8-x}Gd_{0.2}Mg_xNi_{3.3}Co_{0.3}Al_{0.1}$	0 to 0.15	up	up	up	–	–	[137]
Ca	$La_{0.67}Mg_{0.33-x}Ca_xNi_{2.75}Co_{0.25}$	0 to 0.05	–	up	up	–	–	[138]
Al	$La_{0.75}Mg_{0.25}Ni_{3.5-x}Co_{0.2}Al_x$	0 to 0.09	down	down	up	–	–	[139]
Co	$LaNi_{3.2-x}Mn_{0.3}Co_x$	0.2 to 0.8	down	–	up	up	–	[140]
Co	$La_{0.7}Zr_{0.1}Mg_{0.2}Ni_{3.4-x}Co_xFe_{0.1}$	0.15 to 0.25	down	up	up	–	–	[141]
Co	$La_{0.55}Pr_{0.05}Nd_{0.15}Mg_{0.25}Ni_{3.5-x}Co_xAl_{0.25}$	0 to 0.3	up	up	same	–	–	[142]
Co + Al	$La_{0.45}Pr_{0.135}Nd_{0.315}Mg_{0.1}Ni_{3.9}Al_{0.2}$	0 to 0.1	down	up	up	–	–	[143]
Co + Al	$La_2MgMn_{0.3}Ni_{8.7-x}(Co_{0.5}Al_{0.5})_x$	0 to 2	down	up	up	–	–	[144]
Co + Al	$La_{0.55}Pr_{0.05}Nd_{0.15}Mg_{0.25}Ni_{3.5}(Co_{0.5}Al_{0.5})_x$	0 to 0.3	up	–	up	up	–	[145]
Al	$LaNi_{3.8-x}Al_x$	0 to 0.4	up then down	–	–	–	Improves activation	[146]
Mn	$(La_{0.8}Nd_{0.2})_2Mg(Ni_{0.9-x}Co_{0.1}Mn_x)_9$	0 to 0.1	up	–	up	–	–	[147]
Mn	$La_{0.78}Mg_{0.22}(Ni_{0.9-x}Co_{0.1}Mn_x)_3$	0 to 0.01	down	up	down	–	–	[148]
Cu	$LaMg_2Ni_{9-x}Cu_x$	0 to 9	down	–	–	–	–	[149]
Si	$La_{0.8}Mg_{0.2}Ni_{3.3}Co_{0.52}Si_x$	0 to 0.1	down	up	up	–	–	[107]
Ni	$CeMn_{0.25}Al_{0.25}Ni_{1.5+x}$	0 to 1.1	up	–	–	–	–	[150]
H_2O_2 in electrolyte	$Nd_{18.8}Mg_{2.5}Ni_{75.1}Al_{3.6}$	–	up		up	up	–	[151]

Melt-spin	$La_{0.75-x}Zr_xMg_{0.25}Ni_{3.2}Co_{0.2}Al_{0.1}$	0 to 0.2	–	up	–	–	–	[132]
Melt-spin	La_2MgNi_9	–	–	–	–	–	Improves Mg-homogeneity	[152]
Ball milling	$La_{0.7}Mg_{0.3}Ni_{2.8}Co_{0.5-x}Fe_x$	0 to 0.5	up	–	up	–	–	[153]
NiCuP plating	$La_{0.88}Mg_{0.12}Ni_{2.95}Mn_{0.10}Co_{0.55}Al_{0.10}$	–	up	up	–	up	–	[154]
Spark plasma sintering	$La_{0.85}Mg_{0.15}Ni_{3.8}$	–	same	–	up	–	–	[155]
Polyaniline plating	$La_{0.8}Mg_{0.2}Ni_{2.7}Mn_{0.1}Co_{0.55}Al_{0.1}$	–	–	up	up	–	–	[156]
Magnetic annealing	$La_{0.67}Mg_{0.33}Ni_{3.0}$	–	up	up	up	–	–	[157]
Chemical coprecipitation + metal reduction-diffusion	$La_{0.67}Mg_{0.33}Ni_{3.0}$	–	–	–	–	–	Produces multi-phase structure	[158]

Ti-Ni-Based Alloys

The Ti-Ni-based system was the first MH alloy used in NiMH batteries in the early 1970 s. Although its development was interrupted by the fast growth of AB_5 alloy, the Ti-Ni system remains a popular research topic due to its low cost, high hydrogen storage capacity, and fast activation. Two types of Ti-Ni binary alloys, TiNi and Ti_2Ni, are able to absorb large amounts of hydrogen and are candidates as negative electrode material in NiMH batteries. TiNi alloy exhibits polymorphism, which is the basis of its outstanding shape memory property that is fully utilized in many industrial applications. Upon cooling, TiNi transforms from a B2 cubic structure (austenite) to a B19' monoclinic structure (martensite). Early electrochemical studies on TiNi alloy demonstrated great activation performance and an electrochemical discharge capacity of 210 to 250 $mAh{\cdot}g^{-1}$ [159,160]. Compared to TiNi, Ti_2Ni MH alloy had higher hydrogen storage capacity due to its higher content of hydride forming Ti in the formulation, but its electrochemical capacity was only 170 $mAh{\cdot}g^{-1}$ with its stronger metal-hydrogen bond strength, an indicator of dehydriding difficulty [160]. By combining TiNi and Ti_2Ni phases, the electrochemical capacity increases up to 320 $mAh{\cdot}g^{-1}$ [160,161] with the assistance of synergetic effect between the two phases. During hydrogen desorption, TiNi, which has better desorption kinetics, first dehydrides and contributes to the overall electrochemical capacity; then, the hydrogen stored in Ti_2Ni phase is transferred internally into the dehydrided portion of TiNi and discharged through TiNi. Without the assistance of TiNi phase, the hydrogen stored in Ti_2Ni phase cannot be released due to the stronger meta-hydrogen bond strength. However, the cycle stability of such a biphasic system suffers due to the corrosion and oxidation of Ti_2Ni [159,162].

Most research efforts on TiNi alloys focused on elemental modifications on both A- and B-sites. Different material manufacturing procedures, such as mechanical alloying, annealing, and different quench rates, were introduced to study the effect of structure on overall electrochemical properties. Emami *et al.* investigated the influence of Pd by partially replacing Ni [163,164]. While the unit cell of TiNi is enlarged by the larger-sized Pd, the electrochemical capacity is reduced since the Pd-substitution increases the stability of TiNi intermetallic alloy and decreases the stability of their hydrides according to the electronic calculations. Recently, a study on substituting Ni in TiNi with various modifiers revealed that Fe, Co, and Cr increase the electrochemical capacity up to 400 $mAh{\cdot}g^{-1}$ with good activation performance [165]. Zhao *et al.* fabricated amorphous Ti_2Ni alloy by solid-state sintering and ball milling [166], and the resulting material shows improved cycle stability but also lower capacity compared to crystalline Ti_2Ni. Annealing treatment was then proposed in addition to sintering and ball milling, which results in a thicker oxide layer on the surface of amorphous Ti_2Ni and improves both capacity and charge retention of the alloy [167]. In order to increase the capacity, Zhao *et al.* developed another fabrication method for Ti_2Ni: induction melting, ball milling, and annealing [162]. The product is amorphous and nanocrystalline in nature and demonstrates reasonable capacity and good cycle stability; however, the highest reported capacity of 363 $mAh{\cdot}g^{-1}$ for Ti_2Ni MH alloy is obtained at higher temperatures while cycle stability suffers. Zr-substitution for Ti was used as another way to potentially increase capacity and improve cycle stability of amorphous Ti_2Ni [168]. Zr was reported to destabilize amorphous Ti_2Ni phase and promote TiNi phase when partially substituting Ti in ball-milled Ti_2Ni from elemental powders [169]. An amorphous Ti_3Ni_2 alloy was prepared to combine advantages from TiNi and Ti_2Ni, and although its capacity was lower than the crystalline form, cycle stability was much improved [170].

Ti-based icosahedral quasicrystal MH alloy, which has a crystallographically disallowed five-fold rotational symmetry, has only recently been studied for NiMH battery applications. The icosahedral phase is believed to contain much higher densities of tetrahedral interstitial sites compared to normal crystals [171]. Since hydrogen atoms enter favorably into tetrahedral sites, the icosahedral phase can absorb a large amount of hydrogen. Two recent studies, with and without milling after melt-spin, were performed varying the Zr/Ni ratio in $Ti_{45}Zr_{38}Ni_{17}$[172,173]. All alloys consisted of 100% or close to 100% icosahedral phase and demonstrated that higher levels of Ni increase capacity up to 86 to 88 mAh·g^{-1}; however, this value is dramatically lower than the theoretical capacity [172,173]. By changing the Zr/Ti ratio in $Ti_{45}Zr_{30}Ni_{25}$, it was shown that higher Zr deteriorates capacity. Also with annealing, an alloy that has been arc-melted and mechanically alloyed has higher capacity (up to 130 mAh·g^{-1}) compared to an unannealed amorphous alloy [174]. Hu *et al.* improved capacity up to 278 mAh·g^{-1} with decent cycle stability by adding Mn in a melt-spun TiVNi-based alloy [175]. The effect of Sc-addition was reported by the same group in TiVNi-based quasicrystal [176], with extended cycle life.

Mg-Ni-Based Alloys

Due to Mg's abundance, low cost, light weight, and high hydrogen storage capacity (2200 mAh·g^{-1} theoretical electrochemical capacity [9]), Mg-based MH alloy continues to be an interesting research topic and a strong candidate for the negative electrode material of NiMH batteries. In order to be applicable for room temperature battery operation, a stoichiometry of Mg:Ni close to 1:1 is required. However, the MgNi intermetallic compound does not exist on the Mg-Ni binary phase diagram. Therefore, non-equilibrium fabrication methods, such as mechanical alloying, RF sputtering, laser ablation, and melt-spin, are often used to prepare MgNi alloys. Such material usually has a mixed structure of amorphous and nanocrystalline character. Despite the high capacity that MgNi alloy offers, its sluggish hydriding/dehydriding kinetics and poor corrosion resistance in alkaline media prevent MgNi from use in practical application. Various types of modifications to the MgNi system were studied to improve the overall electrochemical performance, such as replacements of A- (by rare earth, transition metals, or others) and B-sites (by transition metals), combinations of different fabrication procedures, and surface treatments [177].Table 5 illustrates recent elemental substitution efforts on MgNi-based alloy. By varying the ball milling time, Anik *et al.* found that 15 h was sufficient to obtain the amorphous/nanocrystalline state [178,179]; however, 25 h was required to incorporate all Ni into the main MgNi phase. Electrochemical discharge capacity is also influenced by the milling period. The same report demonstrated that capacity increased sharply with up to 15 h milling time, stabilized with up to 25 h milling time, and decreased with further increase in milling time. Based on the role of each constituent element, a series of systematic elemental substitutions was conducted [178,179,180,181]. Anik *et al.* reported an improvement in electrochemical performance with $Mg_{0.80}Ti_{0.15}Al_{0.05}Zr_{0.05}Ni_{0.95}$ alloy (420 mAh·g^{-1} and 90% capacity retaining rate at the 20th cycle) compared to the base MgNi alloy (495 mAh·g^{-1} and 35% capacity retaining rate at the 20th cycle) [178,181]. The pulverization mechanism of MgNi was investigated by in-situ monitoring of hydride-dehydride cycles using acoustic emission technique coupled with electrochemical measurement [182,183]. Unlike most cases where pulverization is caused by volume expansion and contraction with repeated hydrogenation, mechanically alloyed MgNi consists of porous agglomerates made up of many particles cold welded together, which are likely to be easily broken down by the mechanical action of hydrogen bubbles during hydrogen evolution. Hydrogen combustion synthesis was employed as another MgNi fabrication method; and with the help of subsequent ball milling and surface protection from NAFION coating, a 400% increase in capacity could be achieved [184].

Table 5. Summary of recent research on electrochemical property improvement for mechanically alloyed MgNi MH alloys.

Substitution	Alloy formula	Range of x	Capacity	Cycle life	Comment	Reference
Ti	$Mg_{1-x}Ti_xNi$	0 to 0.2	up then down	up	As the Ti/Mg ratio increases, surface charge transfer resistance increases	[178,180]
Ti	$Mg_{0.7}Ti_{0.3}Ni$	–	down	up	–	[185]
Ti	$Mg_{1-x}Ti_xNi$	0 to 0.1	down	up	Reduces pulverization	[186]
Zr	$Mg_{1-x}Zr_xNi$	0 to 0.2	up then down	up	–	[178]
La	$Mg_{0.7}Ti_{0.225}La_{0.075}Ni$	–	down	up	Further improves corrosion resistance	[185]
Al	$Mg_{1-x}Al_xNi$	0 to 0.2	down	up	–	[178]
Al	$Mg_{0.9}Ti_{0.1}NiAl_x$	0 to 0.05	down	up	Reduces pulverization	[187]
B	$Mg_{1-x}B_xNi$	0 to 0.2	down	same	–	[178]
Pd	$Mg_{1-x}Pd_xNi$	0 to 0.2	down	up	Surface charge transfer resistance decreases and then increases	[179]
Pd	$Mg_{0.9}Ti_{0.1}NiAl_{0.05}Pd_x$	0 to 0.1	down	up	Increases HRD	[187]
Pd	$Mg_{50}Ni_{50-x}Pd_x$	0 to 5	down	up	–	[188]
Mg/Ni	$Mg_{0.85+x}Ti_{0.15}Ni_{1.0-x}$	0 to 0.1	up	down	–	[180]

Mg_2Ni is another Mg-Ni system that has been extensively studied for its potential applications in hydrogen storage and NiMH battery. With the higher content of hydride-former in the Mg_2Ni formulation, it is capable of storing a greater amount of hydrogen compared to MgNi; however, Mg_2Ni's discharge kinetics and corrosion resistance are worse. Nanocrystalline/amorphous structure was shown to improve hydriding/dehydriding kinetics [189]; such structure can be achieved by mechanical alloying and melt-spin fabrication techniques or elemental substitution. Although the exact effect of melt-spin cooling rates depends on the type of elemental substitution, a higher cooling rate was frequently found to be beneficial in enhancing electrochemical properties and cycle stability [190,191,192,193,194,195,196,197]. Furthermore, the length of milling time plays an important role in varying the performances of mechanically alloyed materials [198,199].

Melt-spin in magnetic field has been reported to greatly improve both capacity and stability [200]. Zhu*et al.* provided a comparison among the influences of various processing methods on electrochemical characteristics [201]. Table 6 summarizes the recent elemental substitution efforts on the Mg_2Ni system. It should be noted that besides the Mg-Ni MH systems, other Mg-based alloys, such as MgTi [202] and MgAl [203] systems doped with Ni, were also investigated for their electrochemical properties.

Table 6. Summary of recent research on electrochemical property improvement for Mg_2Ni MH alloys. MS: melt-spin, MSM: melt-spin in magnetic field, HCS: hydriding combustion synthesis, BM: ball milling, MSP: magnetron sputtering.

Substitution/ Addition	Process	Alloy formula	Range of *x*	Capacity	HRD	Cycle life	Comment	Reference
Co	MS	$Mg_2Ni_{1-x}Co_x$	0 to 0.4	up	up	up	Promotes amorphous phase	[190,191,194,195]
Mn	MS	$Mg_2Ni_{1-x}Mn_x$	0 to 0.4	up	up then down	up	Promotes amorphous phase	[192,194,195]
Cu	MS	$Mg_2Ni_{1-x}Cu_x$	0 to 0.4	up	up then down	up	-	[193–195]
La	MS	$Mg_{2-x}La_xNi$	0 to 0.2	-	up	up	Promotes amorphous phase	[196,197]
-	MSM	Mg_2Ni	-	up	-	up	-	[200]
Co	HCS+BM	$Mg_{2.1-x}Co_xNi$	0 to 0.1	down	down	up	-	[204]
Cr	HCS+BM	$Mg_{2.1-x}Cr_xNi$	0 to 0.1	down	down	up	-	[204]
Nb	HCS+BM	$Mg_{2.1-x}Nb_xNi$	0 to 0.1	down	down	up	-	[204]
Ti	HCS+BM	$Mg_{2.1-x}Ti_xNi$	0 to 0.1	down	down	up	-	[204]
V	HCS+BM	$Mg_{2.1-x}V_xNi$	0 to 0.1	down	down	up	-	[204]
Al	HCS+BM	$Mg_{2-x}Al_xNi$	0 to 0.7	up then down	-	-	-	[199]
Ti	BM	$Mg_{2-x}Ti_xNi$	0 to 0.5	up	-	up	-	[198]
B	BM	$Mg_{1.5}Ti_{0.3}Zr_{0.1}Al_{0.1}Ni$	-	-	-	-	Compares to others in [205]	[205]
C	BM	$Mg_{1.5}Ti_{0.3}Zr_{0.1}Al_{0.1}Ni$	-	-	-	-	Compares to others in [205]	[205]
Fe	BM	$Mg_{1.5}Ti_{0.3}Zr_{0.1}Al_{0.1}Ni$	-	-	-	-	Compares to others in [205]	[205]
Pd	BM	$Mg_{1.5}Ti_{0.3}Zr_{0.1}Al_{0.1}Ni$	-	-	-	-	Compares to others in [205]	[205]
Al	BM	$Mg_{1.5}Ti_{0.3}Zr_{0.1}Al_{0.1}Ni$	-	-	-	-	Compares to others in [205]	[205]
Al	BM	$Mg_{2-x}Al_xNi$	0 to 0.25	up	-	-	-	[206]
Multiwalled carbon nanotubes	BM	$(MgAl)_2Ni$	-	up	-	-	-	[206]
Al	MSP	$Mg_{2-x}Al_xNi$	0 to 0.3	up then down	-	-	Improves corrosion resistance	[207]

Laves Phase-Related BCC Solid Solutions

"Laves phase-related body-centered-cubic (BCC) solid solution" [208] is an interesting MH alloy with a general formula of AB_x, where A is from Group 4A (mainly Ti), B is from Group 5A, 6A, and 7A (mainly V), and x is between 1 and 6. It has a unique two-phase microstructure composed of a BCC phase and a Laves phase (mostly C14). Microstructure evolution as a function of C14/BCC ratio is constructed based on literature review and presented in Figure 5. During solidification, the high V BCC phase solidifies first and form a 3D framework and the rest of the liquid turns into C14 phase, which is also a complementary 3D framework, and such phenomena produces the following evolution in microstructure with the increase in C14/BCC ratio: C14 starts to appear at the grain boundary (Figure 5a) → sections of C14 phase start to connect and form a 3D network (Figure 5b) → BCC phase forms 3D framework and the cross-section is composed of isolated islands embedded in C14 matrix (Figure 5c) → BCC phase forms fish-bone type of inclusions (Figure 5d). The high density of phase boundaries contributes to the hydrogen storage properties in two ways: the promotion of a synergetic effect between BCC (high hydrogen storage capability) and C14 phases (better absorption kinetics and easy formation due to its brittleness), and the formation of a coherent and catalytic interface in between.

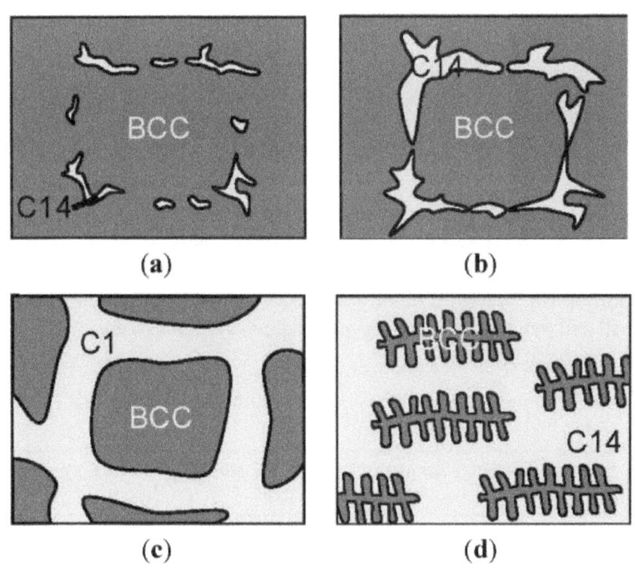

Figure 5. Schematics of microstructure evolution of a series of Laves phase-related body-centered-cubic (BCC) solid solution alloys as the C14 phase abundance increases (from Figure 5a–d).

While Chai *et al.* claimed the best ratio of BCC:C14 is 1:1 for optimized results [209], Guégen *et al.* showed that only a small amount of C14 is needed as a catalyst for the gaseous phase storage [210]. In order to study the correlation between the ratio and electrochemical properties of the alloy, we fabricated a series of Laves/BCC alloys with the Laves phase abundance from 10.5 to 91.6 wt %; their bulk diffusion coefficients and crystallite sizes are plotted in Figure 6. When one phase dominates, the crystallite size is smaller and the diffusion coefficient is larger. Therefore, a small but sufficient amount of secondary phase is preferable in the multi-phase MH alloy system. In addition, a BCC/C14 multi-phase alloy, reported by Wang and Ning, shows an electrochemical capacity of 450 mAh·g^{-1}, but degradation through cycling is severe [211]. The BCC/C14 alloy was remelted with 10 wt % $LaNi_3$ [212] and up to 10 wt % $LaNi_5$ [213,214] by arc melting, and the activation behavior, HRD, low-temperature performance, and cycle life are all improved due to the synergetic effect between main and secondary phases. BCC/Ti-Ni-based multi-phase systems were also fabricated previously [117,215]. The existence of the TiNi secondary phase enhances HRD and cycle life and shows the highest capacity of 470 mAh·g^{-1} [215]. Recently, an anneal-and-quench method [216] has been implemented to prepare BCC/C14 alloys. Furthermore, quasicrystal-included alloys have also been investigated by Liu *et al.* [217,218,219,220,221], and capacity, HRD, and cycle stability at the 30th cycle up to 422 mAh·g^{-1}, 88%, and 81%, respectively, were reported.

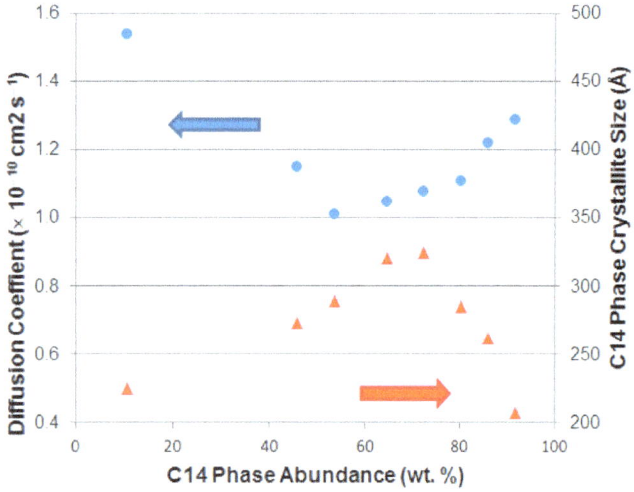

Figure 6. Plots of diffusion constant and crystallite size of the C14 phase determined by FWHM from XRD analysis as functions of C14 phase abundance for a series of Laves phase-related BCC solid solution alloys. The bulk transport is enhanced when the grain size is small and much of the boundary interface is available to contribute to the synergetic effect.

Zr-Ni-Based Alloys

Since the AB_2 intermetallic compound does not exist on the Zr-Ni binary phase diagram, one or more secondary non-Laves Zr_xNi_y phases composed of neighboring phases of AB_2 are often observed when fabricating a Zr-Ni-based MH alloy with an AB_2 composition. Between the working main phases and the catalytic secondary phases, a synergetic effect arises and contributes positively to the overall electrochemical performance. Although these minor phases do not have an appropriate metal-hydrogen bond strength when compared to the AB_2 stoichiometry, it is essential to evaluate various Zr_xNi_y alloys in order to gain better understanding of their roles in AB_2 MH alloys and potentially develop alternative rare earth-free MH alloys for NiMH battery through composition modification. A systematic examination of Zr_8Ni_{21}, Zr_7Ni_{10}, Zr_9Ni_{11}, and ZrNi alloys correlating composition, structure, gaseous phase hydrogen storage, and electrochemical properties was recently reported by Nei *et al.* [77]. Annealing is detrimental to the electrochemical capacity of Zr_xNi_y alloys due to the elimination of minor phases, similar to the annealing effect on AB_2 alloy. Furthermore, the electrochemical capacity maximizes at Zr_7Ni_{10}, which demonstrates that both hydrogen desorption/discharge rate and theoretical maximum hydrogen storage determined by the Zr/Ni ratio influence the electrochemical discharge capacity. Among all, Zr_8Ni_{21} shows the highest HRD and easiest activation. Zr_7Ni_{10} and Zr_8Ni_{21} were selected for further composition modification to improve their electrochemical properties [76,222,223,224,225,226], and the results are summarized in Table 7. One of the drawbacks of AB_2 alloy compared to AB_5 alloy is its lower HRD. For the purpose of increasing HRD in the Zr-based alloys, alloys with higher B/A ratios, such as $ZrNi_5$ [26] and Zr_2Ni_7 [25], become more attractive. Performances of V-modified $ZrNi_5$ and Zr_2Ni_7 are summarized in Table 7. Although their capacities are too low for practical application (≤177 mAh·g^{-1}), their bulk hydrogen diffusion properties are superior to those of existing AB_5, AB_2, and A_2B_7 alloys. With further modification, electrochemical performances of $ZrNi_5$ and Zr_2Ni_7 alloys are expected to improve.

Table 7. Summary of recent works on modification on Zr_xNi_y MH alloys.

Substitution	Alloy formula	Range of *x*	Capacity	HRD	Comment	Reference
Ti	$Ti_xZr_{7-x}Ni_{10}$	0 to 2.5	–	up	Activation becomes easier $Ti_{1.5}Zr_{5.5}Ni_{10}$ has good combination of capacity and HRD, 204 mAh·g^{-1} and 79%	[76]
V	$Ti_{1.5}Zr_{5.5}V_xNi_{10-x}$	0 to 3.0	up	down	Main phase shifts from Zr_7Ni_{10} to C14 $Ti_{1.5}Zr_{5.5}V_{0.5}Ni_{9.5}$ with Zr_7Ni_{10}-predominant structure has good combination of capacity and HRD, 242 mAh·g^{-1} and 80%	[222]
Cr	$Ti_{1.5}Zr_{5.5}V_{0.5}(Cr_xNi_{10-x})_{9.5}$	0.1 to 0.2	down	down	Main phase shifts from Zr_7Ni_{10} to Zr_9Ni_{11} to C14	[223,224]

Mn	$Ti_{1.5}Zr_{5.5}V_{0.5}(Mn_xNi_{10-x})_{9.5}$	0.1 to 0.2	up	up	Main phase shifts from Zr_7Ni_{10} to Zr_9Ni_{11} to C14	[223,224]
Fe	$Ti_{1.5}Zr_{5.5}V_{0.5}(Fe_xNi_{10-x})_{9.5}$	0.1 to 0.2	up	down	Main phase shifts from Zr_7Ni_{10} to C15.	[223,224]
Co	$Ti_{1.5}Zr_{5.5}V_{0.5}(Co_xNi_{10-x})_{9.5}$	0.1 to 0.2	down	down	Main phase shifts from Zr_7Ni_{10} to Zr_9Ni_{11} to C15	[223,224]
Cu	$Ti_{1.5}Zr_{5.5}V_{0.5}(Cu_xNi_{10-x})_{9.5}$	0.1 to 0.2	down	down	Main phase stays Zr_7Ni_{10}	[223,224]
Al	$Ti_{1.5}Zr_{5.5}V_{0.5}(Al_xNi_{10-x})_{9.5}$	0.1 to 0.2	down	down	Main phase shifts from Zr_7Ni_{10} to C14	[223,224]
Mg	$Zr_8Ni_{19}Mg_2$	–	down	down	Main phase shifts from Zr_8Ni_{21} to tetragonal Zr_7Ni_{10}	[225,226]
Al	$Zr_8Ni_{19}Al_2$	–	up	down	Main phase shifts from Zr_8Ni_{21} to tetragonal Zr_7Ni_{10}	[225,226]
Sc	$Zr_8Ni_{19}Sc_2$	–	down	down	Main phase shifts from Zr_8Ni_{21} to tetragonal Zr_7Ni_{10}	[225,226]
V	$Zr_8Ni_{19}V_2$	–	down	down	Main phase shifts from Zr_8Ni_{21} to Zr_2Ni_7	[225,226]
Mn	$Zr_8Ni_{19}Mn_2$	–	down	down	Main phase shifts from Zr_8Ni_{21} to Zr_2Ni_7	[225,226]
Co	$Zr_8Ni_{19}Co_2$	–	down	down	Main phase shifts from Zr_8Ni_{21} to Zr_2Ni_7	[225,226]
Sn	$Zr_8Ni_{19}Sn_2$	–	down	up	Main phase shifts from Zr_8Ni_{21} to Zr_2Ni_7 Annealed $Zr_8Ni_{19}Sn_2$ is Zr_8Ni_{21}-structured	[225,226]
La	$Zr_8Ni_{19}La_2$	–	up	down	Main phase shifts from Zr_8Ni_{21} to orthorhombic Zr_7Ni_{10}	[225,226]
Hf	$Zr_8Ni_{19}Hf_2$	–	up	down	Main phase shifts from Zr_8Ni_{21} to orthorhombic Zr_7Ni_{10}	[225,226]
V	$ZrV_xNi_{4.5-x}$	0 to 0.5	up	down	Main phase shifts from $ZrNi_5$ to monoclinic Zr_2Ni_7	[26]
V	$ZrV_xNi_{3.5-x}$	0 to 0.9	up then down	up	Main phase shifts from monoclinic Zr_2Ni_7 to cubic Zr_2Ni_7	[25]

Other Alloy Systems

It is nearly impossible to simultaneously meet all the requirements for a specific electrochemical application using one MH alloy. For example, an alloy with high capacity usually has lower HRD, and good activation behavior usually indicates shorter cycle life. Therefore, the development of composite MH alloys, which contain two or more hydrogen storage materials/intermetallic compounds/elements, can potentially combine the advantages of constituted alloys. Recent composite alloy studies include BCC related alloys modified by AB_2 (Section 2.6), AB_5 [227], $LaNi_5$[213,214], ZrV_2 [220], A_2B_7-type [217], $LaNi_3$ [212], or other BCC [218], A_2B_7-type alloy modified by AB_5 [228], MgNi alloy modified by Ti(NiCo) [229], Mg_2Ni alloys modified by TiNi, TiFe [230], $(MgMn)_2Ni$ [231], $(MgAl)_2Ni$ [232], Co, or Ti [233], and are summarized in Table 8.

Researchers have ventured out of well-understood MH alloy systems in order to develop a novel material that fulfills all requirements of electrochemical applications. R_6T_{23} systems (R = Gd, Ho, T= Mn, Fe) were recently studied. He *et al.* performed B-site substitution using Co on Ho_6Fe_{23} and found that the alloys have Th_6Mn_{23}-type structure [234]. As the level of Co increases, activation, capacity (443 $mAh \cdot g^{-1}$ at 150 $mA \cdot g^{-1}$), cycle stability, and HRD all improve. Electrochemical results were also reported on the Gd-Co-Mn system, and a promising capacity of 377 $mAh \cdot g^{-1}$ at 150 $mA \cdot g^{-1}$ was obtained [235]. The HRD of this type of alloy, however, is relatively low and in the range of 50% to 60%.

Table 8. Summary of recent research on composite alloys. BM: ball milling, AM: arc melting, ERM: electric resistance melting, IEC: isothermal evaporation casting.

Addition	Process	Base alloy	Addition level	Phase distribution	Capacity	HRD	Cycle life	Reference
$MmNi_{3.99}Al_{0.29}Mn_{0.3}Co_{0.6}$	BM	$Ti_{0.32}Cr_{0.43-x-y}V_{0.25}Fe_xMn_y$	0 to 20 wt %	BCC	up	–	–	[227]
$LaNi_5$	AM	$Ti_{0.10}Zr_{0.15}V_{0.35}Cr_{0.10}Ni_{0.30}$	0 to 10 wt %	BCC+C14+Zr-rich	up then down	up	up	[213,214]
ZrV_2	BM	$Ti_{1.4}V_{0.6}Ni$	0 to 20 wt %	quasicrystal+Ti_2Ni+BCC+C14+C15	up	up	up	[220]
$La_{0.65}Nd_{0.12}Mg_{0.23}Ni_{2.9}Al_{0.1}$	BM	$Ti_{1.4}V_{0.6}Ni$	0 to 20 wt %	quasicrystal+Ti_2Ni+BCC+$LaNi_5$+$PuNi_3$	same	up	down	[217]
$LaNi_3$	AM	$Ti_{0.10}Zr_{0.15}V_{0.35}Cr_{0.10}Ni_{0.30}$	0 to10 wt %	BCC+C14+Zr-rich	up	up	up	[212]
$Ti_{15}Zr_{18}V_{18}Ni_{29}Cr_5Co_7Mn$	BM	$Ti_{1.4}V_{0.6}Ni$	0 to 40 wt %	quasicrystal+Ti_2Ni+BCC+C14	up	up	up	[218]
$La_{0.377}Ce_{0.389}Pr_{0.063}Pr_{0.171}Ni_{3.5}Co_{0.6}Mn_{0.4}Al_{0.5}$	BM	$Mm_{0.80}Mg_{0.20}Ni_{2.56}Co_{0.50}Mn_{0.14}Al_{0.12}$	0 to 30 wt %	$LaNi_5$+La_2Ni_7	down	up then down	up	[228]
$TiNi_{0.56}Co_{0.44}$	BM	MgNi	0 to 50 wt %	Amorphous MgNi	down		up	[229]
TiNi	BM	Mg_2Ni	0 to 100 mol %	TiNi+Mg_2Ni	down			[230]
TiFe	BM	Mg_2Ni	0 to 100 mol %	TiFe+Mg_2Ni	up			[230]
Mg_3MnNi_2	ERM+IEC	Mg_2Ni	0 to 100 mol %	$Mg_2Ni \rightarrow Mg_2Ni+Mg_3MnNi_2 \rightarrow Mg_3MnNi_2$	up		up	[231]
Mg_3AlNi_2	ERM+IEC	Mg_2Ni	0 to 100 mol %	$Mg_2Ni \rightarrow Mg_2Ni+Mg_3AlNi_2 \rightarrow Mg_3AlNi_2$	up		up then down	[232]
Co	BM	Mg_3MnNi_2	0 to 200 mol %	amorphous Mg_3MnNi_2	up		up	[233]
Ti	BM	Mg_3MnNi_2	0 to 200 mol %	amorphous Mg_3MnNi_2	up		up	[233]

CONCLUSIONS

Hydrogen storage alloys for electrochemical application have been extensively studied for many years. We have presented a review of recent research activities on metal hydride alloys for nickel metal hydride battery and also provided an overview of the use of metal hydrides in other electrochemical applications. AB_5 and AB_2 alloys are very well established systems. In order to potentially dominate the future electric vehicle and stationary applications, self-discharge, low-temperature performance, and cycle stability become more important to study, and the trend of recent research reflects the efforts on improving the aforementioned properties. Superlattice A_2B_7-type alloy, which possesses the advantages of both AB_5 and AB_2 and low self-discharge capability, is likely to be the next generation of metal hydride alloy used as the negative electrode material in nickel metal hydride batteries and has attracted much attention. Although the Ti-Ni alloy system is difficult to process and has poorer high-rate performance, its much lower raw material cost makes the system one that merits further studies for improvement. The Mg-Ni alloy system holds great promise in achieving very high capacity, and recent research efforts have concentrated on improving its kinetics and cycle capability for the purpose of practical implementation. Laves phase-related BCC solid solution has high capacity; enhancing its stability is currently the most essential topic. The incorporation of quasicrystals by various fabrication methods remains an interesting subject. Zr-Ni alloy systems were systematically investigated in the last few years. Although their performance might not be satisfactory for electrochemical applications at the present time, further elemental modifications or use as a composite modifier can assist in realizing their potential.

ACKNOWLEDGMENTS

The authors are obliged to the current and former colleagues at Ovonic for sharing the joy and hard work of NiMH battery research. They are: Stanford R. Ovshinsky, Michael A. Fetcenko, Benjamin Reichman, Cristian Fierro, Baoquan Huang, Taihei Ouchi, Su Cronogue, Jun Im, Cheng Tung, Feng Li, William Mays, John Koch, Diana Wong, and Melanie Rainhout. We also appreciate the fruitful discussions with experts outside Ovonic, which include: Koiji Morii from Daido Steel, Hongge Pan from Zhejiang University, Shigekazu Yasuoka and Kazuta Takeno from FDK-Twicell, Lingkun Kong from Shenzhen High Power, Aishen Wu from Shida Battery, Xilin Zhu from REO MH Alloy Co., David Zheng from Gold Peak, Leonid Bendersky from NIST, Simon Ng, Steve Sally, and Gavin Lawes from Wayne State University, and Yi Liu from Oregon State University.

REFERENCES

1. Schmitt, R. Process for manufacturing a negative accumulator electrode for the reversible storage and restitution of hydrogen. U.S. Patent 3,972,726A, 3 August 1976.
2. Willems, J.G. Metal hydride electrodes stability of $LaNi_5$-related compounds. *Philips J. Res.* **1984**, *39*, 1–94.
3. Anani, A.; Visintin, A.; Petrov, K.; Srinivasan, S.; Reilly, J.J.; Johnson, J.R.; Schwarz, R.B.; Desch, P.B. Alloys for hydrogen storage in nickel/hydrogen and nickel/metal hydride batteries. *J. Power Sources* **1994**, *47*, 261–275.
4. Kleperis, J.; Wójcik, G.; Czerwinski, A.; Skowronski, J.; Kopczyk, M.; Beltowska-Brzezinska, M. Electrochemical behavior of metal hydride. *J. Solid State Electrochem.* **2001**, *5*, 229–249.
5. Feng, F.; Geng, M.; Northwood, D.O. Electrochemical behaviour of intermetallic-based metal hydrides used in Ni/metal hydride (MH) batteries: A review. *Int. J. Hydrog. Energy* **2001**, *26*, 725–734.
6. Hong, K. The development of hydrogen storage electrode alloys for nickel hydride batteries.*J. Power Sources* **2001**, *96*, 85–89.
7. Cuevas, F.; Joubert, J.M.; Latroche, M.; Percheron-Guégan, A. Intermetallic compounds as negative electrodes of Ni/MH batteries. *Appl. Phys. A* **2001**, *72*, 225–238.

8. Petrii, O.A.; Levin, E.E. Hydrogen-accumulating materials in electrochemical systems. *Russ. J. General Chem.* **2007**, *77*, 790–796.

9. Zhao, X.; Ma, L. Recent progress in hydrogen storage alloys for nickel/metal hydride secondary batteries. *Int. J. Hydrog. Energy* **2009**, *34*, 4788–4796.

10. Liu, Y.; Pan, H.; Gao, M.; Wang, Q. Advanced hydrogen storage alloys for Ni/MH rechargeable batteries. *J. Mater. Chem.* **2011**, *21*, 4743–4755.

11. Wessells, C.; Ruffo, R.; Huggins, R.A.; Cui, Y. Investigations of the electrochemical stability of aqueous electrolytes for lithium battery applications. *Electrochem. Solid-State Lett.* **2010**, *13*, A59–A61.

12. Nakayama, H.; Nobuhara, K.; Kon, M.; Matsunaga, T. *Electrochemical Properties of Metal Hydrides as Anode for Rechargeable Lithium Ion Batteries*; The Electrochemical Society: Susono, Shizuoka, Japan, 2010; p. 1052.

13. Dong, H.; Kiros, Y.; Noréus, D. An air-metal hydride battery using $MmNi_{3.6}Mn_{0.4}Al_{0.3}Co_{0.7}$in the anode and a perovskite in the cathode. *Int. J. Hydrog. Energy* **2010**, *35*, 4336–4341.

14. Mizutani, M.; Morimitsu, M. *Development of a Metal Hydride/Air Secondary Battery with Multiple Electrodes*; The Electrochemical Society: Las Vegas, NV, USA, 2010; p. 453.

15. Osada, N.; Morimitsu, M. *Cycling Performance of Metal Hydride-Air Rechargeable Battery*; The Electrochemical Society: Las Vegas, NV, USA, 2010; p. 194.

16. Weng, G.; Li, C.V.; Chan, K. *High Efficiency Vanadium-Metal Hydride hybrid Flow Battery: Importance in Ion Transport and Membrane Selectivity*; The Electrochemical Society: Toronto, ON, Canada, 2013; p. 242.

17. Weng, G.; Li, C.V.; Chan, K. *Study of the Electrochemical Behavior of high Voltage Vanadium-Metal hydride Hybrid Flow Battery*; The Electrochemical Society: Toronto, ON, Canada, 2013; p. 484.

18. Weng, G.; Li, C.V.; Chan, K. *Exploring the Role of Ionic Interfaces of the High Voltage Lead Acid-Metal Hydride Hybrid Battery*; The Electrochemical Society: Hilton Hawaiian Village, HI, USA, 2012; p. 372.

19. Purushothama, B.K.; Wainright, J.S. Analysis of pressure variations in a low-pressure nickel-hydrogen battery. Part 2: Cells with metal hydride storage. *J. Power Sources* **2012**, *206*, 421–428. [PubMed]

20. Danko, D.B.; Sylenko, P.M.; Shlapak, A.M.; Khyzhun, O.Y.; Shcherbakova, L.G.; Ershova, O.G.; Solonin, Y.M. Photoelectrochemical cell for water decomposition with a hybrid photoanode and a metal-hydride cathode. *Solar Energy Mater. Solar Cells* **2013**, *114*, 172–178.

21. Beccu, K. Accumulator electrode with capacity for storing hydrogen and method of manufacturing said electrode. U.S. Patent 3,669,745A, 3 June 1972.

22. Willems, J.J.G.; Buschow, K.H.J. From permanent magnets to rechargeable hydride electrodes. *J. Less Common Metals* **1987**, *129*, 13–30.

23. Young, K.; Huang, B.; Regmi, R.K.; Lawes, G.; Liu, Y. Comparisons of metallic clusters imbedded in the surface of AB_2, AB_5, and A_2B_7 alloys. *J. Alloy. Compd.* **2010**, *506*, 831–840.

24. Young, K.; Ouchi, T.; Huang, B. Effects of annealing and stoichiometry to (Nd, Mg)$(Ni, Al)_{3.5}$metal hydride alloys. *J. Power Sources* **2012**, *215*, 152–159.

25. Young, M.; Chang, S.; Young, K.; Nei, J. Hydrogen storage properties of $ZrV_xNi_{3.5-x}$ (x = 0.0–0.9) metal hydride alloys. *J. Alloy. Compd.* **2013**. in press.

26. Young, K.; Young, M.; Chang, S.; Huang, B. Synergetic effects in electrochemical properties of $ZrV_xNi_{4.5-x}$ (x = 0.0, 0.1, 0.2, 0.3, 0.4, and 0.5) metal hydride alloys. *J. Alloy. Compd.* **2013**,*560*, 33–41.

27. Van Vucht, J.H.N.; Kuijpers, F.A.; Burning, H.C.A.M. Reversible room-temperature absorption of large quantities of hydrogen by intermetallic compounds. *Philips Res. Rep.***1970**, *25*, 133–140.

28. Young, K.; Ouchi, T.; Reichman, B.; Koch, J.; Fetcenko, M.A. Improvement in the low-temperature performance of AB_5 metal hydride alloys by Fe-addition. *J. Alloy. Compd.* **2011**,*509*, 7611–7617.

29. Young, K.; Ouchi, T.; Banik, A.; Koch, J.; Fetcenko, M.A.; Bendersky, L.A.; Wang, K.; Vaudin, M. Gas atomization of Cu-modified AB_5 metal hydride alloys. *J. Alloy. Compd.* **2011**, *509*, 4896–4904.

30. Young, K.; Ouchi, T.; Reichman, B.; Koch, J.; Fetcenko, M.A. Effects of Mo additive on the structure and electrochemical properties of low-temperature AB_5 metal hydride alloys. *J. Alloy. Compd.* **2011**, *509*, 3995–4001.

31. Kong, L.; Chen, B.; Young, K.; Koch, J.; Chan, A.; Li, W. Effects of Al- and Mn-contents in the negative MH alloy on the self-discharge and long-term storage properties of Ni/MH battery.*J. Power Sources* **2012**, *213*, 128–139.

32. Young, K.; Huang, B.; Ouchi, T. Studies of Co, Al, and Mn substitutions in $NdNi_5$ metal hydride alloys. *J. Alloy. Compd.* **2012**, *543*, 90–98.

33. Liu, B.; Li, A.; Fan, Y.; Hu, M.; Zhang, B. Phase structure and electrochemical properties of $La_{0.7}Ce_{0.3}Ni_{3.75}Mn_{0.35}Al_{0.15}Cu_{0.75-x}Fe_x$ hydrogen storage alloys. *Trans. Nonferrous Metals Soc. China* **2012**, *22*, 2730–2735.

34. Peng, X.; Liu, B.; Fan, Y.; Ji, L.; Zhang, B.; Zhang, Z. Phase structure and electrochemical hydrogen storage characteristics of $La_{0.7}Ce_{0.3}Ni_{3.85}Mn_{0.8}Cu_{0.4}Fe_{0.15-x}(Fe_{0.43}B_{0.57})_x$ (x = 0–0.15) alloys. *Int. J. Electrochem. Sci.* **2013**, *8*, 3419–3428.

35. Fan, Y.; Liu, B.; Zhang, B.; Ji, L.; Wang, Y.; Zhang, Z. Microstructures and electrochemical properties of $LaNi_{3.55}Co_{0.2-x}Mn_{0.35}Al_{0.15}Cu_{0.75}(Fe_{0.43}B_{0.57})_x$ (x = 0–0.20) hydrogen storage alloys. *Mater. Chem. Phys.* **2013**, *138*, 803–809.

36. Liu, B.; Peng, X.; Fan, Y.; Ji, L.; Zhang, B.; Zhang, Z. Microstructures and electrochemical properties of $LaNi_{3.55}Co_{0.2-x}Mn_{0.35}Al_{0.15}Cu_{0.75}(V_{0.81}Fe_{0.19})_x$ hydrogen storage alloys. *Int. J. Electrochem. Sci.* **2012**, *7*, 11966–11977.

37. Liu, B.; Hu, M.; Fan, Y.; Ji, L.; Zhu, X.; Li, A. Microstructures and electrochemical properties of

$La_{0.7}Ce_{0.3}Ni_{3.75-x}Cu_{0.75}Mn_{0.35}Al_{0.15}(Fe_{0.43}B_{0.57})_x$ hydrogen storage alloys. *Electrochim. Acta* **2012**, *69*, 384–388.

38. Wang, Y.; Fan, Y.; Li, A. Phase structure and electrochemical properties of $La_{0.7}Ce_{0.3}Ni_{3.83-x}Mn_{0.43}Co_{0.25}Al_{0.26}Cu_{0.48}(Fe_{0.43}B_{0.57})_x$ hydrogen storage alloys. *Int. J. Electrochem. Sci.* **2013**, *8*, 5469–5478.

39. Liu, B.; Hu, M.; Zhou, Y.; Li, A.; Ji, L.; Fan, Y.; Zhang, Z. Microstructures and electrochemical characteristics of $La_{0.7}Ce_{0.3}Ni_{3.75-x}Cu_{0.75}Mn_{0.35}Al_{0.15}(V_{0.81}Fe_{0.19})_x$ (x = 0–0.20) hydrogen storage alloys. *J. Alloy. Compd.* **2012**, *544*, 105–110.

40. Peng, X.; Liu, B.; Fan, Y.; Ji, L.; Zhang, B.; Zhang, Z. Microstructures and electrochemical characteristics of $La_{0.7}Ce_{0.3}Ni_{4.2}Mn_{0.9-x}Cu_{0.37}(V_{0.81}Fe_{0.19})_x$ hydrogen storage alloys.*Electrochim.Acta* **2013**, *93*, 207–212.

41. Peng, X.; Liu, B.; Fan, Y.; Zhu, X.; Peng, Q.; Zhang, Z. Microstructures and electrochemical hydrogen storage characteristics of $La_{0.7}Ce_{0.3}Ni_{4.2}Mn_{0.9-x}Cu_{0.37}(Fe_{0.43}B_{0.57})_x$ (x = 0–0.20) alloys. *J. Power Sources* **2013**, *240*, 178–183.

42. Liu, B.; Hu, M.; Ji, L.; Fan, Y.; Wang, Y.; Zhang, Z.; Li, A. Phase structure and electrochemical properties of $La_{0.7}Ce_{0.3}Ni_{3.75}Mn_{0.35}Al_{0.15}Cu_{0.75-x}(Fe_{0.43}B_{0.57})_x$ hydrogen storage alloys. *J. Alloy. Compd.* **2012**, *516*, 53–57.

43. Liu, B.; Hu, M.; Li, A.; Ji, L.; Zhang, B.; Zhu, X. Microstructure and electrochemical characteristics of $La_{0.7}Ce_{0.3}Ni_{3.75}Mn_{0.35}Al_{0.15}Cu_{0.75-x}(V_{0.81}Fe_{0.19})_x$ hydrogen storage alloys.*J. Rare Earths* **2012**, *30*, 769–774.

44. Fan, Y.; Liu, B.; Zhang, B.; Ji, L.; Zhang, Z. Phase structure and electrochemical properties of non-stoichiometric $La_{0.7}Ce_{0.3}(Ni_{3.65}Cu_{0.75}Mn_{0.35}Al_{0.15}(Fe_{0.43}B_{0.57})_{0.10})_x$ hydrogen storage alloys. *J. Rare Earths* **2012**, *30*, 1249–1254.

45. Zhang, B.; Wu, W.; Bian, X.; Tu, G. Investigations on the kinetics properties of the electrochemical reactions of $MlNi_{3.55}Co_{0.75-x}Mn_{0.4}Al_{0.3}(Cu_{0.75}P_{0.25})_x$ (x = 0.0, 0.1, 0.2, 0.3, 0.4) hydrogen storage alloys. *J. Alloy. Compds* **2012**, *538*, 189–192.

46. Zhang, B.; Wu, W.; Bian, X.; Tu, G. Study on microstructure and electrochemical performance of the $MlNi_{3.55}Co_{0.75-x}Mn_{0.4}Al_{0.3}(Cu_{0.75}P_{0.25})_x$ (x = 0–0.5) composite alloys. *J. Power Sources* **2013**, *236*, 80–86.

47. Drulis, H.; Hackemer, A.; Folcik, L.; Giza, K.; Bala, H. ; Gondek, Ł.; Figiel, H. Thermodynamic and electrochemical hydrogenation properties of $LaNi_{5-x}In_x$ alloys. *Int. J. Hydrog. Energy* **2012**, *37*, 15850–15854.

48. Giza, K.; Bala, H.; Drulis, H.; Hackemer, A.; Folcik, L. Gaseous and electrochemical hydrogenation properties of $LaNi_{4.3}(Co, Al)_{0.7-x}In_x$ alloys. *Int. J. Electrochem. Sci.* **2012**, *7*, 9881–9891.

49. Balogun, M.; Wang, Z.; Huang, H.; Liu, W.; Ni, C.; Li, G. Microstructure and electrode performance of $LaNi_{4.1-x}Co_{0.6}Mn_{0.3}Al_x$ hydrogen storage alloys. *J. Guilin Univ. Electron. Technol.* **2012**, *32*, 508–513.

50. Chao, D.; Zhong, C.; Ma, Z.; Yang, F.; Wu, Y.; Zhu, D.; Wu, C.; Chen, Y. Improvement in high-temperature performance of Co-free high-Fe AB_5-type hydrogen storage alloys. *Int. J. Hydrog. Energy* **2012**, *37*, 12375–12383.

51. Balogun, M.; Wang, Z.; Chen, H.; Deng, J.; Yao, Q.; Zhou, H. Effect of Al content on structure and electrochemical properties of $LaNi_{4.4-x}Co_{0.3}Mn_{0.3}Al_x$ hydrogen storage alloys. *Int. J. Hydrog. Energy* **2013**, *39*, 10926–10931.

52. Yang, S.; Han, S.; Li, Y.; Yang, S.; Hu, L. Effect of substituting B for Ni on electrochemical kinetic properties of AB_5-type hydrogen storage alloys for high-power nickel/metal hydride batteries. *Mater. Sci. Eng. B* **2011**, *176*, 231–236.

53. Yang, S.; Han, S.; Song, J.; Li, Y. Influences of molybdenum substitution for cobalt on the phrase structure and electrochemical kinetic properties of AB_5-type hydrogen storage alloys. *J. Rare Earths* **2011**, *29*, 692–697.

54. Srivastava, S.; Upadhyay, R.K. Investigations of AB_5-type negative electrode for nickel-metal hydride cell with regard to electrochemical and microstructural characteristics. *J. Power Sources* **2010**, *195*, 2996–3001.

55. Peng, X.; Liu, B.; Zhou, Y.; Fan, Y.; Yang, Y.; Li, Q. Microstructures and electrochemical hydrogen storage characteristics of $La_{0.65-x}Ce_{0.25-x}Pr_{0.03}Nd_{0.07}Y_{2x}Ni_{3.65}Co_{0.75}Mn_{0.3}Al_{0.3}$ (x= 0–0.04) alloys. *Int. J. Electrochem. Sci.* **2013**, *8*, 2262–2271.

56. Du, Y.; Li, W. Structural and electrochemical properties of annealed $La_{1-x}Y_xNi_{3.55}Mn_{0.4}Al_{0.3}Co_{0.75}$ hydrogen storage alloys. *J. Environ. Sci.* **2011**, *23*, S59–S62.

57. Takasaki, T.; Nishimura, K.; Saito, M.; Fukunaga, H.; Iwaki, T.; Sakai, T. Cobalt-free nickel-metal hydride battery for industrial applications. *J. Alloy. Compd.* 2013. Available online: http://dx.doi.org/10.1016/j.jallcom.2013.01.092 (accessed on 4 July 2013).

58. Su, G.; He, Y.; Liu, K. Effects of pretreatment on $MlNi_{4.00}Co_{0.45}Mn_{0.38}Al_{0.3}$ hydrogen storage alloy powders and the performance of 6 Ah prismatic traction battery. *Int. J. Hydrog. Energy***2012**, *37*, 12384–12392.

59. Tian, X.; Liu, X.; Yao, Z.; Yan, S. Structure and electrochemical properties of rapidly quenched $Mm_{0.3}Ml_{0.7}Ni_{3.55}Co_{0.75}Mn_{0.4}Al_{0.3}$ hydrogen storage alloy. *J. Mater. Eng. Perform.***2013**, *22*, 848–853.

60. Yang, S.; Liu, Z.; Han, S.; Zhang, W.; Song, J. Effects of annealing treatment on the microstructure and electrochemical properties of low-Co hydrogen storage alloys containing Cu and Fe. *Rare Metals* **2011**, *30*, 464–469.

61. Kim, J.; Yamamoto, K.; Yonezawa, S.; Takashima, M. Effects of Ni-PTFE composite plating on AB_5-type hydrogen storage alloy. *Mater. Lett.* **2012**, *82*, 217–219.

62. Cuscueta, D.J.; Corso, H.L.; Arenillas, A.; Martinez, P.S.; Ghilarducci, A.A.; Salva, H.R. Electrochemical effect of carbon nanospheres on an AB_5 alloy. *Int. J. Hydrog. Energy* **2012**, *37*, 14978–14982.

63. Li, X.; Wang, L.; Dong, H.; Song, Y.; Shang, H. Electrochemical hydrogen absorbing properties of graphite/AB_5 alloy composite electrode. *J. Alloy. Compd.* **2012**, *510*, 114–118.

64. Jang, I.; Kalubarme, R.S.; Yang, D.; Kim, T.; Park, C.;

Ryu, H.; Park, C. Mechanism for the degradation of $MmNi_{3.9}Co_{0.6}Mn_{0.3}Al_{0.2}$ electrode and effects of additives on electrode degradation for Ni-MH secondary batteries. *Metals Mater. Int.* **2011**, *17*, 891–897.

65. Zhang, Q.; Su, G.; Li, A.; Liu, K. Electrochemical performances of AB_5-type hydrogen storage alloy modified with Co_3O_4. *Trans. Nonferrous Metals Soc. China* **2011**, *21*, 1428–1434.

66. Su, G.; He, Y.; Liu, K. Effects of Co_3O_4 as additive on the performance of metal hydride electrode and Ni-MH battery. *Int. J. Hydrog. Energy* **2012**, *37*, 11994–12002.

67. Yun, L.; Xing, K.; Feng, H. Impact of hydroxy Ni powders on electrochemical properties of AB_5 hydrogen storage alloy. *Value Eng.* **2013**, *30*, 288–289.

68. Nei, J.; Young, K.; Salley, S.O.; Ng, K.Y.S. Determination of C14/C15 phase abundance in Laves phase alloys. *Mater. Chem. Phys.* **2012**, *136*, 520–527.

69. Boettinger, W.J.; Newbury, D.E.; Wang, K.; Bendersky, L.A.; Chiu, C.; Kanttner, U.R.; Young, K.; Chao, B. Examination of multiphase $(Zr,Ti)(V,Cr,Mn,Ni)_2$ Ni-MH electrode alloys: Part I. Dendritic solidification structure. *Metall. Mater. Trans. A* **2010**, *41*, 2033–2047.

70. Bendersky, L.A.; Wang, K.; Boettinger, W.J.; Newbury, D.E.; Young, K.; Chao, B. Examination of multiphase $(Zr,Ti)(V,Cr,Mn,Ni)_2$ Ni-MH electrode alloys: Part II. Solid-state transformation of the interdendritic B2 phase. *Metall. Mater. Trans. A* **2010**, *41*, 1891–1906.

71. Bendersky, L.A.; Wang, K.; Boettinger, W.J.; Bewbury, D.E.; Young, K.; Chao, B. Microstructural Studies of Multiphase $(Zr,Ti)(V,Cr,Mn,Co,Ni)_2$ Alloys for Ni/MH Negative Electrodes. In *Intermetallic-Based Alloys for Structural and Functional Applications*; Bewlay, P., Kumar, Y., *et al*, Eds.; MRS: Boston, MA, USA, 2010; Volume 1295.

72. Bendersky, L.A.; Wang, K.; Levin, I.; Newbury, D.; Young, K.; Chao, B.; Creuziger, A. $Ti_{12.5}Zr_{21}V_{10}Cr_{8.5}Mn_xCo_{1.5}Ni_{46.5-x}$ AB_2-type metal hydride alloys for electrochemical storage application: Part 1. Structural characteristics. *J. Power Sources* **2012**, *218*, 474–486.

73. Young, K.; Ouchi, T.; Huang, B.; Chao, B.; Fetcenko, M.A.; Bendersky, L.A.; Wang, K.; Chiu, C. The correlation of C14/C15 phase abundance and electrochemical properties in the AB_2alloys. *J. Alloy. Compd.* **2010**, *506*, 841–848.

74. Young, K.; Nei, J.; Ouchi, T.; Fetcenko, M.A. Phase abundances in AB_2 metal hydride alloys and their correlations to various properties. *J. Alloy. Compd.* **2011**, *509*, 2277–2284.

75. Young, K.; Chao, B.; Bendersky, L.A.; Wang, K. $Ti_{12.5}Zr_{21}V_{10}Cr_{8.5}Mn_xCo_{1.5}Ni_{46.5-x}$ AB_2-type metal hydride alloys for electrochemical storage application: Part 2. Hydrogen storage and electrochemical properties. *J. Power Sources* **2012**, *218*, 487–494.

76. Young, K.; Ouchi, T.; Liu, Y.; Reichman, B.; Mays, W.; Fetcenko, M.A. Structural and electrochemical properties of $Ti_xZr_{7-x}Ni_{10}$. *J. Alloy. Compd.* **2009**, *480*, 521–528.

77. Nei, J.; Young, K.; Regmi, R.; Lawes, G.; Salley, S.O.; Ng, K.Y.S. Gaseous phase hydrogen storage and electrochemical properties of Zr_8Ni_{21}, Zr_7Ni_{10}, Zr_9Ni_{11}, and ZrNi metal hydride alloys. *Int. J. Hydrog. Energy* **2012**, *37*, 16042–16055.

78. Young, K.; Ouchi, T.; Yang, J.; Fetcenko, M.A. Studies of off-stoichiometric AB_2 metal hydride alloy: Part 1. Structural characteristics. *Int. J. Hyrdog. Energy* **2011**, *36*, 11137–11145.

79. Young, K.; Nei, J.; Huang, B.; Fetcenko, M.A. Studies of off-stoichiometric AB_2 metal hydride alloy: Part 2. Hydrogen storage and electrochemical properties. *Int. J. Hydrog. Energy* **2011**,*36*, 11146–11154.

80. Young, K.; Ouchi, T.; Fetcenko, M.A. Pressure-composition-temperature hysteresis in C14 Laves phase alloys: Part 1. Simple ternary alloys. *J. Alloy. Compd.* **2009**, *480*, 428–433.

81. Young, K.; Ouchi, T.; Mays, W.; Reichman, B.; Fetcenko, M.A. Pressure-composition-temperature hysteresis in C14 Laves phase alloys: Part 2. Applications in NiMH batteries. *J. Alloy. Compd.* **2009**, *480*, 434–439.

82. Young, K.; Ouchi, T.; Fetcenko, M.A. Pressure-composition-temperature hysteresis in C14 Laves phase alloys: Part 3. Empirical formula. *J. Alloy. Compd.* **2009**, *480*, 440–448.

83. Young, K.; Koch, J.; Ouchi, T.; Banik, A.; Fetcenko, M.A. Study of AB_2 alloy electrodes for Ni/MH battery prepared by centrifugal casting and gas atomization. *J. Alloy. Compd.* **2010**,*496*, 669–677.

84. Young, K.; Ouchi, T.; Banik, A.; Koch, J.; Fetcenko, M.A. Improvement in the electrochemical properties of gas atomized AB_2 metal hydride alloys by hydrogen annealing. *Int. J. Hydrog. Energy* **2011**, *36*, 3547–3555.

85. Young, K.; Ouchi, T.; Huang, B.; Fetcenko, M.A. Effects of B, Fe, Gd, Mg, and C on the structure, hydrogen storage, and electrochemical properties of vanadium-free AB_2 metal hydride alloy. *J. Alloy. Compd.* **2012**, *511*, 242–250.

86. Young, K.; Ouchi, T.; Koch, J.; Fetcenko, M.A. Compositional optimization of vanadium-free hypo-stoichiometric AB_2 metal hydride alloy for Ni/MH battery application. *J. Alloy. Compd.***2012**, *510*, 97–106.

87. Young, K.; Fetcenko, M.A.; Koch, J.; Morii, K.; Shimizu, T. Studies of Sn, Co, Al, and Fe additives in C14/C15 Laves alloys for NiMH battery application by orthogonal arrays. *J. Alloy. Compd.* **2009**, *486*, 559–569.

88. Young, K.; Regmi, R.; Lawes, G.; Ouchi, T.; Reichman, B.; Fetcenko, M.A.; Wu, A. Effects of aluminum substitution in C14-rich multi-component alloys for NiMH battery application. *J. Alloy. Compd.* **2010**, *490*, 282–292.

89. Young, K.; Ouchi, T.; Fetcenko, M.A. Roles of Ni, Cr, Mn, Sn, Co, and Al in C14 Laves phase alloys for NiMH battery application. *J. Alloy. Compd.* **2009**, *476*, 774–781.

90. Young, K.; Ouchi, T.; Reichman, B.; Mays, W.; Regmi, R.; Lawes, G.; Fetcenko, M.A.; Wu, A. Optimization of Co-content in C14 Laves phase multi-component alloys

for NiMH battery application. *J. Alloy. Compd.* **2010**, *489*, 202–210.

91. Young, K.; Ouchi, T.; Huang, B.; Reichman, B.; Fetcenko, M.A. Effect of molybdenum content on structural, gaseous storage, and electrochemical properties of C14-predominant AB_2metal hydride alloys. *J. Power Sources* **2011**, *196*, 8815–8821.

92. Young, K.; Ouchi, T.; Huang, B.; Reichman, B.; Fetcenko, M.A. Studies of copper as a modifier in C14-predominant AB_2 metal hydride alloys. *J. Power Sources* **2012**, *204*, 205–212.

93. Young, K.; Ouchi, T.; Huang, B.; Reichman, B.; Fetcenko, M.A. The structure, hydrogen storage, and electrochemical properties of Fe-doped C14-predominating AB_2 metal hydride alloys. *Int. J. Hydrog. Energy* **2011**, *36*, 12296–12304.

94. Young, K.; Wong, D.F.; Ouchi, T.; Huang, B.; Reichman, B.; Fetcenko, M.A. Effects of La-addition to the structure, hydrogen storage, and electrochemical properties of C14 metal hydride alloys. *J. Alloy. Compd.* Submitted.

95. Young, K.; Ouchi, T.; Koch, J.; Fetcenko, M.A. The role of Mn in C14 Laves phase multi-component alloys for NiMH battery application. *J. Alloy. Compd.* **2009**, *477*, 749–758.

96. Ruiz, F.C.; Peretti, H.A.; Visintin, A.; Triaca, W.E. A study on $ZrCrNiPt_x$ alloys as negative electrode components for NiMH batteries. *Int. J. Hydrog. Energy* **2011**, *36*, 901–906.

97. Young, K.; Ouchi, T.; Huang, B.; Reichman, B.; Blankenship, R. Improvement in −40 °C electrochemical properties of AB_2 metal hydride alloy by silicon incorporation. *J. Alloy. Compd.* **2013**, *575*, 65–72.

98. Young, K.; Fetcenko, M.A.; Ouchi, T.; Li, F.; Koch, J. Effects of Sn-substitution in C14 Laves phase alloys for NiMH battery application. *J. Alloy. Compds* **2009**, *469*, 406–416.

99. Young, K.; Fetcenko, M.A.; Li, F.; Ouchi, T. Structural, thermodynamic, and electrochemical properties of $Ti_xZr_{1-x}(VNiCrMnCoAl)_2$ C14 Laves phase alloys. *J. Alloy. Compd.* **2008**, *464*, 238–247.

100. Young, K.; Fetcenko, M.A.; Li, F.; Ouchi, T.; Koch, J. Effect of vanadium substitution in C14 Laves phase alloys for NiMH battery application. *J. Alloy. Compd.* **2009**, *468*, 482–492.

101. Young, K.; Young, M.; Ouchi, T.; Reichman, B.; Fetcenko, M.A. Improvement in high-rate dischargeability, activation, and low-temperature performance in multi-phase AB_2 alloys by partial substitution of Zr by Y. *J. Power Sources* **2012**, *215*, 279–287.

102. Young, K.; Reichman, B.; Fetcenko, M.A. Electrochemical performance of AB_2 metal hydride alloys measured at −40 °C. J. *Alloy. Compd.* 2013. Available online: http://dx.doi.org/10.1016/j.jallcom.2013.01.125 (accessed on 18 June 2013).

103. Yasuoka, S.; Magari, Y.; Murata, T.; Tanaka, T.; Ishida, J.; Nakamura, H.; Nohma, T.; Kihara, M.; Baba, Y.; Teraoka, H. Development of high-capacity nickel-metal hydride batteries using superlattice hydrogen-absorbing alloys. *J. Power Sources* **2006**, *156*, 662–666.

104. Yasuoka, S.; Kihara, M. Development of high capacity nickel-metal hydride batteries using superlattice hydrogen-absorbing alloys. *Rare Earths* **2006**, *48*, 112–113.

105. Teraoka, H. Development of low self-discharge nickel-metal hydride battery. Available online: http://www.scribd.com/doc/9704685/Teraoka-Article-En (accessed on 18 July 2013).

106. Ching, P.; Liu, S.; Tan, M. The research and development of A_2B_7-type of La_2Ni_7-based MH alloy. *Technol. Trend* **2011**, *14*, 23–24.

107. Zhang, Y.; Chen, L.; Yang, T.; Xu, C.; Ren, H.; Zhao, D. The electrochemical hydrogen storage performances of Si-added La-Mg-Ni-Co-based A_2B_7-type electrode alloys. *Rare Metals*, 2013. Available online: http://dx.doi.org/10.1007/s12598-013-0053-x (accessed on 21 June 2013).

108. Dong, Z.; Ma, L.; Ding, Y.; Wu, Y.; Shen, X.; Wang, L. Effects of sintering temperature on microstructure and electrochemical characteristics of La-Mg-Ni system hydrogen storage alloys. *J. Nanjing Univ. Technol.* **2011**, *33*, 12–15.

109. Liu, J.; Han, S.; Li, Y.; Yang, S.; Shen, W.; Zhang, L.; Zhou, Y. An investigation on phase transformation and electrochemical properties of as-cast and annealed $La_{0.75}Mg_{0.25}Ni_x$ (x = 3.0, 3.3, 3.5, 3.8) alloys. *J. Alloy. Compd.* **2013**, *552*, 119–126.

110. Hu, W.; Denys, R.V.; Nwakwuo, C.C.; Holm, T.; Maehlen, J.P.; Solberg, J.K.; Yartys, V.A. Annealing effect on phase composition and electrochemical properties of the Co-free La_2MgNi_9 anode for Ni-metal hydride batteries. *Electrochim. Acta* **2013**, *96*, 27–33.

111. Huang, T.; Yuan, X.; Yu, J.; Wu, Z.; Han, J.; Sun, G.; Xu, N.; Zhang, Y. Effects of annealing treatment and partial substitution of Cu for Co on phase composition and hydrogen storage performance of $La_{0.7}Mg_{0.3}Ni_{3.2}Co_{0.35}$ alloy. *Int. J. Hydrog. Energy* **2012**, *37*, 1074–1079.

112. Jiang, W.; Mo, X.; Wei, Y.; Zhou, Z.; Guo, J. Effect of annealing treatment on hydrogen storage properties of La-Ti-Mg-Ni-based alloy. *J. Rare Earths* **2012**, *30*, 450–455.

113. Jiang, W.; Qin, C.; Zhu, R.; Guo, J. Annealing effect on hydrogen storage property of Co-free $La_{1.8}Ti_{0.2}MgNi_{8.7}Al_{0.3}$ alloy. *J. Alloy. Compd.* **2013**, *565*, 37–43.

114. Fang, X.; Luo, Y.; Gao, Z.; Zhang, G.; Kang, L. Effects of annealing treatment on microstructure and electrochemical properties of $La_{0.68}Gd_{0.2}Mg_{0.12}Ni_{3.3}Co_{0.3}Al_{0.1}$ hydrogen storage alloys. *J. Funct. Mater.* **2012**, *20*, 2751–2756.

115. Guo, P.; Lin, Y.; Zhao, H.; Guo, S.; Zhao, D. Structure and high-temperature electrochemical properties of $La_{0.6}Nd_{0.15}Mg_{0.25}Ni_{3.3}Si_{0.10}$ hydrogen storage alloys. *J. Rare Earths* **2011**, *29*, 574–579.

116. Li, F.; Young, K.; Ouchi, T.; Fetcenko, M.A. Annealing effects on structural and electrochemical properties of $(LaPrNdZr)_{0.83}Mg_{0.17}(NiCoAlMn)_{3.3}$ alloy. *J. Alloy.*

Compd. **2009**, *471*, 371–377.

117. Quao, Y.Q.; Zhao, M.S.; Zhang, Q.M.; Zhai, J. Crystal structure and some dynamic performances of $Ti_{0.25}V_{0.34}Du_{0.01}Cr_{0.1}Ni_{0.3}$ hydrogen storage electrode. *Chin. Chem. Lett.* **2011**, *22*, 93–96.
118. Luo, Y.; Feng, C.; Teng, X.; Kang, L. Study on electrochemical properties of annealed $R_{0.67}Mg_{0.33}Ni_{3.0}$ (R = La, Ce, Pr, Nd) hydrogen storage alloys. *Rare Metal. Mater. Eng.* **2010**, *39*, 90–95.
119. Zhou, X.; Young, K.; West, J.; Regalado, J.; Cherisol, K. Degradation mechanisms of high-energy bipolar nickel metal hydride battery with AB_5 and A_2B_7 alloys. *J. Alloy. Compd.* **2013**. in press.
120. Liu, Y.; Cao, Y.; Huang, L.; Gao, M.; Pan, H. Rare earth-Mg-Ni-based hydrogen storage alloys as negative electrode materials for Ni/MH batteries. *J. Alloy. Compd.* **2011**, *509*, 675–686.
121. Dong, Z.; Ma, L.; Wu, Y.; Wang, L.; Shen, X. Microstructure and electrochemical hydrogen storage characteristics of $(La_{0.7}Mg_{0.3})_{1-x}Ce_xNi_{2.8}Co_{0.5}$ (x = 0–0.20) electrode alloys. *Int. J. Hydrog. Energy* **2011**, *36*, 3016–3021.
122. Wang, B.; Chen, Y.; Liu, Y. Structure and electrochemical properties of $(La_{1-x}Dy_x)_{0.8}Mg_{0.2}Ni_{3.4}Al_{0.1}$ (x = 0.0–0.20) hydrogen storage alloys. *Int. J. Hydrog. Energy* **2012**, *37*, 9082–9087.
123. Luo, Y.; Jia, Q.; Wu, T.; Kang, L. Influence of rare-earth elements R on phase-structure and electrochemical properties of hydrogen storage alloys (LaRMg)$(NiCoAlZn)_{3.5}$. *J. Lanzhou Univ. Technol.* **2009**, *35*, 1–6.
124. Zhang, Y.; Hou, Z.; Li, B.; Ren, H.; Cai, Y.; Zhao, D. Electrochemical hydrogen storage characteristics of as-cast and annealed $La_{0.8-x}Nd_xMg_{0.2}Ni_{3.15}Co_{0.2}Al_{0.1}Si_{0.05}$ (x = 0–0.4) alloys. *Trans. Nonferrous Metals Soc. China* **2013**, *23*, 1403–1412.
125. Liu, S.; Qin, M.; Qing, P.; Huang, X.; Huang, H.; Guo, J. Study on the hydrogen storage and electrochemical characteristics of $(La_{1-x}Nd_x)_2Mg(Ni_{0.8}Co_{0.15}Mn_{0.05})_9$ (x = 0~0.3) alloys. *Guangxi Sci.* **2011**, *18*, 242–245.
126. Ren, H.; Li, B.; Hou, Z.; Liu, X.; Chen, L.; Zhang, Y. Enhanced electrochemical cycle stability of RE-Mg-Ni system A_2B_7-type alloys by melt spinning and substituting La with Pr. *Adv. Mater. Res.* **2012**, *415–417*, 1603–1607.
127. Zhang, Y.; Hou, Z.; Yang, T.; Zhang, G.; Li, X.; Zhao, D. Structure and electrochemical hydrogen storage characteristics of $La_{0.8-x}Pr_xMg_{0.2}Ni_{3.15}Co_{0.2}Al_{0.1}Si_{0.05}$ (x = 0–0.4) electrode alloys. *J. Cent. South. Univ. Technol.* **2013**, *20*, 1142–1150.
128. Zhang, Y.; Cai, Y.; Zhao, C.; Zhai, T.; Zhang, G.; Zhao, D. Electrochemical performances of the as-melt $La_{0.75-x}M_xMg_{0.25}Ni_{3.2}Co_{0.2}Al_{0.1}$ (M = Pr, Zr; x = 0, 0.2) alloys applied to Ni/metal hydride (MH) battery. *Int. J. Hydrog. Energy* **2012**, *37*, 14590–14597.
129. Li, P.; Hou, Z.; Yang, T.; Shang, H.; Qu, X.; Zhang, Y. Structure and electrochemical hydrogen storage characteristics of the as-cast and annealed $La_{0.8-x}Sm_xMg_{0.2}Ni_{3.15}Co_{0.2}Al_{0.1}Si_{0.05}$ (x = 0–0.4) alloys. *J. Rare Earths* **2012**, *30*, 696–704.
130. Dong, Z.; Wu, Y.; Ma, L.; Shen, X.; Wang, L. Influences of low-Ti substitution for La and Mg on the electrochemical and kinetic characteristics of AB_3-type hydrogen storage alloy electrodes. *Sci. China Technol. Sci.* **2010**, *53*, 242–247.
131. Dong, Z.; Ma, L.; Wang, L.; Wu, Y.; Shen, X. Effects of substituting La with Ti on microstructure and electrochemical properties of AB_3-type hydrogen storage alloys. *J. Nanjing Univ. Technol.* **2011**, *33*, 57–62.
132. Zhang, Y.; Yang, T.; Shang, H.; Chen, L.; Ren, H.; Zhao, D. Electrochemical hydrogen storage kinetics of $La_{0.75-x}Zr_xMg_{0.25}Ni_{3.2}Co_{0.2}Al_{0.1}$ (x = 0–0.2) alloys prepared by melt spinning. *Energy Procedia* **2012**, *16*, 1275–1282.
133. Wei, F.; Li, Li.; Xiang, H.; Li, H.; Wei, F. Phase structure and electrochemical properties of $La_{1.7+x}Mg_{1.3-x}(NiCoMn)_{9.3}$ (x = 0 – 0.4) hydrogen storage alloys. *Trans. Nonferrous Metals Soc. China* **2012**, *22*, 1995–1999.
134. Jiang, L.; Lan, Z.; Li, G.; Guo, J. Influence of magnesium on electrochemical properties of $RE_{3-x}Mg_x(Ni_{0.7}Co_{0.2}Mn_{0.1})_9$ (x = 0.5–1.25) alloy electrodes. *J. Rare Earths* **2012**, *30*, 1255–1259.
135. Gao, Z.; Luo, Y.; Li, R.; Lin, Z.; Kang, L. Phase structures and electrochemical properties of $La_{0.8-x}Gd_{0.2}Mg_xNi_{3.1}Co_{0.3}Al_{0.1}$ hydrogen storage alloys. *J. Power Sources* **2013**, *241*, 509–516.
136. Gao, Z.; Kang, L.; Luo, Y. Influence of magnesium content on structure and performance of A_2B_7-type RE-Mg-Ni Hydrogen storage alloys. *Chem. J. Chin. Univ.* **2012**, *33*, 2035–2042.
137. Guo, X.; Luo, Y.; Gao, Z.; Zhang, G.; Kang, L. The effect of Mg on the microstructure and electrochemical properties of $La_{0.8-x}Gd_{0.2}Mg_xNi_{3.3}Co_{0.3}Al_{0.1}$ (x = 0–0.4) hydrogen storage alloys. *J. Funct. Mater.* **2012**, *18*, 2450–2455.
138. Dong, Z.; Wu, Y.; Ma, L.; Wang, L.; Shen, X.; Wang, L. Microstructure and electrochemical hydrogen storage characteristics of $La_{0.67}Mg_{0.33-x}Ca_xNi_{2.75}Co_{0.25}$ (x = 0–0.15) electrode alloys. *Int. J. Hydrog. Energy* **2011**, *36*, 3050–3055.
139. Dong, X.; Yang, L.; Li, X.; Wang, Q.; Ma, L.; Lin, Y. Effect of substitution of aluminum for nickel on electrochemical properties of $La_{0.75}Mg_{0.25}Ni_{3.5-x}Co_{0.2}Al_x$ hydrogen storage alloys. *J. Rare Earths* **2011**, *29*, 143–149.
140. Qin, M.; Liu, S.; Qing, P.; Lan, Z.; Guo, J. The influence of Co content on $LaNi_{3.2-x}Mn_{0.3}Co_x$ (x = 0.2~0.8) alloy hydrogen storage and electrochemical properties. *Phys. Procedia* **2011**, *22*, 577–583.
141. Lan, Z.; Yan, W.; Qin, C.; Lu, Z.; Jiang, J.; Guo, J. Preparation and electrochemical properties of $La_{0.7}Zr_{0.1}Mg_{0.2}Ni_{3.4-x}Co_xFe_{0.1}$ Alloy. *Guangxi Sci.* **2012**, *19*, 134–138.
142. Qin, M.; He, B.; Qing, P.; Lu, Z.; Liu, Y.; Guo, J. Effect of Co content on electrochemical properties of $La_{0.55}Pr_{0.05}Nd_{0.15}Mg_{0.25}Ni_{3.5-x}Co_xAl_{0.25}$ (x = 0~0.4) hydrogen storage alloy. *Mater. Sci. Eng. Powder Metall.* **2012**, *17*, 160–165.

143. Xiong, K.; Zhu, R.; Ding, Y.; Lan, Z.; Huang, C.; Guo, J. Study on the electrochemical properties of $La_{0.7}Mg_{0.3}Ni_{3.4-x}(Al_{0.3}Co_{0.4})_{0.2+x}$ (x = 0~0.3) hydrogen storage alloy electrodes. *Guangxi Sci.* **2012**, *19*, 57–63.

144. Dong, Z.; Ma, L.; Shen, X.; Wang, L.; Wu, Y.; Wang, L. Cooperative effect of Co and Al on the microstructure and electrochemical properties of AB_3-type hydrogen storage electrode alloys for advanced MH/Ni secondary battery. *Int. J. Hydrog. Energy* **2011**, *36*, 893–900.

145. Qin, M.; Lan, Z.; Ding, Y.; Xiong, K.; Liu, Y.; Guo, J. A study on hydrogen storage and electrochemical properties of $La_{0.55}Pr_{0.05}Nd_{0.15}Mg_{0.25}Ni_{3.5}(Co_{0.5}Al_{0.5})_x$ (x = 0.0, 0.1, 0.3, 0.5) alloys. *J. Rare Earths* **2012**, *30*, 222–227.

146. Wang, W.; Chen, Y.; Li, Q.; Yang, W. Microstructures and electrochemical properties of $LaNi_{3.8-x}Al_x$ hydrogen storage alloys. *J. Rare Earths* **2013**, *31*, 497–501.

147. Qin, M.; Liu, S.; Qing, P.; Lan, Z.; Guo, J. Effect of Mn substitution for Ni on hydrogen storage properties and electrochemical characteristics of $(La_{0.8}Nd_{0.2})_2Mg(Ni_{0.9-x}Co_{0.1}Mn_x)_9$ alloys. *Mater. Rev.* **2011**, *25*, 35–39.

148. Zhang, H.; Wang, B.; Wang, D.; Sui, Y.; He, D.; Xia, B. Effect of Mn content on the structure and electrochemical characteristics of $(La_{0.78}Mg_{0.22})(Ni_{(0.9-x)}Co_{0.1}Mn_x)_{3.5}$ hydrogen storage alloys. *Shanghai Metals* **2010**, *32*, 16–19.

149. Pavlyuk, V.; Pozycka-Sokolowska, E.; Marciniak, B.; Paul-Boncour, V.; Dorogova, M. The structural, magnetic, hydrogenation and electrode properties of $REMg_2Cu_{9-x}Ni_x$ alloys (RE = La, Pr, Tb). *Cent. Eur. J. Chem.* **2011**, *9*, 1133–1142.

150. Hou, C.; Zhao, M.; Li, J.; Huang, L. Phase structure and electrochemical characteristics of $CeMn_{0.25}Al_{0.25}Ni_{1.5+x}$ (x = 0.0, 0.3, 0.5, 0.7, 0.9, 1.1) alloys. *Rare Metal. Mater. Eng.* **2012**, *41*, 100–104.

151. Young, K.; Wu, A.; Qiu, Z.; Tan, J.; Mays, W. Effects of H_2O_2 addition to the cell balance and self-discharge of Ni/MH batteries with AB_5 and A_2B_7 alloys. *Int. J. Hydrog. Energy* **2012**, *37*, 9882–9891.

152. Nwakwuo, C.C.; Holm, T.; Denys, R.V.; Hu, W.; Maehlen, J.P.; Solberg, J.K.; Yartys, V.A. Effect of magnesium content and quenching rate on the phase structure and composition of rapidly solidified La_2MgNi_9 metal hydride battery electrode alloy. *J. Alloy. Compd.* **2013**, *555*, 201–208.

153. Tang, C.; Pan, W.; Qin, Y.; Zhou, H. Preparation and hydrogen storage properties of $La_{0.7}Mg_{0.3}Ni_{2.8}Co_{0.5-x}Fe_x$ series alloys. *Chin. J. Nonferrous Metals* **2011**, *21*, 1373–1379.

154. Yang, S.; Liu, H.; Han, S.; Li, Y.; Shen, W. Effects of electrodeless composite plating Ni-Cu-P on the electrochemical properties of La-Mg-Ni-based hydrogen storage alloy. *Appl. Surf. Sci.* **2013**, *271*, 210–215.

155. Zhang, J.; Villeroy, B.; Knosp, B.; Bernard, P.; Latroche, M. Structural and chemical analyses of the new ternary La_5MgNi_{24} phase synthesized by spark plasma sintering and used as negative electrode material for Ni-MH batteries. *Int. J. Hydrog. Energy* **2012**, *37*, 5225–5233.

156. Shen, W.; Han, S.; Li, Y.; Yang, S.; Miao, Q. Effect of electroplating polyaniline on electrochemical kinetics of La-Mg-Ni-based hydrogen storage alloy. *Appl. Surf. Sci.* **2012**, *258*, 6316–6320.

157. Yang, Z.P.; Li, Q.; Zhao, X.J. Influence of magnetic annealing on electrochemical performance of $La_{0.67}Mg_{0.33}Ni_{3.0}$ hydride electrode alloy. *J. Alloy. Compd.* **2013**, *558*, 99–104.

158. Zhang, S.; Wang, D.; Hou, X.; Huang, X.; Sun, L. Research on a new synthesis technology of La-Mg-Ni system alloys and microstructure of $La_{0.67}Mg_{0.33}Ni_{3.0}$ alloys. *Rare Metal. Mater. Eng.* **2012**, *41*, 2075–2080.

159. Justi, E.W.; Ewe, H.H.; Kalberlah, A.W.; Saridakis, N.M.; Schaefer, M.H. Electrocatalysis in the nickel-titanium system. *Energy Convers.* **1970**, *10*, 146–156.

160. Gutjahr, M.A.; Buchner, H.; Beccu, K.D.; Saeufferer, H. A new type of reversible negative electrode for alkaline storage batteries based on metal alloy hydrides. *Power Sources* **1973**, *4*, 79–91.

161. Beccu, K. Negative electrode of titanium-nickel alloy hydride phases. U.S. Patent 3,824,131, 16 July 1974.

162. Zhao, X.; Li, J.; Yao, Y.; Ma, L.; Shen, X. Electrochemical hydrogen storage properties of non-equilibrium Ti_2Ni alloy. *RSC Adv.* **2012**, *2*, 2149–2153.

163. Emami, H.; Cuevas, F. Hydrogenation properties of shape memory Ti(Ni, Pd) compds. *Intermetallics* **2011**, *19*, 876–886.

164. Emami, H.; Souques, R.; Crivello, J.; Cuevas, F. Electronic and structural influence of Ni by Pd substitution on the hydrogenation properties of TiNi. *J. Solid State Chem.* **2013**, *198*, 475–484.

165. Nei, J.; Young, K.; Wong, D.F. Electrochemical properties of $TiNi_{0.9}X_{0.1}$ (X = Ni, Cr, Mn, Fe, Co, Cu) metal hydride alloys. *J. Alloy. Compd.* Being prepared for submission.

166. Zhao, X.; Ma, L.; Qu, X.; Ding, Y.; Shen, X. Effect of mechanical milling on the structure and electrochemical properties of Ti_2Ni alloy in an alkaline battery. *Energy Fuels* **2009**, *23*, 4678–4682.

167. Zhao, X.; Ma, L.; Yao, Y.; Ding, Y.; Shen, X. Ti_2Ni alloy: A potential candidate for hydrogen storage in nickel/metal hydride secondary batteries. *Energy Environ. Sci.* **2010**, *3*, 1316–1321.

168. Zhao, X.; Zhou, J.; Shen, X.; Yang, M.; Ma, L. Structure and electrochemical hydrogen storage properties of A_2B-type Ti-Zr-Ni alloys. *Int. J. Hydrog. Energy* **2012**, *37*, 5050–5055.

169. Li, X.D.; Elkedim, O.; Nowak, M.; Jurczyk, M.; Chassagnon, R. Structural characterization and electrochemical hydrogen storage properties of $Ti_{2-x}Zr_xNi$ (x = 0, 0.1, 0.2) alloys prepared by mechanical alloying. *Int. J. Hydrog. Energy* **2013**, *38*, 12126–12132.

170. Cao, X.; Ma, L.; Yang, M.; Zhao, X.; Ding, Y. Electrochemical properties of the amorphous Ti_3Ni_2 alloy in Ni/MH batteries. *Rare Metal. Mater. Eng.* **2012**, *41*, 1511–1515.

171. Gibbons, P.C.; Kelton, K.F. *Physical Properties of Quasicrystals*; Stadnik, Z.M., Ed.; Springer: Berlin,

Germany, 1999; p. 403.

172. Takasaki, A.; Kuroda, C.; Lee, S.; Kim, J. Electrochemical properties of $Ti_{45}Zr_{38-x}Ni_{17+x}$ ($0 \leq x \leq 8$) quasicrystals produced by rapid-quenching. *J. Alloy. Compd.* **2011**, *509*, S782–S785.

173. Baster, D.; Takasaki, A.; Kuroda, C.; Hanc, E.; Lee, S.; Swierczek, K.; Szmyd, J.S.; Kim, J.; Molenda, J. Effect of mechanical milling on electrochemical properties of $Ti_{45}Zr_{38-x}Ni_{17+x}$ (x= 0,8) quasicrystals produced by rapid-quenching. *J. Alloy. Compd.* 2003. Available online: http://dx.doi.org/10.1016/j.jallcom.2013.03.272 (accessed on 27 May 2013).

174. Ariga, Y.; Takasaki, A.; Kuroda, C.; Kulka, A. Electrochemical properties of $Ti_{45-x}Zr_{30+x}Ni_{25}$($x$ = −4, 0, 4) quasicrystal and amorphous electrodes produced by mechanical alloying. *J. Alloy. Compd.* **2003**. in press.

175. Hu, W.; Wang, L.; Wang, L. Quinary icosahedral qausicrystalline Ti-V-Ni-Mn-Cr alloy: A novel anode material for Ni-MH rechargeable batteries. *Mater. Lett.* **2011**, *65*, 2868–2871.

176. Hu, W.; Yi, J.; Zheng, B.; Wang, L. Icosahedral qausicrystalline $(Ti_{1.6}V_{0.4}Ni)_{100-x}Sc_x$ alloys: Synthesis, structure and their application in Ni-MH batteries. *J. Solid State Chem.* **2013**, *202*, 1–5.

177. Ma, F.; Liu, X.; Yan, S.; Ao, D.; Ren, Y. Research progress of influence factors on electrochemical performance of Mg-base hydrogen storage alloys. *Metallic Funct. Mater.***2011**, *34*, 67–70.

178. Anik, M.; Özdemir, G.; Küçükdeveci, N.; Baksan, B. Effect of Al, B, Ti and Zr additive elements on the electrochemical hydrogen storage performance of MgNi alloy. *Int. J. Hydrog. Energy* **2011**, *36*, 1568–1577.

179. Anik, M.; Özdemir, G.; Küçükdeveci, N. Electrochemical hydrogen storage characteristics of Mg-Pd-Ni ternary alloys. *Int. J. Hydrog. Energy* **2011**, *36*, 6744–6750.

180. Anik, M. Effect of titanium additive element on the discharging behavior of MgNi alloy electrode. *Int. J. Hydrog. Energy* **2011**, *36*, 15075–15080.

181. Anik, M.; Karanfil, F.; Küçükdeveci, N. Development of the high performance magnesium based hydrogen storage alloy. *Int. J. Hydrog. Energy* **2012**, *37*, 299–308.

182. Etiemble, A.; Idrissi, H. On the decrepitation mechanism of MgNi and $LaNi_5$-based electrodes studied by *in situ* acoustic emission. *J. Power Sources* **2011**, *196*, 5168–5173.

183. Etiemble, A.; Idrissi, H.; Roué, L. *In situ* investigation of the volume change and pulverization of hydride materials for Ni-MH batteries by concomitant generated force and acoustic emission measurements. *J. Power Sources* **2012**, *205*, 500–505.

184. Kim, S.; Chourashiya, M.G.; Park, C.; Park, C. Electrochemical performance of NAFION coated electrodes of hydriding combustion synthesized MgNi based composite hydride.*Mater. Lett.* **2013**, *93*, 81–84.

185. Liu, J. Effect of Ti-La substitution on electrochemical properties of amorphous MgNi-based secondary hydride electrodes. *J. Tianjin Norm. Univ.* **2011**, *31*, 67–70.

186. Etiemble, A.; Idrissi, H. Effect of Ti and Al on the pulverization resistance of MgNi-based metal hydride electrodes evaluated by acoustic emission. *Int. J. Hydrog. Energy* **2013**, *38*, 1136–1144.

187. Etiemble, A.; Rousselot, S.; Guo, W.; Idrissi, H.; Roué, L. Influence of Pd addition on the electrochemical performance of Mg-Ni-Ti-Al-based metal hydride for Ni-MH batteries. *Int. J. Hydrog. Energy* **2013**, *38*, 7169–7177.

188. Santos, S.F.; de Castro, J.F.R.; Ticianelli, E.A. Microstructures and electrode performances of $Mg_{50}Ni_{(50-x)}Pd_x$ alloys. *Cent. Eur. J. Chem.* **2013**, *11*, 485–491.

189. Spassov, T.; Köster, U. Thermal stability and hydriding properties of nanocrystalline melt-spun $Mg_{63}Ni_{30}Y_7$. *J. Alloy. Compd.* **1998**, *279*, 279–286.

190. Zhang, Y.; Song, C.; Ren, H.; Li, Z.; Hu, F.; Zhao, D. Enhanced hydrogen storage kinetics of nanocrystalline and amorphous Mg_2N-type alloy by substituting Ni with Co. *Trans. Nonferrous Metals Soc. China* **2011**, *21*, 2002–2009.

191. Zhang, Y.; Zhao, D.; Ren, H.; Guo, S.; Qi, Y.; Wang, X. Enhanced electrochemical hydrogen storage characteristics of nanocrystalline and amorphous $Mg_{20}Ni_{10-x}Co_x$ (x = 0–4) alloys by melt spinning. *Rare Metal. Mater. Eng.* **2011**, *40*, 1146–1151.

192. Zhang, Y.; LU, K.; Zhao, D.; Guo, S.; Qi, Y.; Wang, X. Electrochemical hydrogen storage characteristics of nanocrystalline and amorphous Mg_2Ni-type alloys prepared by melt-spinning. *Trans. Nonferrous Metals Soc. China* **2011**, *21*, 502–511.

193. Zhang, Y.; Zhang, G.; Yang, T.; Hou, Z.; Guo, S.; Qi, Y.; Zhao, D. Electrochemical and hydrogen absorption/desorption properties of nanocrystalline Mg_2Ni-type alloys prepared by melt spinning. *Rare Metal. Mater. Eng.* **2012**, *41*, 2069–2074.

194. Zhang, Y.; Yang, T.; Shang, H.; Zhao, C.; Xu, C.; Zhao, D. The electrochemical hydrogen storage characteristics of as-spun nanocrystalline and amorphous $Mg_{20}Ni_{10-x}M_x$ (M = Cu,Co,Mn; x = 0–4) alloys. *Rare Metals*, 2013. Available online: http://dx.doi.org/10.1007/s12598-013-0051-z (accessed on 15 July 2013).

195. Zhang, Y.; Zhao, C.; Yang, T.; Shang, H.; Xu, C.; Zhao, D. Comparative study of electrochemical performances of the as-melt $Mg_{20}Ni_{10-x}M_x$ (M = None, Cu, Co, Mn; x = 0, 4) alloys applied to Ni/metal hydride (MH) battery. *J. Alloy. Compd.* **2013**, *555*, 131–137.

196. Zhang, Y.; Qi, Y.; Zhao, D.; Guo, S.; Ma, Z.; Wang, X. An investigation of hydrogen storage kinetics of melt-spun nanocrystalline and amorphous Mg_2Ni-type alloys. *J. Rare Earths* **2011**,*29*, 87–93.

197. Zhang, Y.; Li, B.; Ren, H.; Hou, Z.; Hu, F.; Wang, X. Influences of melt spinning on electrochemical hydrogen storage performance of nanocrystalline and amorphous Mg_2Ni-type alloys. *J. Cent. South. Univ. Technol.* **2011**, *18*, 1825–1832.

198. Huang, L.W.; Elkedim, O.; Nowak, M.; Chassagnon, R.; Jurczyk, M. $Mg_{2-x}Ti_xNi$ (x = 0, 0.5) alloys prepared by mechanical alloying for electrochemical hydrogen

storage: Experiments and first-principles calculations. *Int. J. Hydrog. Energy* **2012**, *37*, 14248–14256.

199. Zhang, J.; Zhu, Y.; Wang, Y.; Pu, Z.; Li, L. Electrochemical hydrogen storage properties of $Mg_{2-x}Al_xNi$ (x = 0, 0.3, 0.5, 0.7) prepared by hydriding combustion synthesis and mechanical milling. *Int. J. Hydrog. Energy* **2012**, *37*, 18140–18147.
200. Jiang, C.; Wang, H.; Wang, Y.; Cheng, X.; Tang, Y.; Liu, Z.; Xie, H. Enhanced electrochemical performance of Mg_2Ni alloy prepared by rapid quenching in magnetic field. *J. Power Sources***2013**, *238*, 257–264.
201. Zhu, Y.; Yang, C.; Zhu, J.; Li, L. Structural and electrochemical hydrogen storage properties of Mg_2Ni-based alloys. *J. Alloy. Compd.* **2011**, *509*, 5309–5314.
202. Rousselot, S.; Guay, D.; Roue, L. Comparative study on the structure and electrochemical hydriding properties of MgTi, $Mg_{0.5}Ni_{0.5}Ti$ and $MgTi_{0.5}Ni_{0.5}$ alloys prepared by high energy ball milling. *J. Power Sources* **2011**, *196*, 1561–1568.
203. Cao, Z.; Wu, D.; Li, C.; Bian, J.; Zhang, K. Effect of Ni addition on electrochemical properties of Mg-Al hydrogen storage alloy. *J. Shenyang Norm. Univ.* **2013**, *31*, 16–20.
204. Yang, C.; Zhu, Y.; Zhang, W.; Zhu, J.; Li, L. Effects of metal doping on electrochemical properties of Mg-based hydrogen storage alloys. *J. Nanjing Univ. Technol.* **2011**, *33*, 52–57.
205. Anik, M. Improvement of the electrochemical hydrogen storage performance of magnesium based alloys by various additive elements. *Int. J. Hydrog. Energy* **2012**, *37*, 1905–1911.
206. Huang, L.W.; Elkedim, O.; Nowak, M.; Jurczyk, M.; Cahssagnon, R.; Meng, D.W. Synergistic effects of multiwalled carbon nanotubes and Al on the electrochemical hydrogen storage properties of Mg_2Ni-type alloy prepared by mechanical alloying. *Int. J. Hydrog. Energy* **2012**,*37*, 1538–1545.
207. Xu, J.; Niu, D.; Fan, Y. Electrochemical hydrogen storage performance of $Mg_{2-x}Al_xNi$ thin films. *J. Power Sources* **2012**, *198*, 383–388.
208. Akiba, E.; Iba, H. Hydrogen absorption by Lases phase related BCC solid solution.*Intermetallics* **1998**, *6*, 461–470.
209. Chai, Y.J.; Zhao, M.S.; Wang, N. Crystal structural and electrochemical properties of $Ti_{0.15}Zr_{0.08}V_{0.35}Cr_{0.10}Ni_{0.30-x}Mn_x$ (x = 0–0.12) alloys. *Mater. Sci. Eng. B* **2008**, *147*, 47–51.
210. Guéguen, A.; Joubert, J.M.; Latroche, M. Influence of the C14 $Ti_{35.4}V_{32.3}Fe_{32.3}$ Laves phase on the hydrogenation properties of the body-centered cubic compd $Ti_{24.5}V_{59.3}Fe_{16.2}$. *J. Alloy. Compd.* **2011**, *509*, 3013–3018.
211. Wang, N.; Chai, Y. Electrochemical properties and reaction mechanism of Ti-V multiphase alloy at high temperature. *Rare Metal. Mater. Eng.* **2011**, *40*, 1519–1521.
212. Wang, Y.; Zhao, M. Electrochemical hydrogen storage characteristics of $Ti_{0.10}Zr_{0.15}V_{0.35}Cr_{0.10}Ni_{0.30}$ – 10% $LaNi_5$ composite and its synergetic effect. *Trans. Nonferrous Metals Soc. China* **2012**, *22*, 2000–2006.
213. Wang, Y.; Zhao, M. Distinct synergistic effect in $Ti_{0.10}Zr_{0.15}V_{0.35}Cr_{0.10}Ni_{0.30}$ + 1.0 wt % $LaNi_5$ hydrogen storage composite electrode. *Int. J. Hydrog. Energy* **2012**, *37*, 3276–3282.
214. Wang, Y.; Zhao, M. Electrochemical characteristics and synergetic effect of $Ti_{0.10}Zr_{0.15}V_{0.35}Cr_{0.10}Ni_{0.30}$ – 10 wt % $LaNi_5$ hydrogen storage composite electrode. *J. Rare Earths* **2012**, *30*, 146–150.
215. Inoue, H.; Koyama, S.; Higuchi, E. Charge-discharge performance of Cr-substituted V-based hydrogen storage alloy negative electrodes for use in nickel-metal hydride batteries.*Electrochim.Acta* **2012**, *59*, 23–31.
216. Hang, Z.; Xiao, X.; Li, S.; Ge, H.; Chen, C.; Chen, L. Influence of heat treatment on the microstructure and hydrogen storage properties of $Ti_{10}V_{77}Cr_6Fe_6Zr$ alloy. *J. Alloy. Compd.***2012**, *529*, 128–133.
217. Liu, W.; Wang, X.; Hu, W.; Kawabe, Y.; Watada, M.; Wang, L. Electrochemical performance of TiVNi-Quasicrystal and AB_3-Type hydrogen storage alloy composite materials. *Int. J. Hydrog. Energy* **2011**, *36*, 616–620.
218. Liu, W.; Zhang, S.; Wang, L. $Ti_{1.4}V_{0.6}Ni$ quasicrystal and its composites with$xV_{18}Ti_{15}Zr_{18}Ni_{29}Cr_5Co_7Mn$ alloy used as negative electrode materials for the nickel-metal hydride (Ni-MH) secondary batteries. *Mater. Lett.* **2012**, *79*, 122–124.
219. Liu, W.; Zhang, S.; Wang, L. Influence of heat treatment on electrochemical properties of $Ti_{1.4}V_{0.6}Ni$ alloy electrode containing icosahedral quasicrystalline phase. *Trans. Nonferrous Metals Soc. China* **2012**, *22*, 3034–3038.
220. Liu, W.; Liang, F.; Zhang, S.; Lin, J.; Wang, X.; Wang, L. Electrochemical properties of Ti-based Quasicrystal and ZrV_2 Laves phase alloy composite materials as negative electrode for Ni-MH secondly batteries. *J. Non-Cryst. Solids* **2012**, *358*, 1846–1849.
221. Liu, W.; Kawabe, Y.; Liang, F.; Okuyama, R.; Lin, J.; Wang, L. A composite based on Fe substituted TiVNi alloy: Synthesis, structure and electrochemical hydrogen storage property. *Intermetallics* **2013**, *34*, 18–22.
222. Young, K.; Ouchi, T.; Fetcenko, M.A.; Mays, W.; Reichman, B. Structural and electrochemical properties of $Ti_{1.5}Zr_{5.5}V_xNi_{10-x}$. *Int. J. Hydrog. Energy* **2009**, *34*, 8695–8706.
223. Young, K.; Ouchi, T.; Huang, B.; Nei, J.; Fetcenko, M.A. Studies of $Ti_{1.5}Zr_{5.5}V_{0.5}(M_xNi_{1-x})_{9.5}$(M = Cr, Mn, Fe, Co, Cu, Al): Part 1. Structural characteristics. *J. Alloy. Compd.* **2010**, *501*, 236–244.
224. Young, K.; Nei, J.; Huang, B.; Ouchi, T.; Fetcenko, M.A. Studies of $Ti_{1.5}Zr_{5.5}V_{0.5}(M_xNi_{1-x})_{9.5}$(M = Cr, Mn, Fe, Co, Cu, Al): Part 2. Hydrogen storage and electrochemical properties. *J. Alloy. Compd.* **2010**, *501*, 245–254.
225. Nei, J.; Young, K.; Salley, S.O.; Ng, K.Y.S. Effects of annealing to $Zr_8Ni_{19}X_2$ (X = Ni, Mg, Al, Sc, V, Mn, Co, Sn, La, and Hf): Structural characteristics. *J. Alloy. Compd.* **2012**, *516*, 144–152.

226. Nei, J.; Young, K.; Salley, S.O.; Ng, K.Y.S. Effects of annealing on $Zr_8Ni_{19}X_2$ (X = Ni, Mg, Al, Sc, V, Mn, Co, Sn, La, and Hf): Hydrogen storage and electrochemical properties. *Int. J. Hydrog. Energy* **2012**, *37*, 8418–8427.

227. Lee, H.; Chourashiya, M.G.; Park, C.; Park, C. Hydrogen storage and electrochemical properties of the $Ti_{0.32}Cr_{0.43-x-y}V_{0.25}Fe_xMn_y$ (x = 0–0.055, y = 0–0.080) alloys and their composites with $MmNi_{3.99}Al_{0.29}Mn_{0.3}Co_{0.6}$ alloy. *J. Alloy. Compd.* **2013**, *566*, 37–42.

228. Huang, H.; Huang, K. Effect of AB_5 alloy on electrochemical properties of $Mm_{0.80}Mg_{0.20}Ni_{2.56}Co_{0.50}Mn_{0.14}Al_{0.12}$ hydrogen storage alloy. *Powder Technol.* **2012**, *221*, 365–370.

229. Huang, H.; Huang, K.; Chen, D.; Liu, S.; Zhuang, S. The electrochemical properties of MgNi-x wt % $TiNi_{0.56}Co_{0.44}$ (x = 0, 10, 30, 50) composite alloys. *J. Mater. Sci.* **2010**, *45*, 1123–1129.

230. Zlatanova, Z.; Spassov, T.; Eggeler, G.; Spassova, M. Synthesis and hydriding/dehydring properties of Mg_2Ni-AB (AB = TiNi or TiFe) nanocomposites. *Int. J. Hydrog. Energy* **2011**, *36*, 7559–7566.

231. Hsu, F.; Lin, C.; Lee, S.; Lin, C.; Bor, H. Effect of Mg_3MnNi_2 on the electrochemical characteristics of Mg_2Ni electrode alloy. *J. Power Sources* **2010**, *195*, 374–379.

232. Lee, S.; Hsu, F.; Chen, W.; Lin, C.; Lin, J. Influence of Mg_3AlNi_2 content on the cycling stability of Mg_2Ni-Mg_3AlNi_2 hydrogen storage alloy electrodes. *Intermetallics* **2011**, *19*, 1953–1958.

233. Lee, S.; Huang, C.; Chou, Y.; Hsu, F. Horng, J. Effects of Co and Ti on the electrode properties of Mg_3MnNi_2 alloy by ball-milling process. *Intermtallics* **2013**, *34*, 122–127.

234. He, W.; Zhao, Y.; Huang, J.; Zhang, Y.; Zeng, L. Electrochemical performance of compds $Ho_6Fe_{23-x}Co_x$ (x = 0, 1, 3). *J. Rare Earths* **2011**, *29*, 124–128.

235. Yao, X.; Ma, R.; Huang, J.; He, J.; He, W.; Zeng, L. Electrochemical performance of Gd-Co-Mn alloys. *J. Rare Earths* **2012**, *30*, 540–544.

CHAPTER 8

ELECTROCHEMICAL AND OPTICAL PROPERTIES OF MAGNESIUM-ALLOY HYDRIDES

Thirugnasambandam G. Manivasagam, Kamil Kiraz and Peter H. L. Notten

Keywords: Hydrogen storage, magnesium hydride, switchable mirrors, bulk powders, thin films, electrochemical hydrogenation, gas phase hydrogenation

INTRODUCTION

The depleting fossil fuel resources and altering climate conditions increases the necessity for alternative sustainable energy technologies. Among the green-house gases, CO_2 emission from burning fossil fuels has been especially identified as the main perpetrator for the undesirable climate change. The CO_2 emission has increased from 20.7 billion tonnes (Gt) in 1990 to 32.5 Gt in 2006 (9.9 Gt of carbon equivalent) [1]. Furthermore, the oil and natural gas reserves are expected to decrease within a few decades. Among the available alternative energy sources such as solar, wind, tidal, geothermal, etc., the hydrogen economy seems advantageous, considering its wide-spread availability, high storage capacity, utilization, efficiency, and environmental compatibility [2]. Hydrogen is therefore expected to play a dominant role in future energy scenarios as a promising energy carrier. It undergoes a clean combustion reaction with oxygen, producing energy and water only, according to

$$H_{2\,(g)} + \frac{1}{2}O_{2\,(g)} \rightarrow H_2O_{(l)} \quad (\Delta H_f^{\circ} = -285.8 \text{ kJ/mol at } 25°C) \qquad (1)$$

On the other hand, for example, in the combustion reaction of the natural gas methane, CO_2 is the dominant byproduct, which obviously is an undesirable green-house gas.

$$CH_{4(g)} + 2O_{2(g)} \rightarrow CO_{2(g)} + 2H_2O_{(g)} \quad (\Delta H_f^{\circ} = -802.3 \text{kJ/mol at } 25\,°C) \qquad (2)$$

The energy per mass of hydrogen (142 MJ kg^{-1}) is favorably much higher than that of other chemical fuels, such as liquid hydrocarbons (47 MJ kg^{-1}) [3,4]. Conventionally, H_2 is generally stored in high-pressure cylinders to increase the volumetric density [5]. Novel high-pressure tanks made of composites are currently being developed which can withstand more than 700 bars [4]. Even though these cylinders can withstand high pressures, their large volume and the risk of explosion limit their practical applications. On the other hand, condensation under cryogenic conditions seems to be attractive due to a significant increase in the volumetric density of up to 70.8 kg/m^3. However, liquid hydrogen containers are open systems and the loss of hydrogen is therefore hard to control. Alternatively, significant density increases can also be achieved by storing hydrogen in the form of metal-hydrides. It is interesting to realize that the density of hydrogen in metal-hydrides is often even higher than in liquid hydrogen. Due to the limited amount of free H_2 gas that is in equilibrium with that stored in the solid, the metal-hydrides offer a relatively safe storage. Table 1 lists the comparison of hydrogen content and energy density information of several storage media.

The reversible hydrogen absorption and desorption capability of numerous metals and intermetallic compounds have already been known for a long period of time [6]. The discovery in the late 1960s at the Dutch Philips Research laboratories showed that the intermetallic compound $LaNi_5$ is

Citation: Thirugnasambandam G. Manivasagam, Kamil Kiraz and Peter H. L. Notten, Electrochemical and Optical Properties of Magnesium-Alloy Hydrides, doi: 10.3390/cryst2041410.

able to absorb a significant amount of hydrogen gas reversibly. Soon after this discovery it was realized that electrodes made of $LaNi_5$ could serve as new electrochemical storage medium and could become a competitive alternative for the cadmium electrode in conventional Ni-Cd batteries. Even though $LaNi_5$ can reversibly absorb and desorb a large amount of hydrogen under ambient conditions, the gravimetric capacity is still relatively low, about 1.2 wt.% of Hydrogen [7]. Considering the cost, availability and gravimetric storage capacity, magnesium is therefore, in principle, a much more suitable candidate with a storage capacity of up to 7.6 wt.% of hydrogen (see Table 1).

Table 1. Hydrogen content and energy density in various media [8].

Medium	Hydrogen Content (wt. %)	Volume Density* (H atoms l^{-1}) (10^{-19})	Energy Density**	
			MJ kg^{-1}	mJ l^{-1}
$H_{2(g)}$ (150 atm)	100.00	0.5	141.90	1.02
$H_{2(l)}$ (−253°C)	100.00	4.2	141.90	9.92
MgH_2	7.65	6.7	9.92	14.32
VH_2	2.10	11.4	-	-
Mg_2NiH_4	3.60	5.9	4.48	11.49
$TiFeH_{1.95}$	1.95	5.5	2.47	13.56
$LaNi_5H_{6.7}$	1.50	7.6	1.94	12.77
$ZrMn_2H_{3.6}$	1.75	6.0	-	-
$ZrMn_2Fe_{0.8}H_{3.4}$	1.38	4.8	-	-

*Does not include container weight or void volumes; ** Refers to the value equivalent to hydrogen in metal hydrides

In comparison to the other metal-hydrides, MgH_2 has a relatively high energy density of 9.92 MJ/kg [8]. Although Mg fulfills major requirements for serving as a novel hydrogen storage material, it cannot be used in practical applications due to its poor (de)sorption kinetics and high (de)hydrogenation reaction temperature (350–400 °C). The poor kinetics for the hydrogenation has been ascribed to two consecutive facts: the dissociation of the hydrogen molecules is hindered by the oxidation of the Mg surface and the hydrogen diffusion is limited in the rutile structure [9]. Despite these drawbacks, the lightweight property, high energy density and low cost make Mg-hydride a potentially promising hydrogen storage material for mobile applications.

Mg-based systems can be modified to improve the sorption properties, such as temperature (thermodynamics) and reaction rate (kinetics). It has been recognised that magnesium can be modified in different ways, i.e., forming nano particles, alloying Mg with transition metals and adding catalysts. This review will focus on the electrochemical and optical properties of the bulk and thin film Mg-TM alloy hydrides.

GAS PHASE *VERSUS* ELECTROCHEMICAL HYDROGEN STORAGE

Many metals and alloys are capable of reversibly absorbing and desorbing large amounts of hydrogen. Loading can be done either by introducing molecular hydrogen gas or reducing water in an electrolyte, as schematically indicated in Figure 1 a and b, respectively. Molecular hydrogen is dissociated at the surface and is subsequently absorbed inside the host lattice while two H atoms recombine to H_2 molecules during the desorption process (Figure 1a).

$$2M + H_2 \leftrightarrow 2MH_{ad} \tag{3}$$

$$MH_{ad} \leftrightarrow MH_{abs} \tag{4}$$

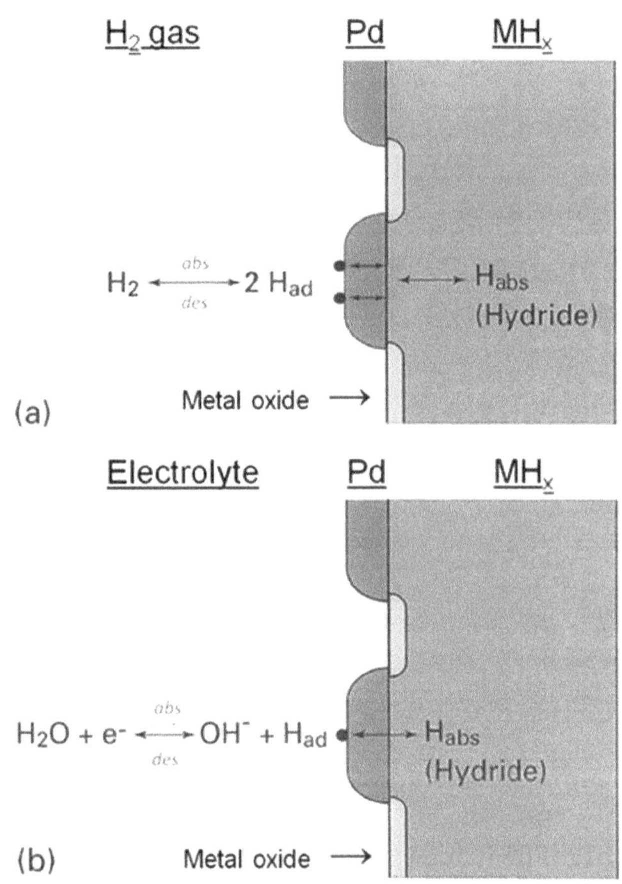

Figure 1. Schematic model for hydrogenation via (**a**) gas phase (**b**) electrolyte.

The thermodynamics of hydride formation via the gas phase can be elegantly described by Pressure–Composition–Isotherms (PCI) as shown in Figure 2. Initially a small amount of hydrogen (H/M < 0.1) is dissolved in the metal lattice during the absorption process and forms a solid solution, which is generally denoted as α-phase. In this hydrogen concentration range the pressure increases as a function of hydrogen content. As the hydrogen concentration increases beyond a certain critical value (H/M > 0.1), the local interaction between hydrogen atoms dominates which leads to the nucleation and growth of a new hydride phase called β-phase, When the two phases coexists the isotherm exhibits a horizontal plateau region and, according to the phase rule, pressure independent characteristics are indeed expected in the plateau region. The length of the plateau is a function of hydrogen content. When all the α-phase has been converted into the β-phase the pressure dependence is again rather strong and the hydrogen pressure raises steeply with hydrogen concentration. The plateau level is experimentally found to be a clear function of the temperature. Beyond the critical temperature (T_c), no clear phase transition occurs anymore and only a solid solution is then present.

The equilibrium pressure depends on the temperature and is related to the change in both enthalpy (ΔH) and entropy (ΔS). The entropy change is mainly associated with the dissociation of hydrogen molecules into adsorbed H-atoms at the surface of the hydride-forming material. As the entropy of the gas is the most dominant parameter, the entropy change is more or less the same for all known hydride systems and is dictated by the hydrogen in the gas phase. The enthalpy term is associated with the bond strength of hydrogen with the host material. The entropy and enthalpy values can be experimentally obtained from the Van 't Hoff relationship, according to

$$ln\left(\frac{p_{eq}}{p_o}\right) = \frac{\Delta H}{RT} - \Delta S_0/R \tag{5}$$

where p_{eq} and p_o are equilibrium and standard pressures, respectively, ΔH is the standard enthalpy change, R the gas constant, T the temperature and ΔS_0 is the entropy change. The operating temperatures of the metal hydrides are determined by the equilibrium pressure and the overall reaction kinetics.

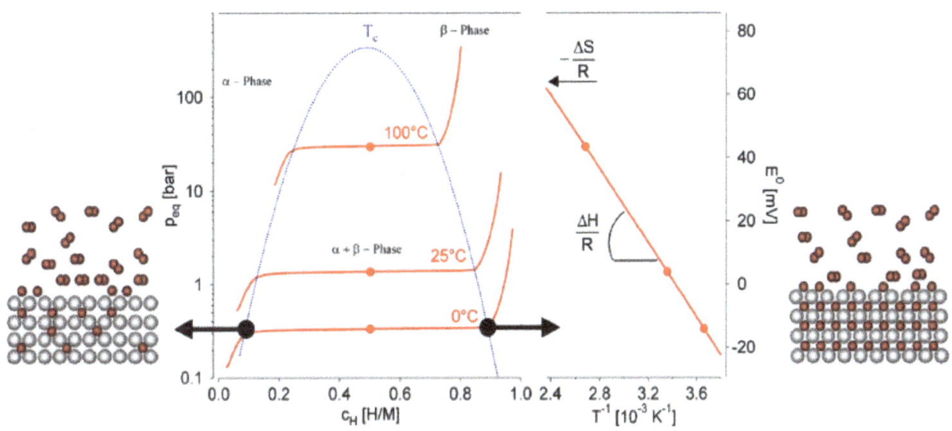

Figure 2. Pressure composition isotherms for hydrogen absorption [10].

Another way to hydrogenate the host metal is by electrochemical reduction of water at a metallic electrode immersed in an alkaline solution, according to

$$M + H_2O + e^- \leftrightarrow MH_{ad} + OH^- \tag{6}$$

This reversible process is also schematically indicated in Figure 1b. Subsequently, the adsorbed hydrogen atoms (H_{ad}) are absorbed by the metal, leading to H_{abs}. In both cases, Pd addition prevents surface oxidation.

$$MH_{ad} \leftrightarrow MH_{abs} \tag{7}$$

By changing the direction of the current, Equations (6) and (7) can be reversed and the absorbed hydrogen can reversibly be extracted from the host electrode. An advantage of this electrochemical method is that the hydrogen content can be precisely calculated from the amount of charge (Q) that is used during electrochemical (de)hydrogenation. Q can be quantified by integrating the current (I) with respect to time (t) according to equation (8)

$$Q = \int_0^{t\ end} |I|dt \tag{8}$$

As every electron corresponds to a single H atom that is inserted or extracted, the number of absorbed/desorbed hydrogen atoms can easily be calculated from Q by using the faraday constant (F)

$$\frac{Q}{F} = mole\ \mathrm{H} \tag{9}$$

The exact hydrogen storage capacity per mass unit can then be calculated with respect to the total mass of the electrode material.

To determine the thermodynamics, the electrochemical absorption and desorption properties in the equilibrium state of the metal-hydride system, the so-called Galvanostatic Intermittent Titration Technique (GITT) is often used. This method is based on collecting equilibrium potential data during partial (dis)charging–resting cycles in which short current pulses are followed by current-off voltage relaxation periods. It has been deduced that the equilibrium potential (E_{eq}) of the fully relaxed metal-hydride electrode can be related to the partial hydrogen pressure P_{H_2}).

$$E_{eq} = \frac{RT}{nF} ln\left(\frac{P_{H2}}{P_{ref}}\right) vs. RHE \tag{10}$$

where P_{ref} is the reference pressure (1 bar) and RHE refers to the reversible hydrogen electrode as reference voltage [11].

APPLICATIONS OF METAL-HYDRIDES

NiMH Battery

Electrochemical hydrogen storage in metal hydrides (MH) is primarily applied in aqueous NiMH secondary batteries. The alternative MH electrode is aimed to replace the Cd-electrode which is widely used in Ni-Cd batteries and is highly toxic. Moreover, MH electrodes reveal significantly improved storage capacities and (dis)charge kinetics.

A schematic representation of a complete NiMH battery, containing a hydride-forming electrode and a nickel electrode, is shown in Figure 3. Both the electrically insulating separator and porous electrodes are impregnated with a strong alkaline solution (concentrated KOH) that provides the ionic conductivity between the two electrodes. The basic electrochemical reactions, occurring at both electrodes during charging and discharging can be represented by equations (11) and (12) respectively.

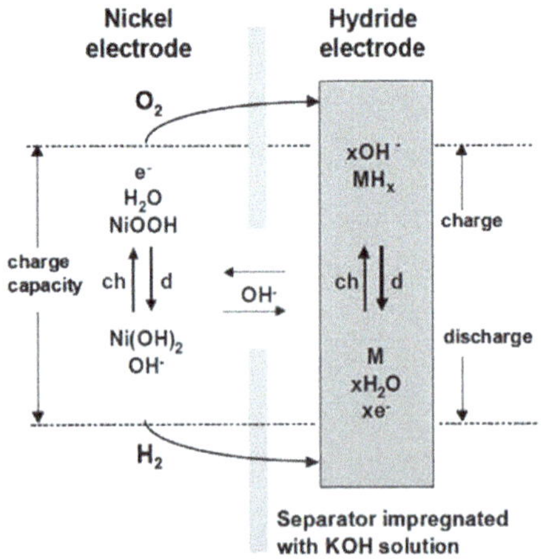

Figure 3. The concept of a sealed rechargeable Ni-metal hydride (MH) battery seen from a schematic point of view [12].

$$\mathrm{Ni(OH)_2 + OH^-} \underset{\text{discharging}}{\overset{\text{charging}}{\rightleftarrows}} \mathrm{NiOOH + H_2O + e^-} \tag{11}$$

$$M + \mathrm{H_2O + e^-} \underset{\text{discharging}}{\overset{\text{charging}}{\rightleftarrows}} MH + \mathrm{OH^-} \tag{12}$$

During charging, divalent Ni(II) is oxidized to the trivalent Ni(III) state and water is reduced to hydrogen atoms which are first adsorbed on the metal electrode, which are then subsequently absorbed by the hydride-forming compound. The reverse reactions take place during discharging. The net effect of this reaction sequence (Equations 11 and 12) is that electrical charge is chemically stored in the electrode in the form of hydrogen. The reaction takes place without any water consumption preventing drying out of the battery unlike Ni-Cd batteries. The hydrogen storage alloy currently used in NiMH batteries is a Mischmetal-based $LaNi_5$ compound. Researchers are still aiming to improve the properties of these materials mainly by focusing on the (dis)charge kinetics, storage capacity and corrosion resistance during over(dis)charging. Currently, the target sector of the NiMH battery technology is mainly mobile applications specifically Hybrid Electrical Vehicles (HEV) in the automotive industry [12].

Electrochromic Applications

Conceptually, 'electrochromism' is the reversible optical change of specific compounds in response to a change in the oxidation state of the involved electro-active species. The terminology was initially introduced by Platt in 1961 [13] for the color change of organic dyes under the influence of a strong electric field. The first extensive study was done by Deb in 1969 [14] in the field of inorganic metal oxide systems (WO_3). Numerous researchers both in industry and academically focused on different classes of electrochromic materials. The introduction of optically active metal-hydride thin films was reported by Huiberts et al. in 1996 and was based on yttrium- and lanthanum hydrides [15]. Metal-hydrides expanded the classes of electrochromic materials and were listed among advanced optically switching inorganic compounds. Unlike most of the electrochromic color changing materials, metal hydrides even showed all three optical states (reflecting, absorbing and transmitting states) during the reversible (de) hydrogenation processes without any color change [16]. In this section, the development of optically active, hydride-forming materials are highlighted.

In the extensive work of Griessen et al., metal-hydride switchable mirrors can be classified into three generations: rare earth (RE) metal hydrides, magnesium – rare earth (MgRE) metal hydrides and magnesium–transition metal (MgTM) hydrides [16]. The leading work of Huiberts et al.focuses on the hydrogenation of rare earth yttrium thin films via the gas phase and in situ optical characterization of the transmission of the sample. The pressure-composition isotherm of a 300 nm Y film capped with a top layer of 15 nm Pd, indicated by black squares, is shown in Figure 4together with the intensity of transmitted light (red circles). The optical transmission has been measured at $\lambda = 635$ nm ($\hbar\omega = 1{,}96$ eV). Both the as-prepared metallic yttrium sample (α phase) and the di-hydrided YH_2 state (β phase) show almost no transmission whereas the tri-hydrided state at $x = 3$ (γ phase) has become optically transparent. Optical switching becomes saturated at composition $YH_{2,86}$ [16].

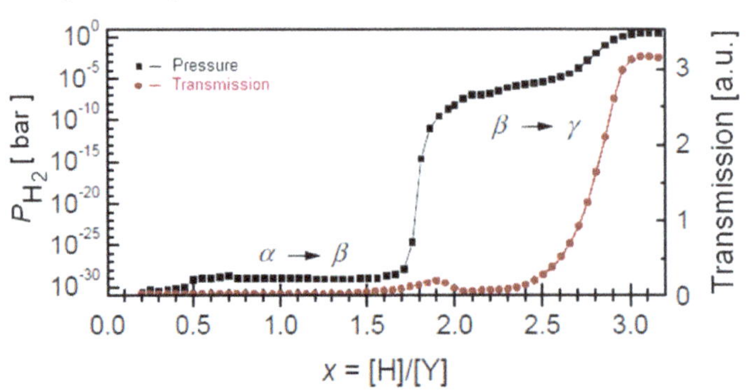

Figure 4. Pressure–Composition isotherm of yttrium film during gas phase hydrogenation (black square) and transmission spectrum depending on the hydrogen concentration (red circle) [16].

The first electrochemical hydrogenation experiments with combined optical analyses of yttrium film proved that optical switching can also be induced electrochemically in alkaline protective media, as is shown in Figure 6 [17]. Here a constant (de)hydrogenation current (1 mA) is applied to a 500 nm thick Y electrode that is capped with a 5 nm thick Pd-catalyst layer. Electrochemical (de)hydrogenation is performed in 6 M KOH electrolyte with Pt counter electrode and Hg/HgO reference electrode. Optical switching is reported to take place in the hydrogen concentration range of $2.7 < x < 2.8$ for YH_x. Electrochemical switching could be accomplished in the order of seconds during charging and minutes during discharging, revealing the slower dehydrogenation kinetics.

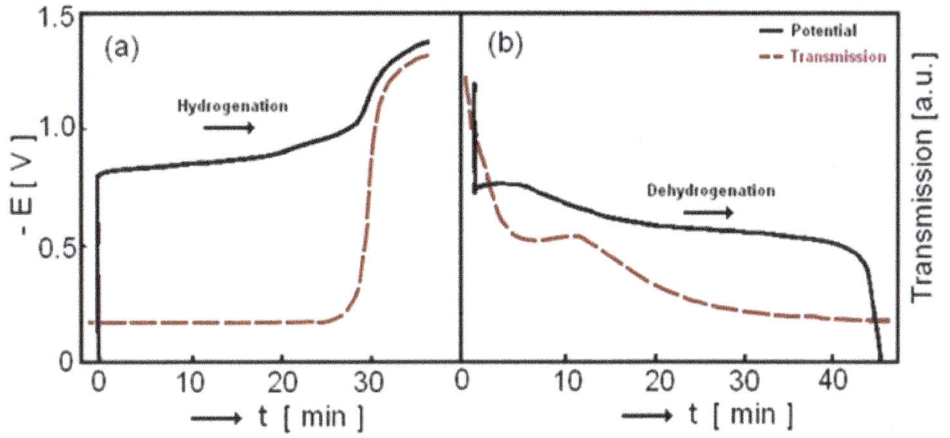

Figure 5. Electrode potential (solid black line) and optical transmission (dashed red line) intensity of yttrium film during electrochemical hydrogenation (a) and dehydrogenation (b) [17].

The second generation Mg-RE switchable mirrors, Gd-Mg alloys with different Gd to Mg ratios were optically examined during gas phase hydrogenation [18]. The effect of alloy composition for $Gd_{1-x}Mg_x$ binary alloys ($0 \leq x \leq 0.9$) on the transmission is reported in Figure 6 for a wide wavelength spectrum in the range of 200 to 1000 nm.

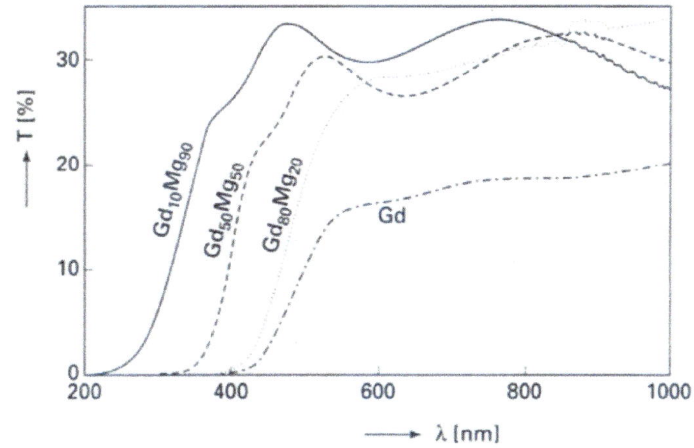

Figure 6. Influence of Gd-Mg composition upon the transmission of $GdMgH_3$ thin films [18].

All samples in Figure 6 were 200 nm films, capped with 10 nm Pd and hydrogenated under a constant pressure of 5 bar of hydrogen at room temperature. In this work, magnesium was intended to be used as alloying element for the rare earth element Gd. Moreover, the color neutrality and transparency for gadolinium-magnesium hydride were remarkable and of significant importance for optical applications. In Figure 7, all three possible optical states can be seen during the transition from metallic GdMg to semiconducting $GdMgH_3$.

Figure 7. Three optical states of Mg-Gd switchable mirror: reflecting (**a**), absorbing (**b**) and transmitting (**c**) states [19].

At the beginning of the new century, magnesium became the material of renewed interest not only from an optical point of view but also from an energy storage aspect. Various transition metals were investigated in order to control the kinetics of the (de)hydrogenation reaction. Niessenet al. showed that metastable alloys can be prepared in thin film form by depositing Mg with Ti, V or Cr, though they are immiscible elements from a thermodynamic point of view, [20]. The hydrogen content of these materials also strongly affects the optical properties, which can be used in hydrogen sensors, smart solar collectors and switchable mirrors. The pioneering work of the third generation of switchable mirrors, Mg-transition metals, is highlighted in Ref. [17] where the reversible optical transition of Mg-rich Nb binary alloys was carried out by Richardson et al. in 2001 [21].

Figure 8 shows the electrochemical (de)hydrogenation of a 50 nm thick Mg-Ni film covered with a 10 nm Pd top layer in 8 M KOH. Nickel is an interesting alloying element, which is thermodynamically expected to destabilize Mg. Optical switching of this material has also been confirmed in the gas phase under a stream of 4% hydrogen in argon or helium. Visual coloration has been reported as pale yellow in the Mg-rich region and deep red in the Ni-rich region in the hydride, transparent, state.

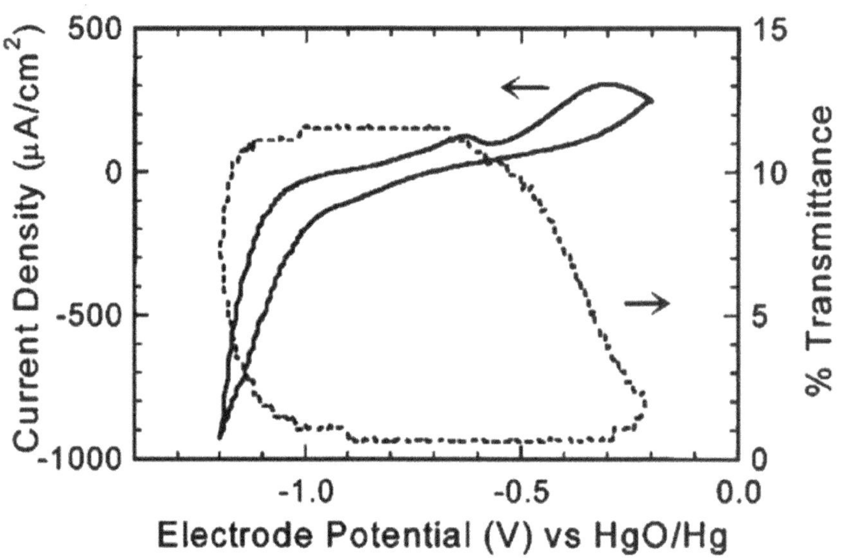

Figure 8. Electrochemical (de)hydrogenation cycle (line) with switching optical response (dashed) between reflective and transparent states of Mg-Ni alloy and its hydride respectively [21].

In order to understand the main concepts behind the different colorations, the optical states and visual responses during the metal-to-metal hydride transition, the energy band gap between the valence band and conduction band of the individual metal-hydride must be considered.

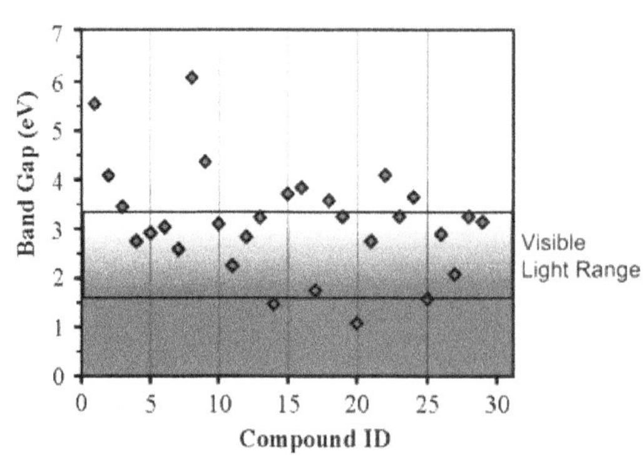

(ID 1-5) MgH_2, $LiMgH_3$, $NaMgH_3$, $KMgH_3$, $RbMgH_3$; (ID 6-10) $CsMgH_3$, $MgAlH_5$, MgB_2H_8, $MgAl_2H_8$, $Ba_6Mg_7H_{26}$; (ID 11-15) $BaMgH_4$, $Ca_4Mg_3H_{14}$, $Ca_{19}Mg_8H_{54}$, $Ca_4Mg_4FeH_{63}$, $CaMgNiH_4$; (ID 16-20) Cs_2MgH_4, Cs_3MgH_5, K_2MgH_4, $LiMg_2RuH_6$, $MgRuH_4$; (ID 21-25) Mg_3ReH_7, Rb_2MgH_4, Rb_3MgH_5, $Rb_4Mg_3H_{10}$, $SrMg_2FeH_8$; (ID 26-29) $SrMgH_4$, $SrMgNiH_4$, $Yb_4Mg_3H_{14}$, $Sr_2Mg_3H_{10}$.

Figure 9. Fundamental band gaps of complex magnesium hydrides [22].

For instance, MgH_2 is a colorless insulator with a band-gap of E_g = 5.16 eV. The optical band-gap of Mg-Ni alloy (Mg_2NiH_4), on the other hand, varies between 2.5 and 2.8 eV, depending on the presence of Mg rich areas which has been attributed to MgH_2 segregation. Since the band-gap energy of these materials seems to be an important and determining factor, an overview of the band gaps is shown in Figure 9 and was prepared by Karazhanov et al. [22].According to the band-gap (E_g) theory, all incident light is transmitted when E_g of the material is greater than the energy of the visible light which ranges from 390 nm violet (3.2 eV) to 760 nm red (1.6 eV). When E_g value is in the range of the energy of the visible light ($1.6 < E_g < 3.2$ eV), color change is expected, depending on the absorbed and emitted light wavelengths (gradient region). Finally, when E_g is less than the energy of the visible light, all the white light energy is absorbed and emitted, which results in a zero transparency (dark region). Figure 9 illustrates the basic idea about the expected optical response bands for magnesium containing metal-hydride compounds.

MAGNESIUM-BASED BINARY ALLOYS

The importance of light-weight materials in hydrogen storage technology is obvious. Magnesium-based alloys offer this potential. Therefore, lots of effort has been put into this research over a long period of time. For example, the extremely poor (de)hydrogenation kinetics of pure Mg have been significantly improved by the discovery of nano-structured alloying materials. In particular, ball-milling was successfully adopted by several research groups to produce interesting storage materials.

It has been shown that alloying Mg with transition metals has also a significant impact on the thermodynamics and kinetics of the hydride-formation reaction. Mg_2Ni was the first alloy that was extensively studied in this respect [23,24,25]. Even though, this alloy has a better performance in terms of thermodynamics and kinetics, the hydrogen content in Mg_2NiH_4 is only about 3.6 wt.%, which is less than half of that of pure MgH_2.

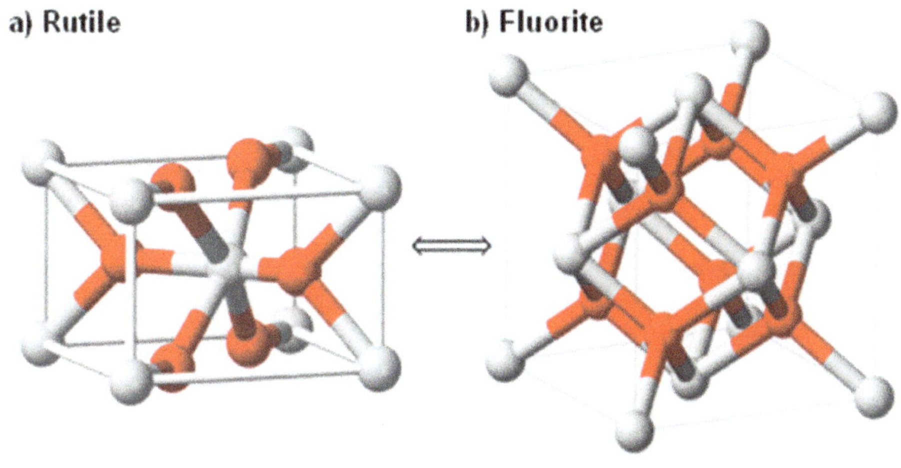

Figure 10. Rutile MgH_2 (a) and fluorite $Mg_xTM_{(1-x)}H_2$ (b).

Under standard conditions, magnesium-dihydride has the rutile structure (see MgH_2 Figure 10a) with space group P42/mnm (136). Magnesium and hydrogen atoms are located in 2a and 4fWyckoff positions, respectively. Each magnesium atom is coordinated octahedrally by hydrogen atoms. Mainly, the hydrides of the transition metals in the 4th and 5th row of the 3B to 6B groups crystallize in a fluorite structure ($MgTMH_2$ see Figure 10 b) with space group Fm3m (225) and transition metal atoms in 4a and hydrogen atoms in 8c Wyckoff positions [26]. Alloying these transition metals helps to preserve the fluorite structure in $Mg_xTM_{(1-x)}H_2$ while MgH_2 has the rutile structure which is denser. Consequently, the hydrogen mobility is limited whereas $Mg_xTM_{(1-x)}H_2$ ($x < 0.8$) adopts the fluorite structure, which is more open for hydrogen transportation.

The hydrogen storage capacity of these fluorite-structured compounds was shown to have a much larger reversible electrochemical storage capacity with respect to Mischmetal-based AB_5compounds in both thin film and bulk form. Comparing the extracted amount of charge of a fully hydrogenated AB_5-type hydride-forming electrode (curve (a) in Figure 11) with that of a fluorite-stabilized magnesium-based compound (curve (b) in Figure 11) reveals that the latter has about four times higher storage capacity. In addition, the rates at which hydrogen can be electrochemically loaded and removed in fluorite-structured Mg-alloys were substantial, making these fluorite-stabilized materials very interesting.

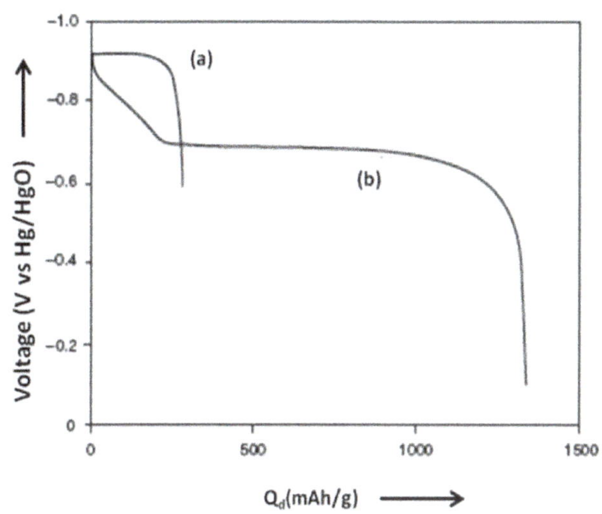

Figure 11. Comparison between (**a**) AB_5 type compound and (**b**) $Mg_{0.72}Sc_{0.28}(Pd_{0.012}Rh_{0.012})$.

Magnesium Scandium Alloys

The specific fluorite-type of magnesium-based compounds alloyed with scandium has been studied electrochemically with compositions ranging from 50 to 100 at.%, both in thin film and bulk forms. The thin film electrodes were prepared with a thickness of 200 nm on quartz substrates capped with a 10 nm Pd layer whereas the bulk powders were prepared through arc melting in a molybdenum crucible $Mg_xSc_{1-x}Pd_{0.24}$. The XRD results of the thin films confirm the formation of HCP solid solution irrespective of the composition. This is far from the equilibrium state reported in the phase diagram, where a Cs-Cl type solid solution is found between x = 50 − 65 at.% Mg and single phase α-Mg is only formed beyond 80 at.% of Mg. The XRD results of bulk powders, on the other hand, confirm the equilibrium phase diagram.

However, the electrochemical properties of Mg-Sc thin films and bulk materials are very similar. Both thin film and bulk materials show similar trends in storage capacity and the shape of the discharge curves are also very similar for the various compositions (Figure 12). The maximum storage capacity was reached at a composition of 80 at.% Mg and was 1740 mAh/g for thin films, corresponding to 6.5 wt.% of Hydrogen and 1495 mAh/g (5.6 wt.%) for bulk materials. The rate capability of the thin films is, however, much better than that of the bulk materials due to shorter diffusion lengths. [27,28].

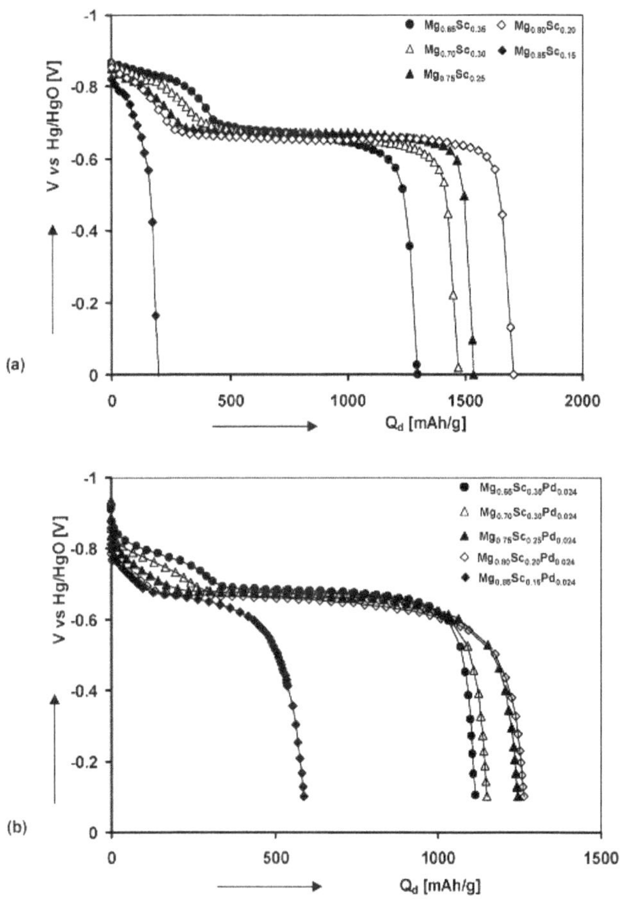

Figure 12. Galvanostatic discharge curves of: thin films (a) and bulk materials (b); the current density was 1000 mA/g for the thin films and 50 mA/g for the bulk materials [28].

It was found that the favorable discharge kinetics was retained up to 80 at.% magnesium. The improvement of the (de)hydriding kinetics compared to pure magnesium has been attributed to the more open fluorite structure of scandium-hydride instead of the more compact rutile structure of the conventional magnesium-hydrides. The plateau pressure is not a function of composition and almost equal to the equilibrium pressure of pure MgH_2. Diffraction data obtained from the bulk system elucidates that the fully charged hydride is a single-phase fcc-structured hydride and no phase segregation into MgH_2 and ScH_2 has taken place [28].

Recently, the crystal structure of a palladium-containing $Mg_{65}Sc_{35}$ compound has been studied upon hydrogenation by neutron diffraction [29]. The metallic phase adopts a pseudo-CsCl-type structure. When hydrogenated, the structure changes from pseudo-CsCl to fcc involving an elongation along the c-axis and shrinkage along the a-axis (Figure 13). Upon deuteration, a two-phase domain is subsequently observed between 0.85 and 1.55 [D/M] with two hydrides in equilibrium. Above this value, a pressure increase becomes apparent in the isotherm, leading to fully occupied tetrahedral sites and partially filled octahedral sites [30].

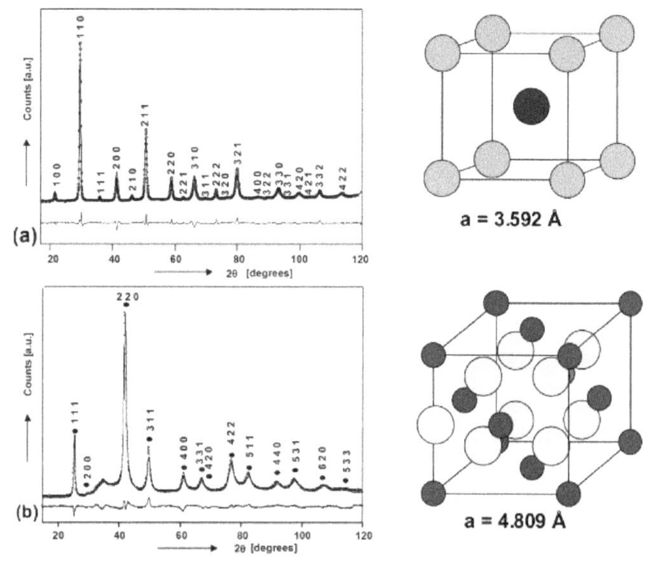

Figure 13. Neutron diffraction of $Mg_{0.65}Sc_{0.35}$ (a) before deuterium loading (b) after deuterium loading [30].

It has been argued that interstitial hydrogen is highly mobile, due to the large and partially occupied octahedral sites that that favorably influence the transportation properties. Using proton nuclear magnetic resonance (NMR), the crystallography has been proven to be responsible for the beneficial transportation properties of the fluorite-structured magnesium compounds. The frequency hopping rate for the fluorite-structured compounds at the relevant temperatures was found to be many orders of magnitude higher than that for the rutile-structured material. At any given temperature, the hopping rate of $Mg_{(1-y)}Sc_yH_x$ is faster than in ScH_2 and much faster than in the ionic MgH_2, but still slower than in $LaNi_5H_{6.8}$ (Figure 14) [31].

First principle calculations were done on $Mg_{(1-y)}Sc_y$ alloys by Pauw et al. [32]. Figure 15 shows that the decrease in the formation enthalpy with increasing Sc-content is more pronounced for the fluorite structure than for the rutile structure. The crossing point of the two curves is located around 20% Sc (y = 0.20 in Figure 16), which is the exact composition where the electrochemical results showed the maximum rate capability and hence electrochemical storage capacity. The first principle calculations are in good agreement with these experimental findings.

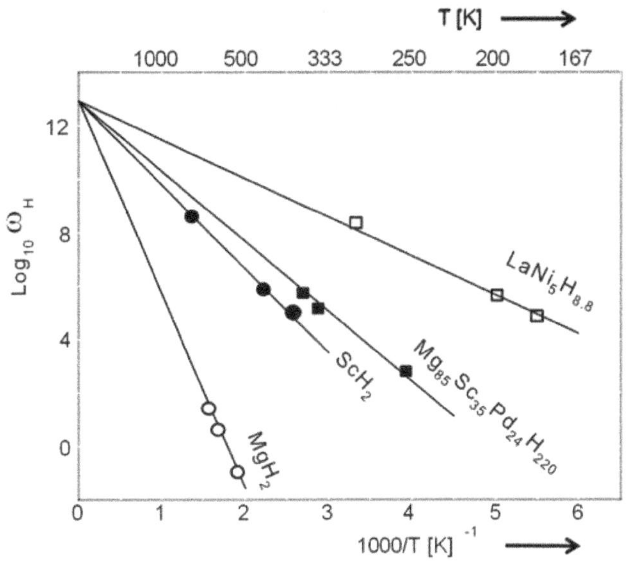

Figure 14. Relaxation map showing rate ω_H of H hopping as a function of reciprocal temperature in MgH_2, ScH_2, $Mg_{(1-y)}Sc_yH_x$, and $LaNi_5H_{6.8}$ [31].

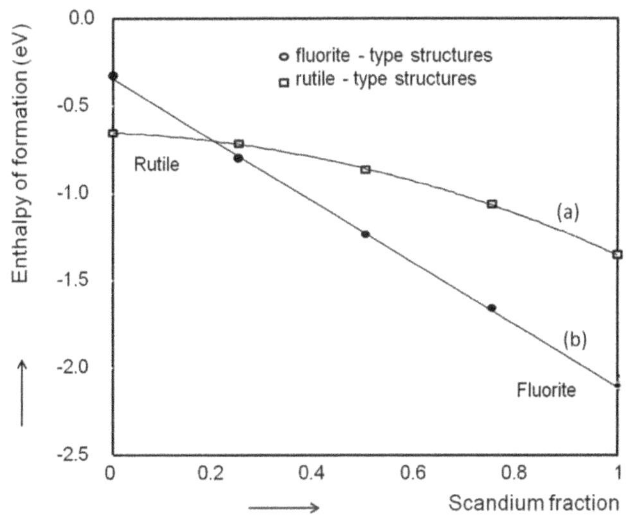

Figure 15. Enthalpies of formation of $Mg_{1-y}Sc_yH_2$. (**a**) Rutile (**b**) fluorite [32].

Magnesium Titanium Alloys

Since scandium is rather expensive, attempts to find cheaper substitutes were recently undertaken. Titanium dihydrides are also promising candidates because of having the same fluorite structure as that of scandium dihydride. Experiments performed with $Mg_{0.80}Ti_{0.20}$ thin film electrodes indeed showed a similar drastic improvement of the discharge (dehydrogenation) kinetics compared to pure magnesium hydride (MgH_2). A more elaborate study on the Mg–Ti system (Figure 16) revealed that the composition dependence of the hydrogen storage capacity was quite similar to that of the Mg-Sc system, also showing a maximum capacity around 80 at.% magnesium [33].

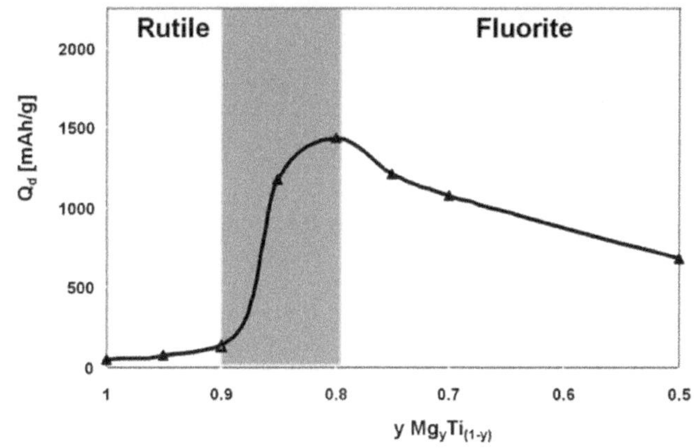

Figure 16. Electrochemical determined reversible electrochemical capacity at room temperature as a function of magnesium content in Mg-Ti thin film electrodes.

Pauw et al. [32] also investigated the formation enthalpies of metal hydrides in both the rutile and fluorite structure. The calculations showed that the addition of Ti induces a remarkable change in the crystal structure compared to pure MgH_2. The formation enthalpy of rutile $Mg_{0.50}Ti_{0.50}H_2$ is −0.421 eV/H_2, which is 22.7 kJ mol^{-1} less negative than that of pure MgH_2. The rutile structure with composition $Mg_{0.25}Ti_{0.75}H_2$ is also less stable than MgH_2, which means that Ti destabilizes the rutile structure over almost the entire compositional range. The formation enthalpy of the fluorite structure shows a tendency towards more negative values over the entire compositional range, the same as for the Mg-Sc system. The fluorite structure becomes the most stable phase around y = 0.20, which is also very similar to Mg-Sc and these theoretical results are therefore in good agreement with the experimental results. The crystal structure of the resulting hydrides seems to play a crucial role in determining the (de)hydrogenation kinetics for this interesting new class of hydrogen storage materials [32].

In situ electrochemical XRD was carried out by Vermeulen et al. to investigate the crystallographic phase transformation during the electrochemical hydrogenation of the MgTi alloys [34]. While hydrogenating $Mg_{0.90}Ti_{0.10}$ thin films electrochemically, the hexagonal reflections decrease with increasing hydrogen content (Figure 17 a). Simultaneously, the bct reflections evolve, indicating the formation of a body centered tetragonal phase which is the same as for pure MgH_2. On the other hand, hydrogenation of $Mg_{0.70}Ti_{0.30}$ showed an increase in the fcc phase (Figure 17 b). Substantial electrochemical storage capacities of more than 1500 mAh g^{-1} have been reported for the Mg-Ti system at room temperature. This corresponds to a reversible hydrogen content of about 6 wt.%. In addition, it has been shown that these meta-stable materials cannot only be produced by thin film evaporation and sputtering techniques but also by mechanical alloying.

Several attempts have been made by researchers to produce Mg-Ti alloys by mechanical alloying. Rousselot et al. investigated the phase transformation during electrochemical hydrogenation [35]. The meta-stable hcp $Mg_{0.50}Ti_{0.50}$ alloy was formed through mechanical alloying. XRD was performed on the electrochemically hydrogenated alloy. Upon cycling, the intensity of the hcp phase decreases and a new peak arises, which corresponds to hydrided fcc phase. The fcc phase is irreversibly formed and does not disappear after dehydrogenation, instead a slight decrease in the lattice constant was observed upon discharge. Kalisvaart et al. reported the formation of two fcc phases of the Mg-Ti alloy when the milling process was carried out with stearic acid as process control agent [36,37]. Among the two fcc phases only one was concluded to be electrochemically active. The storage capacity of this bulk material was reported to be 500 mAh/g.

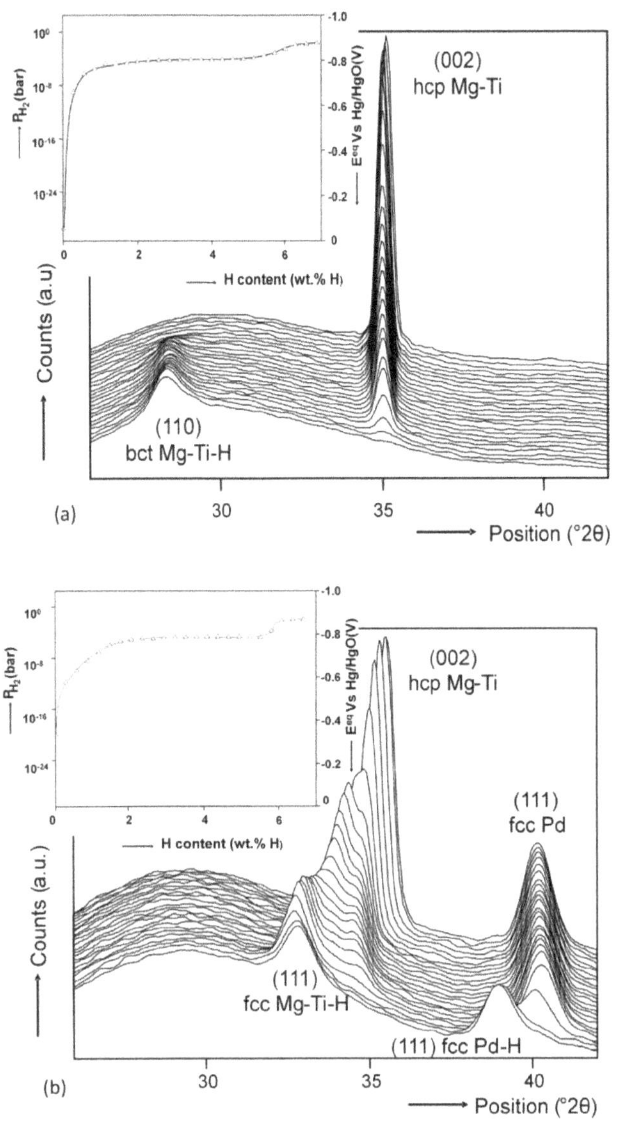

Figure 17. In-situ electrochemical XRD measurements of $Mg_{0.90}Ti_{0.10}$ (**a**) and $Mg_{0.70}Ti_{0.30}$ (**b**) during hydrogenation [34].

Kyoi et al. synthesized Mg-Ti-H compound at 600 °C in a high-pressure anvil cell above 8 GPa by mixing MgH_2 and $TiH_{1.9}$ [38]. The metal atom structure was found to be close to the Ca_7Ge type fcc structure. This is in line with the findings of Vermeulen et al [34]. It could be concluded from in situ XRD studies that the hydrogen atoms occupy the tetrahedral sites.

The Mg-Ti alloy has a positive heat of mixing and must therefore be considered as an immiscible system [39]. The alloy can therefore, in principle, not be synthesized under equilibrium conditions and thus the only way to produce these meta-stable materials is by making use of non-equilibrium preparation methods. The stability of the meta-stable alloys is still questionable at high temperature and pressure conditions. Liang et al. reported an extended solubility of Ti in Mg, during mechanical alloying of Mg and Ti [40]. In the case of Mg with 20 at.% Ti, about 12.5 at.% Ti is dissolved in the Mg lattice when the mechanical alloying process reaches a stable state, while the rest remains as fine particles of size 50–150 nm in diameter. Dissolution of Ti in the Mg lattice causes a decrease in unit cell volume. The phase segregation begins at high temperature and the supersaturated Mg-Ti alloys start to decompose at 200 °C. Hydrogenation is an exothermic reaction and enhances the decomposition process.

Srinivasan et al. investigated the structure of bulk magnesium-titanium deuteride $Mg_{0.65}Ti_{0.35}D_x$ prepared via mechanical alloying and deuterium gas-phase loading at 75 bars and 175 °C [41]. The structural analysis was made with XRD, NMR and neutron diffraction. Deuterium loading causes the formation of both interdispersed rutile $MgD_2 – TiD_y$ nano-domains and a separate fluorite TiD_z phase. Exchange NMR technique indicates complete deuterium exchange between the MgD_2 and TiD_y phase within one second. Gas phase hydrogenation of the Mg-Ti alloys at higher temperatures increases the risk of phase segregation.

Even though the kinetics of the fluorite-structured compounds is faster compared to the MgH_2, the thermodynamic properties are not improved by the addition of these transition metals. For instance, the dehydrogenation enthalpy of $Mg_xSc_{(1-x)}H_2$ and $Mg_xTi_{(1-x)}H_2$ is higher compared to MgH_2. Thus, high temperature is required to dehydrogenate these compounds in case of gas phase loading. However, only Mg-Sc alloys are stable with respect to decomposition into its elemental metallic form. Since scandium is expensive, the next optional metal Ti is preferred considering the storage capacity compared to Cr and V. Due to the high stability of the Mg-Ti hydrides and low thermal stability of Mg-Ti alloy, the addition of a third element into the binary alloys is required to stabilize the host materials and, consequently, to destabilize the alloy hydrides. Careful selection of the third alloying element could assist in improving the thermodynamic properties while maintaining the storage capacities.

Mg-Based Ternary Alloys

Increasing the plateau pressure of the metal-hydrogen system can be accomplished by destabilization of the metal hydride. Miedema proposed a mathematical model to calculate the effect on the thermodynamic stability of an additional element(s) to the metallic compound [42,43]. The Miedema model was used by Vermeulen et al. to screen elements that stabilize the metallic Mg-Ti system [44]. Al and Si were selected to destabilize the Mg-Ti alloy, due to their advantageous lightweight mass to maintain the high gravimetric capacity of the overall Mg-Ti alloy. The isothermal curves of the $Mg_{0.69}Ti_{0.21}Al_{0.10}$ alloy show two plateau regions, a plateau up to 4 wt% of Hydrogen at 10^{-6} bars and a second sloping plateau which lies at higher pressures (Figure 18).

Addition of higher Al content in the Mg-Ti lattice leads to a decrease in the reversible storage capacity. If only 10 at.% is incorporated in Mg-Ti alloy with a high Mg-content, it does not seem to significantly affect the reversible capacity. The declining capacity at high Al content could be due to kinetic limitations imposed by the material. Additionally, Si also clearly affects the isotherms compared to Mg-Ti alloys. The length of the first plateau decreases, when P_{H_2} increases steeply and soon reaches a sloping plateau. Since Si is heavier than Al, the storage capacity is lower than that of the Al-containing ternary alloys. The occurrence of the second plateau was also observed in other systems, such as $Mg_2CuAl_{0.375}$, $LaCo_6$, Ti-Fe and $NdCo_5$. Interestingly, the Mg-Ti-Al-H system shows a plateau near atmospheric pressure and temperature.

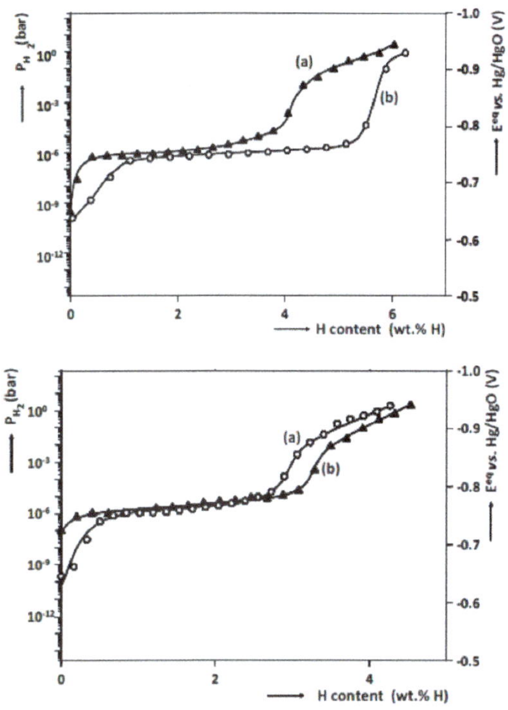

Figure 18. Electrochemically determined dehydrogenation isotherms of 200 nm; (i-a) $Mg_{0.69}Ti_{0.21}Al_{0.10}$ and (i-b) $Mg_{0.80}Ti_{0.20}$ thin films with a 10 nm Pd top-coat (ii-a) $Mg_{0.55}Ti_{0.35}Si_{0.10}$ and (ii-b) $Mg_{0.69}Ti_{0.21}Si_{0.10}$ films capped with 10 nm Pd [44].

HYDROGENOGRAPHY

The introduction of hydrogenography by Griessen et al. in 2007 was an interesting approach to analyze gradient alloy compositions during the (de)hydrogenation cycles of alloy thin films [45]. This combinatorial method enables optically monitoring of the (de)hydrogenation of numerous compositions of alloying metals simultaneously under the same experimental conditions. Controlling the metal alloy composition is the key step in the hydrogenography technique. Off-centered, co-sputterred two or more metals generate a series of heterogeneously deposited gradient compositions on the same substrate. Then, the different optical responses that are generated by multiple alloy compositions with respect to the hydrogen content are spatially analyzed via a CCD camera. Hydrogenography can be used as an in situ screening test to quickly find the optimal composition which performs faster optical conversion or higher transmission during hydrogen (un)loading. Depending on the research of interest, all optical responses (transparent, absorbent and reflecting states) with respect to the hydrogen content can be studied.

For instance, by applying combinatorial hydrogenography, multi-compositional optical analyses can be performed in a single run by focusing on the spatially defined heterogeneous metal-hydride composition, as can be seen in Figure 19. In (a) and (b) binary $Mg_{(1-y)}Ti_y$ alloys ($0.11 \leq y \leq 0.4$) and in (c) ternary gradient alloys with composition $Mg_yNi_zTi_{(1-y-z)}$ ($0.4 \leq y \leq 0.9$ and $0 \leq y \leq 0.5$) are analyzed by combined gas phase hydrogenation and in situ hydrogenography. The gradually varying optical responses upon hydrogenation can clearly be seen with respect to the different binary and ternary alloy compositions, indicating that the hydrogenation properties are strongly dependent on the materials composition [45].

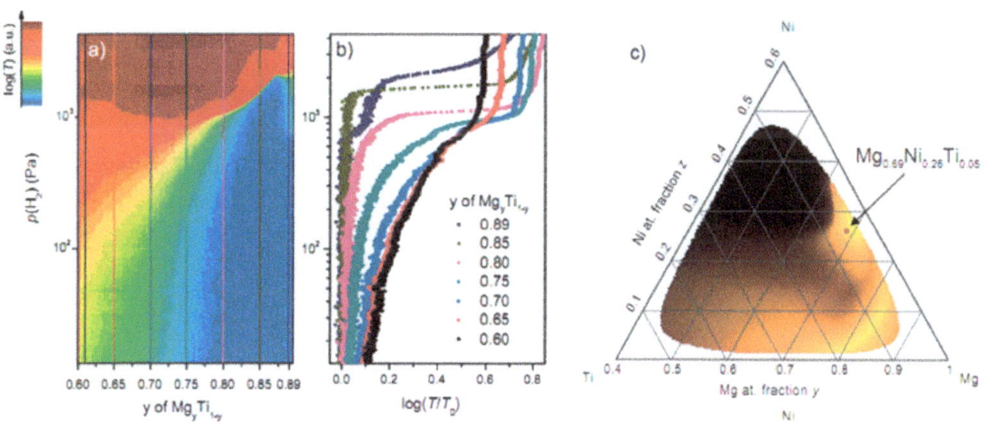

Figure 19. Pressure-optical transmission isotherms of (**a**) continuously gradient $Mg_yTi_{(1-y)}$; (**b**) discrete compositions of $Mg_yTi_{(1-y)}$; and (**c**) phase image of ternary gradient Mg-Ni-Ti alloy hydrides. [45]

CONCLUSIONS

Metal-hydrides are promising candidates in terms of energy density, lightweight, safety and reversibility. Specifically, intensive work has been done upon magnesium-based alloy systems (up to 7.6 wt% of hydrogen storage capacity) to improve hydrogen storage properties. Electrochemical hydrogenation, which is widely analyzed in this work, does not only enable reversible, precise (dis)charging under ambient temperature and pressure but also combines fundamental studies of potential battery and fuel cell applications. Conceptually, stability modification of the metal/metal-hydride systems has been achieved by adding a second or even a third element to Mg. Binary $Mg_xTM_{(1-x)}$(TM = Sc,Ti) possesses better kinetics compared to pure MgH_2. However, these alloy-hydrides suffer from relatively poor thermal stability. Thus, the necessity of further alloying with foreign elements is emerging in order to adjust the thermal characteristics of the alloy-hydrides. Both

Al and Si were found to be suitable ternary additives to binary alloys for achieving better thermodynamic properties.

On the other hand, magnesium-hydride potentially is an interesting candidate for switchable smart mirror applications due to its high band-gap energy of E_g = 5.6 eV. The hydrides change from metal-to-semiconductor, which results in electrochromic and optical responses of the thin films as a function of the hydrogen content. Pioneering results showed that magnesium, alloyed with transition metals like gadolinium, titanium and nickel exhibits a conversion from the metallic reflective state to the semiconducting transparent state.

Besides these promising results, these state-of-the-art materials are still far from meeting the goals for practical applications. Therefore, continued research work is needed to further improve the hydrogenation properties, such as reaction rate, thermal stability and cyclic behavior of the metal-hydrides.

REFERENCES

1. Guo, Z.X.; Shang, C.; Aguey-Zinsou, K.F. Materials challenges for hydrogen storage. J. Eur. Ceram. Soc. **2008**, 28, 1467–1473.
2. Momirlan, M.; Veziroglu, T.N. The properties of hydrogen as fuel tomorrow in sustainable energy system for a cleaner planet. Int. J. Hydrogen Energy **2005**, 30, 795–802.
3. Züttel, A.; Borgschulte, A.; Schlapbach, L. Hydrogen as A Future Energy Carrier; Wiley-VCH Verlag GmbH & Co. KGaA: Weinheim, Germany, 2008; p. 5.
4. Schlapbach, L.; Züttel, A. Hydrogen-storage materials for mobile applications. Nature **2001**,414, 353–358.
5. Ross, D.K. Hydrogen storage: The major technological barrier to the development of hydrogen fuel cell cars. Vacuum **2006**, 80, 1084–1089.
6. Fukai, Y. The Metal-Hydrogen System, Basic Bulk Properties. In Springer Series in Materials Science; Springer: Berlin, Germany, 1993; pp. 55–88.
7. Van Vucht, J.H.N.; Kujipers, F.A.; Bruning, H.C.A.M. Reversible room-temperature absorption of large quantites of hydrogen by intermetallic compounds. Philips Res. Rep.**1970**, 25, 133–140.
8. Selvam, P.; Viswanathan, B.; Swamy, C.S.; Srinivasan, V. Magnesium and magnesium alloy hydrides. Int. J. Hydrogen Energy **1986**, 11, 169–192.
9. Luz, Z.; Genossar, J.; Rudman, P.S. Identification of the diffusing atom in MgH_2. J. Less Common Metals **1980**, 73, 113–118.
10. Züttel, A. Materials for hydrogen storage. Mater. Today **2003**, 6, 24–33.
11. Willems, J.J.G. Metal Hydride Electrodes Stability of $LaNi_5$-Related Compounds; Philips Research Laboratories: Eindhoven, The Netherlands, 1984; Volume 39, pp. 1–94.
12. Notten, P.H.L. Rechargeable Nickel-Metal hydride Batteries: A Successful New Concept. InInterstitial Intermetallic Alloys; Grandjean, F., Long, G.L., Buschow, K.H.J., Eds.; Kluwer Academic Publishers: Dordrecht, The Netherlands, 1995; Volume 281, p. 151.
13. Platt, J.R. Electrochromism, a Possible Change of Color Producible in Dyes by an Electric Field. J. Chem. Phys. **1961**, 34, 862–863.
14. Deb, S.K. A Novel electrophotographic system. Appl. Opt. **1969**, 8, 192–195.
15. Huiberts, J.N.; Griessen, R.; Rector, J.H.; Wijngaarden, R.J.; Dekker, J.P.; de Groot, D.G.; Koeman, N.J. Yttrium and lanthanum hydride films with switchable optical properties.Nature **1996**, 380, 231–234.
16. Griessen, R.; Giebels, I.A.M.E.; Dam, B. Optical Properties of Metal-Hydrides: Switchable Mirrors. Available online: http://www.nat.vu.nl/en/Images/ReviewSwitchableMirrors10AUG04_tcm69-85550.pdf (accessed on 28 November 2011).
17. Notten, P.H.L.; Kremers, M.; Griessen, R. Optical switching of Y-hydride thin film electrodes. J. Electrochem. Soc. **1996**, 143, 3348–3353.
18. Van der Sluis, P.M.; Ouwerkerk, M.; Duine, P.A. Optical switching based on magnesium lanthanide alloy hydrides. Appl. Phys. Lett. **1997**, 70, 3356–3358.
19. Griessen, R.; van der Sluis, P. Schaltbare Spiegel-Elektronenkorrelationen in der Anwendung. Physik Unserer Zeit **2001**, 32, 76–83.
20. Niessen, R.A.H.; Notten, P.H.L. Electrochemical hydrogen storage characteristics of thin film MgX (X=Sc, Ti, V, Cr) compounds. Electrochem. Solid StateLett. **2005**, 8, A534–A538.
21. Richardson, T.J.; Slack, J.L.; Armitage, R.D.; Kostecki, R.; Farangis, B.; Rubin, M.D. Switchable mirrors based on nickel-magnesium films. Appl. Phys. Lett. **2001**, 78, 3047–3049.
22. Karazhanov, S.Zh.; Ulyashin, A.G.; Vajeeston, P.; Ravindran, P. Hydrides as materials for semiconductor electronics. Philos. Mag. **2008**, 88, 2461–2476.
23. Reilly, J.J.; Wiswall, R.H. Reaction of hydrogen with alloys of magnesium and nickel and the formation of Mg_2NiH_4. Inorg. Chem. **1968**, 7, 2254–2256.
24. Janot, R.; Aymard, L.; Rougier, A.; Nazri, G.; Tarascon, J. Enhanced hydrogen sorption capacities and kinetics of Mg_2Ni alloys by ball-milling with carbon and Pd coating. J. Mater. Res. **2003**, 18, 1749–1752.
25. Zhang, S.G.; Hara, Y.; Suda, S.; Morikawa, T.; Inoue, H.; Iwakura, C. Physicochemical and electrochemical hydriding-dehydriding characteristics of amorphous $MgNi_x$ (x = 1.0, 1.5, 2.0) alloys prepared by mechanical alloying. J. Solid State Electrochem. **2001**, 5, 23–28.
26. Süleyman, Er.; Tiwari, D.; de Wijs, G.A.; Brocks, G. Tunable Hydrogen Storage in Magnesium-Transition Metal Compounds: First-Principles Calculations. Phys. Rev. B **2009**,79, 1–8.
27. Niessen, R.A.H.; Notten, P.H.L. Hydrogen storage in thin film magnesium-scandium alloys.J Alloy Compd. **2005**, 404-406, 457–460.
28. Kalisvaart, W.P.; Niessen, R.A.H.; Notten, P.H.L. Electrochemical hydrogen storage in MgSc alloys: A comparative study between thin films and bulk materials. J. Alloy Compd. **2006**, 417, 280–291.
29. Latroche, M.; Kalisvaart, P.; Notten, P.H.L. Crystal

structure of $Mg_{0.65}Sc_{0.35}D_x$ deuterides studied by X-ray and neutron powder diffraction. J. Solid State Chem. **2006**, 179, 3024–3032.

30. Kalisvaart, W.P.; Latroche, M.; Cuevas, F.; Notten, P.H.L. In situ neutron diffraction study on Pd-doped $Mg_{0.65}Sc_{0.35}$ electrode material. J. Solid State Chem. **2008**, 181, 1141–1148.
31. Conradi, M.S.; Mendenhall, M.P.; Ivancic, T.M.; Carl, E.A.; Browning, C.D.; Notten, P.H.L.; Kalisvaart, W.P.; Magusin, P.C.M.M.; Bowman, R.C., Jr.; Hwang, S.; Adolphi, N.L. NMR to determine rates of motion and structures in metal-hydrides. J. Alloy Compd. **2007**, 446-447, 499–503.
32. Pauw, B.R.; Kalisvaart, W.P.; Tao, S.X.; Koper, M.T.M.; Jansen, A.P.J.; Notten, P.H.L. Cubic MgH_2 stabilized by alloying with transition metals: A density functional theory study. Acta Mater. **2008**, 56, 2948–2954.
33. Vermeulen, P.; Niessen, R.A.H.; Notten, P.H.L. Hydrogen storage in metastable $Mg_yTi_{(1-y)}$thin films. Electrochem. Commun. **2006**, 8, 27–32.
34. Vermeulen, P.; Wondergem, H.J.; Graat, P.C.J.; Borsa, D.M.; Schreuders, H.; Dam, B.; Griessen, R.; Notten, P.H.L. In situ electrochemical XRD study of (de) hydrogenation of $Mg_yTi_{(100-y)}$ thin films. J. Mater. Chem. **2008**, 18, 3680–3687.
35. Rousselot, S.; Bichat, M.P.; Guay, D.; Roué, L. Structure and electrochemical behaviour of metastable $Mg_{50}Ti_{50}$ alloy prepared by ball milling. J. Power Sources **2008**, 175, 621–624.
36. Kalisvaart, W.P.; Notten, P.H.L. Mechanical alloying and electrochemical hydrogen storage of Mg-based systems. J. Mater. Res. **2008**, 23, 2179–2187.
37. Kalisvaart, W.P.; Wondergem, H.J.; Bakker, F.; Notten, P.H.L. Mg-Ti based materials for electrochemical hydrogen storage. J. Mater. Res. **2007**, 22, 1640–1649.
38. Kyoi, D.; Sato, T.; Rönnebro, E.; Kitamura, N.; Ueda, A.; Ito, M.; Katsuyama, S.; Hara, S.; Noréus, D.; Sakai, T. A new ternary magnesium-titanium hydride Mg_7TiH_x with hydrogen desorption properties better than both binary magnesium and titanium hydrides. J. Alloy Compd. **2004**, 372, 213–217.
39. De Boer, F.R.; Boom, R.; Mattens, W.C.M.; Miedema, A.R.; Niessen, A.K. Cohesion in Metals; North-Holland: Amsterdam, The Netherlands, 1988; p. 127.
40. Liang, G.; Schulz, R. Synthesis of Mg-Ti alloy by mechanical alloying. J. Mater. Sci. **2003**, 38, 1179–1184.
41. Srinivasan, S.; Magusin, P.C.M.M.; Kalisvaart, W.P.; Notten, P.H.L.; Cuevas, F.; Latroche, M.; van Santen, R.A. Nanostructures of $Mg_{0.65}Ti_{0.35}D_x$ studied with x-ray diffraction, neutron diffraction, and magic-angle-spinning ^2H NMR spectroscopy. Phys. Rev. B **2010**, 81, 1–10.
42. Miedema, A.R.; de Châtel, P.F.; de Boer, F.R. Cohesion in Alloys-fundamentals of a Semi-empirical model. Physica B **1980**, 100, 1–28.
43. Miedema, A.R. The electronegativity parameter for transition metals: Heat of formation and charge transfer of alloys. J. Less Common Metals **1973**, 32, 117–136.
44. Vermeulen, P.; van Thiel, E.F.M.J.; Notten, P.H.L. Ternary MgTiX-alloys: A promising route towards low-temperature, high-capacity, hydrogen-storage materials. Chem. Eur. J. **2007**, 13, 9892–9898.
45. Gremaud, R.; Broedersz, C.P.; Borsa, D.M.; Borgschulte, A.; Mauron, P.; Schreuders, H.; Rector, J.H.; Dam, B.; Griessen, R. Hydrogenography: An optical combinatorial method to find new light-weight hydrogen-storage materials. Adv. Mater. **2007**, 19, 2813–2817.

CHAPTER 9

HIGH TEMPERATURE METAL HYDRIDES AS HEAT STORAGE MATERIALS FOR SOLAR AND RELATED APPLICATIONS

Michael Felderhoff and Borislav Bogdanović

Keywords: Hydrogen storage, heat storage, magnesium hydride, Mg_2FeH_6

INTRODUCTION

The paper on hand deals with a chemical-based method for thermal solar energy storage. Materials which are appropriate for this purpose are chemical compounds of metals, metal alloys or intermetallic compounds and hydrogen known as *metal hydrides* (MH_n, Equation 1). The majority of metals, metal alloys and intermetallic compounds react directly with gaseous hydrogen (Equation 1 from left to right) to form metal hydrides in an exothermal way, that is, with release of heat (ΔH). If the formation of a metal hydride under certain conditions is a reversible reaction ($\rightleftharpoons$), then the metal hydride can by a heat supply upon decomposition into the metal and hydrogen. Upon thermal dissociation of a metal hydride by heat absorption (Equation 1, from right to left), the thermal energy is transformed into chemical energy. Expressed in another way: thermal energy can be reversibly stored as the heat of reaction of reversible chemical reactions. Among the methods for heat storage, sensible and thermochemical heat storage, the thermochemical method offers in general the higher amount of energy stored per mass of storage material (Table 1) [1, 2].

Table 1. Frequently used solid and liquid materials for the storage of sensible heat [1]

	M.p. [°C]	Temperature range [°C]	ρ [g cm^{-3}]	c [kJ kg^{-1} K^{-1}]
Water	0	0 - 100	0.98 [a]	4.19
Thermo-oil	-70 / -10	up to 350 [b]	0.87 [a]	2.1
53 KNO_3/40 $NaNO_2$/ 7 $NaNO_3$ [c]	142	up to 450	1.85 [a]	1.3
Na	98		0.84 [d]	1.26
Cast iron	1150 – 1300		7.2 [d]	0.54
Aluminium	660		2.7 [d]	0.92
Fire clay			2.1-2.6 [d]	1.0
Al_2O_3	1700		3.0 [d]	1.0
MgO	1700		3.0 [d]	1.0
Stone			1.9-2.6 [d]	0.8-0.9
Concrete				0.9

[a] Average density in the temperature range given. [b] some thermooils can be used up to 390 °C. [c] wt.%. [d] at room temperature.

The special feature of metal hydrides as reversible heat storage systems is that upon their thermal dissociation the liberated hydrogen is simultaneously a fuel with the highest known gravimetric energy density (usually expressed in MJ·kg^{-1}). The gravimetric energy density of hydrogen [1, 3] is about three times higher than that of gasoline. Hydrogen in liquid form is therefore used as a rocket fuel. A further remarkable feature of hydrogen as a fuel is that upon its combustion at high temperatures in air [4] as well as upon "cold combustion" in low temperature fuel cells, only pure water is produced. Thus, from ecological standpoint hydrogen is an ideally clean, albeit a secondary energy carrier. The last attribute means that hydrogen has first to be produced under consumption of primary energy sources which are at present primarily fossile fuels or, otherwise, nuclear energy (via

Citation: Michael Felderhoff and Borislav Bogdanović, High Temperature Metal Hydrides as Heat Storage Materials for Solar and Related Applications, doi: 10.3390/ijms10010325.

water electrolysis) or, in the future, the expected solar energy. From the above statements it follows that the reversible metal hydride/metal systems (Equation 1) are to be considered as energy carriers in a double sense - namely as heat and as hydrogen storage systems.

$$M + n/2\,H_2 \underset{\text{heat absorption (heat storage)}}{\overset{\text{heat evolution (heat recovery)}}{\rightleftharpoons}} MH_n + \Delta H \quad (1)$$

M = metal, metal alloy, intermetallic compound

In the heat storage application, the heat of reaction liberated in the hydride formation process serves as useful heat. Thereby hydrogen is confined in a closed system and is preserved during a great number of heat storage cycles. In the other case – as a hydrogen store – the hydrogen liberated from the hydride is present in an open system, and after displacement from the hydride it is irreversibly burned to water.

The principle of heat and Hydrogen Storage by the MgH_2/Mg system

The magnesium hydride/magnesium system is particularly suitable for both purposes, because in MgH_2 the reversible hydrogen storage capacity is the highest among reversible binary hydrides (7.6 wt.%) and because upon the reaction of magnesium with hydrogen (Equation 2) a relatively large amount of heat (75 kJ·mol^{-1} H_2, (0.9 kWh kg-1Mg) at a temperature level between 200 and 500 °C is set free. Herewith an extremely important region for heat storage becomes available, since according to the Carnot principle, heat energy at this level can be transformed with high efficiency into other energy forms like mechanical energy and hence in electricity. We are dealing here mainly with the application of magnesium hydride as a heat storage medium. The MgH_2/Mg system (Equation 2), however, can also be applied for hydrogen storage. Upon gradually heating up of MgH_2 in a closed system the compound will dissociate giving Mg and H_2 and the hydrogen pressure above the remaining MgH_2 will increase exponentially following the van't Hoff law (Figure 1).

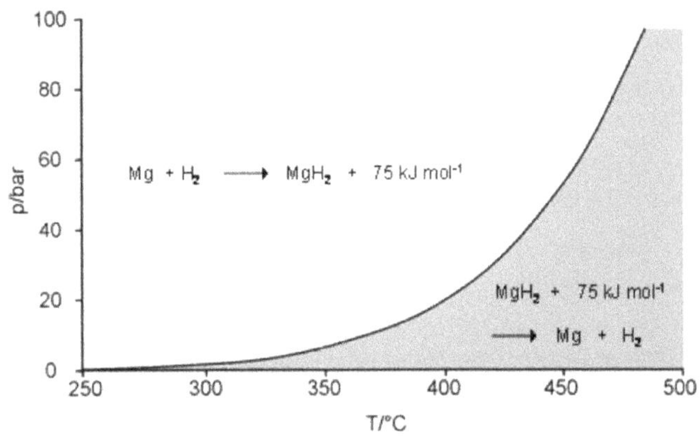

Figure 1. Dissociation pressure curve of MgH_2.

$$MgH_2 + 75\ \text{kJ mol}^{-1} \underset{\text{heat recovery}}{\overset{\text{heat storage}}{\rightleftharpoons}} Mg + H_2 \quad (2)$$

T = 250 – 500 °C

The knowledge of the H_2 dissociation (equilibrium) pressure as a function of temperature is of fundamental importance for the operation of the MgH_2 heat and hydrogen storage system. At a given H_2 pressure the course of the curve in Figure 1 determines which minimal temperature that is definitely necessary in order to liberate the hydrogen from MgH_2 under a given H_2 pressure (f.i. 450 °C for 40 bar of H_2 pressure); and conversely, upon charging of Mg with hydrogen under given pressure conditions the temperature of the store will correspond to the hydrogen charging pressure. Thus, by choice of the hydrogen pressure it is possible to determine at which temperature level the stored heat will be delivered. The diagram in the Figure 1 shows additionally under which temperature and pressure conditions the system will absorb or desorb heat (shaded part - heat absorption; clear part - heat desorption). Although the diagram in the Figure 1 indicates under which conditions the reaction can go in one or in the other direction, it does not give any information if at all and how rapidly the reaction will take place. In the early [5] and in the more recent literature [6] it is often stated that the reaction of magnesium with hydrogen is notoriously sluggish and therefore hardly useful for heat storage [5]. For an optimal thermochemical heat storage material it is required that the chemical reaction be much more rapid than the rate of heat transport, so that the dynamics of the system will depend mainly on the rate of the heat transport. Additionally, it is required that the reaction in both directions is any times repeatable and that the material shows no sign of fatigue. For accelleration of chemical reactions, apart from increase of temperature, and the effect of the microstructure of particles [6], the general answer is catalysis.

At the beginning of our research work on metal hydrides there was the experimental finding that the reaction of magnesium with hydrogen in solution can be catalyzed [7]. During the past years the research in this field has led to MgH_2/Mg materials which fulfill the above mentioned criteria: satisfactory kinetics combined with stability upon cyclic hydrogen charging and discharging. The first method for the preparation of the so-called "active MgH_2/Mg materials" (Equation 3) [1, 8] consists in treatment of commercial Mg powder (f.i. 270 mesh) with hydrogen in an organic solvent (i.e. toluene) in the presence of a small amount of the soluble organometallic nickel complex bis(cyclooctadiene-1,5-Ni(0) [9] at ambient temperature and normal pressure. The Ni-complex is hereby instantaneously hydrogenated to cyclooctane and metallic nickel and the latter, in a fine dispersion, is precipitated on the surface of the Mg particles and acts as a catalyst for hydrogenation and dehydrogenation of the former. The second method [8] consists of intimately mixing of the Mg powder with 4–10 wt.% of fine commercial Ni powder. The activity of the material prepared according to the second method is somewhat lower than that of the first method. Cycle stability of the "standard materials" is presented in Ref. [8] (Figures 4, 5 and 9 therein). The Mg powders doped with Ni in these ways and fabricated several times on a 10 kg scale were shown in numerous tests to be suitable materials for heat and hydrogen storage and were employed by us and by others in several storage installation prototypes [1, 8, 10–12], three of which are presented in Section 3. Already after the first hydrogenation the materials are fully active and remain active during hundreds of hydrogen charging-discharging cycles. A more recent 1,000 cycle test using both materials is shown in Figure 2 [13]. Some restrictions at high temperatures are discussed further in Section 4.

Mg (2 wt.%Ni) ---, Mg (5 wt.% Ni) --- [13].

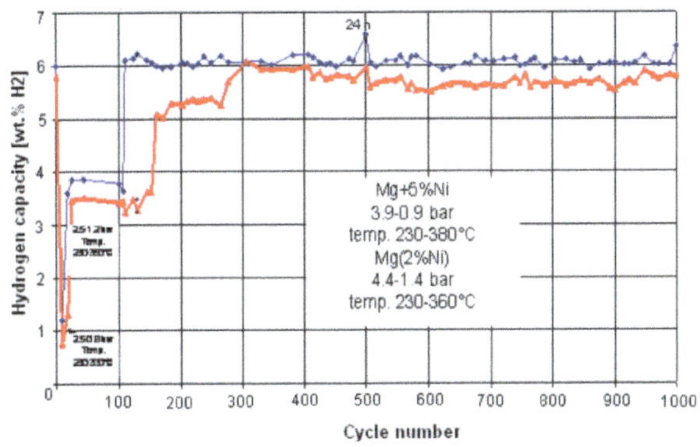

Figure 2. 1000 cycle test of Ni-doped MgH_2, time for a full cycle 3 h.

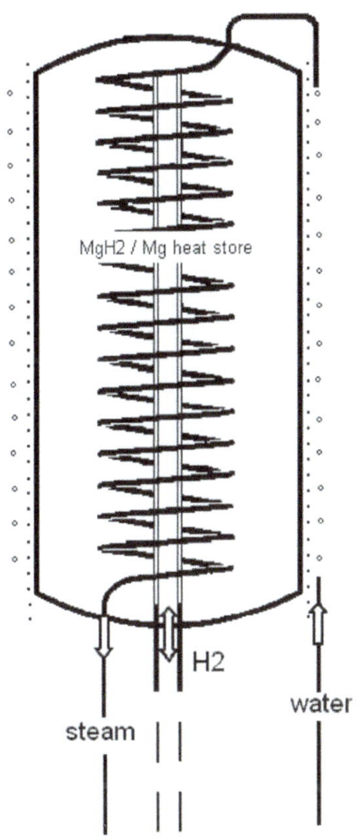

Figure 4. The process steam generator based on MgH_2 [17, 18].

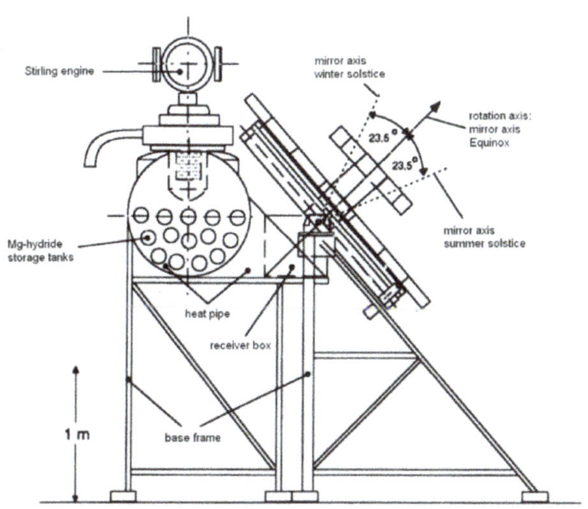

Figure 5. Cross section of the model of the solar power station, with MgH_2 heat storage units and fix-focus solar concentrator.

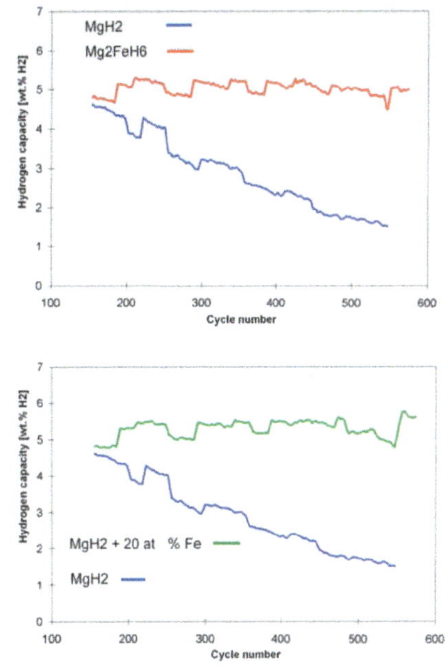

Figure 9. Hydrogen storage capacity of MgH_2, Mg_2FeH_6 (top) and of the mixed Mg_2FeH_6-MgH_2 system [22]. (Conditions for re/dehydrogenations: 482/533 °C, 80/86 bar, 1.5/1.5 h).

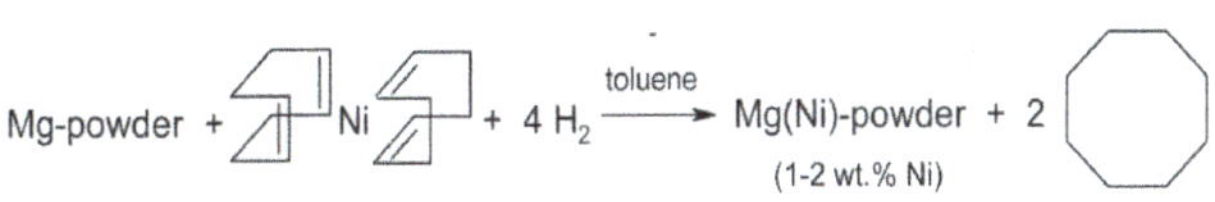

(3)

As already mentioned (Equation 2), hydrogen is liberated in the endothermic process of feeding heat (75 kJ mol^{-1} Mg) to MgH_2, which until the reverse process of heat recovery must be temporarily stored. A heat storage system based on MgH_2-Mg (Figure 3) has to be constructed as a closed system consisting of a MgH_2 pressure container and an additional pressure container for the temporary storage of the displaced hydrogen (Part A). As schematically represented (Figure 3, Part A), the magnesium hydride container can be designed so to be much smaller than the pressure container for gaseous hydrogen, because of the much higher hydrogen concentration in the form of magnesium hydride. In the form of MgH_2 hydrogen attains a concentration corresponding to a hydrogen pressure of 700 – 800 bar. Up to ~150 °C MgH_2 has only a very low dissociation pressure, which means that hydrogen storage in this form is comparably a secure form for hydrogen storage.

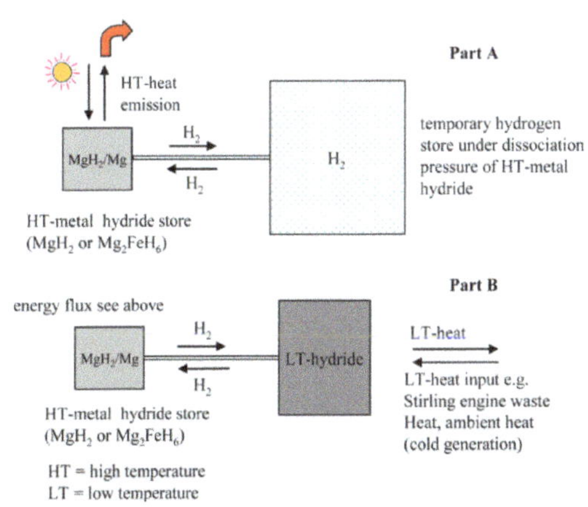

Figure 3. Schematical representation of construction of MgH_2/Mg heat stores. Part A: temporary storage of hydrogen in a pressure container; part B: temporary storage of hydrogen in a low-temperature metal hydride.

In comparison to the first version (Figure 3, Part A), a much more compact construction can and has been realized, when for the temporary storage of hydrogen in place of a pressure container is applied an at low temperature hydrogen absorbing alloy (Figure 3, Part B) which with hydrogen forms a so-called low temperature (LT) metal hydride. A LT hydride is typically a hydride on Fe-Ti or La-Ni-basis [14] or, in future possibly on alanate basis [15].

Using LT hydrides for temporary storage of hydrogen, hydrogen is available at ambient or even below ambient trerperature. Apart from the compact construction, such a heat store (Figure 3, Part B) includes possibilities which are of great importance for the rational use of solar and other energy sources:

- absorption of hydrogen by the LT-alloy produces LT-heat (f.i. 30–70°C) which is typically ≈ 1/3 of the MgH_2 heat amount (75 kJ mol^{-1}); this waste heat can be used for example for warm water supply;
- the desorption of hydrogen from the side of the LT hydride takes place with withdrawal of heat from the surroundings, that is with production of cold, that can be used for production of ice or for climatisation.
- altogether a closed circular system of two (or more) conjoined reversible metal hydride systems which operate at different temperature levels represent a so-called hydride chemical heat pump, which is applicable for different types of heat and energy transformations [16].

EXPERIMENTAL RESULTS CONCERNING THE MGH_2-MG SYSTEM

A Process Steam Generator *[17, 18]*

In the following three installations which are designed for different fields of application and which have been realized as laboratory prototypes are described. These were primarily engineering tasks which were tackled at first by engineers by mathematical modelling (simulations) and calculations.

Unlike engineers, chemists prefer an experimental approach. In this sense the first experimental heat accumulator on magnesium hydride basis was built and tested in our institute. The facility for the production of the process steam was installed in the pilot plant of the institute and was in operation during 1½ years. In this connection the heat transfer problems within the hydride bed (heat pipe effect) was also investigated. The process steam generator is schematically represented in Figure 4, the isolated cylindric steel pressure vessel is filled with 14.5 kg of Ni-doped Mg-powder.

In the center of the vessel is installed a sintered metal tube for supply and removal of hydrogen. Around the vessel is wound a steel tube which is connected to a source for pressurized water. The tube is further lead into the inner of the vessel where it is helically embedded in the MgH_2/Mg bulk. Hydrogenation of the Ni-doped storage material is carried out directly in the pressure vessel. For heat desorption the pressure vessel is electrically heated from outside, resulting in MgH_2dissociation. The hydrogen which escapes is temporary stored in a bunch of six commercial hydrogen pressure containers. As soon as the heating is interrupted and the temperature of the heat accumulator begins to decrease the hydrogen flows back from the pressure containers to the Mg/MgH_2 vessel where the HT heat is regenerated and produces in the spiral tube superheated water vapor. Some technical data about the heat store unit are given in the Table 2.

Table 2. Technical Data of the MgH_2-Heat Store Unit [17, 18]

Amount of storage material (Ni-doped Mg)	14.5 kg
Volume of the pressure vessel	19.4 dm^3
Weight of the pressure vessel	26 kg
Bulk density	0.75 g cm^{-3}
Maximum operational pressure	50 bar
Maximum operational temperature	450 °C
Total weight	40 kg
Amount of stored heat / total weight	0.25 kWh kg^{-1}
Maximum power output	4 kW
Amount of water to be vaporized at the power output of 4 kW	6 kg h^{-1}
Maximum steam temperature	ca. 400 °C
Maximum steam pressure	40 bar

The heat accumulator had the ability to produce 4 kW of heat, up to 90% capacity at the same temperature level. A coupled unit of a MgH_2/Mg heat store with a hydrogen pressure container (Figure 3, Part A) can also serve as a heat buffer for intermittent HT heat sources.

A model of a thermochemical solar power plant *[1, 8, 11, 13, 19, 20]*

The MgH_2/Mg-heat accumulator, first used to demonstrate the feasibility for production of "process vapor", is clearly applicable for storage of industrial heat and at this temperature level (350–450°C) of solar heat in particular. According to the "solar concept", the combination of a solar-thermal power generator with a thermochemical MgH_2 heat accumulator should enable generation of power also in the absence of sunshine, that is after sundown and during the passage of clouds. Within the frame of a publicly sponsored joint project between the both Max-Planck-Institutes in Mülheim, the IKE Institute of the Stuttgart University and the HTC Solar Company in Lörrach a first model of a small solar power station was built. Figure 5 shows a cross section of the installation. The main components were a "fixed focus concentrator" for solar radiation [11], a cavity radiation receiver, a heat pipe system for heat transfer, a Stirling engine, a MgH_2 store, a hydrogen pressure or LT hydride store. For supply of the whole installation with solar heat a solar mirror collector with 6.5 m² area of a light weight type construction was built. The concentrated solar heat is used to drive the Stirling engine and produce power. Simultaneously, a more or less part of the solar heat is absorbed by the MgH_2 heat store and the released hydrogen streams in the pressure container or, alternatively, in the LT hydride store. In periods without or weak solar irradiation the MgH_2/Mg system cools down and the hydrogen flows back to the MgH_2 store and produces there high temperature heat and via Stirling engine electric power. The model of the solar power plant was tested under laboratory [10] and in limited extent also under solar test field conditions [11] and in this way the ability of the system to function was established. Shortcomings of the installation were realized and the measures to address them proposed.

A solar cooking- and cooling-device [12, 19]

As a third example for the magnesium hydride application for heat storage and transformation a laboratory model of a solar cooking-and-cooling device was built. The device consists of a MgH_2-heat store unit filled with 4.4 kg (~3 KWh heat) of the Ni-doped Mg-material and equipped with a cooking plate. The unit is linked to a LT metal hydride store unit with ~30 kg of a LT metall alloy ($MmNi_{4.22}Fe_{0.78}$) coupled to a refrigerator and a water reservoir. As a cooling cabinet a commercial ice-box of 113 L capacity is used. The solar mirror in this case has an area of 2.5 m^2 and is made out of a fibrous plastic material.

During solar irradiation hours solar heat is concentrated by means of the fixed-focus concentrator and directly used for cooking. Simultaneously, hydrogen from the MgH_2-store is desorbed and absorbed in the LT alloy, thus performing the task of heat storage. The LT heat of 30–40°C is absorbed by a water reservoir. In order to be able to cook in the absence of solar irradiation the hydrogen is allowed to stream in the reverse direction, so that the heat of MgH_2 formation provides heat for cooking. During the hydrogen desorption from the LT hydride, the desorption heat is removed from the surroundings and the resulting cold with the help of a water-glycol circle system is transferred to the ice-box in which ice is produced.

A complete cycle of an experimentally performed cooking- and cooling process is described in the following. At the beginning of such a cycle hydrogen is located in the LT hydride store. Upon opening of the valve between the two stores, the hydrogen streams to the HT-store resulting in a heat evolution of around 300 °C which remains during 5–6 h at this level and is usable for cooking. At the same time the LTH store itself cools down to -10 °C; ice is formed in the ice-box and the temperature there remains at about 0 °C. In this time 0.9 KWh of cold are produced and 35% of this amount are allotted for the production of ice. After a rest of 7 h during the night, during which the hydrogen supply is disconnected, through solar irradiation, at 350–450 °C, hydrogen is displaced from MgH_2 and absorbed in the LT alloy with formation of warm water (30–40 °C). The device is now ready for a new cooking-cooling cycle. Laboratory tests were satisfactory, but solar (field) tests have not yet been done. As operational areas can be envisaged regions with a lot of sunshine but without energy infrastructure. Without any consumption of fuels or electric power it should be possible to cook and to refrigerate food at any day or night time.

NI-DOPED VERSUS UNDOPED MGH_2-MG MATERIALS FOR HIGH TEMPERATURE HEAT OR HYDROGEN STORAGE [21]

The design and construction of the above described appliances for energy and heat storage on MgH_2/Mg basis are more or less tasks for engineers. For chemists there still remains the challenge of improving the properties of metal hydride storage materials and to develop new ones. So, for instance, the question arises which is the highest application temperature level for functioning of MgH_2/Mg heat storage system. The question is of great importance, because according to the Carnot principle the efficiency of heat engines – be they small Stirling engines or steam turbines – increases with the temperature level of the fed heat. In order to remain with our example: the process steam of 500 °C is more efficient for the production of power than the same heat quantity at 400 °C or lower. If under this aspect we look again at temperature dependence of the MgH_2dissociation pressure (Figure 1), it can be seen that at the temperature of, f.i. 500 °C, the hydrogen pressure in the system is around 100 bar. Results of a 650 cycle test with mechanically Ni-doped Mg-powder (4 wt.%) under severe conditions are shown in Figure 6 [21].

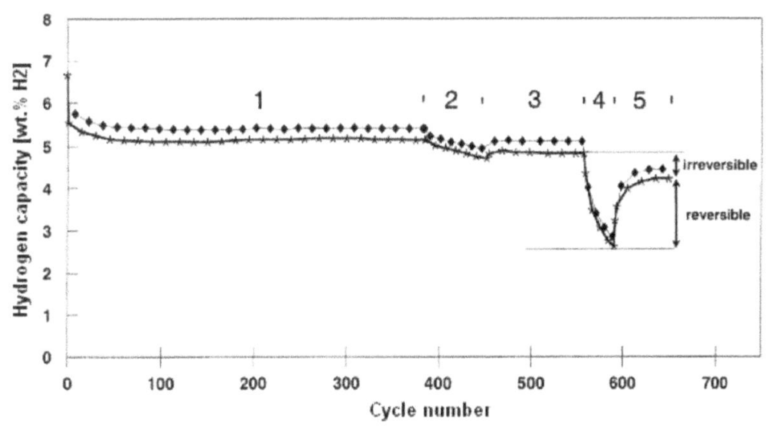

Figure 6. A 650 cycle test performed with mechanically Ni-doped (4 wt.%) 270 mesh Mg powder, demonstrating reversible and irreversible hydrogen capacity losses under severe conditions; —*— and —♦—, 45 and 135 min hydrogenation times, respectively [21].

Under a high dehydrogenation temperature (section 2, 480 °C) the hydrogen capacity and with it, the amount of stored heat, decreases. An investigation of this phenomenon showed that a part of these capacity losses can be recovered under milder conditions (Sections 2/3, 4/5, "reversible capacity losses"). A part of the capacity losses, however, is permanent (Sections 1/3, 3/5, "irreversible capacity losses"). Additionally, it could be shown that capacity losses upon cyclisation are higher with Ni-doped material than with undoped Mg-powder. The consequence is therefore that for heat storage at a high temperature level (> ~440 °C), it is of advantage to use plain Mg-powder in place of a Ni-doped one. The cause for this fall-off in performance at higher temperatures, as we presently know it, is a morphological one – namely the sintering of Mg particles at temperatures above 450 °C, which is more pronounced in a two (Mg-Ni) than in an one component system (Mg) [21].

THE MG-FE-H SYSTEM AND ITS POTENTIAL FOR THERMOCHEMICAL THERMAL ENERGY STORAGE [22]

Mg_2FeH_6 AS REVERSIBLE TERNARY HYDRIDE SYSTEM

A way out of the dilemma discussed in Section 4 was to look for alternative metal hydride/metal systems in which such sintering phenomena do not occur. To overcome this problem magnesium iron hydride Mg_2FeH_6 can be used for heat storage. The compound was discovered and the structure elucidated in 1984 by Yvon *et al.* after heating of 2:1 mixtures of 2 Mg and Fe powders at 450 °C under 20–120 bar H_2 pressure [23].

As it can be seen in Figure 7, Mg_2FeH_6 is composed of Mg^{2+} cations and FeH_6^{4-} anions in the form of octahedrons. In the center of the octahedrons is located a Fe atom, which is surrounded by six covalently bound H atoms. Alternatively to the high pressure method, Mg_2FeH_6 can be synthesized by reactive mechanical alloying of Fe and Mg elemental powders in a hydrogen atmosphere at room temperature [24–26]. The gravimetric hydrogen content of Mg_2FeH_6 is 5.5 wt.%. The volumetric hydrogen density reaches 150 kg m^{-3} which is double of the density of liquid hydrogen (70 kg m^{-3}). The advantage of Mg_2FeH_6 as a heat storage material in comparison to MgH_2 (Table 3) is its higher stability and hence a lower hydrogen dissociation pressure (Figure 8). So, for example, the dissociation pressure of Mg_2FeH_6 at 500 °C (66 bar) - is about 1/3 lower than that of MgH_2 at the same temperature (100 bar). This gives for the construction of pressure containers for Mg_2FeH_6 an advantage versus containers for MgH_2.

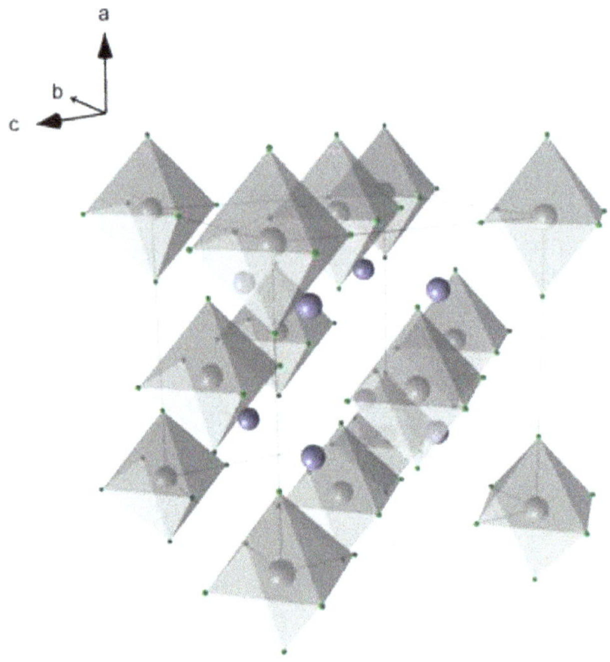

Figure 7. Crystal unit cell of Mg_2FeH_6. Mg-atoms are shown in blue, Fe-atoms are located in the centers of the octahedrons.

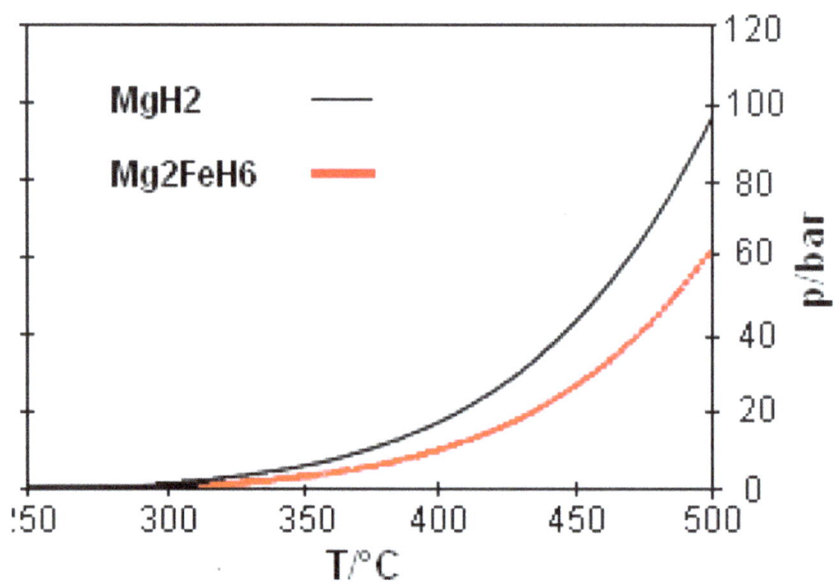

Figure 8. Temperature dependence of the dissociation pressure of Mg_2FeH_6 in comparision with that of MgH_2 [22].

Table 3. Comparision of important values for heat storage using MgH_2 and Mg_2FeH_6 [22]

	MgH_2	Mg_2FeH_6
Reaction enthaply ΔH [kJ mol^{-1}]	74	77.4
Hydrogen content calc. [wt.%]	7.66	5.47
Hydrogen content expt. [wt.%]	6	5
Theoretical density of the hydride [g cm^{-3}]	1.42	2.74
Experimentally realized density [g cm^{-3}]	0.8	1.22
Heat storage density (calc. weight) [kJ kg^{-1}]	2.814	2.106
Heat storage density (expt. weight) [kJ kg^{-1}]	2.204	1.921
Heat storage density (calc. volume) [kJ dm^{-3}]	3.996	5.768
Heat storage density (expt. volume) [kJ dm^{-3}]	1.763	2.344

Numerous cycle tests done with the Mg-Fe-H system prove that the system is superior under high temperature and pressure conditions with respect to cycle stability in comparison to the pure MgH_2-Mg system (Figure 9, top) [22]. While the latter system under different HT conditions already looses capacity after 200–250 cycles (due to sintering), the Mg-Fe-H-systems, under comparable conditions, show no signs of loss of capacity and reaction rates even after 600 or more cycles. Even under very severe conditions (510–590 °C/138–149 bar) cycle stabilities with a capacity of more than 5 wt.% H_2 are achieved. Interestingly, the improved cycle stability has been shown to exist also using less than stoichiometric amounts of Fe with respect to Mg (Figure 9, bottom). In view of the possible technical application of these systems for heat storage in future especially of solar heat in the temperature region around 500 °C or higher, this finding appears to be of importance [22]. However, further investigation of cycling properties of the mixed MgH_2/Mg – Mg_2FeH_6/2Mg Fe hydride sytems is necessary.

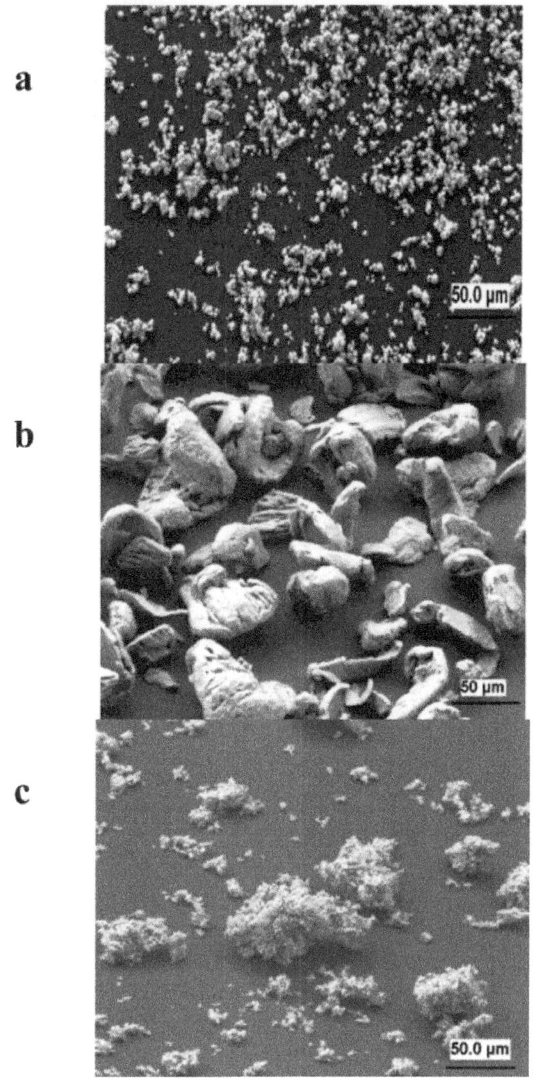

Figure 10. Raster electron images from top to the bottom: a) Fe-metal, b) Mg-flakes, c) Mg_2FeH_6 after 600 cycles.

Electron Microscopy Investigations of the Mg_2FeH_6 Materials

The cause for the improvement of thermochemical storage properties of the Mg-Fe-H system with respect to the MgH_2 system could be partially elucidated via electron microscopy investigations on Mg-Fe-H samples in the hydrogenated and dehydrogenated state. Raster electron-microscopy images of the starting Fe and Mg powders (Fe 5–10 µm and Mg particles 50–100 µm) and of Mg_2FeH_6 after 600 hydrogenation/dehydrogenation cycles at high temperatures are shown in Figures 10a–c. A high resolution transmission electron microscopy (TEM) micrograph of Mg_2FeH_6 is reproduced in Figure 11 (top). At 1:500,000 magnification one can already perceive the atomic net plains of the crystalline magnesium-iron hydride. The material is perfectly homogeneous. On the lower part of the Figure (Figure 11, bottom) can be seen a TEM photo of the dehydrogenated material: in this case black and colorless regions can be discerned. By energy dispersive X-ray (EDX) analyses it could be proved that the black regions are almost pure iron and the colorless transparent ones are pure magnesium. Upon each dehydrogenation the two metals separate from each other.

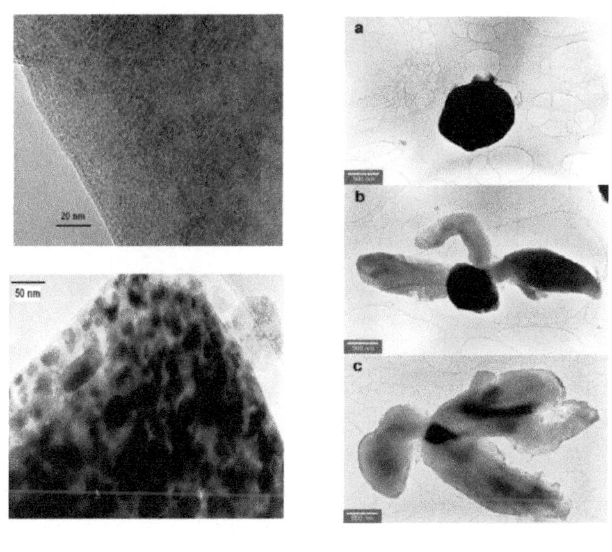

Figure 11. Left: HR-TEM-pictures of Mg_2FeH_6 (top) and dehydrogenated material (2Mg + Fe) (bottom); right: TEM micrographs of different steps of Mg_2FeH_6 formation are recorded. Dark regions of the particles are Fe regions and the lighter consist of Mg_2FeH_6. (a) Initial stage of the Mg_2FeH_6formation; (b) vermicular excresence of Mg_2FeH_6 out of the surface of an iron seed; (c) final stage of the Mg_2FeH_6formation [22].

The size of the regions are in the range of 20–50 nm. Apparently, due to this extremely fine dispersion it is possible that after each hydrogenation step the homogeneous Mg_2FeH_6 phase is formed. In the hydrogenation step a transfer of (probably) magnesium atoms must take place. With the help of TEM it was possible to identify different stages of Mg_2FeH_6 formation from Fe, Mg and H_2 and which takes place on the surface of Fe grains (Figure 11); for details see Ref. [22]. It appears plausible that this special circumstance - the separation of metals in two immiscible phases upon dehydrogenation and the merging to one phase upon hydrogenation - counteracts agglomeration of particles and thus, is one of the probable causes of cyclic stability.

A PICTURE OF A THERMOCHEMICAL STORAGE OF SOLAR HEAT ON THE MGH_2/MG BASIS

Comparision of the MgH_2/Mg system with the existing heat storage system of Andasol 1

Andasol 1 is a solar thermal power plant located in Andalucia (Spain) with a 50 MW steam turbine for electricity production. To run the plant during the dark a heat storage system is installed for additional 7.5 h electricity production at night. The storage system consists of two insulated storage tanks each 14 m high and 36 m in diameter (Figure 12). They are filled with 28,000 tons of a molten salt mixture of sodium and potassium nitrate. The tanks are kept at different temperatures, the cold tank at 260 °C and the hot tank at 390 °C. To charge heat, the "cold" molten salt picks up heat in a heat exchanger from the 400 °C hot oil of the solar field and stores in the hot tank. For discharging the process is reversed [27]. The overall storable heat amount in this system is 1,000 MWh [28].

Figure 12. 50 MW Solar power plant Andasol 1 with solar thermal energy storage system [27], photo with permission from Solar Millennium.

As mentioned in Section 2 the amount of heat released upon reaction of 1 kg of Mg with H_2 is 0.9 kWh. Thus it can be roughly estimated that an amount of about 1,100 tons of Mg-metal would be able to store the same amount of heat as the Andasol storage system. In addition to the storage of the Mg/MgH_2 mixture, a second container is necessary for the temporary storage of the hydrogen released during the dissociation of MgH_2 (heat absorption Figure 3, Part A). The most simple option is the storage of hydrogen gas under own dissociation pressure of MgH_2. As an example we assume here working temperatures of the heat storage between 370 and 400 °C. At 370 °C the dissociation pressure of MgH_2 is 10 bar and at 400 °C roughly 20 bar (Figure 1) [29]. In this pressure and temperature ranges, a simple calculation shows [30], a volume of 109,000 m^3 of H_2gas resulting from decomposition of MgH_2 must be stored. This sounds a lot, but, as shown below, gas storage of this order of magnitude has been standard gas technology for more than hundred years.

Options for the storage of large volumes of hydrogen gas

Hydrogen gas storage in containers

Natural gas storage under elevated pressures in spherical gas tanks is a widely used method. A system with a diameter up to 40 m and a maximum gas pressure of roughly 10 bar for natural gas has been constructed in Wuppertal, Germany and is in use since the 1950's. The volume of this storage vessel is 55,000 m^3. Pipe containers are another simple method for the storage of natural gas up to pressures of 100 bar. A similar construction can also be used for the storage of hydrogen.

Hydrogen storage in gas pipelines

Large volumes of H_2 can be stored in existing hydrogen pipelines. A hydrogen pipeline can possibly be situated near a thermal solar power plant.

Hydrogen gas storage in geological formations

The storage of natural gas in underground salt caverns is also a widely used method to overcome fluctations in the demand for the gas [31]. Recently, several studies have shown that underground caverns can be used for the storage of hydrogen under elevated pressures and with extremely low leak rates [32, 33]. Underground hydrogen storage in salt caverns is situated, f.i. in Teesside, Great Britain. The salt caverns are in a depth of 350 m with an overall volume of 70,000 m^3. The hydrogen gas is under a constant pressure of 45 bar, which is the result of brine replacement. These values are close to those of a conceptual storage system based on a MgH_2/Mg storage system for a 50 MW solar thermal power plant. On condition is that a suitable geological formation can be found close to the power plant, but underground hydrogen storage option appears to be a cheap and simple storage method for hydrogen released from MgH_2.

CONCLUSIONS

In the face of todays' urgent climate, energy and economical problems we are at present witnessing intensified planning and building of thermal solar power plants [34]. To enable the plants to function overnight storage of HT heat is necessary, which in practice usually involves storage of sensible heat by means of molten salts. As chemists we propose that, as presented in Section 6, the "thermochemical option" based on HT metal hydrides should also be taken into consideration.

On the basis of the present short overview one can easily recognize that in the research field of metal hydrides as energy storage systems both at high and low temperatures there remains yet a lot to be discovered. Currently the search for low temperature reversible hydride systems with higher than at present known H_2 storage capacity, as required for heat pumps and production of cold, but first and foremost as hydrogen storage materials for fuel-cell driven cars [15, 35], seems especially worthy of efforts at the moment.

ACKNOWLEDGMENTS

The assistance and intellectual input from B. Spliethoff and K. Schlichte over several years of research is gratefully acknowledged!

REFERENCES

1. Bogdanović, B; Ritter, A; Spliethoff, B. Active MgH_2-Mg systems for reversible chemical energy storage. *Angew. Chem. Int. Ed* **1990**, *29*, 223–234.
2. Beckmann, G; Gilli, PV. *Thermal Energy Storage*; Springer-Verlag: Wien, New York, 1984; p. 57, Table 2.4.
3. Buchner, H. *Energiespeicherung in Metallhydriden*; Springer Verlag: Wien, New York, 1982.
4. The formation of NO_x in application of H_2 for engines with internal combustion can be practically eliminated using lean gas mixtures (λ λ 2); taken from [3], p. 126.
5. Rummel, W. Heat storage in magnesium-hydrogen system. *Siemens Forsch. Entwicklungsber***1978**, *7*, 44.
6. Zaluska, A; Zaluski, B; Ström-Olsen, JO. Nanocrystalline magnesium for hydrogen storage. *J. Alloys Comp* **1999**, *288*, 217–225.
7. Bogdanović, B; Liao, S; Schwickardi, M; Sikorsky, P; Spliethoff, B. Catalytic synthesis of magnesium hydride under mild conditions. *Angew. Chem. Int. Ed. Engl* **1980**, *19*, 818–819.
8. Bogdanović, B; Hartwig, Th; Spliethoff, B. The development, testing and optimization of energy storage materials based on the MgH_2-Mg system. *Int. J. Hydrogen Energy* **1993**, *18*, 575–589.
9. Bogdanović, B; Kröner, M; Wilke, G. Olefin-Komplexe des Nickels (0). *Liebigs Ann. Chem***1966**, *699*, 1–23.
10. Wierse, M; Groll, M. Development, construction and testing of a thermo-chemical energy store on the MgH_2 basis for solar power plants and other energy supplying systems (in German), a joint project of the German BMFT, 0328939 B, 1995; Wierse, M.; Groll, M. System performance of a solar thermal power station with a thermochemical energy storage. Proceedings of the 11th Hydrogen Energy Conference, Stuttgart, Germany, 23-28/7/1996; Veziroglu, TN, Winter, CJ, Baselt, JP, Kreysa, G, Eds.; Dechema: Frankfurt, Germany, 1996; pp. 1997–2003.
11. Kleinwächter, J; Mitzel, M. dto., a joint project of the German BMFT, 0328993 A. 1996.
12. Steiner, D; Wierse, M; Groll, M. Development of a solar cooking/cooling unit with a thermochemical energy store based on metal hydrides. Proceedings of the 11th Hydrogen Energy Conference, Stuttgart, Germany, 23-28/7/1996; Veziroglu, TN, Winter, CJ, Baselt, JP, Kreysa, G, Eds.; Dechema: Frankfurt, Germany, 1996; pp. 2005–2010.
13. Reiser, A; Bogdanović, B; Schlichte, K. The application og Mg-based metal hydrides as heat energy storage systems. *Int. J. Hydrogen Energy* **2000**, *25*, 425–430.
14. Bowman, RC; Fultz, B. Metallic hydrides I: Hydrogen storage and other gas-phase applications. *MRS Bulletin* **2002**, *27*, 688–693.
15. Schüth, F; Bogdanović, B; Felderhoff, M. Light metal hydrides and complex hydrides for hydrogen storage. *Chem. Commun* **2004**, 2249–2258.
16. Dantzer, P. Metal-hydride technology: A critical review. In *Hydrogen in Metals III, Topics in Applied Physics.*; Wipf, H, Ed.; Springer-Verlag: Berlin, Heidelberg, Germany, 1997; Volume 73.
17. Bogdanović, B; Ritter, A; Spliethoff, B; Straßburger, K. A process steam generator based on the high temperature magnesium hydride/magnesium heat storage system. *Int. J. Hydrogen Energy* **1995**, *20*, 811–822.
18. Straßburger, K. *Wärmetransformation und thermische Energiespeicherung durch das aktive MgH_2/Mg-System*; . Thesis, University Essen: Essen, Germany, 1999.
19. Bogdanović, B; Spliethoff, B; Ritter, A. The magnesium hydride system for heat storage and cooling. *Ztschr. Physik. Chem. Neue Folge* **1989**, *164*, 1497–1508.
20. Bogdanović, B; Ritter, A; Spliethoff, B; Straßburger, K. Development, optimisation and testing of energy storage materials on the magnesium hydride magnesium basis (in German), a joint project of the German Federal Ministry for Research and Technology (BMFT), 0328939 C. 1992.
21. Bogdanović, B; Hofmann, H; Neuy, A; Reiser, A; Schlichte, K; Spliethoff, B; Wessel, S. Ni-doped versus undoped Mg-MgH_2 materials for high temperature heat or hydrogen storage. *J Alloys Compd* **1999**, *292*, 57–71.
22. Bogdanović, B; Reiser, A; Schlichte, K; Spliethoff, B; Tesche, B. Thermodynamics and dynamics of the Mg-Fe H system and its potential for thermochemical thermal energy storage. *J. Alloys Comp* **2002**, *345*, 77–89.
23. Didisheim, JJ; Zolliker, P; Yvon, K; Fischer, P; Schefer, J; Gubelmann, M; Williams, AF. Dimagnesium Iron(II) Hydride, Mg_2FeH_6, Containing Octahedral FeH_6^{4-} Anions. *Inorg. Chem* **1984**, *23*, 1953–1957.
24. Huot, J; Boily, S; Akiba, E; Schulz, R. Direct synthesis of Mg_2FeH_6 by mechanical alloying. *J. Alloys Compd* **1998**, *280*, 306–309.
25. Gennari, FC; Castro, FJ; Andrade Gamboa, JJ. Synthesis of Mg_2FeH_6 by reactive mechanical alloying: formation and decomposition properties. *J. Alloys Compd* **2002**, *339*, 261–267.
26. Varin, RA; Li, S; Calka, A; Wexler, D. Formation and environmental stability of nanocrystalline and amorphous hydrides in the 2Mg-Fe mixture processed by controlled reactive mechanical alloying (CRMA). *J. Alloys Compd* **2004**, *373*, 270–286.
27. Fairly, P. Largest Solar Thermal Power Plant to Start Up.
28. .Bogdanović, B; Bohmhammel, K; Christ, B; Reiser, A; Schlichte, K; Vehlen, K; Wolf, U. Thermodynamic

investigation of the magnesium hydrogen system. *J. Alloys Compd* **1999**, *282*, 84–92.

29. The amount of 1,100 tons of Mg, when converted into MgH_2, would upon dissociation deliver 1.09 × 10^6 Nm^3 of H_2 or 109,000 m^3 at 10 bar pressure. Thus, for temporary storage of hydrogen a storage tank of 109,000 m^3 capacity is necessary. We envisage that at the beginning of a heat storage cycle the tank is filled with 10 bar of hydrogen. Upon heating of the storage unit filled with 1,100 tons of MgH_2 from 370 to 400 °C, the unit, upon complete dissociation of MgH_2 would release 109,000 m^3 of hydrogen, which when introduced into the storage tank would result in a pressure increase from 10 to 20 bar (Figure 1). During night the heat store unit will cool down and the hydrogen will flow back to the heat store, producing there 1,000 MWh of heat at the 400 to 370 °C temperature level. The pressure in the system will drop from 20 to 10 bar, thus restoring the initial state. It should be mentioned that the described system is self-controlled, depending only of the amount of heat output or input to the storing unit. Certainly, the presented is a very crude outline of a heat storage cycle, since during the cycle ubiquitous heat losses have to be taken into account. On the other hand, by assuming higher than at presently used working temperatures and pressures for the heat storage cycle, much smaller volumes for the hydrogen storage tank can be calculated.
30. Sedlacek, R. Underground Gas Storage in Europe. *Erdöl, Erdgas, Kohle* **1999**, *115*, 537–540.
31. Leighty, W. Running the world on renewables: Hydrogen transmission pipelines and firming geologic storage. *Int. J. Energy Res* **2008**, *32*, 408–426.
32. Crotogino, F; Hamelmann, R. Wasserstoff-Speicherung in Salzkavernen zur Glättung des Windstromangebotes. 14. Symposium zur Nutzung regenerativer Energiequellen und Wasserstofftechnik, Stralsund, Germany, Nov. 2007. *Energieperspektiven 2*, 2008.
33. .Züttel, A, Borgschulte, A, Schlapbach, L, Eds.; *Hydrogen as a Future Energy Carrier*; Wiley-VCH: Weinheim, Germany, 2008.

CHAPTER 10

EFFECTS OF VANADIUM/NICKEL CONTENTS IN LAVES PHASE-RELATED BODY-CENTERED-CUBIC SOLID SOLUTION METAL HYDRIDE ALLOYS

Kwo-hsiung Young, Diana F. Wong and Jean Nei

Keywords: Hydrogen absorbing alloys, metal hydride (MH) electrode, Laves phase alloys, body-centered-cubic alloys, phase segregation

INTRODUCTION

"Laves phase-related body-centered-cubic (BCC) solid solution" is a family of metal hydride (MH) alloys composed mainly of a BCC phase and a Laves phase (mostly C14), which can be used as the negative electrode active material for nickel/metal hydride (Ni/MH) batteries [1,2]. Other minor phases, such as C15, TiNi, Ti_2Ni, and VNi can also be included. Both V and TiCr intermetallic alloys have the same BCC crystal structure, which enable a wide composition solubility range for BCC phase in the Ti-V-Cr ternary phase diagram [3]. Alloys with sole BCC structure can have a hydrogen storage capacity higher than 3.5 wt% [4,5], but the electrochemical performance is poor because of severe cracking and passivation on the surface [5]. Adding Ni into the alloy formula promotes the formation of a C14 phase [6]. The introduction of Laves phases promotes good absorption kinetics [7], easy formation due to its brittleness [8,9,10], and high surface catalytic activity through a synergetic effect [11]. Many works have been reported in this family of alloys that partially replace Ti, V, and Cr with other transition metals for improvements in both hydrogen storage [9,12,13,14,15,16,17,18,19] and electrochemical applications [20,21,22,23]. Our contributions to the field include the introduction of a high hydrogen pressure activation process [24], an optimization of annealing conditions (900 °C for 12 h) [25], and a study examining the contributions of the constituent elements in Laves phase-related BCC solid solutions. The contributions of the elements are summarized as: removal of Fe and decrease in V-content in exchange for higher Ni-content to improve both the electrochemical capacity and high-rate dischargeability (HRD); elimination of the C14 phase by removal of Zr contributing to a reduced discharge capacity, a prolonged activation period, and a less catalytic surface for electrochemical reaction; and the TiNi phase was also found to contribute positively to the bulk diffusion of hydrogen while hindering the surface electrochemical reaction. In the same study, an alloy (P17) achieved a full capacity of 390.8 $mAh·g^{-1}$ at a small discharge rate of 4 $mA·g^{-1}$; however, the capacity decreases to 365.4 $mAh·g^{-1}$ when the discharge rate was increased to 50 $mAh·g^{-1}$ [26]. The capacity degradation with increased rate in P17 is worse than that in the conventional AB_5 MH alloys (from 336 $mAh·g^{-1}$ to 329 $mAh·g^{-1}$) under the same test condition [27]. The HRD is very important for the high-power applications for Ni/MH battery, such as in hybrid electrical vehicles and power tools. Previously, Ni has been known to increase the HRD of the MH alloys [22,28,29,30,31,32]. In order to further improve the HRD performance, the vanadium/nickel content in alloy P17 was adjusted and the results are presented here.

EXPERIMENTAL SECTION

Ingot sample was prepared by an arc-melting, which includes a non-consumable tungsten electrode, a water-cooled copper tray and a continuous argon flow. A piece of sacrificial titanium

Citation: Kwo-hsiung Young, Diana F. Wong and Jean Nei, Effects of Vanadium/Nickel Contents in Laves Phase-Related Body-Centered-Cubic Solid Solution Metal Hydride Alloy, doi: 10.3390/batteries1010034.

underwent a few melting-cooling cycles to reduce the residual oxygen concentration in the system before each run. Each 12 g ingot was re-melted and turned over several times to ensure uniformity in chemical composition. Following previously developed annealing conditions, each ingot was annealed in vacuum at 900 °C for 12-h [25]. A chemical composition for each ingot was analyzed using a Varian Liberty 100 inductively coupled plasma-optical emission spectrometer (ICP-OES, Agilent Technologies, Santa Clara, CA, USA). A Philips X'Pert Pro X-ray diffractometer (XRD, Philips, Amsterdam, Netherlands) was used to study the microstructure, and a JEOL-JSM6320F scanning electron microscope (SEM, JEOL, Tokyo, Japan) with energy dispersive spectroscopy (EDS) capability was used to study the phase distribution and composition. Gaseous phase hydrogen storage characteristics for each sample were measured using a Suzuki-Shokan multi-channel pressure-concentration-temperature (PCT, Suzuki Shokan, Tokyo, Japan) system. In the PCT analysis, each sample (a single piece of ingot with a newly cleaved surface and a weight of about 2 g) was first activated by a 2-h thermal cycle (between 300 °C and 30 °C) at 5 MPa H_2pressure. PCT isotherms at 30 and 60 °C were then measured. Magnetic susceptibility was measured using a Digital Measurement Systems Model 880 vibrating sample magnetometer (MicroSense, Lowell, MA, USA).

RESULTS AND DISCUSSION

Alloy Composition

Eight alloys with a general formula of $Ti_{15.6}Zr_{2.1}V_{44-x}Cr_{11.2}Mn_{6.9}Co_{1.4}Ni_{18.5+x}Al_{0.3}$, $x = 0, 4, 6, 8, 10, 12, 14$, and 16, were prepared by arc melting for this study. Design compositions of these alloys are listed in Table 1. Alloys P17 ($x = 0$) and P22 ($x = 4$) have been previously studied. P17 shows good balance between capacities and HRD [26], and P22 was the subject of the annealing condition matrix study [25]. The new compositions extend the study of V-content *versus* Ni-content. ICP results from the ingot samples are listed in Table 1. Other than a small but consistent deficiency in Zr (due to formation of ZrO_2 as oxygen scavenger), the ICP results are very close to the target compositions. The measured B/A ratios are slightly higher than the designed values due to the loss in Zr.

Table 1. Design compositions and inductively coupled plasma (ICP) results for alloys in this study. All numbers are in at%

Alloy		Ti	Zr	V	Cr	Mn	Co	Ni	Al	B/A
P17	Design	15.6	2.1	44.0	11.2	6.9	1.4	18.5	0.3	4.65
	ICP	15.6	2	44.1	11.3	6.4	1.4	18.9	0.3	4.68
P22	Design	15.6	2.1	40.0	11.2	6.9	1.4	22.5	0.3	4.65
	ICP	15.5	1.8	41.0	11.2	6.5	1.4	22.2	0.4	4.78
P23	Design	15.6	2.1	38.0	11.2	6.9	1.4	24.5	0.3	4.65
	ICP	15.2	1.7	38.7	11.4	6.8	1.4	24.3	0.4	4.91
P24	Design	15.6	2.1	36.0	11.2	6.9	1.4	26.5	0.3	4.65
	ICP	15.5	1.8	36.0	11.3	6.4	1.5	27.1	0.4	4.78
P25	Design	15.6	2.1	34.0	11.2	6.9	1.4	28.5	0.3	4.65
	ICP	15.5	1.8	34.4	11.4	6.6	1.4	28.4	0.4	4.77
P26	Design	15.6	2.1	32.0	11.2	6.9	1.4	30.5	0.3	4.65
	ICP	15.5	1.7	33.5	10.8	6.8	1.5	29.8	0.4	4.81
P27	Design	15.6	2.1	30.0	11.2	6.9	1.4	32.5	0.3	4.65
	ICP	15.2	1.8	30.6	11.7	6.6	1.4	32.2	0.4	4.88
P28	Design	15.6	2.1	28.0	11.2	6.9	1.4	34.5	0.3	4.65
	ICP	15.3	1.9	28.1	11.4	6.6	1.4	34.9	0.4	4.81

X-Ray Diffractometer Structure Analysis

XRD patterns of the eight alloys are shown in Figure 1. Five sets of diffraction peaks are observed: a BCC, two Laves phases (C14 and C15), a TiNi and a σ-VNi. As the Ni-content in the alloy increases, the BCC peaks shift to higher angles, C14 phase is replaced by C15 phase, and σ-VNi phase starts to appear. σ-VNi phase is in a tetragonal structure, and the lattice constants are a = 8.966 Å and c = 4.641 Å with a composition at $V_{0.6}Ni_{0.4}$ [33]. It has wide solubility in V from 57.5 at% to 75 at% [34]. The V-Ni-Cr ternary phase diagram indicates an extension of σ-VNi phase from 0% to about 40% Cr [35].

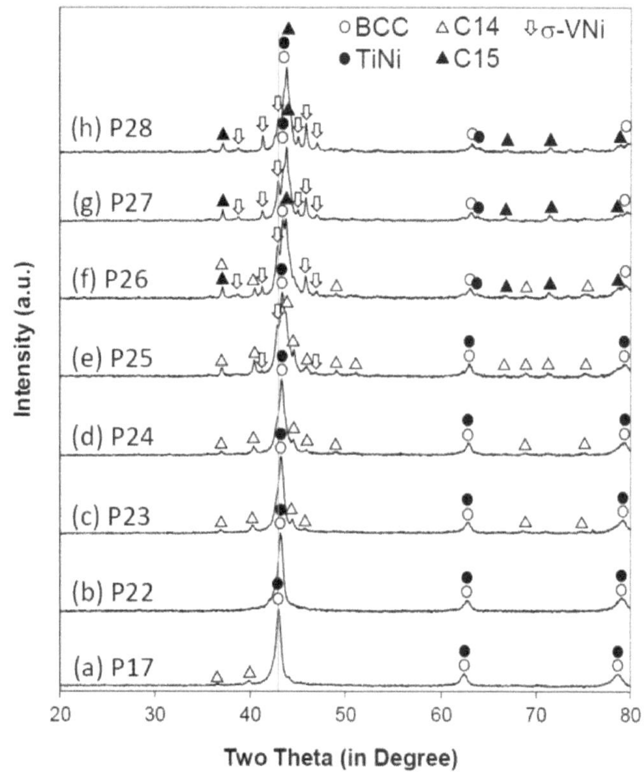

Figure 1. X-ray diffractometer (XRD) patterns using Cu-Kα as the radiation source for alloys: (**a**) P17; (**b**) P22; (**c**) P23; (**d**) P24; (**e**) P25; (**f**) P26; (**g**) P27; and (**h**) P28. A vertical line is used to indicate the shift in body-centered-cubic (BCC) peaks.

Lattice constants, crystallite sizes, and phase abundances of all five constituent phases calculated from the XRD patterns are listed in Table 2. The variations in lattice constants from BCC, C14, and TiNi phases are plotted in Figure 2. As the amount of Ni increases, the lattice constant a in BCC first remains constant and then decreases, both a and c in C14 phase decrease at about the same rate yielding a slight decrease in c/a ratio, while a in TiNi sees a discontinuous two-segment decrease. The changes in lattice constants for C15 and VNi are less prominent. Lattice parameter a in BCC phase decreases from 2.977 Å to 2.946 Å as the Ni-content in the alloy increases, which is further away from the optimized value of 3.042 Å corresponding to a maximized hydrogen storage capacity [36].

Table 2. Lattice parameters, unit cell volumes, phase abundances, and crystallite sizes of phases derived from XRD analysis. *XS* denotes crystallite size

Alloy	BCC			C14					
	a (Å)	*XS* (Å)	Abundance (wt%)	*a* (Å)	*c* (Å)	*c*/*a*	Unit cell volume (Å^3)	*XS* (Å)	Abundance (wt%)
P17	2.977	171	52.8	4.912	8.010	1.631	167.3	215	13.2
P22	2.974	57	50.7	4.890	7.968	1.629	165.0	275	17.1
P23	2.973	86	57.5	4.878	7.936	1.627	163.5	311	19.4
P24	2.974	72	47.0	4.865	7.915	1.627	162.2	255	20.3
P25	2.954	86	61.2	4.858	7.895	1.625	161.4	287	22.1
P26	2.950	71	55.2	4.844	7.883	1.627	160.2	272	11.1
P27	2.948	68	47.8	-	-	-	-	-	-
P28	2.946	73	34.2	-	-	-	-	-	-

Alloy	C15			TiNi			VNi			
	a (Å)	*XS* (Å)	Abundance (wt%)	*a* (Å)	*XS* (Å)	Abundance (wt%)	*a* (Å)	*c* (Å)	*XS* (Å)	Abundance (wt%)
P17	-	-	-	2.993	170	34.0	-	-	-	-
P22	-	-	-	2.962	175	32.4	-	-	-	-
P23	-	-	-	2.959	273	23.0	-	-	-	-
P24	-	-	-	2.958	204	32.6	-	-	-	-
P25	-	-	-	2.899	105	8.9	9.009	4.686	184	7.7
P26	6.859	637	15.1	2.905	228	4.5	9.010	4.692	219	14.1
P27	6.860	405	26.6	2.912	127	13.0	9.017	4.644	250	12.6
P28	6.858	347	24.7	2.912	115	25.5	9.012	4.642	344	15.7

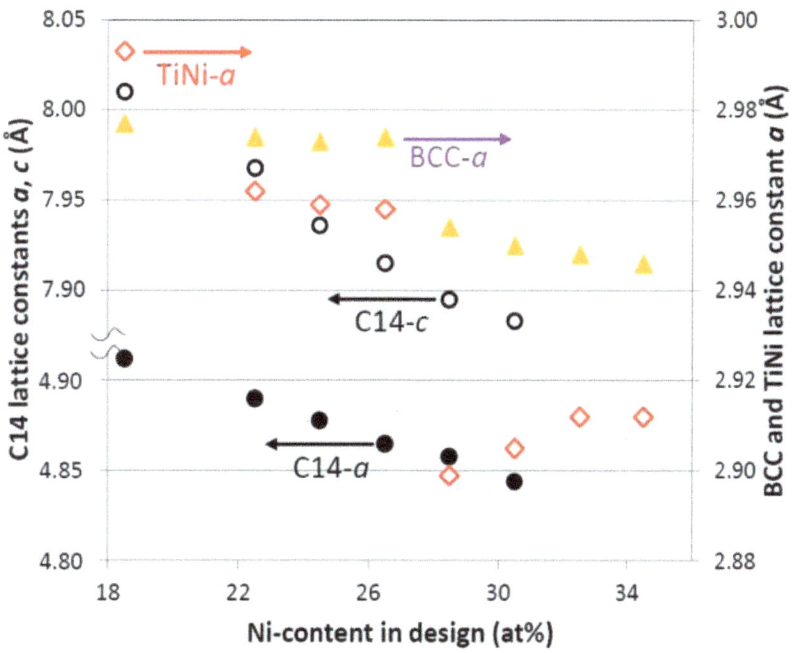

Figure 2. Lattice constants *a* and *c* from C14 phase, and lattice constants *a* from BCC and TiNi phases as functions of Ni-content in design.

Phase abundance and crystallite sizes of each phase are listed in Table 2. These values were obtained from full pattern fitting of the XRD data using the Rietveld method and Jade 9 Software (KS Analytical System, Aubrey, TX, USA).

The evolution of phase abundance is plotted in Figure 3. C14 Laves phase abundance first increases and then decreases as it is replaced by another Laves phase—C15 phase. The total Laves phase abundance generally increases, but begins to taper and decrease as it is changed to C15 phase. TiNi phase abundance varies widely as it first decreases, then increases, decreases, and finally increases again. BCC phase constitutes about 50–60 wt% and then drops to below 40 wt%. σ-VNi-phase appears in alloys with higher Ni-content, and the abundance further increases. The crystallite sizes (*XS*) of all five constituent phases have different trends. As the Ni-content in the alloy increases at the expense of V, the *XS* of BCC and C15 phases decrease, the *XS* of C14 and TiNi phases first increase and then decrease, and in VNi phase, the *XS* increases.

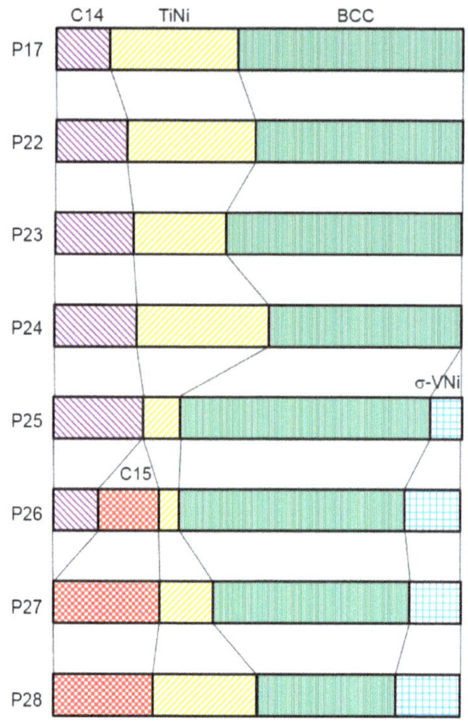

Figure 3. Evolution in the phase abundances of constituent phases from the full-pattern analysis of XRD data.

Scanning Electron Microscope/Energy Dispersive Spectroscopy Microstructure Analysis

Microstructures for this series of alloys were studied using SEM, and the resulting back-scattering electron images (BEI) are presented in Figure 4.

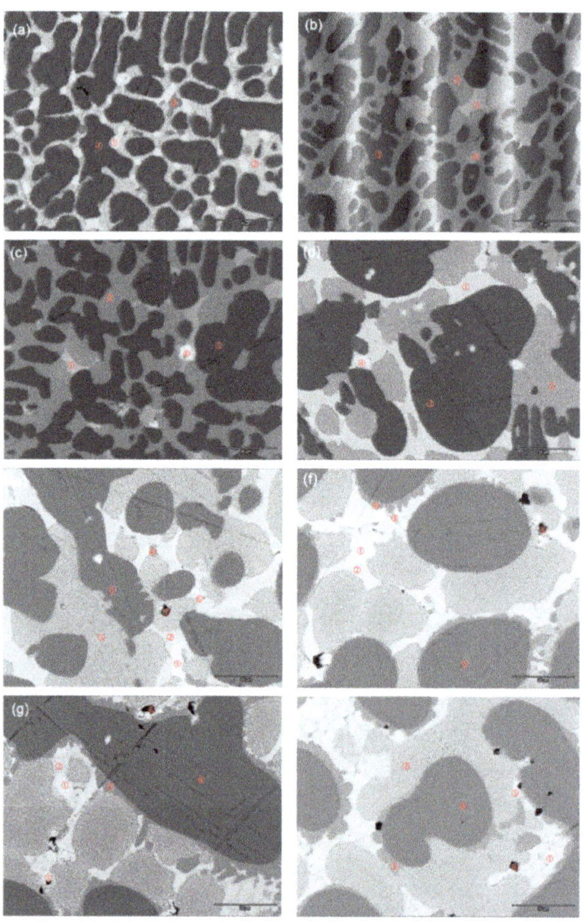

Figure 4. Scanning electron microscope (SEM) back-scattering electron images (BEI) for alloys: (**a**) P17; (**b**) P22; (**c**) P23; (**d**) P24; (**e**) P25; (**f**) P26; (**g**) P27; and (**h**) P28. Chemical compositions in the numbered areas measured by EDS are listed in Table 3.

Table 3. Energy dispersive spectroscopy (EDS) composition data from select spots in Figure 3. All numbers are in at%. B/A ratios for C14 and C15 phases were calculated assuming V is in the B-site while those in TiNi and VNi phases were calculated assuming V in the A-site. Compositions in the main BCC phase are highlighted in **bold**

Location	Zr	Ti	V	Cr	Mn	Fe	Co	Ni	Al	B/A	*e/a*	Phase
Figure 4a-1	16.2	16.3	7.8	4.4	5.9	0.0	1.8	47.5	0.1	2.08	7.28	C14
Figure 4a-2	4.7	36.4	7.8	0.8	3.6	0.0	2.0	44.4	0.3	1.04	-	TiNi
Figure 4a-3	6.2	48.5	11.7	1.2	2.3	0.0	1.8	28.3	0.1	0.51	-	Ti_2Ni
Figure 4a-4	**0.1**	**5.5**	**63.5**	**16.9**	**7.4**	**0.0**	**0.8**	**5.6**	**0.2**	-	-	BCC
Figure 4b-1	14.3	17.3	9.4	0.6	4.8	0.0	0.8	52.5	0.3	2.16	7.44	C14
Figure 4b-2	2.8	34.3	7.3	0.9	4.7	0.1	2.7	46.4	0.8	1.25	-	TiNi
Figure 4b-3	**0.1**	**3.0**	**60.7**	**17.7**	**11.8**	**0.0**	**0.9**	**5.7**	**0.1**	-	-	BCC
Figure 4b-4	87.2	3.0	3.1	0.5	0.8	0.1	0.4	4.8	0.1	-	-	Zr
Figure 4c-1	12.0	18.6	12.4	0.7	3.5	0.0	0.9	51.6	0.3	2.27	7.38	C14
Figure 4c-2	2.4	34.5	8.4	1.0	3.7	0.0	2.3	47.2	0.5	1.21	-	TiNi
Figure 4c-3	**0.1**	**4.5**	**58.1**	**18.2**	**8.6**	**0.0**	**1.1**	**9.3**	**0.1**	-	-	BCC
Figure 4c-4	93.9	1.1	2.5	0.4	0.3	0.1	0.1	1.5	0.1	-	-	Zr
Figure 4d-1	10.6	17.9	16.5	2.5	4.7	0.0	1.4	46.2	0.2	2.51	7.20	C14
Figure 4d-2	1.9	33.9	8.6	1.1	3.5	0.0	2.4	47.6	1.0	1.25	-	TiNi
Figure 4d-3	**0.0**	**2.3**	**60.0**	**21.1**	**8.3**	**0.0**	**1.0**	**7.1**	**0.2**	-	-	BCC
Figure 4d-4	94.7	0.7	1.8	0.4	0.1	0.0	0.2	1.9	0.2	-	-	Zr
Figure 4e-1	10.8	16.0	14.2	1.8	4.2	0.0	1.0	51.7	0.3	2.73	7.45	C15
Figure 4e-2	8.5	17.7	18.2	2.5	5.0	0.0	1.5	46.2	0.3	2.81	7.22	C14
Figure 4e-3	1.6	33.8	8.6	1.1	4.1	0.1	2.5	47.5	0.8	1.28	-	TiNi
Figure 4e-4	0.2	7.0	48.2	6.6	5.2	0.1	1.3	31.3	0.1	0.81	-	VNi
Figure 4e-5	**0.0**	**2.4**	**58.7**	**21.8**	**8.5**	**0.0**	**1.0**	**7.4**	**0.1**	-	-	BCC
Figure 4e-6	89.0	3.2	2.4	0.4	0.2	0.0	0.2	4.5	0.1	-	-	Zr
Figure 4e-7	4.7	69.5	7.8	1.3	1.6	0.0	0.6	14.1	0.3	-	-	TiO_2
Figure 4f-1	10.5	16.2	13.3	1.9	4.4	0.0	1.0	52.5	0.2	2.75	7.50	C15
Figure 4f-2	8.9	17.3	15.7	2.5	5.1	0.0	1.4	48.8	0.2	2.81	7.35	C14
Figure 4f-3	1.4	30.7	10.3	1.1	5.1	0.0	2.0	48.6	0.8	1.36	-	TiNi
Figure 4f-4	0.2	6.8	48.4	7.1	6.8	0.0	1.5	29.0	0.1	0.80	-	VNi
Figure 4f-5	**0.0**	**2.1**	**58.3**	**21.6**	**8.7**	**0.0**	**1.1**	**8.0**	**0.1**	-	-	BCC
Figure 4f-6	6.0	79.0	8.3	0.1	0.5	0.0	0.2	6.0	0.0	-	-	TiO_2
Figure 4g-1	10.4	16.5	11.9	2.1	5.2	0.1	1.2	52.4	0.3	2.72	7.53	C15
Figure 4g-2	4.5	30.2	9.2	1.0	5.0	0.0	1.0	48.8	0.2	1.28	-	TiNi
Figure 4g-3	0.1	6.6	46.2	9.4	6.6	0.0	1.8	29.3	0.1	0.89	-	VNi
Figure 4g-4	**0.0**	**1.8**	**55.4**	**25.3**	**8.6**	**0.0**	**1.1**	**7.6**	**0.1**	-	-	BCC
Figure 4g-5	85.5	5.1	2.1	0.5	0.6	0.2	0.1	5.8	0.1	-	-	Zr
Figure 4g-6	3.5	82.8	7.6	0.1	0.6	0.0	0.1	5.2	0.1	-	-	TiO_2
Figure 4h-1	10.2	16.2	11.8	2.4	5.5	0.0	1.1	52.5	0.3	2.79	7.53	C15
Figure 4h-2	1.3	30.1	9.1	1.0	6.3	0.0	2.3	49.0	0.9	1.47	-	TiNi
Figure 4h-3	0.1	6.6	43.5	11.6	6.8	0.2	1.8	29.2	0.2	0.99	-	VNi
Figure 4h-4	**0.0**	**1.7**	**53.0**	**27.4**	**8.3**	**0.0**	**1.2**	**8.3**	**0.1**	-	-	BCC
Figure 4h-5	86.1	3.9	2.2	0.4	0.7	0.0	0.2	6.2	0.2	-	-	Zr
Figure 4h-6	2.0	90.6	3.3	0.1	0.5	0.0	0.2	3.3	0.1	-	-	TiO_2

EDS analysis was used to study the chemical compositions of several areas showing different contrasts and the results are summarized in Table 3. In the case of BCC-C14 dual-phase alloy, during cooling of the melt, BCC phase with high V-content solidifies first to form a three-dimensional framework while the rest of the liquid solidifies into Laves phase as the alloy cools further. The Laves phase itself is also a three-dimensional framework (Figure 4 in [37]). This interlacing dual-framework can be seen in the alloys in the current study, but the framework has less alignment. The main BCC phase is shown in a darker contrast due to its low Zr- and Ni-contents (lower average atomic weight). The grain size of the main BCC phase increases but the crystallite size decreases (as seen from the XRD analysis) as the Ni-content in the alloy increases. The majority of the secondary phases is in the BCC phase grain boundary, and it evolves from TiNi + C14 + Ti_2Ni (P17, very small amount and not detectable by XRD) to TiNi + C14 + Zr (P22 − P24), TiNi + C14 + C15 + VNi + Zr + TiO_2 (P25, P26), and TiNi + C15 + VNi + Zr + TiO_2 (P27, P28). The average electron density (*e/a*) calculated from the measured composition was used to distinguish the C14 and C15 Laves phases. The *e/a* values are listed in Table 3, and in this case, the C14/C15 threshold falls at a value of 7.45 [38,39]. The VNi phase starts at the center of the secondary phase region (Figure 4e-4) and moves to the boundary between BCC main phase and other secondary phases (TiNi and Laves). It is clear from the V-Ni phase diagram that the cooling sequence is first BCC phase, followed by σ-VNi phase and finally ending with a VNi_2 (*oI*6 structure) + σ-VNi mixture [34]. In our multi-element case, the final VNi_2 is replaced by a combination of TiNi and Laves phases.

The atomic percentages of a few key elements in BCC, TiNi, and C14 phases examined by SEM/EDS are plotted in Figure 5 as functions of Ni-content in the overall alloy design. Since the sampling volume of one spot in the EDS analysis is very limited (up to a few microns in diameter), the results may not be very representative and should be interpreted with caution. In the main BCC phase, as the Ni-content increases at the expense of V, the V-content decreases, the Cr-content increases, and the Mn- and Ni-contents remain at about the same values (Figure 5a). Because Cr is smaller than V, this decreases the lattice constant in BCC phase. V is a hydride former, and a smaller unit cell with less V-content has direct impacts on several properties: the metal-hydrogen bond strength will be weaker (better HRD), the plateau pressure will be higher, and the storage capacity will be lower. In the TiNi phase, as the Ni-content in design increases, the Ni-content increases, the Ti-content decreases, and the V- and Mn-contents remain relatively unchanged (Figure 5b). Because Ni is smaller than Ti, this results in a smaller unit cell, which is also beneficial in the TiNi catalytic phase for easier hydrogen diffusion. In the C14 phase, as the Ni-content in design increases, the Zr-content decreases, the V-content increases, and Ti and Ni remain unchanged. The decrease in Zr is due to the fact that Laves phase contains most of the Zr and its abundance in Laves phase increases with the increase in Ni. The overall Zr-content is fixed; therefore, the Zr-content in

Laves phase is diluted and the heavily over-stoichiometry (B/A > 2.0) invites more V into the Laves phase. Both the lattice constants *a* and *c* decrease as more Zr (larger) is replaced by V (smaller). The relatively constant *c*/*a* indicates that most of the V sits in the A-site since a relatively large substitution in the B-site will cause an anisotropic enlargement in the unit cell [40,41]. The shrinkage in catalytic C14 phase unit cell also facilitates hydrogen diffusion in the alloy bulk.

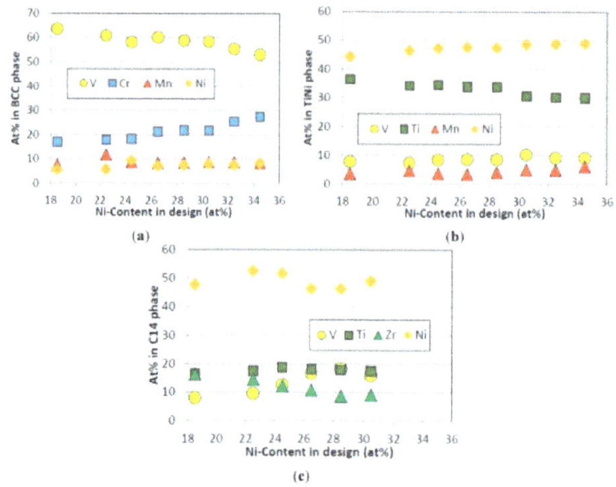

Figure 5. Atomic percentages of key elements in: (**a**) BCC; (**b**) TiNi; and (**c**) C14 phases as functions of Ni-content in design.

Gaseous Phase Study

Gaseous phase hydrogen storage properties of the alloys were studied by PCT. All samples are activated with one thermal cycle in the presence of hydrogen. Consequent measurements did not change the PCT characteristic significantly. The resulting absorption and desorption isotherms measured at 30 °C and 60 °C are shown in Figure 6. Information obtained from the PCT study is summarized in Table 4. A two-phase characteristic can be observed in alloys P22–P24. The heat of hydride formation of each of the phases can be estimated from the atomic percentages and the heats of hydride formation (ΔH_h) for the three main hydride-former atoms, *i.e.*, Zr ($\Delta H_h = -163$ kJ·mol H_2^{-1} for ZrH_2), Ti ($\Delta H_h = -124$ kJ·mol H_2^{-1} for TiH_2), and V ($\Delta H_h = -34$ kJ·mol H_2^{-1} for VH_2) [42]. For example, the calculated ΔH_h for TiNi will be −62 kJ·mol H_2^{-1}, which is close to the experimental result of −60 kJ·mol H_2^{-1} [43]. The such-obtained ΔH_h for the constituent phases for P22 are −25.5 (BCC), −49.6 (TiNi), and −46.8 kJ·mol H_2^{-1} (C14). Therefore, the main plateau at higher equilibrium pressure in the PCT isotherm for P22 alloy is attributed to the main BCC phase, and the second plateau, which occurs at lower hydrogen concentration and lower equilibrium pressure, is attributed to the combination of TiNi and C14 secondary phases. The same estimation is performed on alloy P26, and the results are −22.4 (BCC), −25.2 (VNi), −43.8 (TiNi), −41.3 (C14), and −41.7 kJ·mol H_2^{-1} (C15). The first two phases (BCC and VNi) are considered to be storage phases, which are responsible for the plateau in the PCT isotherm, and the other three phases with higher Ni-content are the catalytic phases that do not contribute directly to the hydrogen storage capacity.

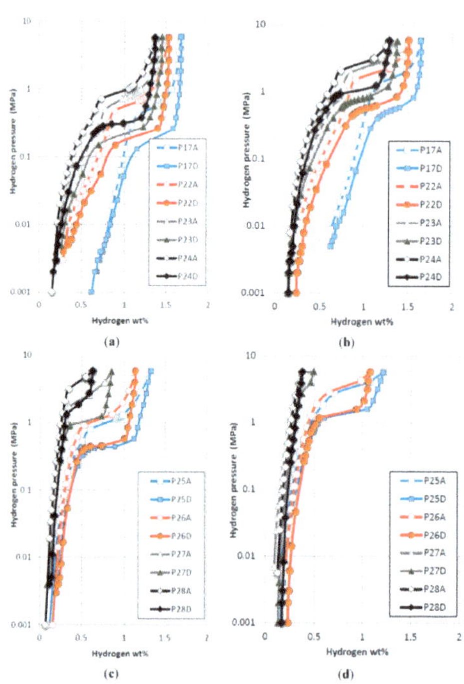

Figure 6. (**a**) 30 °C and (**b**) 60 °C pressure-concentration-temperature (PCT) isotherms of alloys P17, P22, P23, P24; and (**c**) 30 °C and (**d**) 60 °C of alloys P25, P26, P27, and P28. Open and solid symbols are for absorption and desorption curves, respectively.

Table 4. Summary of gaseous phase and thermodynamic properties. The ΔH and ΔS were calculated from the equilibrium pressure plateaus measured at 30, 60, and 90 °C with the error bars of about ±2%

Alloy No.	Desorption pressure @ 30 °C (MPa)	Desorption pressure @ 60 °C (MPa)	Maximum capacity @ 30 °C (wt%)	Reversible capacity @ 30 °C (wt%)	PCT hysteresis @ 30 °C	PCT hysteresis @ 60 °C	$-\Delta H$ (kJ·mol^{-1})	$-\Delta S$ (J·mol^{-1}·K^{-1})
P17	0.17	0.53	1.68	1.01	1.06	1.10	30.9	107
P22	0.21	0.60	1.53	1.24	1.14	1.13	29.9	105
P23	0.22	0.79	1.45	1.25	1.20	1.06	35.3	123
P24	0.32	1.07	1.38	1.20	1.10	0.86	34.2	122
P25	0.45	1.42	1.33	1.16	0.77	0.82	32.1	119
P26	0.43	1.36	1.14	0.93	0.79	0.84	32.4	119
P27	1.06	3.20	0.85	0.75	0.79	0.46	31.0	122
P28	1.90	-	0.63	0.51	0.76	-	-	-

As the Ni-content in the alloy increases, the plateau pressure in the BCC phase increases monotonically (due to the decrease in the BCC unit cell volume), the maximum storage capacity decreases, and the reversible storage first increases (due to the increase in abundance of catalytic C14 phase and the shrinkage in unit cell volumes for BCC, TiNi, and C14 phases) and then decreases (due to the reduction in the maximum capacity). Hysteresis of the PCT isotherm is defined as $\ln(P_a/P_d)$, where P_a and P_d are the absorption and desorption equilibrium pressures at the main plateau (BCC phase), respectively. In general, the hysteresis measured at 30 °C divides the alloys into two groups: alloys with relatively low Ni-content (P17, P22–P24) that show higher hysteresis than those in the other group (P25–P28). PCT hysteresis represents the energy required to elastically distort the lattice at the metal/hydride interface [44]. The first alloys (P17, P22–P24) also have distinctively larger unit cells (Figure 2) and also are free of C15 and σ-VNi phases. The connection between PCT hysteresis and structural properties of these alloys requires further investigation.

In order to study the thermodynamic properties of these alloys, equilibrium pressure in the BCC plateau at 30 °C and 60 °C were used to estimate the ΔH_h and entropy (ΔS) by the equation:

$$\Delta G = \Delta H_h - T\Delta S = RT\ln P \quad (1)$$

where R is the ideal gas constant and T is the absolute temperature. Results of these calculations are listed in Table 4. The values may not be accurate and can only be used for comparison among these alloys. As the Ni-content in the design increases, both ΔH_h and ΔS decrease first and then increase. The decrease in ΔH_h of the first three alloys (P17, P22 and P23) is due to the reduction in the TiNi phase abundance reducing the proximity effects from the catalytic phases [45]. Further increase in ΔH_h is due to the shrinking unit cell volume in BCC phase [46]. ΔS is an indication of how far the MH system is from a perfect, ordered state. The theoretical value of ΔS is the entropy of hydrogen gas, which is close to –135 J mol^{-1} K^{-1}. Alloy P23 shows the lowest ΔH_h and ΔS, and this indicates that it has the least interaction between the main storage phase and the catalytic phase, acting as a single BCC phase MH alloy.

Electrochemical Measurement

The electrochemical discharge capacity of each alloy was measured in a flooded-cell configuration against a partially pre-charged $Ni(OH)_2$ positive electrode. No alkaline pretreatment was applied before the half-cell measurement. For the activation behavior study, each sample electrode was charged at a constant current density of 50 mA·g^{-1} for 10 h and then discharged at a current density of 50 mA·g^{-1} followed by two pulls at 12 mA·g^{-1} and 4 mA·g^{-1}. The results of the low-rate (full) capacity and the HRD_{50} (defined as the ratio in capacities obtained between 50 mA·g^{-1} and 4 mA·g^{-1}) for the first thirteen cycles are plotted in Figure 7. From the comparison, it is easy to see that the activation in low-rate capacity is similar for all alloys (Figure 7a), but the activation in HRD (more related to surface oxidation and pulverization) is easier in alloys with lower Ni-content and becomes more difficult with the increase in Ni-content (Figure 7b). Capacities measured at the 4th cycle are listed in Table 5 together with other important electrochemical and magnetic properties of the alloys in this study. The capacities measured at the 4th cycle are also plotted in Figure 8 with the gaseous phase capacity converted to a theoretical electrochemical capacity by

$$1\text{ wt\% of } H_2 = 268\text{ mAh·g}^{-1} \quad (2)$$

As the Ni-content in the design increases, both capacities measured at 50 mA·g^{-1} and 4 mA·g^{-1}rates decrease monotonically, which is similar to the decrease in the maximum storage capacities measured in the gaseous phase. In a conventional MH alloy, the electrochemical capacities measured at different rates fall between the boundaries set by the equivalent maximum and reversible hydrogen storage capacities [29,47,48,49,50]. In the current study, the electrochemical capacities drop below the lower bound set by the gaseous phase reversible capacity for alloys P24–P28. It is due to the relatively high equilibrium pressure at room temperature (RT) of these alloys (>0.6 MPa) preventing a full charge under the open flooded half-cell configuration. The electrochemical discharge capacities of these alloys should be higher in a sealed cell where the cell pressure can be kept as high as 2.8 MPa.

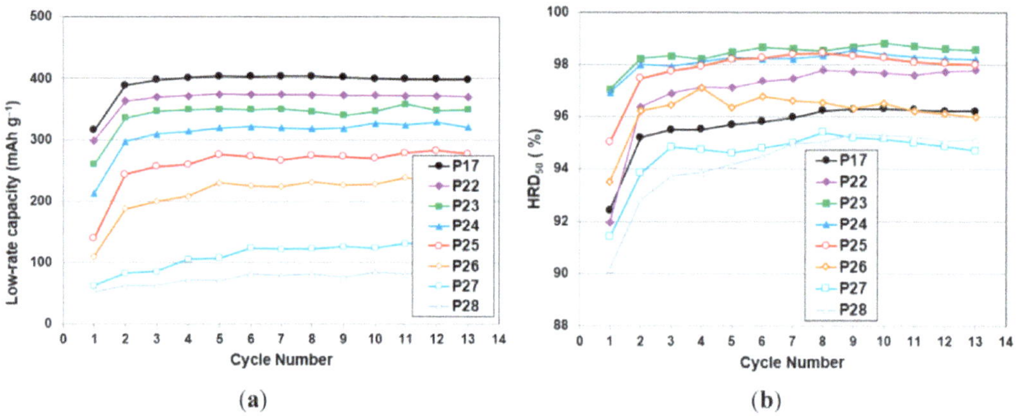

Figure 7. Activation behavior for the first 13 cycles of (**a**) low-rate capacity and (**b**) HRD_{50} (ratio of capacities at 50 mA·g^{-1}and 4 mA·g^{-1} discharge rates).

Table 5. Summary of electrochemical and magnetic properties. High-rate dischargeability $(HRD)_{200}$ is the ratio of capacities measured at 200 mA·g^{-1} and 4 mA·g^{-1}. RT: room temperature

Alloy No.	4th cycle capacity @ 50 mA·g^{-1} (mAh·g^{-1})	4th cycle capacity @ 4 mA·g^{-1} (mAh·g^{-1})	HRD_{200}	Activation cycle to reach maximum capacity	Diffusion coefficient, D @RT (10^{-10} cm^2·s^{-1})	Exchange current I_o @RT (mA·g^{-1})	Ms (emu·g^{-1})	$H_{1/2}$ (kOe)	Open circuit voltage (V)
P17	382.7	400.7	0.815	6	1.69	36.5	0.39	0.11	1.302
P22	360.4	371.0	0.873	5	1.67	25.2	1.40	0.11	1.306
P23	342.6	348.8	0.920	5	2.95	25.5	0.62	0.11	1.326
P24	308.1	314.1	0.939	6	2.85	20.0	0.92	0.10	1.330
P25	253.9	259.2	0.965	5	1.58	18.9	1.03	0.10	1.312
P26	201.6	207.7	0.913	5	1.22	12.8	1.81	0.11	1.288
P27	100.1	105.6	-	6	1.39	12.3	1.42	0.14	1.264
P28	68.0	72.4	-	6	1.46	16.3	1.40	0.12	1.332

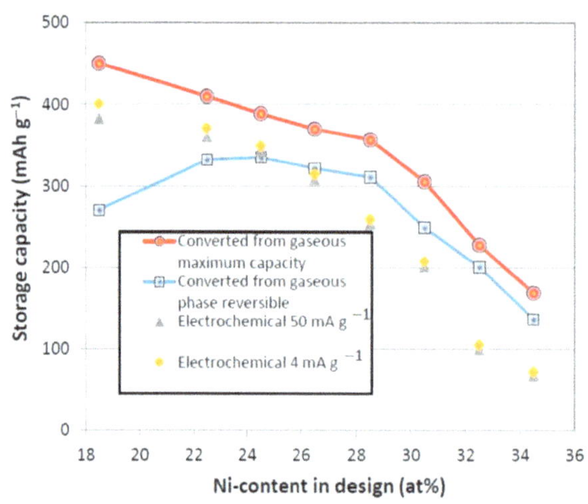

Figure 8. Plot of gaseous phase and electrochemical storage capacities as functions of Ni-content in design.

Half-cell HRD_{200} of each alloy, defined as the ratio of discharge capacities measured at 200 mA·g^{-1} and 4 mA·g^{-1}, is listed in Table 5. As the Ni-content in the alloy increases, HRD_{200} increases in the beginning, maximizes at alloy P25 and then decreases. When the discharge capacities at different discharge rates are compared against a commercially available standard AB_5 MH alloy with a HRD_{200} of 0.958 (Figure 9), only alloy P25 shows better rate-capability, but its capacities are too low. We find that although BCC-TiNi-Laves MH alloys have higher discharge capacities at a lower rate (<200 mA·g^{-1}), their high-rate capacities are not as good as in conventional AB_5 MH alloys.

In order to further study the inferior HRD capability in BCC-TiNi-Laves MH alloys compared to AB_5 MH alloys, both the bulk diffusion coefficient (D) and the surface exchange current (I_o) of the alloys under the current study were measured electrochemically. The details of both parameters' measurements were previously reported [51], and the values are listed in Table 5. As the Ni-content in design increases, the D value increases and then decreases, while I_o decreases.

The evolution of the HRD is more similar to the trend in D than in I_o, and therefore we conclude that the high-rate performance in this series of alloys is more related to the bulk diffusion than to the surface reaction. The increase in D initially is due to the shrinking unit cells for BCC, TiNi, and C14 phases (Table 2). In a comparison study, C15 phase shows faster hydrogen bulk diffusion than C14 phase [52]; therefore, the subsequent decrease in D with higher Ni-content is most likely related to the formation of σ-VNi phase. CALPHAD thermodynamic calculations show that hydrogen permeability is reduced with higher Ni-content in V-Ni alloys [53]. Compared to the D value from a typical AB_5 MH alloy (2.55×10^{-10} $cm^2 \cdot s^{-1}$ [54]), alloys P23 and P24 have faster bulk hydrogen diffusion. The reason the BCC-TiNi-Laves alloys have lower HRDs despite better bulk diffusion is because of their relatively low I_o values (Table 5) compared to AB_5 MH alloy (43.2 $mA \cdot g^{-1}$, [54]). When the abundances of the catalytic secondary phases (Laves and TiNi) are plotted against the I_o values as shown in Figure 10, it is clear that while Laves phases (C14 and C15) appear beneficial, TiNi phase appears detrimental to the surface electrochemical reaction. Similar findings regarding TiNi phase hindering the HRD performance of Laves phase based AB_2 MH alloys were reported before [55].

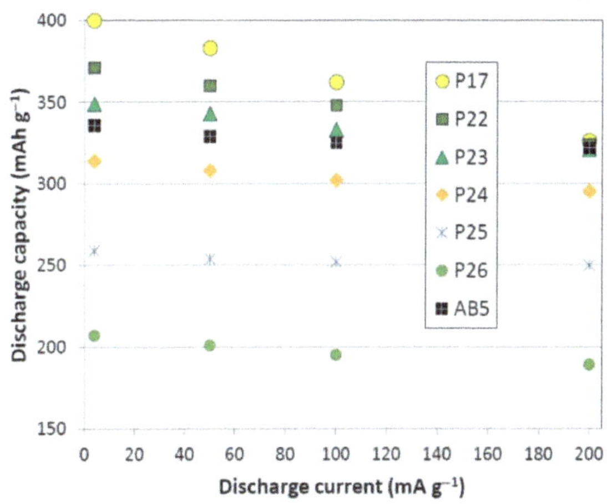

Figure 9. Rate-dependent discharge capacity for alloys P17, P22–P26 and a commercially available AB_5 MH alloy.

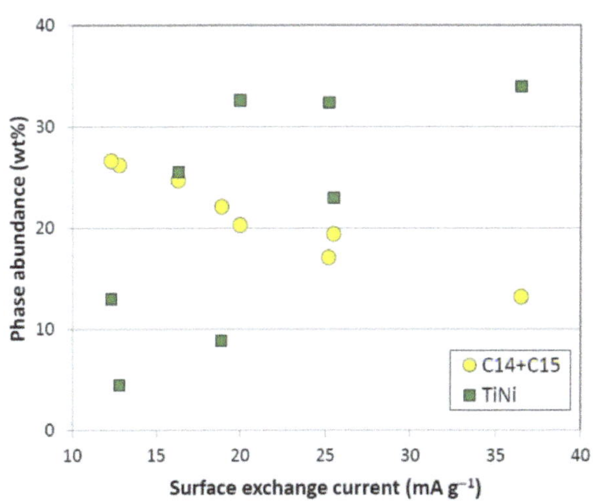

Figure 10. Plots of Laves (C14 + C15) and TiNi phase abundances *vs.* the surface exchange current.

The lower surface reaction currents measured in the current series of alloys were further investigated by magnetic susceptibility measurement. Details of the background and the experimental method are published in [56]. Metallic Ni is an active catalyst for the water splitting and recombination reactions that contribute to the I_o in this electrochemical system, and this technique allows us to obtain the saturated magnetic susceptibility (M_S), a quantification of the amount of surface metallic Ni (the product of preferential oxidation), and the magnetic field strength at one-half of the M_S value ($H_{1/2}$), a measure of the average reciprocal number of Ni atoms in a metallic cluster. M_S and $H_{1/2}$ values for the alloys in this study are listed in Table 5. While the $H_{1/2}$ values of these alloys are very similar, the trend of M_S values with the increase in Ni-content is not obvious. When these values are compared to values from an AB_5 MH alloy (M_S = 0.434 $emu \cdot g^{-1}$and $H_{1/2}$= 0.17 kOe [54]), we find that all alloys except P17 have higher M_S values. These BCC-TiNi-Laves alloys have higher weight percentage of metallic Ni in their activated surfaces and should have better HRD [57]. The reason for the lower I_o values in this current series of alloys may be due to their relatively larger Ni cluster sizes, resulting in less surface area, which is indicated by smaller $H_{1/2}$ values. Systematic transmission electron microscope work is necessary here to continue the investigation of the metallic inclusions, particularly their size and shape, embedded in the activated surface oxide layer of these alloys.

The open circuit voltage (OCV) of the electrochemical cells comprising a negative electrode from each alloy and a $Ni(OH)_2$/NiOOH positive electrode is listed in Table 5. OCV is important for the power density of the cell and is usually proportional to the logarithm of equilibrium pressure, typical of most MH alloys. In the previous study in a series of BCC-C14 alloys, the OCV is inversely proportional to the desorption equilibrium pressure [26]. In the current series of alloys, the plateau pressure increases with the increase in Ni-content, but the OCV first increases and then decreases except for the last alloy P28, which shows very small discharge capacity due to insufficient charging in the open-air configuration. The evolution of OCV in this study is similar to the trends in HRD and maximizes with alloy P24.

CONCLUSIONS

The Ni-content in a base alloy, P17, was increased with the intention to improve the electrochemical high-rate performance in the negative electrode active material for (Ni/MH) batteries. Although the HRD improved with the increase in Ni-content at the expense of V, the discharge capacity was deteriorated. An alloy, P23, with a composition of $Ti_{15.6}Zr_{2.1}V_{36}Cr_{11.2}Mn_{6.9}Co_{1.4}Ni_{26.5}Al_{0.3}$ shows a good balance between capacity and high-rate capability. The increase in Ni-content shifts the Laves phase from C14 into C15 because of the increase in the average electron density in the phase. The increase in Ni-content also promotes a new σ-VNi phase, which has a similar hydrogen-metal bond strength as the main BCC phase. The new phase is associated with reduced bulk diffusion of hydrogen. Further improvement in high-rate capability in this family of alloys should maintain the Ni-level at 22.5 at% to 26.5 at% and increase the content of other C14-promoters, such as Zr and Hf.

ACKNOWLEDGMENTS

This work is financially supported by Advanced Research Projects Agency-Energy (ARPA-E) under the robust affordable next generation EV-storage (RANGE) program (DE-AR0000386). The authors thank these colleagues for technical assistance: Simon Ng, Shuli Yan, Shiuan Chang (Wayne State University), Michael A. Fetcenko, Baoquan Huang, Tiejun Meng, David Pawlik, Allen Chan, Ryan J. Blankenship and Su Cronogue (BASF-Ovonic).

REFERENCES

1. Iba, H.; Akiba, E. Hydrogen absorption and microstructure in BCC alloys with C14-type Lave phase. *J. Jpn. Inst. Met. Mater.* **1994**, *58*, 1225–1232.
2. Akiba, E.; Iba, H. Hydrogen absorption by Laves phase related BCC solid solution. *Intermetallics* **1998**, *6*, 461–470.
3. Enomoto, M. The Cr-Ti-V system. *J. Phase Equilibria* **1992**, *13*, 195–200.
4. Inoue, H.; Arai, S.; Iwakura, C. Crystallographic and electrochemical characterization of $TiV_{4-x}Ni_x$ alloys for nickel-metal hydride batteries. *Electrochim. Acta* **1996**, *41*, 937–939.
5. Yu, X.; Wu, Z.; Xia, B.; Xu, N. A Ti-V-based bcc phase alloy for use as metal hydride electrode with high discharge capacity. *J. Chem. Phys.* **2004**, *121*, 987–990.
6. Young, K.-H.; Ouchi, T.; Huang, B.; Nei, J. Structure, hydrogen storage, and electrochemical properties of body-centered-cubic $Ti_{40}V_{30}Cr_{15}Mn_{13}X_2$ alloys (X = B, Si, Mn, Ni, Zr, Nb, Mo, and La). *Batteries* **2015**, *1*, 74–90.
7. Chen, N.; Li, R.; Zhu, Y.; Liu, Y.; Pan, H. Electrochemical hydrogenation and dehydrogenation mechanisms of the Ti-V base multiphase hydrogen storage electrode alloy. *Acta Metall. Sin.* **2004**, *40*, 1200–1204.
8. Iba, H.; Akiba, E. The relation between microstructure and hydrogen absorbing property in Laves phase-solid solution multiphase alloys. *J. Alloy. Compd.* **1995**, *231*, 508–512.
9. Rönnebro, E.; Noréus, D.; Sakai, T.; Tsukahara, M. Structural studies of a new Laves phase alloy $(Hf,Ti)(Ni,V)_2$ and its very stable hydride. *J. Alloy. Compd.* **1995**, *231*, 90–94.
10. Tsukahara, M.; Takahashi, K.; Mishima, T.; Isomura, A.; Sakai, T. V-based solid solution alloys with Laves phase network: Hydrogen absorption properties and microstructure. *J. Alloy. Compd.* **1996**, *236*, 151–155.
11. Qiu, S.; Chu, H.; Zhang, Y.; Sun, D.; Song, X.; Sun, L.; Fen, X. Electrochemical kinetics and its temperature dependence behaviors of $Ti_{0.17}Zr_{0.08}V_{0.35}Cr_{0.10}Ni_{0.30}$ alloy electrode. *J. Alloy. Compd.* **2009**, *471*, 453–456.
12. Iba, H.; Akiba, E. Hydrogen absorption and modulated structure in Ti–V–Mn alloys. *J. Alloy. Compd.* **1997**, *253–254*, 21–24.
13. Shashikala, K.; Banerjee, S.; Kumar, A.; Pai, M.R.; Pillai, C.G.S. Improvement of hydrogen storage properties of TiCrV alloy by Zr substitution for Ti. *Int. J. Hydrog. Energy* **2009**, *34*, 6684–6689.
14. Yang, X.; Li, J.; Zhang, T.; Hu, R.; Xue, X.; Fu, H. Role of defect structure on hydrogenation properties of $Zr_{0.9}Ti_{0.1}V_2$ alloy. *Int. J. Hydrog. Energy* **2011**, *36*, 9318–9323.
15. Huot, J.; Akiba, E.; Ogura, T.; Ishido, Y. Crystal structure, phase abundance and electrode performance of Laves phase compounds (Zr, A)$V_{0.5}Ni_{1.1}Mn_{0.2}Fe_{0.2}$ (A ≡ Ti, Nb or Hf). *J. Alloy. Compd.* **1995**, *218*, 101–109.
16. Qiu, S.-J.; Chu, H.-L.; Zhang, J.; Zhang, Y.; Sun, L.-X.; Xu, F.; Sun, D.-L.; Ouyang, L.-Z.; Zhu, M.; Grolier, J.-P.E.; *et al.* Effect of La partial substitution for Zr on the structural and electrochemical properties of $Ti_{0.17}Zr_{0.08-x}La_xV_{0.35}Cr_{0.1}Ni_{0.3}$ (x = 0–0.04) electrode alloys. *Int. J. Hydrog. Energy* **2009**, *34*, 7246–7252.
17. Huang, Z.; Cuevas, F.; Liu, X.; Jiang, L.; Wang, S.; Latroche, M.; Du, J. Effects of Si addition on the microstructure and the hydrogen storage properties of $Ti_{26.5}V_{45}Fe_{8.5}Cr_{20}Ce_{0.5}$ BCC solid solution alloys. *Int. J. Hydrog. Energy* **2009**, *34*, 9385–9392.
18. Kamegawa, A.; Shirasaki, K.; Tamura, T.; Kuriiwa, T.; Takamura, H.; Okada, M. Crystal structure and protium absorption properties of Ti–Cr–X alloys. *Mater. Trans.* **2002**, *43*, 470–473.
19. Liu, Y.; Zhang, S.; Li, R.; Gao, M.; Zhong, K.; Miao, H.; Pen, H. Electrochemical performances of the Pd-added Ti-V-based hydrogen storage alloys. *Int. J. Hydrog. Energy* **2008**, *33*, 728–734.
20. Kuriiwa, T.; Tamura, T.; Amemiya, T.; Fuda, T.; Kamegawa, A.; Takamura, H.; Okada, M. New V-based alloys with high protium absorption and desorption capacity. *J. Alloy. Compd.* **1999**, *293–295*, 433–436.
21. Yu, J.; Liu, B.; Cho, K.; Lee, J. The effects of partial substitution of Mn by Cr on the electrochemical cycle life of Ti-Zr-V-Mn-Ni alloy electrodes of a Ni/MH battery. *J. Alloy. Compd.* **1998**, *278*, 283–290.
22. Lee, H.-H.; Lee, K.-Y.; Lee, J.-Y. The Ti-based metal hydride electrode for Ni–MH rechargeable batteries. *J. Alloy. Compd.* **1996**, *239*, 63–70.
23. Kim, J.; Paik, C.; Cho, W.; Cho, B.; Yun, K.; Kim, S. Corrosion behaviour of $Zr_{1-x}Ti_xV_{0.6}Ni_{1.2}M_{0.2}$ (M = Ni, Cr, Mn) AB_2-type metal hydride alloys in alkaline solution. *J. Power Sources* **1998**, *75*, 1–8.
24. Young, K.; Wong, D.; Wang, L. Effect of Ti/Cr content on the microstructures and hydrogen storage properties of Laves phase-related body-centered-cubic solid solution alloys. *J. Alloy. Compd.* **2015**, *622*, 885–893.
25. Young, K.; Ouchi, T.; Nei, J.; Wang, L. Annealing effects on Laves phase-related body-centered-cubic solid solution metal hydride alloys. *J. Alloy. Compd.* **2016**, *654*, 216–225.
26. Young, K.; Ouchi, T.; Nei, J.; Meng, T. Effects of Cr, Zr, V, Mn, Fe, and Co to the hydride properties of Laves phase-related body-centered-cubic solid solution alloys. *J. Power Sources* **2015**, *281*, 164–172.
27. Young, K.; Ouchi, T.; Reichman, B.; Koch, J.; Fetcenko, M.A. Effects of Mo additive on the structure and electrochemical properties of low-temperature AB_5 metal hydride alloys. *J. Alloy. Compd.* **2011**, *509*, 3995–4001.

28. Chai, Y.; Zhao, M. Structure and electrochemical properties of $Ti_{0.25}V_{0.35}Cr_{0.40-x}Ni_x$ (x = 0.05–0.40) solid solution alloys. *Int. J. Hydrog. Energy* **2005**, *30*, 279–283.

29. Young, K.; Fetcenko, M.A.; Li, F.; Ouchi, T. Structural, thermodynamic, and electrochemical properties of $Ti_xZr_{1-x}(VNiCrMnCoAl)_2$ C14 Laves phase alloys. *J. Alloy. Compd.* **2008**, *464*, 238–247.

30. Young, K.; Ouchi, T.; Fetcenko, M.A. Roles of Ni, Cr, Mn, Sn, Co, and Al in C14 Laves phase alloys for NiMH battery application. *J. Alloy. Compd.* **2009**, *476*, 774–781.

31. Li, R.; Pan, H.; Gao, M.; Zhu, Y.; Liu, Y.; Jin, Q.; Lei, Y. Structural and electrochemical properties of hydrogen storage alloys $Ti_{0.8}Zr_{0.2}V_{2.7}Mn_{0.5}Cr_{0.8}Ni_x$ (x = 1.50–2.25). *J. Alloy. Compd.* **2004**, *373*, 223–230.

32. Pan, H.; Li, R.; Gao, M.; Liu, Y.; Lei, Y.; Wang, Q. Effects of Ni on the structural and electrochemical properties of Ti-V-based hydrogen storage alloys. *Int. J. Hydrog. Energy* **2006**, *31*, 1188–1195.

33. Pearson, W.; Christian, J. The structure of the σ phase in vanadium-nickel alloys. *Acta Crystallogr.* **1952**, *5*, 157–162.

34. Massalski, T.; Okamoto, H.; Subramanian, P.; Kacprzak, L. *Binary Alloy Phase Diagram*, 2nd ed.; ASM International: Materials Park: Ohio, USA, 1990; p. 2880.

35. Kodentzov, A.A.; Dunaev, S.F.; Slusarenko, E.M. Determination of the phase diagram of the V-Ni-Cr system using diffusion couples and equilibrated alloys. *J. Less Common Met.* **1987**, *135*, 15–24.

36. Yan, Y.; Chen, Y.; Liang, H.; Zhou, X.; Wu, C.; Tao, M.; Pang, L. Hydrogen storage properties of V–Ti–Cr–Fe alloys. *J. Alloy. Compd.* **2008**, *454*, 427–431.

37. Young, K.; Nei, J.; Wong, D.; Wang, L. Structural, hydrogen storage, and electrochemical properties of Laves-phase related body-centered-cubic solid solution metal hydride alloys. *Int. J. Hydrog. Energy* **2014**, *39*, 21489–21499.

38. Johnston, R.L.; Hoffmann, R. Structure-bonding relationships in the Laves phases. *Z. Anorg. Allg. Chem.* **1992**, *616*, 105–120.

39. Nei, J.; Young, K.; Salley, S.O.; Ng, K. Determination of C14/C15 phase abundance in Laves phase alloys. *Mater. Chem. Phys.* **2012**, *136*, 520–527.

40. Young, K.; Fetcenko, M.A.; Li, F.; Ouchi, T.; Koch, J. Effect of vanadium substitution in C14 Laves phase alloys for NiMH battery application. *J. Alloy. Compd.* **2009**, *468*, 482–492.

41. Young, K.; Ouchi, T.; Huang, B.; Reichman, B.; Fetcenko, M.A. Effect of molybdenum content on structure, gaseous storage, and electrochemical properties of C14-predominant AB_2 metal hydride alloys. *J. Power Sources* **2011**, *196*, 8815–8821.

42. Young, K. Metal Hydrides. In *Elsevier Reference Module in Chemistry, Molecular Sciences and Chemical Engineering*; Elsevier B.V.: Waltham, MA, USA, 2013.

43. Griessen, R.; Riesterer, T. Heat of Formation Models. In *Hydrogen in Intermetallic Compounds I*; Schlapbach, L., Ed.; Springer-Verlag: Berlin, Germany, 1988; p. 273.

44. Balasubramaniam, R. Hysteresis in metal-hydrogen systems. *J. Alloy. Compd.* **1997**, *253–254*, 203–206.

45. Wong, D.; Young, K.; Nei, J.; Wang, L.; Ng, K. Effects of Nd-addition on the structural, hydrogen storage, and electrochemical properties of C14 metal hydride alloys. *J. Alloy. Compd.* **2015**, *647*, 507–518.

46. Nakano, H.; Wakao, S. Substitution effect of elements in Zr-based alloys with Laves phase for nickel-hydride battery. *J. Alloy. Compd.* **1995**, *231*, 587–593.

47. Nei, J.; Young, K.; Sally, S.O.; Ng, K. Effects of annealing on $Zr_8Ni_{19}X_2$ (X = Ni, Mg, Al, Sc, V, Mn, Co, Sn, La, and Hf): Hydrogen storage and electrochemical properties. *Int. J. Hydrog. Energy* **2012**, *37*, 8418–8427.

48. Sun, D.; Jiang, J.; Lei, Y.; Liu, W.; Wu, J.; Wang, Q.; Yang, G. Effects of measurement factor on electrochemical capacity of some hydrogen storage alloys. *Mater. Sci. Eng. B* **1995**, *30*, 19–22.

49. Young, K.; Nei, J.; Huang, B.; Ouchi, T.; Fetcenko, M.A. Studies of $Ti_{1.5}Zr_{5.5}V_{0.5}(M_xNi_{1-x})_{9.5}$(M = Cr, Mn, Fe, Co, Cu, Al): Part 2. Hydrogen storage and electrochemical properties. *J. Alloy. Compd.* **2010**, *501*, 245–254.

50. Young, K.; Chao, B.; Huang, B.; Nei, J. Studies on the hydrogen storage characteristic of $La_{1-x}Ce_x(NiCoMnAlCuSiZr)_{5.7}$ with a B2 secondary phase. *J. Alloy. Compd.* **2014**, *585*, 760–770.

51. Li, F.; Young, K.; Ouchi, T.; Fetcenko, M.A. Annealing effects on structural and electrochemical properties of $(LaPrNdZr)_{0.83}Mg_{0.17}(NiCoAlMn)_{3.3}$ alloy. *J. Alloy. Compd.***2009**, *471*, 371–377.

52. Young, K.; Ouchi, T.; Huang, B.; Chao, B.; Fetcenko, M.A.; Bendersky, L.A.; Wang, K.; Chiu, C. The correlation of C14/C15 phase abundance and electrochemical properties in the AB_2alloys. *J. Alloy. Compd.* **2010**, *506*, 841–848.

53. Shim, J.H.; Ko, W.S.; Kim, K.H.; Lee, H.S.; Lee, Y.S.; Suh, J.Y.; Choa, Y.W.; Lee, B.-J. Prediction of hydrogen permeability in V-Al and V-Ni alloys. *J. Membr. Sci.* **2014**, *430*, 234–241.

54. Young, K.; Nei, J. The current status of hydrogen storage alloy development for electrochemical applications. *Materials* **2013**, *6*, 4574–4608.

55. Young, K.; Nei, J.; Ouchi, T.; Fetcenko, M.A. Phase abundances in AB_2 metal hydride alloys and their correlations to various properties. *J. Alloy. Compd.* **2011**, *509*, 2277–2284.

56. Young, K.; Wong, D.; Wang, L.; Nei, J.; Ouchi, T.; Yasuoka, S. Mn in misch-metal based superlattice metal hydride alloy—Part 1 structural, hydrogen storage and electrochemical properties. *J. Power Sources* **2015**, *277*, 426–432.

57. Young, K.; Huang, B.; Regmi, R.K.; Lawes, G.; Liu, Y. Comparisons of metallic clusters imbedded in the surface oxide of AB_2, AB_5, and A_2B_7 alloys. *J. Alloy. Compd.* **2010**, *506*, 831–840.

CHAPTER 11

EFFECT OF MAGNESIUM FLUORIDE ON HYDROGENATION PROPERTIES OF MAGNESIUM HYDRIDE

Pragya Jain, Viney Dixit, Ankur Jain, Onkar N. Srivastava and Jacques Huot

Keywords: Magnesium hydride, magnesium fluoride, hydrogen storage, metal hydrides, kinetics

INTRODUCTION

Magnesium hydride is a potential candidate for hydrogen storage because of its high gravimetric and volumetric capacities. Pure magnesium's low environmental impact and abundant availability makes it very attractive for hydrogen storage application. However, high working temperature and slow kinetics limit its potential as hydrogen storage material for practical applications. Therefore, research is required to circumvent these difficulties and make MgH_2 a viable hydrogen storage material. Nano-structuring of MgH_2 is one of the most adopted methods to improve the hydrogenation performance [1]. However, this method has a limitation to achieve the nanocrystalline size (<5 nm) required for destabilization of MgH_2 [2].

Further improvements in sorption behaviour have been achieved by adding a wide variety of pure transition metals [3,4], their oxides [5] and halides [6,7,8]. The remarkable catalytic effect of transition metal oxide, Nb_2O_5 has been well reported. However, during cycling at elevated temperatures, reduction of Nb_2O_5 occurs with augmentation of MgO content [5]. Later, it was found that some transition metal halides, such as FeF_3, $CrCl_3$, NiF_2, $NbCl_5$ and $TiCl_3$ possess better catalytic activity than pure metals or their oxides [6,7,8]. In the case of halides, Malka *et al.* [9] showed that fluorides are better catalysts than chlorides for MgH_2. Addition of transition metal fluorides during milling helps to lower the hydrogen release temperature and increase the rate of hydrogen uptake by MgH_2. It has been shown by different groups [7,8,9] that during milling of MgH_2 with transition metal fluorides, the formed MgF_2 phase replaces the original oxide layer and provides a reactive and protective fluorinated surface for hydrogen uptake. This compound possesses high affinity with hydrogen because of the F-anion, which weakens the Mg-H bonding and improves the sorption properties [9].

However, not much work has been done on direct use of MgF_2 as an additive for MgH_2. Ivanov *et al.* [10] reported that addition of 5 wt % MgF_2 to pure Mg during milling leads to 5 wt % hydrogen absorption in over 20 h but has an insignificant effect on dehydrogenation kinetics of MgH_2. Loss in absorption capacity from the second cycle onwards was also observed. Recently, Ma *et al.* [11] investigated the catalytic effects of MgF_2 and TiH_2 to understand the kinetic improvements obtained when MgH_2 was ball milled with 4 mol % TiF_3. They reported that sole addition of 6 mol % MgF_2 has negligible catalytic effect on MgH_2 at an operating temperature of 150 °C.

The limited and inconsistent results attained on catalytic effect of MgF_2 on MgH_2 shows that more work needs to be done to understand this system both from hydrogenation and material perspective. The present work is aimed to investigate the microstructural, morphological and hydrogenation behaviour of MgH_2 when MgF_2 is used as additive.

Citation: Pragya Jain, Viney Dixit, Ankur Jain, Onkar N. Srivastava and Jacques Huot, Effect of Magnesium Fluoride on Hydrogenation Properties of Magnesium Hydride, doi:10.3390/en81112330.

RESULTS

Comparison of Undoped and 5 mol % MgF_2 *Doped MgH_2 at 335 °C*

The X-ray diffraction (XRD) patterns of MgH_2 without and with 5 mol % MgF_2 prepared by 1 h ball milling are shown in Figure 1. For the undoped sample, the diffraction pattern peaks are associated with main phase of β-MgH_2 and some unreacted Mg. There is no evidence of the metastable γ-MgH_2 phase. This is due to the short milling time and low milling intensity. A broad peak centered at 43° is attributed to MgO. The crystallite size of β-MgH_2 is evaluated from Rietveld refinement to be 23.4 ± 0.3 nm. Milling with MgF_2 additive was even more effective for reduction of crystallite size of MgH_2 that was evaluated as 10.9 ± 0.3 nm. During milling there is physical interaction between the different species during the repeating collisions. Therefore, the MgF_2 has also some mechanical effect on MgH_2. Unfortunately, the alloying of brittle-brittle system is poorly understood [12]. However, from the present experiment it seems that addition of small amount of MgF_2 improves crystallite size reduction but the exact mechanism is still unclear.

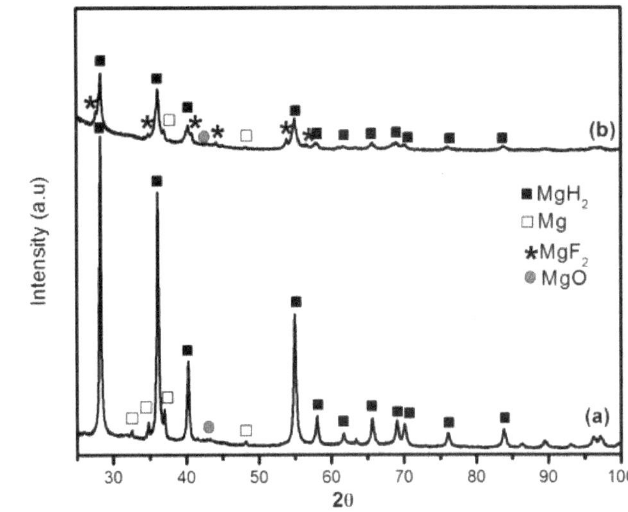

Figure 1. XRD patterns of ball milled samples milled 1 h: (**a**) pure MgH_2 and (**b**) MgH_2 + 5 mol % MgF_2.

In practical applications, desorption will be performed under a pressure of at least 100 kPa of hydrogen. However, in order to study the behaviour of MgH_2–MgF_2 system, we decided to fully dehydride the samples after ball milling. Therefore, after milling the samples were completely desorbed at 335 °C under dynamic vacuum before investigating their hydrogenation properties. Representative hydrogenation and dehydrogenation characteristics are shown in Figure 2.

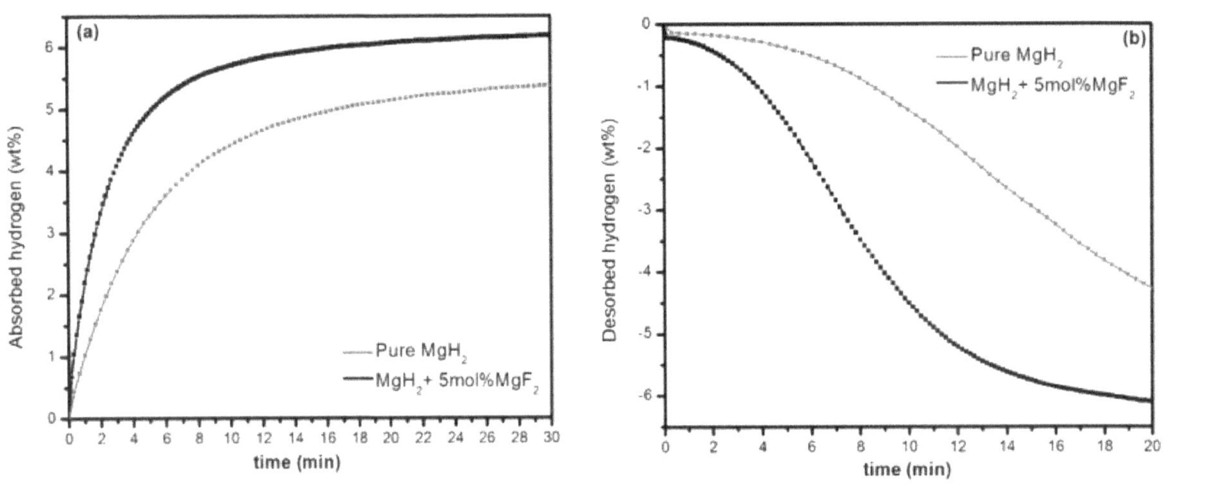

Figure 2. Hydrogen sorption kinetics at 335 °C of 1 h milled MgH_2 without and with 5 mol % MgF_2. (**a**) First absorption under 1000 kPa H_2; (**b**) desorption under 100 kPa H_2.

It is observed that at 335 °C under 1000 kPa H_2 pressure MgH_2 + 5 mol % MgF_2 system absorbs 6.2 wt % hydrogen in 30 min in comparison to only 5.3 wt % absorption by pure MgH_2. This shows a large improvement in absorption capacity is achieved, yielding 92% of the theoretical capacity in comparison to 70% for the pure MgH_2. In addition, significant improvement in desorption kinetics is achieved with complete desorption of the hydride phase in less than 20 min in presence of MgF_2. Thus, the beneficial effect of MgF_2 is clearly evident on the hydriding/dehydriding aspect of MgH_2.

Figure 3 shows the diffraction patterns of the doped sample in its desorbed and reabsorbed states. The desorbed pattern shows a small amount of un-desorbed MgH_2. The interesting fact is that MgF_2 is still present in the sample. This could be expected because it is known that for MgH_2-transition metal (TM) fluoride systems, milling or dehydrogenation induces the formation of $MgF_2$and TM hydride [4,11]. Thus, MgF_2 is a stable compound and does not react to form MgH_2.

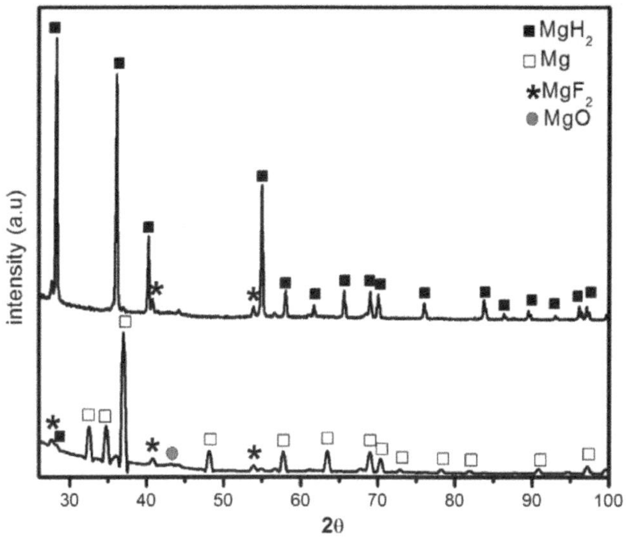

Figure 3. XRD patterns of MgH_2 + 5 mol % MgF_2 (**a**) after desorption at 335 °C under 100 kPa H_2 and (**b**) after re-hydrogenation at 335 °C under 1000 kPa H_2.

This is confirmed by the diffraction pattern of fully hydrided sample. The phases present are MgH_2 and MgF_2 along with small amount of unreacted Mg. Compared to the patterns of Figure 1we see that the peaks of patterns of Figure 3 are not as broad, implying that the crystallite size increased. From Rietveld analysis we found that the crystallite size of Mg in the dehydrided pattern is 49.1 ± 0.8 nm while the crystallite size of MgH_2 in the reabsorbed pattern is 64 ± 2 nm. This shows that there is grain growth compared to the as-milled sample. This may be due to the high temperature of hydrogenation and also because of desorption/absorption itself.

Figure 4 shows the SEM images of MgH_2 + 5 mol % MgF_2 composite in (Figure 4a) desorbed state and (Figure 4b) after re-hydrogenation at 335 °C in comparison with pure MgH_2 (Figure 4c). The images show that ball milling with additive leads to effective decrease in particle size. In addition, energy dispersive X-ray (EDX) mapping done at higher magnification (Figure 5) shows that agglomerates consist of smaller MgH_2 particles and additive.

Elemental mapping made on MgH_2 + 5 mol % MgF_2 in both the desorbed state (Figure 5a) and re-hydrogenated state (Figure 5b) confirms homogenous distribution of MgF_2.

High energy milling leads to uniform dispersion of MgF_2 phase in MgH_2 matrix which may act as a catalytic layer and contributes in improving sorption properties. Chemical analysis performed by EDX spectroscopy during transmission electron microscopy (TEM) investigation of the desorbed sample gave the average atomic composition of different elements as 9.8% O, 14.7% F and 75.7% Mg which is very close to the nominal composition (86% Mg and 14% F). The presence of oxygen in EDX pattern in comparison to its small trace in XRD pattern could be due to small crystallite size of MgO making it peak difficult to distinguish from the background. A similar EDX investigation was performed on a sample that has been submitted to five dehydrogenation/hydrogenation cycles. Because abundances vary from point to point, we average over four different localisations. We found that, after cycling, the atomic composition of different elements was 15% ± 4% O, 11% ± 4% F and 74% ± 6% Mg. Within experimental error, these values are similar to the ones before cycling. However, this may be an indication that cycling induces a loss of MgF_2 and increase of MgO. Typical TEM micrographs presented in Figure 6 shows the morphology of the desorbed sample.

The image shows presence of large number of particles agglomerated together with no visibility clear particle boundaries. These observations are quite similar to those reported recently by Grzech *et al.* [13]. High resolution pictures taken over region-1 in Figure 6 and its corresponding selected area electron diffraction (SAED) patterns shows reflections at *d*-spacing 2.45, 1.90, 1.60, 1.36 and 1.22 Å which are characteristic of Mg (101), (102), (110), (112) and (202) planes respectively along with reflections at d values 2.10 and 1.48 Å, corresponding to MgO (200) and (220) planes. Thus, the surface consists of small crystallites of MgO (forming well defined ring and represented by red rings) surrounding the large crystallites of Mg (seen as discontinuous spots and represented by white rings). While the multiple SAED patterns acquired from the region-2 were well indexed as a mixture of large crystallite Mg and small crystallite of MgF_2 (seen as well-defined rings and colored yellow). The absence of oxide in region-2 is evidence that presence of fluoride limits MgO only to the surface. Both structural and morphological studies support the presence of MgF_2 phase even after complete hydrogen absorption/desorption cycle at 335 °C.

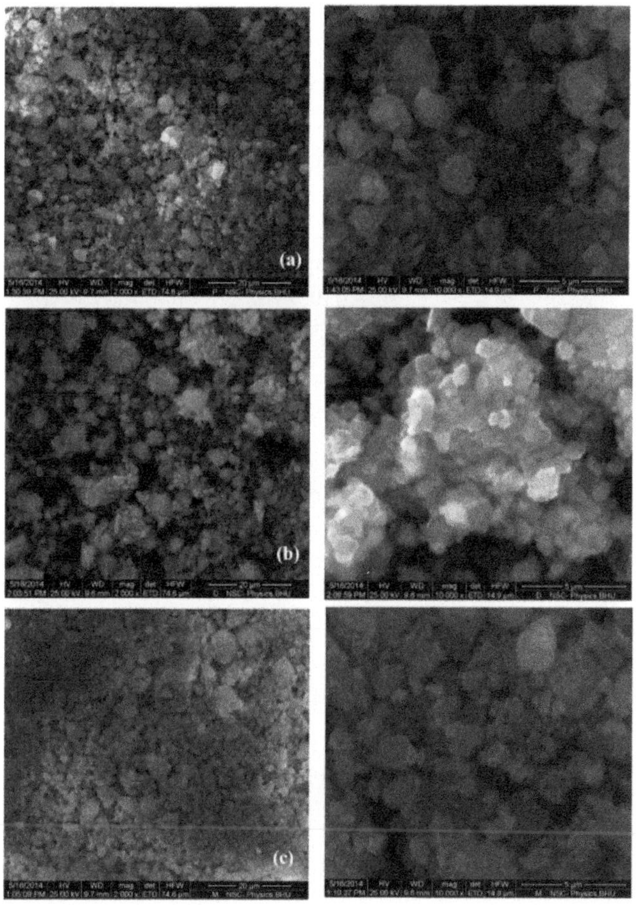

Figure 4. SEM images for MgH_2 + 5 mol % MgF_2 in (**a**) desorbed state and (**b**) re-hydrogenated state in comparison to (**c**) pure MgH_2.

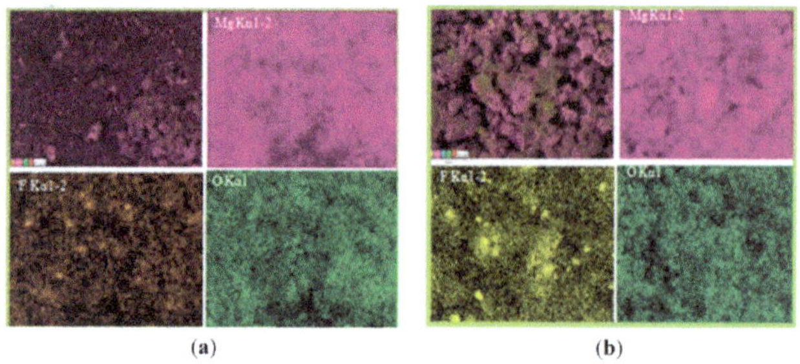

Figure 5. Elemental mapping showing particle morphology and distribution of 1 h milled MgH_2 + 5 mol % MgF_2: (**a**) after desorption and (**b**) after re-hydrogenation at 335 °C.

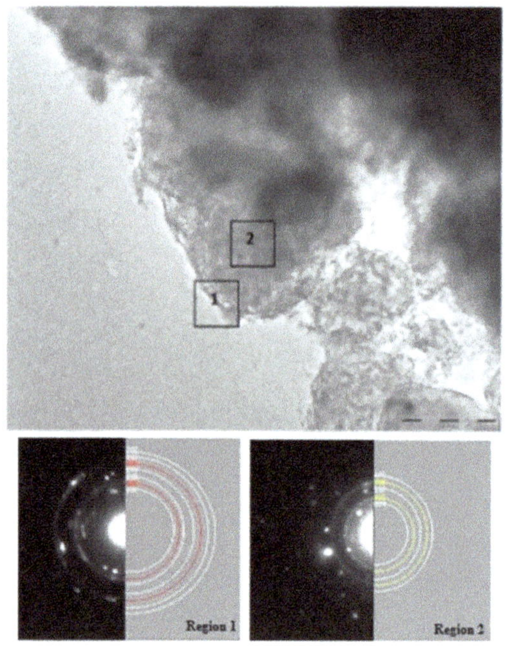

Figure 6. Transmission electron microscopy (TEM) micrograph of MgH_2 + 5 mol % MgF_2 sample after desorption at 335 °C with selected area electron diffraction (SAED) patterns and simulations. Region 1 is composed of Mg (**white rings**) covered with MgO layer (**red rings**) in simulated data while Region 2 shows diffraction rings corresponding to Mg (**white rings**) and MgF_2 (**yellow rings**).

Hydrogenation Characteristics of MgH_2 + *5 mol % MgF_2 at Lower Temperatures*

The catalytic effect of 5 mol % MgF_2 on hydrogen sorption properties of MgH_2 was further investigated at lower temperatures. Figure 7 shows the absorption kinetics at 335, 310, 285 and 145 °C under 1000 kPa of hydrogen.

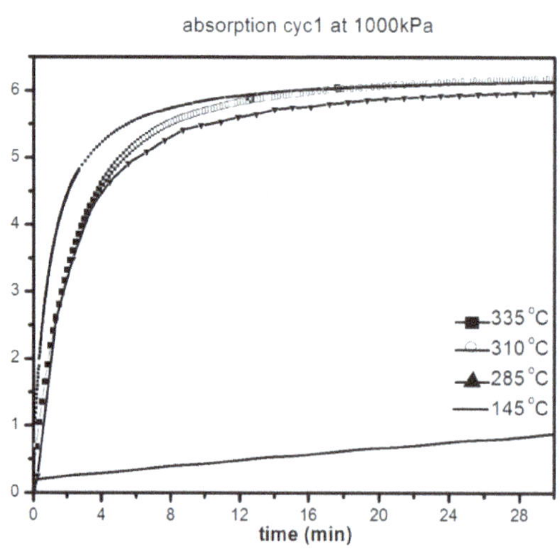

Figure 7. First absorption under 1000 kPa H_2 at different temperatures of 1 h milled MgH_2 + 5 mol % MgF_2. The insert is a compete absorption curve at 145 °C.

It should be pointed out that the samples were initially desorbed at 335 °C in order to ensure that full desorption was achieved before all absorption measurements. We notice only a slight loss in absorption capacity with reduction of temperature from 335 °C (6.2 wt % H_2) to 285 °C (5.8 wt % H_2). As seen in Figure 7, there was slight loss in kinetics and capacity in the temperature range 335–285 °C with the material reaching its complete capacity in less than 30 min. However, at 145 °C, the kinetics are much slower, but after 20 h, a capacity of 5.5 wt % is reached, as shown in Figure 7b. Desorption kinetic under 100 kPa H_2 at 285, 310 and 335 °C are shown in Figure 8.

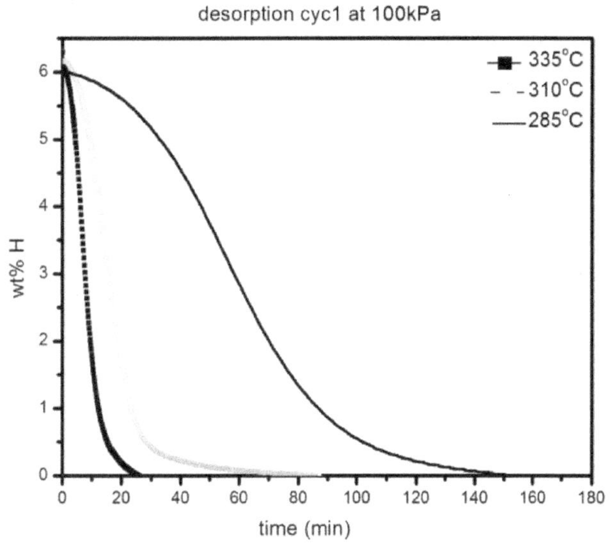

Figure 8. Desorption under 100 kPa H_2 at different temperatures of 1 h milled MgH_2 + 5 mol % MgF_2.

As expected, the kinetics are getting slower as temperature decreases but is still relatively fast even at 285 °C were complete desorption takes place in less than 3 h. These results reveal that even by sole addition of alkaline metal fluorides, improvements in hydrogenation characteristics of magnesium hydride can be achieved.

Cyclic Stability of MgH_2 + *5 mol % MgF_2 at T = 310 °C*

Micro structural results have confirmed that MgF_2 phase does not decompose and no new phase formation occurs during hydrogen absorption/desorption measurements for MgH_2 + 5 mol % MgF_2system. Therefore, cyclic performance of catalyzed magnesium hydride was examined at moderate operating temperature of 310 °C at pressure of 1000 kPa (for absorption) and 10 kPa (for desorption) to evaluate performance stability. Prior to the measurements the sample was completely desorbed at 335 °C. Figure 9 shows that the absorption capacity goes down from 5.9 wt % in first cycle to 5.6 wt % in the 10th cycle.

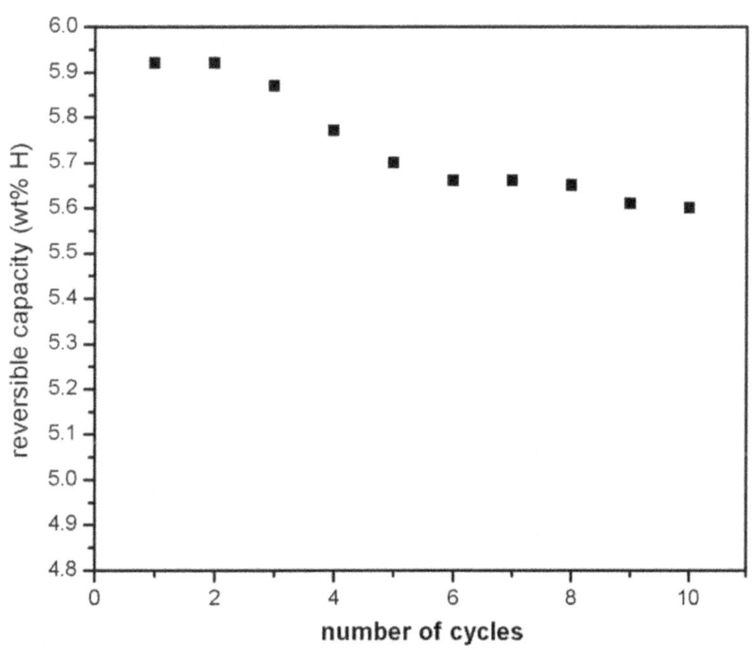

Figure 9. Hydrogen absorption kinetics at 310 °C under 1000 kPa of hydrogen of ball milled MgH_2 + 5 mol % MgF_2 for 10 cycles.

The observed loss of 0.3 wt % in capacity was reached within the first five cycles and thereafter the maximum capacity achieved by the system is more or less stabilized. These results show that magnesium hydride exhibits good hydrogen storage capacity and cyclic stability when magnesium fluoride is used as catalyst in comparison to the use of transition metal fluoride like NbF_5 or $ZrF_4$where sharp decline in storage capacity was observed by Malka *et al.* [14] in the first 10 hydrogenation cycles recorded at 325 °C. X-ray diffraction patterns of the sample taken after 1st and 10th desorption cycle are presented in Figure 10. It shows that the β-MgH_2 phase and the catalytic material remain intact while small increase in content of MgO occurs. Thus, the growth of the MgO layer is mostly responsible for an observed loss in capacity.

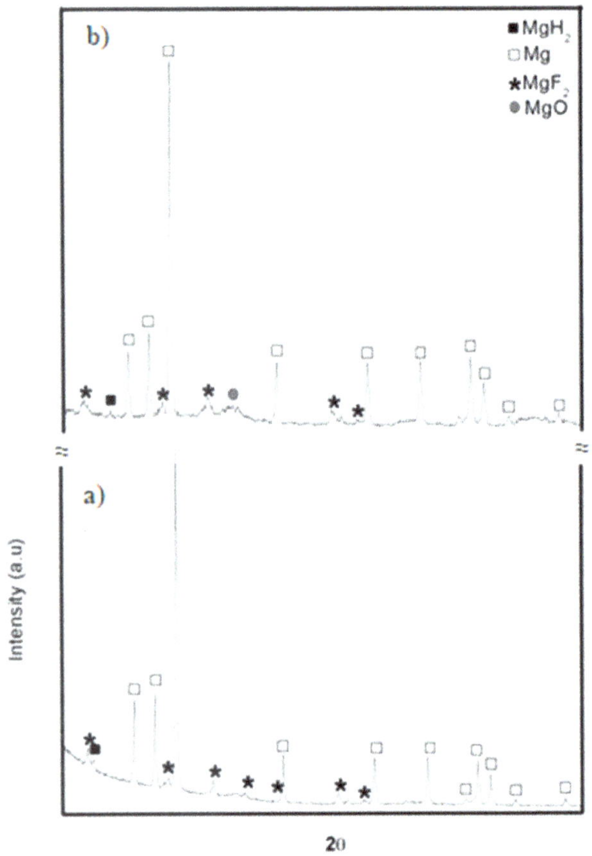

Figure 10. X-ray diffraction patterns of ball milled MgH_2 + 5 mol % MgF_2 after (**a**) one desorption and (**b**) 10 desorption cycles at 310 °C.

DISCUSSION

The structural and hydrogenation results suggest that hydrogen absorption/desorption kinetics of MgF_2 doped MgH_2 is relatively slower than that attained with transition metal fluorides (TmF = TiF_3, ZrF_4, NbF_5, TaF_5). This could be explained by the presence of only one catalytically active phase in the present case (MgF_2) while two active phases are present when transition metal fluorides are used as additives. More explicitly, upon dehydrogenation the following reaction takes place in the present case.

$$MgH_2 + MgF_2 \rightarrow Mg + MgF_2 + H_2 \quad (1)$$

While, as reported by Ma *et al.* [11], when transition metal fluoride is added the reaction taking place is:

$$3MgH_2 + 2TiF_3 \rightarrow 3\,MgF_2 + 2TiH_2 + H_2 \quad (2)$$

As TiH_2 possess more negative enthalpy formation (−136 kJ/mol) than MgH_2 (−75 kJ/mol) it will remain as a stable phase during desorption of MgH_2 in later cycles [13]. Furthermore, presence and concentration of TiH_2 phase would increase on multiple absorption/desorption cycling, which results in reduction of overall storage capacity. In addition, the transition metal fluoride is very sensitive to atmospheric conditions. Ball milled MgH_2 + TM-fluoride samples require oxygen and moisture level to be less than 0.1 ppm for obtaining good hydrogenation results [7,9,11,15]. Additionally, even in the desorbed state the material is pyrophoric, which makes it difficult to handle. Moreover, transition metals are much heavier than magnesium thereby increasing the mass of entire system. It thus seems more practical to use MgF_2 as a doping agent to increase the hydrogenation/dehydrogenation kinetics than transition metal fluorides.

EXPERIMENTAL SECTION

The starting materials MgH_2 (99.8% purity) and MgF_2 (99.9% purity) purchased from Alfa Aesar (Ward Hill, MA, USA) were vacuum annealed for few hours at 80 °C before using them for experiments. Afterwards, MgH_2 powder with 2, 5, and 10 mol % MgF_2, was milled under Ar atmosphere using Fritsch P4 planetary mill (Idar-Oberstein, Germany) with ball to powder ratio of 50:1 at a crucible rotation speed of 220 rpm. Milling was done for 60 min with 15 min rest after every 15 min of milling. The final milled products were handled in a glove box with oxygen and moisture level below 0.1 ppm. Initial hydrogen desorption curves taken at 335 °C under 100 kPa H_2pressure showed that with 2 mol % catalytic material the kinetics was too slow which could be improved by increasing the additive content to 5 mol %. Further increase in concentration of catalytic material to 10 mol % didn't cause any significant change in kinetics. Therefore, MgF_2concentration was restricted to 5 mol % for further investigation.

The hydrogenation characteristics were measured on homemade Sievert-type apparatus and the cyclic studies were made on an automated-four channel apparatus called Multi Channel Hydride Evaluation System from Advanced Materials Corporation, Petersburg, VA, USA. Approximately 400 mg of powder was placed in a sample cell and completely desorbed under dynamic vacuum at 335 °C prior to any measurement. Thereafter all measurements were made under 1000 kPa H_2pressure for absorption and 100 kPa H_2 pressure for desorption at temperatures ranging from 335 °C to 145 °C. X-ray diffraction was performed using Bruker D8 Focus X-Ray apparatus (Bruker, Madison, WI, USA) with CuKα radiation. Phase abundances were evaluated from Rietveld method using Topas software [16]. Small quantity of milled MgH_2 + 5 mol % MgF_2 (a) after desorption at 325 °C and (b) after rehydrogenation under 1000 kPa H_2 was characterized for morphological studies with chemical analysis using JEOL JSM-5500 scanning electron microscope (JEOL, Tokyo, Japan). The sample was filled in air tight bottles and taken to SEM-EDX lab were they were slightly exposed to air for loading in SEM chamber. TEM analysis was performed on FEI: Technai 20G^{2}electron microscope (FEI, Hillsboro, OR, USA), operating at 200 kV accelerating voltage. TEM samples were prepared by dry dispersion of the powder onto a carbon substrate supported by copper TEM grid. This was done in an argon glove box before the TEM session, and the prepared sample was sealed by covering with parafilm tape to be carried to TEM lab. The sample was exposed to air for short duration during loading onto the TEM holder. Thus, partially transformed samples were characterised using scanning and transmission electron microscopy.

CONCLUSIONS

This investigation showed that magnesium fluoride could significantly influence the hydrogen sorption properties of magnesium hydride. It has been shown that MgF_2 additive acts as a catalyst for MgH_2, thereby improving its hydrogenation/dehydrogenation kinetics. These kinetic improvements are due to the presence of chemically stable MgF_2 powder well mixed in MgH_2matrix and MgO layer being limited only to the surface. Cyclic stability reveals that 5 mol % MgF_2helps to accelerate the reversible kinetics of MgH_2 with higher capacity in comparison to other transition metal fluoride catalysts. This is probably due to the persistence of MgF_2 phase during hydrogen cycling. These results suggest that owing to its fast sorption properties, low sensitivity to atmospheric conditions and easy handling ability, this material can be used in applications where operation at relatively high temperature is not considered a significant issue.

ACKNOWLEDGMENTS

Pragya Jain would like to thank University Grant Commission (UGC) for the Dr. D.S. Kothari post-doctoral fellowship to carry out this work.

REFERENCES

1. Jain, I.P.; Lal, C.; Jain, A. Hydrogen storage in Mg: A most promising material. *Int. J. Hydrog. Energy* **2010**, *35*, 5133–5144.
2. Wu, Z.; Allendorf, M.D.; Grossman, J.C. Quantum Monte Carlo Simulation of Nanoscale MgH_2 Cluster Thermodynamics. *J. Am. Chem. Soc.* **2009**, *131*, 13918–13919.
3. Dornheim, M.; Eigen, N.; Barkhordarian, G.; Klassen, T.; Bormann, R. Tailoring Hydrogen Storage Materials Towards Application. *Adv. Eng. Mater.* **2006**, *8*, 377–385.
4. Hanada, N.; Ichikawa, T.; Fujii, H. Catalytic effect of nanoparticle 3d-transition metals on hydrogen storage properties in magnesium hydride MgH_2 prepared by mechanical milling. *J. Phys. Chem. B* **2005**, *109*, 7188–7194.
5. Aguey-Zinsou, K.F.; Ares Fernandez, J.R.; Klassen, T.; Bormann, R. Effect of Nb_2O_5 on MgH_2properties during mechanical milling. *Int. J. Hydrog. Energy* **2007**, *32*, 2400–2407.
6. Ma, L.P.; Wang, P.; Cheng, H.M. Improving hydrogen sorption kinetics of MgH_2 by mechanical milling with TiF_3. *J. Alloys Compd.* **2007**, *432*, L1–L4.
7. Yavari, A.R.; LeMoulec, A.; de Castro, F.R.; Deledda, S.; Friedrichs, O.; Botta, W.J.; Vaughan, G.; Klassen, T.; Fernandez, A.; Kvick, Å. Improvement in H-sorption kinetics of MgH_2powders by using Fe nanoparticles generated by reactive FeF_3 addition. *Scr. Mater.* **2005**, *52*, 719–724.
8. Deledda, S.; Borissova, A.; Poinsignon, C.; Botta, W.J.; Dornheim, M.; Klassen, T. H-sorption in MgH_2 nanocomposites containing Fe or Ni with fluorine. *J. Alloys Compd.* **2005**.
9. Malka, I.E.; Pisarek, M.; Czujko, T.; Bystrzycki, J. A study of the ZrF_4, NbF_5, TaF_5, and $TiCl_3$influences on the MgH_2 sorption properties. *Int. J. Hydrog. Energy* **2011**, *36*, 12909–12917.
10. Ivanov, E.; Konstanchuk, I.; Bokhonov, B.; Boldyrev, V. Hydrogen interaction with mechanically alloyed magnesium-salt composite materials. *J. Alloys Compd.* **2003**, *359*, 320–325.
11. Ma, L.-P.; Wang, P.; Cheng, H.-M. Hydrogen sorption kinetics of MgH_2 catalyzed with titanium compounds. *Int. J. Hydrog. Energy* **2010**, *35*, 3046–3050.
12. Soni, P.R. *Mechanical Alloying: Fundamentals and Applications*; Cambridge International Science Publishing: Cambridge, UK, 2000; p. 160.
13. Grzech, A.; Lafont, U.; Magusin, P.; Mulder, F.M. Microscopic Study of TiF_3 as Hydrogen Storage Catalyst for MgH_2. *J. Phys. Chem. C* **2012**, *116*, 26027–26035.
14. Malka, I.E.; Bystrzycki, J.; Płociński, T.; Czujko, T. Microstructure and hydrogen storage capacity of magnesium hydride with zirconium and niobium fluoride additives after cyclic loading. *J. Alloys Compd.* **2011**, *509*, S616–S620.
15. Jin, S.-A.; Shim, J.-H.; Cho, Y.W.; Yi, K.-W. Dehydrogenation and hydrogenation characteristics of MgH_2 with transition metal fluorides. *J. Power Sources* **2007**, *172*, 859–862.
16. *Topas V4: General Profile and Structure Analysis Software for Powder Diffraction Data*; Bruker_AXS: Karlsruhe, Germany, 2008.

CHAPTER 12

EFFECT OF COLD ROLLING ON THE HYDROGEN DESORPTION BEHAVIOR OF BINARY METAL HYDRIDE POWDERS UNDER MICROWAVE IRRADIATION

Ivaldete da Silva Dupim, Sydney Ferreira Santos and Jacques Huot

Keywords: Metal hydrides, microwave heating, hydrogen desorption, cold rolling, magnesium hydride, titanium hydride, zirconium hydride

INTRODUCTION

The use of microwave energy in material synthesis and processing has been widely investigated over the years. This approach has been successfully used in the synthesis of micrometric and nanometric particles, microwave sintering, surface modification, *etc.* [1,2,3,4,5,6]. The most investigated use of microwaves in materials processing and synthesis is the microwave sintering of oxides, carbides, nitrides, silicides, *etc.* [7].

The microwave heating of metals was considered a minor subject for a long time since the penetration depth of electromagnetic fields into metals is quite small (up to few micrometers), limiting the heating reached in bulk metals. For detailed information on the electromagnetic field interaction with metals during microwave irradiation, the interested reader may refer to the review article by Yoshikawa [8]. However, this scenario changed markedly in 1999 when Roy *et al.* [9] reported for the first time the microwave heating and sintering of a porous metal compact, extending the applicability of microwave irradiation in materials processing. Subsequently, the interest in the microwave assisted processing of metals, alloys and metal matrix composites became reality [10,11,12,13].

Nowadays, the search for clean energy sources is a major target worldwide and hydrogen appears as a promising energy vector candidate [14,15,16]. Conversely, a number of important technological problems must to be solved in order to allow the large scale utilization of hydrogen in practical energy systems. One of the most challenging issues to be solved is finding safe and reliable technologies for hydrogen storage. Metal hydrides are interesting materials for solid state hydrogen storage, and several hydride systems and different materials processing routes have been extensively investigated [14,15,16,17,18,19 ,20,21,22,23,24].

Despite of the abovementioned growing interest in the microwave heating of metallic materials, until now only few papers addressing the microwave irradiation in metal hydrides have been published. Luo *et al.* [25] reported that titanium sintered parts obtained by the microwave irradiation of TiH_2 showed higher densities, finer residual pores and better mechanical properties than those obtained from hydride-dehydrided titanium powder. Li *et al.* [26] employed microwave sintering to process Mg-La-Ni hydrogen storage alloys from elemental powders and successfully obtained multi-phase alloys. Microwave assisted synthesis of hydrides was also performed by Tapia-Ruiz *et al.* [3] to produce the Li_4NH hydride by solid state reaction between Li_3N and LiH compounds. This reaction is complex, presenting possible alternative pathways and side reactions with atmosphere or container material contaminations. These authors reported success in producing single phase Li_4NH by microwave synthesis over a much shorter period of time than necessary for conventional heating methods.

Citation: Ivaldete da Silva Dupim, Sydney Ferreira Santos and Jacques Huot, Effect of Cold Rolling on the Hydrogen Desorption Behavior of Binary Metal Hydride Powders under Microwave Irradiation, doi:10.3390/met5042021.

Hydrogen desorption promoted by microwave heating was also reported in a few papers. Nakamori *et al.* [27] investigated the effect of microwave irradiation on some metal hydrides and complex hydrides. These authors reported that some of the investigated transition metal hydrides presented small hydrogen desorption (TiH_2 and $VH_{0.81}$), as indicated by shifts in diffraction peaks positions of the hydride phases, whereas others did not indicate any hydrogen release (ZrH_2 and $LaH_{2.48}$). Similar behavior was observed for complex hydrides, where $LiBH_4$ showed partial desorption whereas $NaBH_4$ and KBH_4 do not show any evidence of desorption by microwave irradiation. The authors concluded that microwave penetration depth and conduction losses are controlling factors on microwave heating and hydrogen desorption for the investigated hydrides. Krishnan [28] also studied the effects of microwave irradiation in $NaAlH_4$ and $TiCl_2$ doped $NaAlH_4$. The authors concluded that, when the metallic phase Al is present in the starting material, microwave exposure leads to the production of reversible Na_3AlH_6 and Al phases. However, when the sample does not contain an Al phase then microwave irradiation produces an amorphization of $NaAlH_4$. Zhang *et al.* [29] proposed the use of a Ni-coated honeycomb ceramic monolith (HCM) as susceptor to promote fast heating and hydrogen desorption in MgH_2 and other hydrides. Awad *et al.* [30] investigated the effect of different types of carbon (carbon fibers, graphite and diamond) on the hydrogen desorption behavior of Mg + C nanocomposites obtained by ball-milling and reported that more effective desorption was obtained by increasing the milling time, the carbon type content and microwave power. Among the investigated composites, the MgH_2 + 10% CF showed the larger hydrogen release, converting about 90% of the initial MgH_2 in Mg.

A possible application of the microwave desorption is for Zircaloy recycling as mentioned by Dupim *et al.* [31]. As for absorption, to our knowledge absorption under microwave radiation has not been reported yet. The reason may be the necessity of having a sample holder that has to support hydrogen pressure while being permeable to microwaves, which increases the safety issue. In the present investigation only desorption was investigated.

In this work, we investigate the effect of cold rolling on the hydrogen desorption behavior of binary hydride powders during microwave irradiation. This investigation was motivated by the reports concerning the positive effects of cold rolling on the hydrogen desorption kinetics in several hydride systems under conventional heating [31,32,33,34,35,36] and also the lack of investigations focusing the effect of cold rolling on the hydrogen desorption of hydrides during microwave heating.

EXPERIMENTAL SECTION

MgH_2, TiH_2, and ZrH_2 powders having respectively 98%, 99%, and 98.5% purity (metal basis), were used in microwave desorption experiments. Cold rolling was performed by placing the hydride powder between two aluminum plates and rolling it 5 times using a Durston 130 laboratory rolling mill (Durston, High Wycombe, Buckinghamshire, UK). The cold rolling was performed in air.

For microwave irradiation, a domestic microwave oven (Danby model DMW 749SS, Danby, Guelph, ON, Canada) specially adapted for these experiments was used. This microwave oven has a power of 700 W and frequency of 2.5 GHz. A quartz tube with one side dome closed and the other open (200 mm in length and 10 mm diameter) was used as sample holder and a stainless steel cylinder (diameter 0.8 mm) was used as susceptor. The open extremity of this tube was linked to a vacuum pump for dynamic evacuation during the experiments. The level of vacuum was about 10^{-2} torr. The mass of the hydride powder was about 200 mg. A schematic representation of the sample holder used for these experiments is shown in Figure 1. The investigated samples (as-received and cold-rolled) were placed into the quartz tube inside an argon filled glove-box. Irradiation experiments were performed with a fixed duration of 30 min.

The morphology of the metal hydrides was observed by scanning electron microscopy (SEM) using a Philips XL30 microscope (FEI, Hillsboro, Oregon, USA). Working distance and accelerating voltage were varied according to the operating mode of the microscope (backscattered or secondary electron). The actual values are indicated on the respective micrographs. The investigated samples were characterized by X-ray diffraction using a Bruker D8 Focus diffractometer (Cu-Kα radiation), Bruker, Madison, WI, USA. Rietveld refinement was performed using the Topas software (Bruker, Madison, WI, USA) [37].

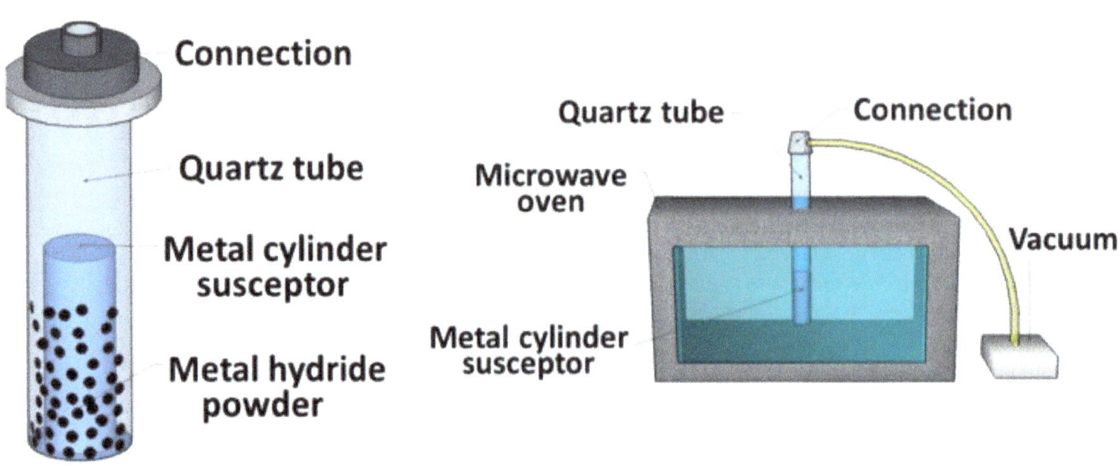

Figure 1. Schematic representation of the sample holder used for microwave irradiation experiments.

RESULTS AND DISCUSSION

Electron Microscopy

Figure 2 shows the morphologies of hydride powders investigated by scanning electron microscopy images obtained with a secondary electrons detector (SE-SEM). Figure 2a–c correspond to MgH_2, TiH_2, and ZrH_2 as-received powders, respectively. It is clear that all the samples show irregular particle morphology, with relatively widespread particle sizes. Moreover, it is noticeable that ZrH_2 powder has the larger mean particle size, followed by MgH_2 and TiH_2.

The effect of cold rolling on the morphology of the hydride powders can be observed in Figure 2d–f for MgH_2, TiH_2, and ZrH_2 respectively. All the rolled samples presented a morphology of large plate-like agglomerates composed of very fine particles, with particles sizes in the range of sub-microns to few microns. The strong particle size refinement promoted by cold rolling on the hydride powders can be clearly seen. The particle size refinement is a positive feature for microwave heating since the microwave penetration depth in metals is small. Moreover, the decrease in particle size allows the metal powder to reach higher temperatures as demonstrated by Mondal *et al.* [38] in the case of microwave irradiation of Cu powder.

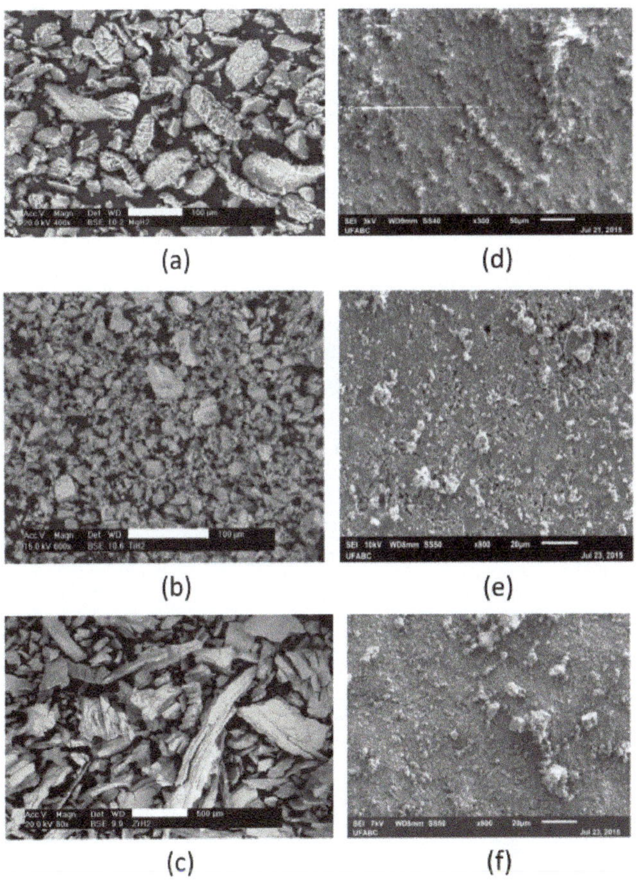

Figure 2. Morphology of metal hydride powders observed by scanning electron microscopy (SEM). Images **(a)**, **(b)**, and **(c)** correspond to as-received MgH_2, TiH_2, and ZrH_2, respectively, whereas **(d)**, **(e)**, and **(f)** correspond to coldrolled MgH_2, TiH_2, and ZrH_2, respectively.

X-ray Diffraction: Cold Rolling Effects

The X-ray diffraction (XRD) patterns of as-received and cold rolled samples of MgH_2, TiH_2, and ZrH_2 are shown in Figure 3, Figure 4 and Figure 5, respectively.

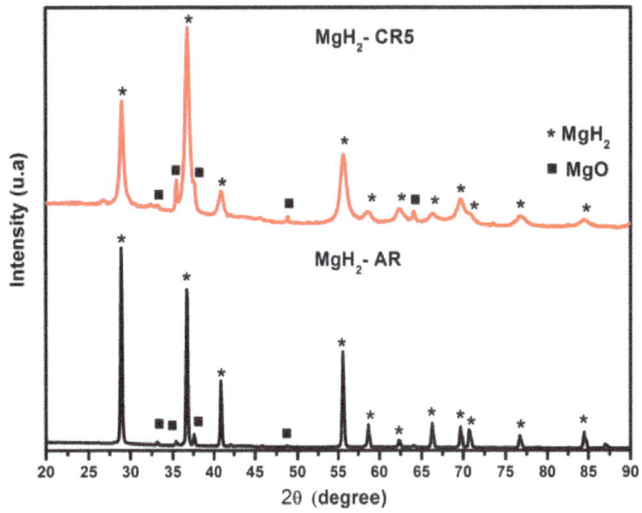

Figure 3. X-ray diffraction patterns of as-received and cold rolled (5 passes) MgH_2 powders. (CR): cold rolled; (AR): as-received.

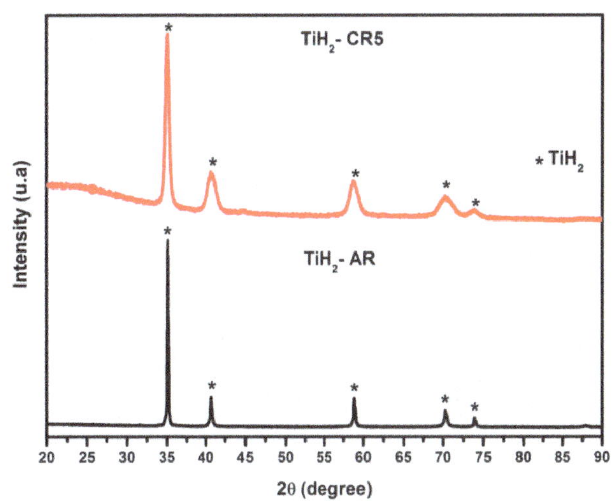

Figure 4. X-ray diffraction patterns of as-received and cold rolled (5 passes) TiH_2 powders.

The XRD patterns of MgH_2 shown in Figure 3 indicate the predominance of the hydride phase with some observable peaks of magnesium. Cold rolling caused broadening of the diffraction peaks which can be ascribed to lattice defects and crystallite size reduction. A quantitative analysis performed using Rietveld refinement indicated the same amount of Mg and MgH_2 in the as-received and cold rolled samples. However, the crystallite size was greatly reduced by cold rolling, going from 114 nm in the as-received sample to 14 nm in the cold rolled sample.

In the case of TiH_2, the X-ray patterns presented in Figure 4 indicate that cold rolling greatly broadened the Braggs peaks. However, contrary to the results found with MgH_2 where the broadening was due to reduction of crystallite size, here the crystallite size remains essentially constant (60 nm) but the microstrain is increased by an order of magnitude from 0.02% in the as-received sample to 0.4% in the cold rolled sample.

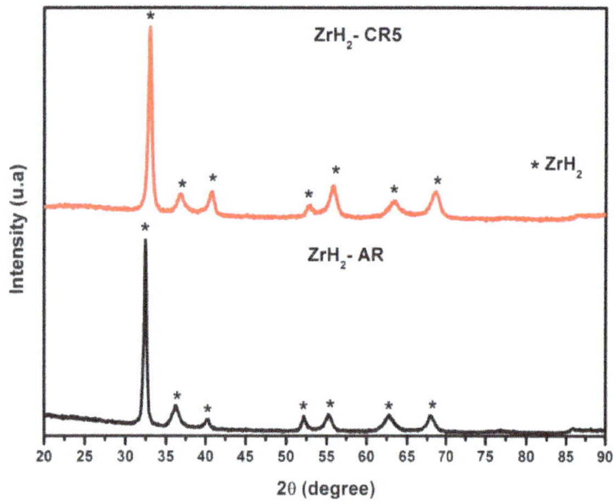

Figure 5. X-ray diffraction patterns of as-received and cold rolled (5 passes) ZrH_2 powders.

As shown in Figure 5, the X-ray patterns indicate that, in ZrH_2 sample, no other phases are present beside the hydride phase. It should be noticed that the peak broadening induced by cold rolling is not as important as for MgH_2 and TiH_2. In fact, Rietveld analysis indicated that, comparing as-received and cold rolled samples, the crystallite size goes from 29 nm to 20 nm while the microstrain goes from 0.2% to 0.3%.

X-ray Diffraction: Microwave Irradiation Effects

The X-ray diffraction patterns of as-received and cold rolled samples after microwave irradiation of MgH_2, TiH_2, and ZrH_2 are shown in Figure 6, Figure 7 and Figure 8, respectively. In the case of MgH_2 (Figure 6), both as-received and cold rolled samples were fully dehydrided by microwave irradiation. Nakamori *et al.* [27] investigated the effect of microwave irradiation in several hydrides and reported the ineffectiveness of that technique for desorption of MgH_2. In their case, the results may be explained by the limited increases of temperature they recorded. However, for ZrH_2 and TiH_2 relatively high temperatures were reached (more than 700 K and 800 K, respectively) and minimal hydrogen desorption was recorded. Therefore, the microwave heating and resulting dehydriding observed in our experiment might be ascribed to the use of a microwave susceptor (metallic cylinder). In our work, the metallic cylinder used as susceptor ensured a close contact with the hydride powder which remained confined between the susceptor and the sample holder wall ensuring an effective heating of the hydride powder. The efficiency on the use of susceptors for microwave irradiation and hydrogen desorption in MgH_2 was demonstrated by Zhang *et al.*[29] who designed a honeycomb ceramic monolith susceptor for this purpose. Awad *et al.* [30] prepared MgH_2 composites adding different types of carbon compounds (carbon fiber, graphite, and diamond, which are good microwave absorbers) during ball-milling and observed the effectiveness of this approach in increasing the hydride decomposition during irradiation.

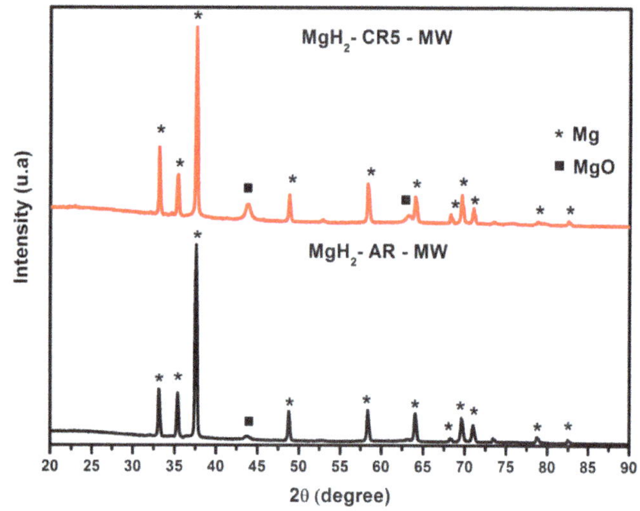

Figure 6. X-ray diffraction patterns of as-received and cold rolled (5 passes) MgH_2 powders after microwave irradiation.

The XRD patterns of TiH_2 samples after microwave irradiation are shown in Figure 7. In both samples (as-received and cold rolled), the hydride decomposition was not complete. The positive effect of cold rolling on the hydrogen desorption is clear from the relative intensities of the diffraction peaks of hydrides and Ti metal phases. From Rietveld refinement we found that, after microwave irradiation, the as-received sample was made of 93 wt.% of titanium hydride (TiH_2 and $TiH_{1.5}$) and only 7 wt.% of titanium. In the case of cold rolled sample, after microwave irradiation the proportion of TiH_2 was 46 wt.%, the rest being titanium.

Nakamori *et al.* [27] highlighted that TiH_2 is suitable for microwave heating (microwave absorber) but showed limited hydrogen desorption. This behavior was ascribed to a limited electromagnetic penetration depth in TiH_2, deduced by these authors as 11 μm for their experimental condition, leading to hydride decomposition only close to the surface. In our work, the hydrogen desorption from the TiH_2 sample was much higher than that reported in Nakamori's work. These results demonstrated the improvement in hydrogen desorption from TiH_2 promoted by cold rolling.

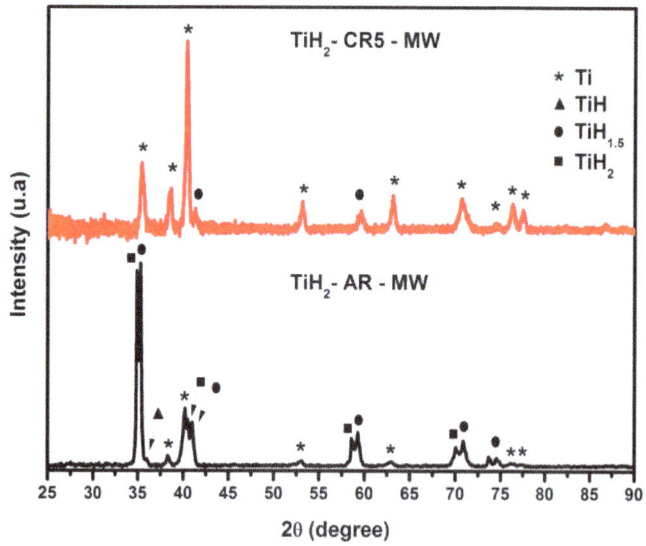

Figure 7. X-ray diffraction patterns of as-received and cold rolled (5 passes) TiH_2 powders after microwave irradiation.

Cold rolling was most effective in the case of ZrH_2. As shown in Figure 8, the pattern for the as-received sample display only peaks during to the hydride phase which means that microwave heating was totally ineffective. On the contrary, the pattern of the cold rolled sample did not show any ZrH_2 peaks but instead only showed peaks that could be indexed to the Zr metal phase and zirconium oxide. Considering the high thermodynamic stability of ZrH_2, the obtained result is very interesting and may open new possibilities for fast hydride and dehydride Zr and its alloys, which is an interesting topic in several areas, such as the nuclear industry.

As for TiH_2, the absence of dehydrogenation for the as-received sample may be due to by the small electromagnetic field penetration depth in Zr and the high thermodynamic stability of the ZrH_2 phase which prevented the hydride decomposition. By cold rolling, the particle size is reduced and defects are generated which contribute positively to the effectivity of microwave heating. Concerning the positive effect of cold rolling on hydrogen desorption, similar behavior has been observed for the conventional heating of cold rolled hydride powders [31,39,40]. This behavior has been attributed to the decrease in crystallite size and crystal defects generated during rolling which could act as fast diffusion paths and heterogeneous nucleating sites for diffusional phase transformations. Similar behavior may be expected for microwave heating. Particularly concerning ZrH_2, beneficial effect of cold rolling for improving the hydrogen desorption kinetics in conventionally heated ZrH_2 was reported in our previous investigation on the hydriding and dehydriding behavior of Zircaloy chips [32].

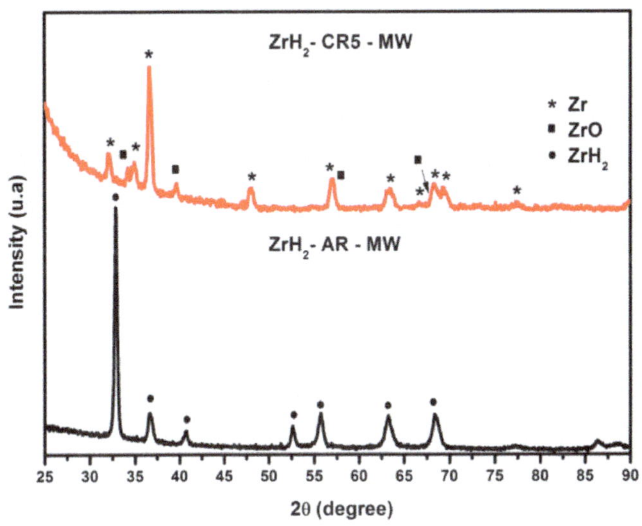

Figure 8. X-ray diffraction patterns of as-received and cold rolled (5 passes) ZrH_2 powders after microwave irradiation.

It should be noticed that after microwave irradiation in a few cases there was formation of oxides while in other cases no oxides were formed. For example, MgH_2 in as-received and cold rolled states presented formation of oxides after microwave irradiation while for TiH_2 no oxides were formed. The explanation may be that when a hydride phase subsisted, it prevented oxidation of the material. As a proof we see that for ZrH_2 the as-received sample after microwave irradiation did not desorbed at all and there were no oxides formed while the cold rolled sample fully desorbed and there is the presence of zirconium oxide.

From these experiments and by comparing with the results of Nakamori *et al.* [27], it is clear that the use of a susceptor greatly improved the effectiveness of microwave irradiation. A critical measurement that is missing in this investigation is the temperature during the reaction. This lack of recording temperature was due to the interaction between microwave and thermocouple. We are planning to solve this problem in a future investigation. Nevertheless, as MgH_2 decomposed to magnesium powder after microwave irradiation, we know that melting temperature was not reached, which means that in our experiments the temperature reached was probably not so different than in Nakamori's case.

CONCLUSIONS

Microwave irradiation in binary metal hydrides was investigated. In the case of MgH_2 the feasibility of microwave heating was observed and its complete dehydriding was made possible by using a susceptor in a simple experimental setup. Despite the low penetration depth of the electromagnetic field in TiH_2, previously reported in the literature, extensive desorption was achieved after cold rolling this hydride powder. As ZrH_2 is much more stable than TiH_2, decomposition of as-received ZrH_2 was even more difficult than TiH_2 under our experimental conditions. On the other hand, full desorption of ZrH_2 was reached after cold rolling this sample. Therefore, for both TiH_2 and ZrH_2 cold rolling of the hydride powders was shown to be a feasible approach for improving the hydrogen desorption. The beneficial effect of cold rolling on the hydride decomposition is attributed to a combination of microstructural and morphological changes promoted by this processing route, which comprises particle size reduction, crystallite size reduction, and increase of residual stress. In the case of particle size reduction, this allows a more effective heating of the particles by approximating the size of the particles to the penetration depth of the electromagnetic field in metallic powders. Moreover, residual stress and crystallite size reduction can increase the number on nucleation sites and accelerate hydrogen diffusivity contributing to effective hydrogen desorption from the hydride powders.

ACKNOWLEDGMENTS

ISD acknowledges the fellowship from the "Programme des Futurs Leaders dans les Ameriques (PFLA)" which is managed by the "Bureau Canadien de l'Education Internationale (BCEI)", Canada. SFS acknowledges FAPESP for visiting researcher fellowship (Grant 14/06127-5) and CNPq grants (558112/2010-2; 461034/2014-0; and 309202/2014-0).

REFERENCES

1. Das, S.; Mukhopadhyay, A.K.; Datta, S.; Basu, D. Prospects of microwave processing: An overview. *Bull. Mater. Sci.* **2009**, *32*, 1–13.
2. Alsharaeh, E.H.; Othman, A.A.; Aldosari, M.A. Microwave irradiation effect on the dispersion and thermal stability of RGO nanosheets within a polystyrene matrix. *Materials***2014**, *7*, 5212–5224.
3. Tapia-Ruiz, N.; Sorbie, N.; Vaché, N.; Hoang, T.K.A.; Gregory, D.H. Rapid microwave synthesis,

characterization and reactivity of litium nitride hydride, Li_4NH. *Materials* **2013**, *6*, 5410–5426.

4. Inamoto, M.; Kurihara, H.; Yajima, T. Vanadium pentoxide-based composite synthesized using microwave water plasma for cathode material in rechargeable magnesium batteries. *Materials* **2013**, *6*, 4514–4522.
5. Seetharaman, S.; Subramanian, J.; Tun, K.S.; Hamouda, A.S.; Gupta, M. Synthesis and characterization of nano boron nitride reinforced magnesium composites produced by the microwave sintering method. *Materials* **2013**, *6*, 1940–1955.
6. Yu, P.; Stephani, G.; Luo, S.D.; Goehler, H.; Qian, M. Microwave-assisted fabrication of titanium hollow spheres with tailored shell structures for varios potential applications. *Mater. Lett.* **2012**, *86*, 84–87.
7. Mondal, A.; Agrawal, D.; Upadhyaya, A. Microwave sintering of refractory metals/alloys: W, Mo, W-Cu, W-Ni-Cu and W-Ni-Fe alloys. *J. Microw. Power Electromagn. Energy* **2010**, *44*, 28–44.
8. Yoshikawa, N. Fundamentals and applications of microwave heating of metals. *J. Microw. Power Electromagn. Energy* **2010**, *44*, 4–13.
9. Roy, R.; Agrawal, D.K.; Chang, J.P.; Gedevanishvili, S. Full sintering of powdered metals using microwaves. *Nature* **1999**, *399*, 668–670.
10. Mondal, A.; Upadhyaya, A.; Agrawal, D. Microwave sintering of W-18Cu and W-7Ni3Cu alloys. *J. Microw. Power Electromagn. Energy* **2009**, *43*, 11–16.
11. Takayama, S.; Link, G.; Miksch, M.; Sato, M.; Ichikawa, J.; Thumm, M. Millimetre wave effects on sintering behavior of metal powder compacts. *Powder Metall.* **2006**, *48*, 274–280.
12. Rybakov, K.I.; Semenov, V.E.; Egorov, S.V.; Eremeev, A.G.; Plotnikov, I.V.; Bykov, Y.V. Microwave heating of conductive powder materials. *J. Appl. Phys.* **2006**.
13. Jayalakshmi, S.; Sahu, S.; Sankaranarayanan, S.; Gupta, S.; Gupta, M. Development of novel Mg-$Ni_{60}Nb_{40}$ amorphous particle reinforced composites with enhanced hardness and compressive response. *Mater. Des.* **2014**, *53*, 849–855.
14. Schlapbach, L.; Züttel, A. Hydrogen-storage materials for mobile applications. *Nature* **2001**, *414*, 353–358.
15. Huot, J. Metal hydrides. In *Handbook of Hydrogen Storage: New Materials for Future Energy Storage*, 1st ed.; Hirscher, M., Ed.; Willey—VCH: Darmstadt, Germany, 2010; pp. 81–116.
16. Zaluska, A.; Zaluski, L.; Strom-Olsen, J.O. Nanocrystalline magnesium for hydrogen storage. *J. Alloys Compd.* **1999**, *288*, 217–225.
17. Chen, F.; Tao, Z.; Liang, J.; Chen, J. Efficient hydrogen storage with the combination of lightweight Mg/MgH_2 and nanostructures. *Chem. Commun.* **2012**, *48*, 7334–7343.
18. Santos, S.F.; Ishikawa, T.T.; Botta, W.J.; Huot, J. MgH_2 + FeNb nanocomposites for hydrogen storage. *Mater. Chem. Phys.* **2014**, *147*, 557–562.
19. Huot, J.; Ravnsbaek, D.B.; Zhang, J.; Cuevas, F.; Latroche, M.; Jensen, T.R. Mechanochemical synthesis of hydrogen storage materials. *Prog. Mater. Sci.* **2013**, *58*, 30–75.
20. Santos, S.F.; Huot, J. Hydrogen storage in Ti-Mn-(FeV) BCC alloys. *J. Alloys Compd.* **2009**, *480*, 5–9.
21. Sakintuna, B.; Lamari-Darkrim, F.; Hirscher, M. Metal hydride materials for solid hydrogen storage: A review. *Int. J. Hydrog. Energy* **2007**, *32*, 1121–1140.
22. Emami, H.; Edalati, K.; Matsuda, J.; Akiba, E.; Horita, Z. Hydrogen storage performance of TiFe after processing by ball milling. *Acta Mater.* **2015**, *88*, 190–195.
23. Shao, H.; Xin, G.; Zheng, J.; Li, X.; Akiba, E. Nanotechnology in Mg-based materials for hydrogen storage. *Nano Energy* **2012**, *1*, 590–601.
24. Jeon, K.-J.; Moon, H.R.; Ruminski, A.M.; Jiang, B.; Kisielowski, C.; Bardhan, R.; Urban, J.J. Air-stable magnesium nanocomposites provide rapid and high-capacity hydrogen storage without using heavy-metal catalysts. *Nat. Mater.* **2011**, *10*, 286–290.
25. Luo, S.D.; Yang, Y.F.; Schaffer, G.B.; Qian, M. Novel fabrication of titanium by pure microwave radiation of titanium hydride powder. *Scr. Mater.* **2013**, *69*, 69–72.
26. Li, Q.; Pan, Y.; Leng, H.; Chou, K. Structure and properties of Mg-La-Ni ternary hydrogen storage alloys by microwave-assisted activation synthesis. *Int. J. Hydrog. Energy* **2014**, *39*, 14247–14254.
27. Nakamori, Y.; Matsuo, M.; Yamada, K.; Tsutaoka, T.; Orimo, S. Effect of microwave irradiation on metal hydrides and complex hydrides. *J. Alloys Compd.* **2007**, *446–447*, 698–702.
28. Krishnan, R.; Agrawal, D.; Dobbins, T. Microwave irradiation effects on reversible hydrogen desorption in sodium aluminum hydrides ($NaAlH_4$). *J. Alloys Compd.* **2009**, *470*, 250–255.
29. Zhang, H.; Geerlings, H.; Lin, J.; Chin, W.S. Rapid microwave hydrogen release from $MgH_2$and other hydrides. *Int. J. Hydrog. Energy* **2011**, *36*, 7580–7586.
30. Awad, A.S.; Tayed, T.; Nakl, M.; Zakhour, M.; Ourane, B.; Troedec, M.L.; Bobet, J.-L. Effect of carbon types (graphite, CFs and diamond) on the hydrogen desorption of Mg-C powder mixtures under microwave irradiation. *J. Alloys Compd.* **2014**, *607*, 223–229.
31. Dupim, I.S.; Moreira, J.M.L.; Huot, J.; Santos, S.F. Effect of cold rolling on the hydrogen absorption and desorption kinetics of Zircaloy-4. *Mater. Chem. Phys.* **2015**, *155*, 241–245.
32. Amira, S.; Santos, S.F.; Huot, J. Hydrogen sorption properties of Ti-Cr alloys synthesized by ball milling and cold rolling. *Intermetallics* **2010**, *18*, 140–144.
33. Huot, J.; Skryabina, N.Y.; Fruchart, D. Application of severe plastic deformation techniques to magnesium for enhanced hydrogen sorption properties. *Metals* **2012**, *2*, 329–343.
34. Huot, J. Nanocrystalline metal hydrides obtained by severe plastic deformations. *Metals***2012**, *2*, 22–40.
35. Jain, P.; Lang, J.; Skryabina, N.; Fruchart, D.; Santos, S.F.; Binder, K.; Klassen, T.; Huot, J. MgH_2 as dopant for improved activation of commercial Mg ingot. *J. Alloys Compd.* **2013**, *575*, 364–369.

36. Zhang, L.T.; Ito, K.; Vasudevan, V.K.; Yamaguchi, M. The effect of cold-rolling on the hydrogen absorption/desorption behavior of Ti-22Al-27Nb alloys. *Mater. Sci. Eng. A* **2002**, *329–331*, 362–366.

37. Bruker AXS. *TOPAS V4: General profile and structure analysis software for powder diffraction data*; Bruker AXS: Karlsruhe, Germany, 2008.

38. Mondal, A.; Shukla, A.; Upadhyaya, A.; Agrawal, D. Effect of porosity and particle size on microwave heating of copper. *Sci. Sinter.* **2010**, *42*, 169–182.

39. Lang, J.; Huot, J. A new approach to the processing of metal hydrides. *J. Alloys Compd.* **2011**, *509*, L18–L22.

40. Lang, J.; Eagles, M.; Conradi, M.S.; Huot, J. Hydrogenation rate limiting step, diffusion and thermal conductivity in cold rolled magnesium hydride. *J. Alloys Compd.* **2014**, *583*, 116–120

CHAPTER 13

INCREASING HYDROGEN DENSITY WITH THE CATION-ANION PAIR BH^{4-}-NH^{4+} IN PEROVSKITE-TYPE $NH_4Ca(BH_4)_3$

Pascal Schouwink, Fabrice Morelle, Yolanda Sadikin, Yaroslav Filinchuk and Radovan Černý

Keywords: Metal borohydride, perovskite, ammonium, ammonia-borane, protic-hydridic, hydrogen storage

INTRODUCTION

Complex hydrides material design has undergone major progress since the first reports suggesting $LiBH_4$ as a solid state hydrogen storage material [1]. Semi-empirical chemical concepts combining the stabilities of the less stable and more stable borohydrides [2], for instance, have given way to more profound concepts that exploit different kinds of weak interaction [3]. Recently, we reported on a large family of mixed-metal borohydrides crystallizing in the perovskite structure type, the latter being a prime example of functional materials design [4]. As a very stable structure type, the perovskite form allows one to study the behavior of a large range of different metals in a borohydride environment, but also to apply an immense toolbox developed throughout decades of metal-oxide perovskite design. The most simple concept comprises cation-substitution on the cuboctahedral *A*- and octahedral *B*-sites of a material generalized as $AB(BH_4)_3$. A logical further step is to explore anion-mixing on the ligand site. It was shown that the borohydride perovskite host readily takes up halide anions on the borohydride site [4]. In this context, it comes to mind to test the nitrogen-based hydrogen molecules, which are currently being vigorously investigated as means to tailor hydrogen release temperatures [3]. Recently we described interesting lattice instabilities in the metal boroyhdride perovskites where supposedly close homopolar dihydrogen contacts between adjacent borohydride anions stabilize lattice distortions in the high-temperature phase of $AB(BH_4)_3$, while these are most commonly destabilized by temperature in metal oxide or halide perovskites [4].

Aside from ammonium borohydride NH_4BH_4 itself [5,6,7], the ammonium cation has been rather neglected by the hydrogen storage community, and so far considered only in conjunction with higher boranes such as $NH_4B_3H_8$ [8], $(NH_4)_2B_{10}H_{10}$ or $(NH_4)_2B_{12}H_{12}$ [9] which impede the borazine evolution during thermolysis thanks to their stable borane cages, but never as a building block in a metal-based material. Attempts to tailor the thermal stability of NH_4BH_4, which decomposes slowly at room temperature (RT), have also been carried out to develop it as solid state hydrogen storage material [7,10].

The underlying concept of incorporating NH_4^+ was to increase the gravimetric hydrogen capacity ρ_m of perovskite-type $KCa(BH_4)_3$ by monovalent cation substitution on the K-site or divalent cation substitution on the Ca-site. Attempts to achieve this by the substitution K^+-Na^+ or Ca^{2+}-Mg^{2+} fail, as the hypothetical compounds $NaCa(BH_4)_3$, $KMg(BH_4)_3$ or $NaMg(BH_4)_3$ all come to lie outside the experimentally established stability field of the structure type within borohydrides [4]. The ionic radius of monovalent NH_4^+ (1.43 Å), on the other hand, is close to that of K^+ (1.33 Å) and ammonium easily substitutes for potassium in many metal halides. The calculated tolerance and octahedral factors suggest $NH_4Ca(BH_4)_3$ to be a stable compound. Its thermal stability and structure were recently reported in [4], accomplishing

Citation: Pascal Schouwink, Fabrice Morelle, Yolanda Sadikin, Yaroslav Filinchuk and Radovan Černý, Increasing Hydrogen Density with the Cation-Anion Pair BH4−-NH4+ in Perovskite-Type NH4Ca(BH4)3, doi: 10.3390/en8088286.

the long-term RT-stabilization of ammonium borohydride NH_4BH_4, with the extreme hydrogen density of 24.5 wt % H_2 in a metal ammonium borohydride, $NH_4Ca(BH_4)_3$, with a high density of ρ_m = 15.7 wt % H_2. Herein, we modify the reported synthesis based on ammonium chloride NH_4Cl to produce chlorine-free samples, taking up this study and focus on the actual decomposition mechanism of $NH_4Ca(BH_4)_3$ *via* an unknown ammonia-borane complex, which is obtained as a single phase sample thanks to the modified procedure and whose structure is solved on the basis of synchrotron X-ray powder diffraction. The vividly applied concept to tailor hydrogen release properties, the hetero-polar contact between protic and hydridic hydrogen, is thus extended for the first time to a cation-anion pair, NH_4^+-BH_4^-, in a metal borohydride. The presented concept of cation-substitution by ammonium is easily extended to a large number of previously reported mixed-metal borohydrides [11].

RESULTS AND DISCUSSION

The substitution of K^+ for NH_4^+ results in a stable compound $NH_4Ca(BH_4)_3$, which is presently investigated by means of two different synthetic mechanochemistry routes using different sources of the ammonium molecule. While the first employs a metathesis reaction based on ammonium chloride, in the second, chlorine-free, approach, NH_4BH_4 is loaded at low temperatures in the milling jar and reacted directly with $Ca(BH_4)_2$ in an addition reaction. The Cl-approach results in better crystallized compounds and a less complex phase composition (of novel phases) in as-milled samples, which is why the decomposition is discussed for these samples, despite the chlorine-content. On the other hand, the main decomposition product occurring during thermolysis of both samples is obtained as a well crystallized single phase composition in the NH_4BH_4-based one, when milled in the presence of a solvent (ether). The crystal structure of the main decomposition product is hence solved and discussed on the basis of these data.

Formation of $NH_4Ca(BH_4)_3$

In samples synthesized from NH_4Cl, perovskite-type $NH_4Ca(BH_4)_3$ is formed according to a general metathesis reaction used for borohydride-perovskites, developed for cases when one of the precursors cannot be provided as a borohydride:

$$NH_4Cl(s) + LiBH_4(s) + Ca(BH_4)_2(s) \rightarrow NH_4Ca(BH_4)_3(s) + LiCl(s) \quad (1)$$

A number of different nominal NH_4Cl:$Ca(BH_4)_2$:$LiBH_4$ compositions were investigated, $NH_4Ca(BH_4)_3$ forming in the less NH_4Cl-rich mixtures 1:1:1, 1:2:1 and 1:2:2, but neither in 2:2:1 nor 3:1:3. At least three other phases with very low decomposition temperatures were observed as impurities or the main phase in 2:2:1 and 3:1:3 mixtures in as-milled samples prepared from the chloride, presumably containing NH_4^+ as indicated by their low decomposition temperatures, but these could not be identified and will not be further discussed presently.

The formation of $NH_4Ca(BH_4)_3$ from the borohydrides, *i.e.*, $Ca(BH_4)_2$ and NH_4BH_4, was attempted in various trials. To this end milling jars were preliminarily cooled to 243 K, in order to stabilize NH_4BH_4 (unstable at RT) in a ratio 1:1 next to solid $Ca(BH_4)_2$. Soft milling (30 min at 500 rpm) of this mixture resulted in a multiphase powder with an unsatisfactory phase yield of $NH_4Ca(BH_4)$ (black curve in Figure 1). The low yield may be owed to partial decomposition of NH_4BH_4, since the jars were not cooled during milling and the inside-temperature reached approximately 310 K.

The procedure was subsequently modified to a solvent-assisted protocol where diethyl ether was added to the mixture and the inside-temperature was kept below 270 K, resulting in a significantly improved phase yield of $NH_4Ca(BH_4)_3$ (blue curve in Figure 1). At temperature below RT, and neglecting impurities, the formation from the borohydrides can hence ideally be approximated as:

$$NH_4BH_4(s) + Ca(BH_4)_2(s) \xrightarrow{ether} NH_4Ca(BH_4)_3(s) \quad (2)$$

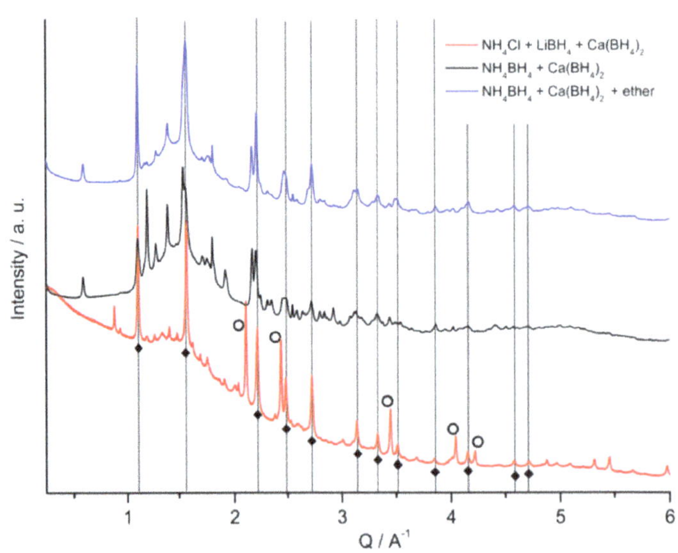

Figure 1. Room temperature synchrotron X-ray powder diffraction patterns shown for chloride-based (bottom curve, mixture 1:1:1), dry NH_4BH_4-ballmilled (middle) and ether-NH_4BH_4-ballmilled samples (top). Squares and vertical lines mark peak position of $NH_4Ca(BH_4)_3$, open circles correspond to LiCl.

Despite differences in the phase composition with respect to impurities the perovskite-type borohydride $NH_4Ca(BH_4)_3$ is the main phase in both the presently discussed chloride (1:1:1) and ether-based approaches. Furthermore, the fact that the lattice parameter *a* is identical within error (5.655(3) and 5.651(5) Å at 313 K for $NH_4Ca(BH_4)_3$ produced by chloride- and ether-based ball milling) as indicated by identical Bragg peak position in Figure 1 demonstrates the efficiency of Reaction (1) in terms of its capability to produce a borohydride-pure product without chloride on the anion-site, showing that the metathesis is completed in such an approach.

Thermolysis of $NH_4Ca(BH_4)_3$

The ammonium metal-borohydride perovskite $NH_4Ca(BH_4)_3$ starts decomposing at approximately 380 K in both, chloride and chloride-free samples. The respective temperature-dependent *in-situ* synchrotron diffraction data are shown for the chloride sample in Figure 2. The suggested decomposition mechanism for the chloride synthesis involves the formation of β-$Ca(BH_4)_2$ and $LiBH_4$ *via* an intermediate decomposition product, which occurs as an impurity already in some as milled mixtures in both approaches (Supplementary Figure S1 and Figure S2). This decomposition intermediate is an unreported AB-complex of $Ca(BH_4)_2$ (AB = NH_3BH_3) and is discussed in detail below.

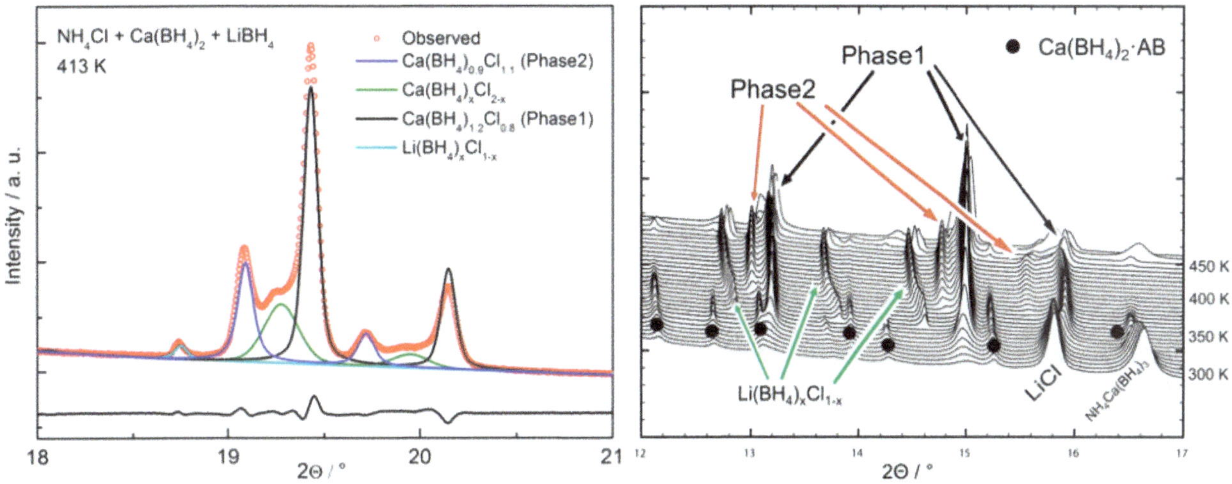

(a) (b)

Figure 2. (**a**) Selected part of Rietveld plot at 413 K for the chloride-based mixture 1:1:1. Three different compositions of $Ca(BH_4)_xCl_{2-x}$ were modeled in addition to $Li(BH_4)_xCl_{1-x}$. The difference curve to the fit is shown on the bottom; (**b**) Close-up on *in-situ* diffraction data showing crystallization and chlorine uptake by $Ca(BH_4)_xCl_{2-x}$. The models of $Ca(BH_4)_{0.9}Cl_{1.1}$ and $Ca(BH_4)_{1.2}Cl_{0.8}$ at 413 K (left) correspond to Phase 1 and Phase 2 in the temperature ramp.

As temperature is increased, calcium borohydride then proceeds to take up chlorine according to $Ca(BH_4)_{2-x}Cl_x$ in Cl-mixtures, as discussed below. Interestingly, chloride seems to stabilize the β-phase of $Ca(BH_4)_2$, which has previously been observed [12], whereas in ether-milled samples the phase composition *post-decomposition* is a mixture of β- and γ-phases. Ideally, the decomposition mechanism for the chloride-free protocol consists in inverting Reaction (2), *i.e.*, the stored NH_4BH_4would be released. In reality, however, the decomposition involves the intermediate 1, *i.e.*, the novel AB-complex $Ca(BH_4)_2$·AB. Hence, while $NH_4Ca(BH_4)_3$ may be considered a means of storing (irreversibly) ammonium borohydride in the solid state, it cannot be released. The idealized decomposition mechanism of $NH_4Ca(BH_4)_3$ can therefore be written as:

$$NH_4Ca(BH_4)_3(s) \rightarrow Ca(BH_4)_2{\cdot}NH_3BH_3(s) + H_2(g) \qquad (3)$$

While this reaction is stoichiometric and observed in the chloride free samples, the situation is more complex when the presence of chlorine is taken into account, leading to multiple processes releasing and absorbing chlorine which are outlined below.

Figure 3 shows the previously published [4] analysis of volatile decomposition products with the corresponding thermal and gravimetric signals, measured on the chloride-sample. Notably, there is no diborane released during the whole thermal decomposition of $NH_4Ca(BH_4)_3$, which in turn does release one molecule of hydrogen around 380 K (observed on Cl sample) and transforms to $Ca(BH_4)_2$·AB, in both chloride and ether milled samples, according to Reaction (3).

In chloride-based samples the decomposition proceeds directly to the AB-complex. This gas release is readily identified as the first mass loss in Figure 3, where mass spectrometry (top panel) shows that the released product's composition is nearly pure hydrogen (very small signal of NH_3at 380 K). $Ca(BH_4)_2$·AB then decomposes to release additional hydrogen above 400 K, accompanied by a minor amount of ammonia NH_3 (Figure 3), similar to previously reported reaction mechanisms for AB complexes involving recovery of β-$Ca(BH_4)_2$ [13,14]. This corresponds to the second mass loss and is visible as a continuous evolution in the spectrometric detection of H_2. The decomposition temperature of the AB complex compares quite well with those of the first two reported members of the metal borohydride AB family [15].

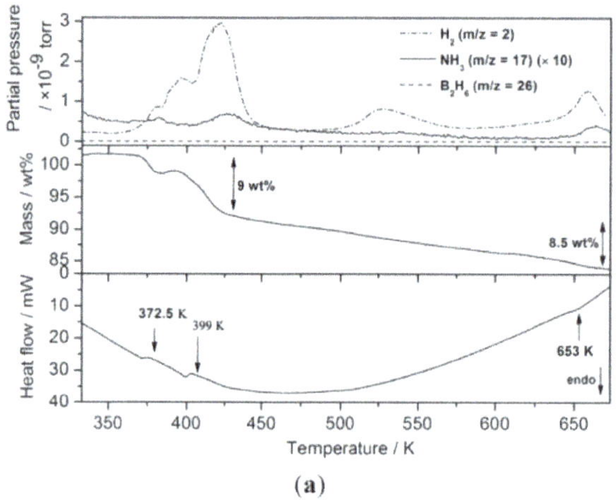

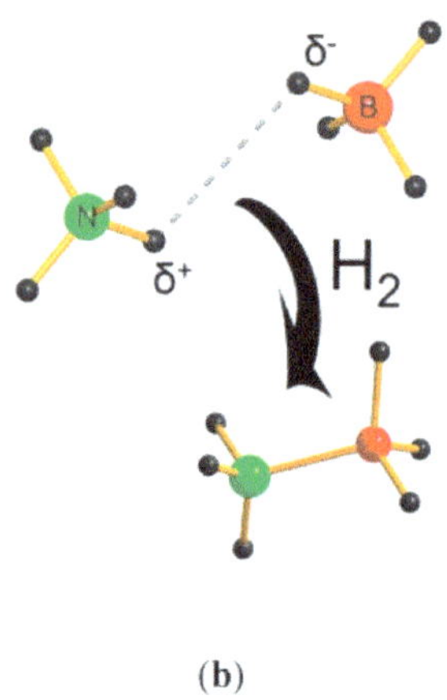

Figure 3. Simultaneous gas release and thermal analysis of the decomposition of $NH_4Ca(BH_4)_3$ under nitrogen flow (heating rate 5 K/min, chloride sample, reproduced from [4]). (**a**) Results from mass-spectrometry (MS), thermal gravimetric analysis (TGA) and differential scanning calorimetry (DSC), the signal corresponding to NH_3 has been multiplied ×10; (**b**) Close heteronuclear dihydrogen contact between protic hydrogen pertaining to the ammonium cation and hydridic hydrogen from the borohydride anion facilitates hydrogen releases via AB.

Chloride Substitution in β-$Ca(BH_4)_2$

In the following we briefly discuss the release and uptake of chlorine by different phases during thermal decomposition of the mixture NH_4Cl: $Ca(BH_4)_2$:$LiBH_4$ 1:1:1 and referring to the *in-situ*diffraction data shown in Figure 2. Chloride substitution has recently been investigated by ball-milling $CaCl_2 + Ca(BH_4)_2$ [16]. Therein, anion-substitution is achieved by heating as-milled mixtures of unreacted precursors, which results in halide-substituted β-$Ca(BH_4)_{2-x}Cl_x$. The authors report different compositions with discrete lattice parameters depending on the nominal composition of precursors in the respective sample. The chlorine uptake by β-$Ca(BH_4)_2$ during the decomposition of $NH_4Ca(BH_4)_3$ is fundamentally different, since it is not $CaCl_2$ acting as a chlorine-source. In this case, the chlorine is provided by LiCl, which is formed during the metathesis Reaction (1). Concurring with the decomposition of the ammonium perovskite LiCl continuously transforms into hexagonal HT-phase $Li(BH_4)_xCl_{1-x}$, which crystallizes together with halide-substituted β-$Ca(BH_4)_2$ (Figure 2) according to Reaction (4):

$$\beta\text{-}Ca(BH_4)_2 + n LiCl \rightarrow Ca(BH_4)_x Cl_{2-x} + n Li(BH_4)_{n-x} Cl_{1-n-x} \quad (4)$$

While the dissolution of chloride is known to stabilize the hexagonal HT-phase of LiBH4, we observe a hitherto unreported behavior of the HT-phase (β) of $Ca(BH_4)_2$. A detailed examination of the *in-situ* diffraction data provided in Figure 2 is essential to the following discussion. At approximately 350 K the remnant $Ca(BH_4)_2$ not reacted but amorphized during ball-milling starts recrystallizing, accompanying the formation and crystallization of the AB-complex $Ca(BH_4)_2$·AB. Despite the temperature increase the lattice parameters of β-$Ca(BH_4)_2$ decrease as crystallinity increases (see Bragg signal at 15° shifting to higher angles with temperature, in Figure 2), due to chlorine-uptake originating from LiCl, while the Li is used to form $Li(BH_4)_xCl_{1-x}$. At temperatures of approximately 400 K, coinciding with the decomposition of the AB-complex in the chloride-based sample, halide-substituted β-$Ca(BH_4)_2$ abruptly forms two phases of well-defined chemical composition and discrete lattice parameters (labeled Phase1 and Phase2 in Figure 2), accompanied by a third phase of intermediate composition and much broader diffraction peaks. Rietveld refinement was performed on data collected at 413 K and including 3 different β-$Ca(BH_4)_xCl_{2-x}$compositions. The result of the refinement is shown in Figure 2. Two compositions $Ca(BH_4)_{0.9}Cl_{1.1}$and $Ca(BH_4)_{1.2}Cl_{0.8}$ could be refined reliably corresponding to unit cell volumes of 194.8(9) and 185.3(7) Å^3, respectively. the evolution of the cell volumes at higher temperature is shown in Supplementary Figure S3. These values are far below that of the reported unit cell volume of pure homoleptic β-$Ca(BH_4)_2$ at 305 K, 208.1 Å^3 [17], in agreement with the smaller anion radius of Cl^-. The anion-composition of the third component corresponding to the broad signals in Figure 2could not be determined reproducibly but is thought to lie between the two determined stoichiometries. The line width is approximately a factor 2.5 larger than in both other compositions. This may be due to a continuous distribution of chlorine dissolution on the BH_4-site. But also a shorter coherence length could be at the origin of the signal-width, suggesting intermediate chemical compositions on very short length scales, for instance at the grain-boundaries between compositions $Ca(BH_4)_{0.9}Cl_{1.1}$ and $Ca(BH_4)_{1.2}Cl_{0.8}$. To our knowledge the coexistence of various discrete anion-compositions has not been reported for β-$Ca(BH_4)_2$, despite anion-substitution having been quite well studied in this compound [12,16].

Ether-Samples

In the samples which were prepared by ball milling solidified NH_4BH_4 in ether at low temperatures the decomposition mechanism is not influenced by the kinetics of chlorine-exchange between the involved pseudobinary compounds β-$Ca(BH_4)_2$ and $LiBH_4$.

While the transformation of $NH_4Ca(BH_4)_3$ to the novel AB-complex is observed as in the chloride-sample, the decomposition now involves a further intermediate **2** (Supplementary Figure S2) that has not been characterized (T-interval 370–385 K, given the vicinity to the solved structure of $Ca(BH_4)_2{\cdot}AB$ we assume **2** to be a further AB-complex, possibly a phase transformation then takes place at 380 K), but indexed to an orthorhombic cell of a = 11.602, b = 14.339, c = 14. 296 Å, V = 2378 Å^3. This phase does not occur in chloride-syntheses. It is unlikely that there is a hydrogen loss occurring between intermediate 2 and $Ca(BH_4)_2{\cdot}AB$, since the reported product of the first two decomposition steps of NH_4BH_4 is the diammoniate of diborane $[(NH_3)_2BH_2](BH_4)$, which itself is an isomer in equilibrium with AB [18], and no intermediate deprotonated molecule exists. Given the small stability field of approximately 5 K and the vicinity to the stability field of $Ca(BH_4)_2{\cdot}AB$, intermediate 2 may be a further AB-derivative of different stoichiometry, or merely a superstructure to $Ca(BH_4)_2{\cdot}AB$, related by a polymorphic transformation.

In ether milled mixtures $Ca(BH_4)_2{\cdot}AB$ begins to decompose to β- and γ-$Ca(BH_4)_2$ at 400 K, proceeding until 430 K (Supplementary Figure S2), which is in surprising agreement with the decomposition-onset of the previously reported AB-complex of $Ca(BH_4)_2$, determined at 398 K, and ending at 428 K [14]. Hence, while the second decomposition step follows the established mechanism of AB-complexes, the perovskite adds an additional hydrogen release prior to this.

We have not identified further decomposition products. The previously reported AB-complexed $Ca(BH_4)_2$ decomposes to its constituents. AB melts at 377 K therefore its possible existence cannot be determined from the diffraction patterns. A difference in structure between the two (discussed below) could suggest a slightly modified decomposition pathway, the reported AB-complex [14] possibly showing a greater predisposition to dissociate to $Ca(BH_4)_2$ and AB due to its structural topology, which is built of defined layers of both precursors, while $Ca(BH_4)_2{\cdot}AB$ reported herein contains AB molecules whose nearest neighbor interactions are considered to be weaker with respect to those taking place in AB itself.

Crystal Structure of $NH_4Ca(BH_4)_3$

In the ammonium calcium borohydride perovskite the NH_4^+ cation occupies the A-site of the perovskite structure, where Ca is the B-cation and the BH_4^- anion is the ligand. We have chosen to model the structure in an ordered manner in a subgroup symmetry which allows for an identical unit cell metric to the primitive perovskite unit. A fully disordered structure can be modelled in Pm-3m, however the fit is not better. The ordered description is supported, albeit not fully justified, by X-ray diffraction data collected down to temperatures of 100 K which did not show a change in diffraction pattern that would suggest a phase transition due to ordering. All reported phase transitions in borohydride perovskites occur above RT without any further events down to 100 K [4,19].

Furthermore, the dominant weak interaction in $NH_4Ca(BH_4)_3$ is of attractive nature and structurally points towards an ordered model. Future neutron studies, as planned also for other metal borohydride perovskites may however result in a correction of this space group symmetry. Our structural description currently is in P-43m, which is a subgroup to Pm-3m, and allows maximizing homopolar dihydrogen contacts BH···HB between hydride anions, which are constrained to lower values by symmetry in other subgroups to Pm-3m, as for instance P432. One boron position and one nitrogen position provide orientation between the molecular species that are rotated by 90° around the principal cubic direction with respect to each other (Figure 4).

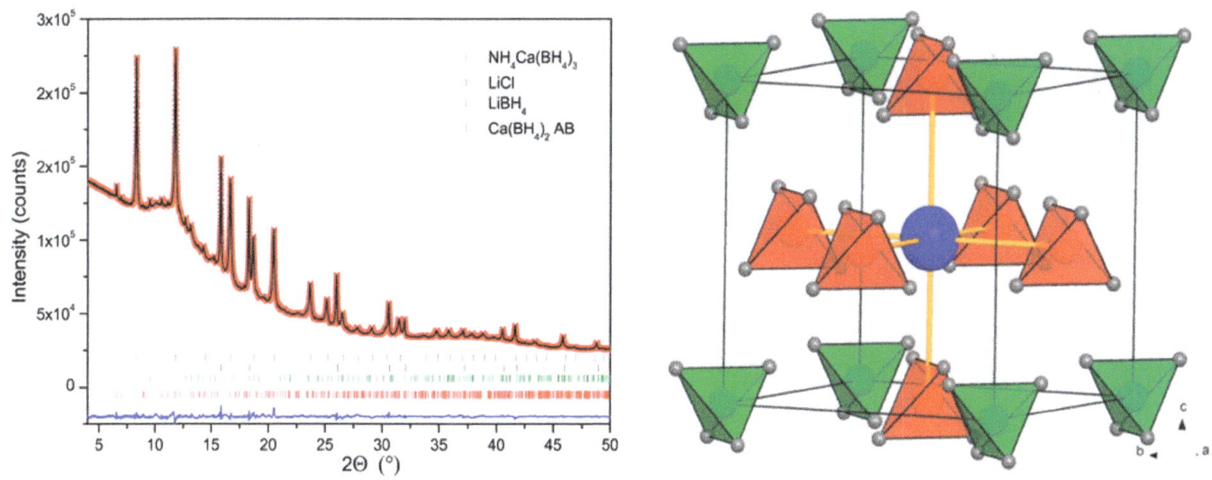

(a) (b)

Figure 4. (**a**) Rietveld plot for the chloride-based synthesis of $NH_4Ca(BH_4)_3$, refined at 298 K; and (**b**) structural model in space group P-43m. NH_4^+ green, BH_4^- red, Ca blue.

Crystal Structure of $Ca(BH_4)_2{\cdot}NH_3BH_3$

The dehydrogenation product of the ammonium borohydride perovskite is the calcium borohydride ammonia-borane complex $Ca(BH_4)_2{\cdot}AB$, which contains one AB molecule per $Ca(BH_4)_2$ unit and whose structure was solved *ab-initio* on a single phase sample from diffraction data collected at 414 K (Figure 5) during the decomposition of ether-milled $NH_4Ca(BH_4)_3$.

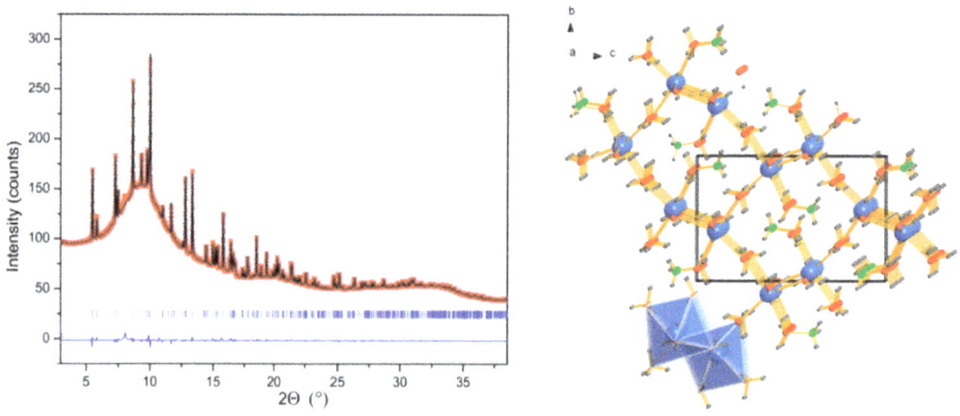

(a) (b)

Figure 5. (**a**) Rietveld plot for the ether-milled sample, refined at 414 K, showing a pure phase composition of $Ca(BH_4)_2{\cdot}AB$; (**b**) Corresponding structural model. N green, B red, H grey, Ca blue.

Interestingly, an AB complex of $Ca(BH_4)_2$ has previously been reported, which contains two AB molecules per unit cell [14] and was observed not during thermal composition, but obtained as an adduct by ball-milling NH_3BH_3 with $Ca(BH_4)_2$. The structure therein was described as containing layers of AB alternating with layers (square lattice) of borohydride stacked along the long axis, *b*.

A close inspection of the structural model of $Ca(BH_4)_2{\cdot}AB$ (Figure 6) reveals that the borohydride layers formed by a square lattice are conserved, however buckled by 90°. A further relationship is found in α-$Ca(BH_4)_2$-like ribbons, running parallel to the *a*-axis in $Ca(BH_4)_2{\cdot}AB$. These $Ca(BH_4)_2$-ribbons are highlighted in the grey box in Figure 6. In this description the NH_3BH_3 molecules form a further one-dimensional structural element (zig-zag chain), which alternates with $Ca(BH_4)_2$ribbons In this context, the relationship between the AB-complex $Ca(BH_4)_2{\cdot}AB$ and the reported $Ca(BH_4)_2{\cdot}2AB$ may therefore be described best as a reduction in dimensionality of the structural elements of the precursors in the former.

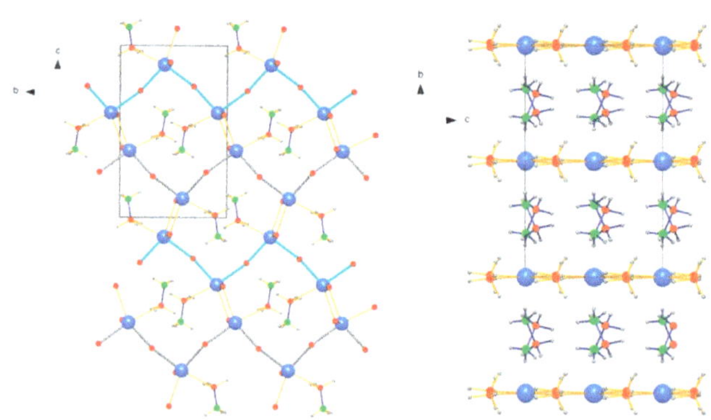

(a) (b)

Figure 6. (**a**) Comparison of the crystal structure of $Ca(BH_4)_2{\cdot}AB$; and (**b**) the reported $Ca(BH_4)_2{\cdot}2AB$-complex [14]. On the left, the buckled square lattice of $Ca(BH_4)_2$ layers is drawn in light-blue and grey. N green, B red, H grey, Ca blue. H atoms have been omitted on BH_4 groups for sake of clarity on the left.

In α-$Ca(BH_4)_2$ the square lattice (layers) of Ca-BH_4 conserved in both AB-complexes is oriented perpendicular to the *b*-axis, while the ribbons conserved only in $Ca(BH_4)_2{\cdot}AB$ connect these layers and run along the *a*-axis. If the buckled Ca-BH_4 square lattice (blue and grey sheets in Figure 6) is flattened out the α-$Ca(BH_4)_2$ structure is recovered in $Ca(BH_4)_2{\cdot}AB$, however with 50% of the ribbons replaced by zig-zag chains of NH_3BH_3. There is therefore no major reconstruction of the α-$Ca(BH_4)_2$ structure in this AB-complex.

The B-N lengths in $Ca(BH_4)_2{\cdot}AB$ at 414 K amount to 1.613(5) Å, significantly longer than those in the reported $Ca(BH_4)_2{\cdot}2AB$ (1.546 Å at 298 K) and very close to those in solid NH_3BH_3 (1.592 Å) [13,14]. The intermolecular distances between hydride and proton ions BH···HN pertaining to adjacent AB molecules amount to 2.05 Å, as opposed to 1.736 Å in the reported $Ca(BH_4)_2{\cdot}2AB$, and are slightly larger to those in AB (2.02 Å). This attests to a larger degree of interaction to BH_4groups as opposed to $Ca(BH_4)_2{\cdot}2AB$, which could lead to a slightly modified decomposition mechanism. The closest refined BH···HN distance between NH_3BH_3 and adjacent hydridic H^- of BH_4 groups is 2.02 Å, which is in quite good agreement with the values of 1.986 and 2.037 Å reported for $Ca(BH_4)_2{\cdot}2AB$. The main difference hence lies in intermolecular AB contacts, which are significantly larger in $Ca(BH_4)_2{\cdot}AB$. Nevertheless, we need to recall that the determination of interatomic hydrogen distances from X-ray powder diffraction data are severely biased by the application of restraints during refinement, and the determined values need not be very accurate.

EXPERIMENTAL SECTION

Synthesis

$Ca(BH_4)_2$, $LiBH_4$ and NH_4Cl were used as purchased from Sigma-Aldrich (Buchs, Switzerland). Chloride based samples were prepared by high-energy ball milling (mechanochemistry) in a Pulverisette P7 planetary ball mill (Fritsch, Idar-Oberstein, Germany). The ball-to-powder mass ratio was approximately 50, 60 milling cycles at 600 rpm of 2 min each were interrupted by 5 min cooling breaks to prevent heating and powder agglomeration on the walls of the milling jar. For the second approach based on NH_4BH_4, the latter was prepared according to ref. [20] with some minor modifications. Milling was performed in the Fritsch Pulverisette P7 planetary ball mill in a stainless steel vial cooled down to approximatively 240 K prior to milling. The milling speed was set to 500 rpm and the ball to powder mass ratio was approximately 50 for both dry and liquid assisted (500 μL (350 mg) of Et_2O at 500 mg of sample mass) milling. The temperature of the milling jar was not kept low during milling. In the case of dry milling, the milling time was 30 min and the temperature of the jar was around 35 °C (308 K) at the end of the milling. In the case of the ether assisted milling, the milling time was of 5 min, resulting in a much smaller temperature increase during milling. The ether was removed by evacuating the sample directly after the milling.

Synchrotron Radiation X-ray Powder Diffraction (SR-XPD)

Synchrotron radiation powder X-ray diffraction (SR-PXD) data were collected at the Swiss Norwegian Beamlines of the European Synchrotron Radiation Facility (ESRF, Grenoble, France) and at the Materials Science Beamline of the Swiss Light Source (Paul Scherrer Institute, Villigen, Switzerland) 0.6888 and 0.8271 Å, respectively, as calibrated by a Si standard. Samples were loaded into 0.5 mm borosilicate capillaries. Structures were solved in direct space using the software FOX [21] and treating BH_4 as a rigid body. After structure solution the respective models were refined with Fullprof [22] or TOPAS [23].

CONCLUSIONS

$NH_4Ca(BH_4)_3$ is to our knowledge the first perovskite-type inorganic compound that contains two inherently dynamic hydrogen-rich molecules, both on the anion and cation sites. The usage of ammonium NH_4^+ as A-site cation allows stabilizing ammonium borohydride for the first time at room temperature and pressure. This describes an approach that is not restricted to perovskite-type borohydrides and the incorporation of NH_4^+ may be generalized to search for new hydrogen-rich compounds. Besides the chemical benefit to hydrogen storage applications, the ammonium cation also enriches the spectrum of structural tools by a further parameter, providing, next to the non-spherical borohydride anion, a cation with tetrahedral charge distribution. The combination of non-spherical ions with chemically distinct hydrogen species may be exploited in complex hydrides to design polar structures.

$NH_4Ca(BH_4)_3$ decomposes at 380 K to new ammonia-borane complex $Ca(BH_4)_2 \cdot NH_3BH_3$ and releases one hydrogen molecule in the process. A major drawback of boron and nitrogen-based hydrogen storage systems is the possible evolution of ammonia NH_3 and diborane B_2H_6, which both lead to the failure of fuel cells. Neither of them are observed during the hydrogen release of $NH_4Ca(BH_4)_3$, while minor NH_3 evolution occurs during decomposition of $Ca(BH_4)_2 \cdot NH_3BH_3$, similar to previously reported AB-complexes of metal-borohydrides. $NH_4Ca(BH_4)_3$ is the first compound of a series of novel metal borohydrides which is expected to grow quickly thanks to a concept that is easily applied to a large and growing number of mixed-metal borohydride salts that contain cations similar in size and thus readily replaced by NH_4^+. For instance, replacing NH_4^+ for K in $K_3Mg(BH_4)_5$ [24] or $KAl(BH_4)_4$ [25,26] would result in hydrogen densities of 21.2 and 19 wt%, which is already very close to NH_4BH_4 itself. Generally, the decomposition of such proposed compounds is assumed to be dominated by the transformation of NH_4^+ and BH_4^- to AB-complexes. While this mechanism is bound to release pure hydrogen, the subsequent decomposition of AB-complexes poses the known problems of rehydrogenation and fuel cell contamination.

ACKNOWLEDGMENTS

This work was supported by the Swiss National Science Foundation under project 200020_149218, by the Fonds National de la Recherche Scientifique (FNRS, Belgium) via instruments CC, PDR, EQP and by FRIA (Fonds pour la formation à la recherche dans l'Industrie et l'Agriculture, Belgium) funding FM's PhD works. The authors thank SNBL at the ESRF and SLS at the PSI for the beamtime allocation and Dmitry Chernyshov (SNBL), Vadim Dyadkin (SNBL) and Nicola Casati (SLS) for their assistance.

AUTHOR CONTRIBUTIONS

P.S. and Y.S. performed the chloride based synthesis, F.M. performed the ammonium borohydride based synthesis. The synchrotron X-ray powder diffraction data were measured by P.S., Y.S., F.M., Y.F. and R.C. The diffraction data were analyzed by P.S. and R.C., crystal structures were solved by P.S. The manuscript has been written by P.S., and was commented by all authors.

REFERENCES

1. Schlapbach, L.; Züttel, A. Hydrogen-storage materials for mobile applications. *Nature* 2001, *414*, 353–358.
2. Schrauzer, G.N. Ueber ein periodensystem der metallboranate. *Naturwissenschaften* 1955, *42*, 438.
3. Jepsen, L.H.; Ley, M.B.; Su-Lee, Y.; Cho, Y.W.; Dornheim, M.; Jensen, J.O.; Filinchuk, Y.; Jorgensen, J.E.; Besenbacher, F.; Jensen, T.R. Boron-nitrogen based hydrides and reactive composites for hydrogen storage. *Mater. Today* 2014, *17*, 129–135.
4. Schouwink, P.; Ley, M.B.; Tissot, A.; Hagemann, H.; Jensen, T.R.; Smrčok, L.; Černý, R. Structure and properties of a new class of complex hydride perovskite materials. *Nat. Commun.* 2014, *5*, 5706.
5. Dixon, D.A.; Gutowski, M. Thermodynamic Properties of Molecular Borane Amines and the $[BH_4^-][NH_4^+]$ Salt for Chemical Hydrogen Storage Systems from *ab Initio* Electronic Structure Theory. *J. Phys. Chem. A* 2005, *109*, 5129–5139.
6. Karkamkar, A.; Kathmann, S.M.; Schenter, G.K.; Heldebrant, D.J.; Hess, N.; Gutowski, M.; Autrey, T. Thermodynamic and Structural Investigations of Ammonium Borohydride, a Solid with a Highest Content of Thermodynamically and Kinetically Accessible Hydrogen. *Chem. Mater.* 2009, *21*, 4356–4358.
7. Flacau, R.; Ratcliffe, C.I.; Desgreniers, S.; Yao, Y.; Klug, D.D.; Pallister, P.; Moudrakowski, I.L.; Ripmeester, J.A. Structure and dynamics of ammonium borohydride. *Chem. Commun.*2010, *46*, 9164–9166.
8. Huang, Z.; Chen, X.; Yisgedu, T.; Meyers, E.A.; Shore, S.G.; Zhao, J.-C. Ammonium Octahydrotriborate ($NH_4B_3H_8$): New Synthesis, Structure, and Hydrolytic Hydrogen Release. *Inorg. Chem.* 2011, *50*, 3738–3742.

9. Yisgedu, T.B.; Huang, Z.; Chen, X.; Lingam, H.K.; King, G.; Highley, A.; Maharrey, S.; Woodward, P.M.; Behrens, R.; Shore, S.G.; *et al.* The structural characterization of $(NH_4)_2B_{10}H_{10}$ and thermal decomposition studies of $(NH_4)_2B_{10}H_{10}$ and $(NH_4)_2B_{12}H_{12}$. *Int. J. Hydrog. Energy* 2012, *37*, 4267–4273.

10. Nielsen, T.K.; Karkamkar, A.; Bowden, M.; Besenbacher, F.; Jensen, T.R.; Autrey, T. Methods to stabilize and destabilize ammonium borohydride. *Dalton Trans.* 2013, *42*, 680–687.

11. Rude, L.H.; Nielsen, T.K.; Ravnsbæk, D.B.; Bösenberg, U.; Ley, M.B.; Richter, B.; Arnbjerg, L.M.; Dornheim, M.; Filinchuk, Y.; Besenbacher, F.; *et al.* Tailoring properties of borohydrides for hydrogen storage: A review. *Phys. Status Solidi* 2011, *208*, 1754–1773.

12. Rude, L.H.; Filinchuk, Y.; Sørby, M.H.; Hauback, B.C.; Besenbacher, F.; Jensen, T.R. Anion Substitution in $Ca(BH_4)_2$-CaI_2: Synthesis, Structure and Stability of Three New Compounds. *J. Phys. Chem. C* 2011, *115*, 7768–7777.

13. Xiong, Z.; Keong Yong, C.; Wu, G.; Chen, P.; Shaw, W.; Karkamkar, A.; Autrey, T.; Owen Jones, M.; Johnson, S.R.; Edwards, P.P.; *et al.* High-capacity hydrogen storage in lithium and sodium amidoboranes. *Nat. Mater.* 2007, *7*, 138–141.

14. Wu, H.; Zhou, W.; Yildirim, H. Alkali and Alkaline-Earth Metal Amidoboranes: Structure, Crystal Chemistry, and Hydrogen Storage Properties. *J. Am. Chem. Soc.* 2008, *130*, 14834–14839.

15. Wu, H.; Zhou, W.; Pinkerton, F.E.; Meyer, M.S.; Srinivas, G.; Yildirim, T.; Udovic, T.J.; Rush, J.J. A new family of metal borohydride ammonia borane complexes: Synthesis, structures, and hydrogen storage properties. *J. Mater. Chem.* 2010, *20*, 6550–6556.

16. Grove, H.; Rude, L.H.; Jensen, T.R.; Corno, M.; Ugliengo, P.; Baricco, M.; Sørby, M.H.; Hauback, B.C. Halide substitution in $Ca(BH_4)_2$. *RSC Adv.* 2014, *4*, 4736–4742.

17. Finlinchuk, Y.; Rönnebro, E.; Chandra, D. Crystal structures and phase transformations in $Ca(BH_4)_2$. *Acta Mater.* 2009, *57*, 732–738.

18. Bowden, M.; Autrey, T. Characterization and mechanistic studies of the dehydrogenation of NH_xBH_x materials. *Curr. Opin. Solid State Mater. Sci.* 2011, *15*, 73–79.

19. Schouwink, P.; Hagemann, H.; Embs, J.P.; D'Anna, V.; Černý, R. Di-hydrogen contact induced lattice instabilities and structural dynamics in complex hydride perovskites. *J. Phys. Condens. Matt.* 2015, *27*, 265403.

20. Parry, R.W.; Schultz, D.R.; Girardot, P.R. The Preparation and Properties of Hexamminecobalt (III) Borohydride, Hexamminechromium (III) borohydride and Ammonium Borohydride. *J. Am. Chem. Soc.* 1958, *1*, 1–3.

21. Favre-Nicolin, V.; Černý, R. FOX, "free objects for crystallography": A modular approach to *ab initio* structure determination from powder diffraction. *J. Appl. Cryst.* 2002, *35*, 734–743.

22. Rodriguez-Carvajal, J. *FULLPROF SUITE*; LLB Sacley & LCSIM: Rennes, France, 2003.

23. Coelho, A.A. TOPAS-Academic V5. 2012. Available online: http://www.topas-academic.net/(accessed on 15 February 2013).

24. Schouwink, P.; D'Anna, V.; Ley, M.B.; Lawson Daku, L.M.; Richter, B.; Jensen, T.R.; Hagemann, H.; Černý, R. Bimetallic Borohydrides in the System $M(BH_4)_2$-KBH_4 (M = Mg, Mn): On the Structural Diversity. *J. Phys. Chem. C* 2012, *116*, 10829–10840.

25. Dovgaliuk, I.; Ban, V.; Sadikin, Y.; Černý, R.; Aranda, L.; Casati, N.; Devillers, M.; Filinchuk, Y. The first halide-free bimetallic aluminum borohydride: Synthesis, structure, stability, and decomposition pathway. *J. Phys. Chem. C* 2014, *118*, 145–153.

26. Knight, D.A.; Zidan, R.; Lascola, R.; Mohtadi, R.; Ling, C.; Sivasubramanian, P.; Kaduk, J.A.; Hwang, S.-J.; Samanta, D.; Jena, P. Synthesis, Characterization, and Atomistic Modeling of Stabilized Highly Pyrophoric $Al(BH_4)_3$ via the Formation of the Hypersalt $K[Al(BH_4)_4]$. *J. Phys. Chem. C* 2013, *117*, 19905–19915.

CHAPTER 14

METAL HYDRIDES: AN INNOVATIVE AND CHALLENGING CONVERSION REACTION ANODE FOR LITHIUM-ION BATTERIES

Luc Aymard, Yassine Oumellal and Jean-Pierre Bonnet

Keywords: Conversion reaction; lithium-ion batteries; metal hydrides

INTRODUCTION

To satisfy the continuously raising need for energy is now a key priority worldwide. The challenge is to obtain environmentally friendly renewable power sources with enhanced electrical energy conversion efficiency at moderate costs. However, these energy sources, such as windmill or solar cells, are intrinsically intermittent and, consequently, need to be associated with efficient energy storage devices in order to provide electricity on demand. With regard to this, lithium-ion (Li-ion) batteries can present an attractive solution, provided that they exhibit sufficient potential and gravimetric/volumetric capacities. Graphite, which is usually used as negative electrode with an intercalation reaction of lithium, is not suitable here due to its intrinsic insufficient specific capacities (370 $A \cdot h \cdot kg^{-1}$, 840 $A \cdot h \cdot L^{-1}$). To overcome these restrictions, new concepts for the negative electrode must be developed, i.e., the Li/graphite intercalation reaction needs to be replaced by either alloying or conversion reactions with lithium. Previously, metal oxides, nitrides, sulfides, phosphides and fluorides were successively investigated as conversion-reaction materials for the negative electrodes of Li-ion batteries [1-4]. In 2008, metal hydrides were proposed for this purpose [5]. Compared to other conversion compounds MgH_2 exhibits remarkable properties such as the lowest polarization value for conversion electrodes (less than 0.2 V) at an average potential of 0.5 V vs Li^+ /Li^0 and a high reversible capacity (1480 $mA \cdot h \cdot g^{-1}$ which is four times that of Li/C electrodes). All these properties make MgH_2 suitable as a material for negative electrodes. MgH_2 reacts with lithium ions in a reversible lithium-driven conversion reaction generating lithium hydride and magnesium metal: $MgH_2 + 2Li^+ + 2e^- \rightleftharpoons Mg +$. Moreover, this conversion reaction is not restricted to MgH_2. It can also be carried out with several different binary and ternary hydrides. The general chemical reaction is then:

$$MH_x + xLi^+ + xe^- \rightleftharpoons M + xLiH.$$

The purpose of this review is to describe the properties of these metal hydrides properties in the reaction vs Li^+ / Li^0 (conversion reaction) with a focus on thermodynamics, involved reaction mechanisms, and some key issues to improve the performance of hydride-based electrodes.

CONVERSION REACTION OF HYDRIDES WITH LITHIUM IONS

Gravimetric and Volumetric Storage Capacity of Hydrides

Figure 1 shows both theoretical gravimetric and volumetric capacities of some binary and ternary hydrides. It can be noticed that the capacities of all hydrides are larger than that of graphite (370 $A \cdot h \cdot kg^{-1}$, 840 $A \cdot h \cdot L^{-1}$). Regarding ternary hydrides, the gravimetric capacities are between 340 $A \cdot h \cdot kg^{-1}$

Citation: Luc Aymard, Yassine Oumellal and Jean-Pierre Bonnet, Metal Hydrides: An Innovative and Challenging Conversion Reaction Anode for Lithium-Ion Batteries, doi:10.3762/bjnano.6.186.

($LaNi_4MnH_5$) and 750 A·h·kg^{-1} (ZrV2H4.9). The highest values are obtained with binary hydrides of light metals, namely 1074 A·h·kg^{-1} and 2038 A·h·kg^{-1} for TiH_2 and MgH_2, respectively. Volumetric capacities are above 2000 A·h·L^{-1} for all hydrides, for instance, 2298 A·h·L^{-1}, 2878 A·h·L^{-1}, 3815 A·h·L^{-1} for $LaNi_4MnH_5$, MgH_2 and TiH_2, respectively. Complex hydrides based on Mg follow this general trend of hydrides (i.e., capacities larger than graphite with Mg_2NiH_4: 963 A·h·kg^{-1}, 2822 A·h·L^{-1}; Mg_2CoH_5: 1191 A·h·kg^{-1}, 3200 A·h·L^{-1}; Mg_2FeH_6: 1456 A·h·kg^{-1}, 3995 A·h·L^{-1}). These large capacities render hydrides as good candidate material for negative electrodes in lithium-ion batteries for stationary as well as mobile applications for which the volumetric capacity plays a key role.

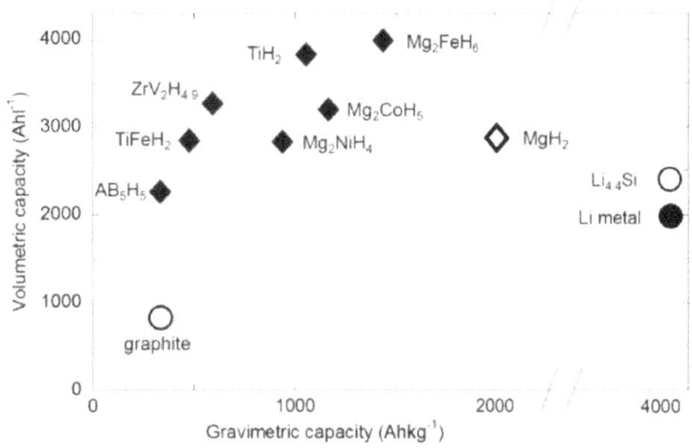

Figure 1: Gravimetric and theoretical volumetric capacities of metals and complex hydrides compared with those of graphite and other materials for negative electrodes. Reproduced with permission from [13]. Copyright 2013 Elsevier.

Thermodynamics of Hydrides

After experimental results on the conversion reaction with MgH_2 /Li [5,6] were reported, other systems that could be involved in the electrochemical conversion process were addressed from thermodynamic rules. The general reaction between hydride and lithium is given in Equation 3, where MH_x is the hydride, M the metal or the intermetallic phase, and x the number of hydrogen atoms. Under standard conditions (p = 1 atm and T = 298 K), the Gibbs free energy of the reaction in Equation 3, $\Delta_r G$ in kJ·mol^{-1}, can be calculated from the values of Gibbs free energy of formation of MH_x and LiH ($\Delta_f G°298$(LiH) and ΔfG°298(MH_x). It corresponds to the sum of reactions (Equation 1 + Equation 2) given as:

$$x\left(\mathrm{Li}+1/2\,\mathrm{H}_2 \to \mathrm{LiH}\right);\quad x\Delta_f G°_{298}(\mathrm{LiH}), \tag{1}$$

$$\mathrm{MH}_x \to \mathrm{M}+x/2\,\mathrm{H}_2;\quad -\Delta_f G°_{298}\left(\mathrm{MH}_x\right), \tag{2}$$

$$\mathrm{MH}_x + x\mathrm{Li} \to \mathrm{M}+x\mathrm{LiH};\quad \Delta_r G°_{298}. \tag{3}$$

The Gibbs free energy of the reaction in Equation 3 is:

$$\Delta_r G°_{298} = \left[\Delta_f G°_{298}(\mathrm{M}) + x\Delta_f G°_{298}(\mathrm{LiH})\right] - \left[\Delta_f G°_{298}(\mathrm{MH}_x) + x\Delta_f G°_{298}(\mathrm{Li})\right]$$

$$\Delta_r G_{298} = \Delta_r G°_{298} = x\Delta_f G°_{298}(\mathrm{LiH}) - \Delta_f G°_{298}(\mathrm{MH}_x).$$

According to the thermodynamics rules, the reaction is feasible for $\Delta_r G°_{298} \le 0$ and spontaneous for $\Delta_r G°_{298} < 0$.

Given the fact that the formation of LiH independent from the type of hydride MHx (common reaction product), another simple criterion to predict the possibility of the conversion process is the Gibbs free energy of formation of the hydride MHx divided by x. This value must be above that of LiH ($\Delta_f G°_{298}(\mathrm{MH}_x)/x > \Delta_f G°_{298}(\mathrm{LiH})$). The knowledge of reaction Gibbs free energy allows for the evaluation of the electromotive force (emf) and of the equilibrium potential of the cell, E, by using the Nernst law: $\Delta_r G° = -x \cdot E \cdot F$ (F: Faraday constant, x: number of electrons involved in the reaction). The equilibrium potential of the cell is deduced from the half reaction (Equation 4 and Equation 5) and the sum reaction (Equation 6).

$$\mathrm{MH}_x + x\mathrm{Li}^+ + xe^- \to \mathrm{M}^0 + x\mathrm{LiH};\quad \Delta G°_{\mathrm{I},298}, \tag{4}$$

$$x\mathrm{Li}^0 \to xe^- + x\mathrm{Li}^+;\quad \Delta G°_{\mathrm{II},298}, \tag{5}$$

$$\mathrm{MH}_x + x\mathrm{Li}^+ \to \mathrm{M}^0 + x\mathrm{LiH};\quad \Delta G°_{\mathrm{III},298}. \tag{6}$$

Hess's law gives:

$$\Delta G°_{\mathrm{III},298} = \Delta G°_{\mathrm{I},298} + \Delta G°_{\mathrm{II},298},$$

$$\Delta G°_{\mathrm{I},298} = -xE°_0\left(\mathrm{MH}_x/\mathrm{M}^0,\mathrm{LiH}\right)F,$$

$$\Delta G°_{\mathrm{II},298} = -xE°_0\left(\mathrm{Li}^+/\mathrm{Li}^0\right)F = 0,$$

the couple $\left(\mathrm{Li}^+/\mathrm{Li}^0\right)$ being used as reference,

$$\Delta G°_{\mathrm{III},298} = \Delta G°_{\mathrm{I},298} = \Delta_r G°_{298},$$

$$\Delta_r G°_{298} = -xEF = -xE°_0\left(\mathrm{MH}_x/\mathrm{M}^0,\mathrm{LiH}\right)F,$$

$$E°_0\left(\mathrm{MH}_x/\mathrm{M}^0,\mathrm{LiH}\right)F = -\Delta_r G°_{298}/xF.$$

With a lithium activity a(Li) = 0, and a lithium ion concentration [Li$^+$] = 1 M ([Li$^+$] inside the electrolyte), the equilibrium potential of the cell (E_{eq}) is:

$$E_{\mathrm{eq}}\left(\mathrm{MH}_x/\mathrm{M}^0,\mathrm{LiH}\right) = E°_0\left(\mathrm{MH}_x/\mathrm{M}^0,\mathrm{LiH}\right)$$

$$E_{\mathrm{eq}}\left(\mathrm{MH}_x/\mathrm{M}^0,\mathrm{LiH}\right) = -\Delta_r G°_{298}/xF.$$

This calculation was applied for different binary and ternary hydrides used to represent the different intermetallic families. The HSC database [7] and literature data [8] were used to obtain the $\Delta_f H$, $\Delta_f G$ and $\Delta_f S$ values.

Binary Hydrides

Figure 2a shows the equilibrium potential of the MHx /Li cell for different binary hydrides M = Y, La, Ba, Ca, Zr, Ti, Na, Cs, Mg. The potential versus Li^+ /Li^0 is positive for Zr, Ti, Na, Cs and Mg and negative for Y, La, Ba and Ca hydrides. It can be clearly concluded from this figure that the conversion reaction is favorable, from a thermodynamics point of view, for Zr, Ti, Na, Cs and Mg hydrides and not possible for Y, La, Ba, Ca. The Gibbs free enthalpy of forma-

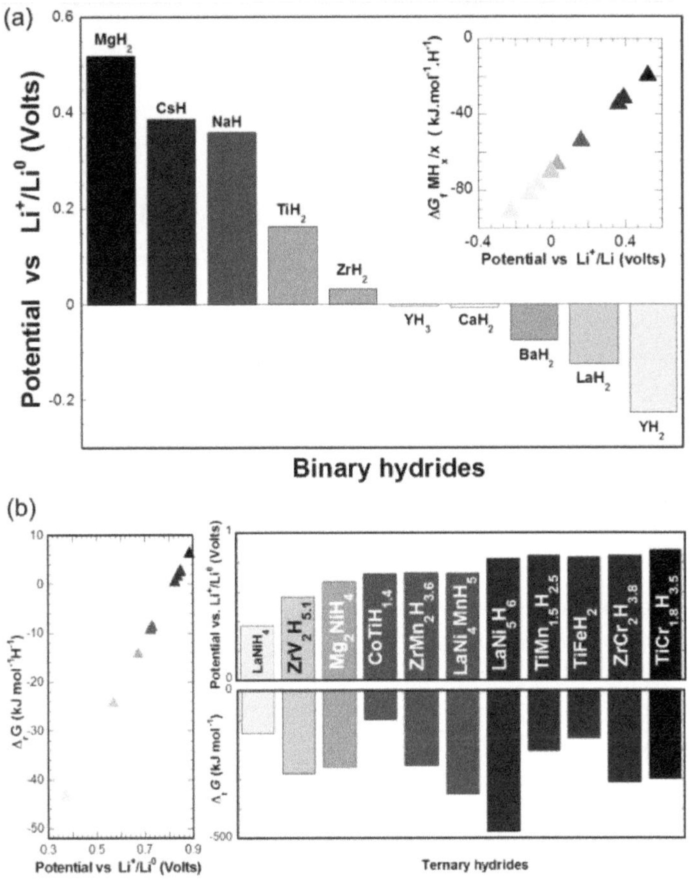

Figure 2: Theoretical equilibrium potential for the MHx/Li cell vs Li^+ /Li^0. a) For binary hydrides M = Y, La, Ba, Ca, Zr, Ti, Na, Cs, Mg. Inset: Gibbs free formation enthalpy of these hydrides as a function of the equilibrium potential. b) For ternary hydrides Mg_2NiH_4, $LaNiH_4$, $CoTiH_{1.4}$, $TiFeH_2$, $ZrV_2H_{5.1}$, $ZrCr_2H_{3.8}$, $ZrMn_2H_{3.6}$, $TiMn_{1.5}H_{2.5}$, $TiCr_{1.8}H_{3.5}$, $LaNi_5H_6$, $LaNi_4MnH_5$ [11]. tion (divided by the amount of substance of hydrogen) of these hydrides is smaller than that of LiH (inset Figure 2a).

Ternary hydrides

In the case of ternary hydrides, equilibrium potentials of the MHx /Li cells were evaluated for different representative compounds ABx of intermetallic families (with x = 0.5, 1, 2 and 5) and their corresponding hydrides (Mg_2NiH_4, $LaNiH_4$, $CoTiH_{1.4}$, $TiFeH_2$, $ZrV_2H_{5.1}$, $ZrCr_2H_{3.8}$, $ZrMn_2H_{3.6}$, $TiMn_{1.5}H_{2.5}$, $TiCr_{1.8}H_{3.5}$, $LaNi_5H_6$, $LaNi_4MnH_5$). . Figure 2b (lower part) shows that the Gibbs free enthalpy of formation value for all these ternary hydrides (per mole hydrogen) is above that of LiH. For these hydrides the equilibrium potential of the cell is positive (Figure 2b) and the conversion reaction can be achieved. Equilibrium potentials are in the range of 0.3–1.0 V vs Li^+ /Li^0 , which is suitable for a negative electrode in Li-ion batteries. The equilibrium potential of the cell can be adjusted for different ABx intermetallic families by varying the site substitutions of A and B [8]. In fact, the plateau pressure of hydride correlates with the lattice cell volume, allows one to change the thermodynamic stability of the hydrides, especially for the families AB_5 and AB_2. Depending of the nature of the hydrides a wide range of hydrogen sorption temperatures from −40 to 300°C gives another argument to a tailor a negative electrode for the desired applications [9].

Comparison between Thermodynamic Prediction and Experience

Figure 3 shows the experimentally measured potential vs Li^+ /Li^0 for electrochemical MH_x /Li cells using binary and ternary hydrides. The potential–capacity curves recorded are in agreement with the assumption based on thermodynamics that the conversion reaction is possible for MgH_2, TiH_2, NaH and ternary hydrides. These discharge curves correspond to the theoretical reaction $MH_x + xLi^+ + xe^- \rightarrow M^0 + xLiH$ and their lengths are in agreement with the number of hydrogen atoms that react with lithium (Figure 3a). For instance the discharge curves of MgH_2, TiH_2 [10], NaH involve two and one lithium respectively for two and one hydrogen [11]. Values superior to the number of hydrogen atoms x can, however, be reached in relation with either a plateau corresponding to the electrolyte decomposition on carbon at 0.8 V or with a metal alloying reaction at low potentials, especially for Mg (0.17 V vs Li^+ /Li^0). The potential–capacities curves are lower than the theoretical equilibrium potential due to internal resistance of the cell and are also not totally flat due to kinetic limitations of the system. The equilibrium potential of the MH_x /Li cells can be obtained by galvanostatic intermittent titration (GITT) in open circuit voltage with, for instance, an experimental value of 0.537 V for the MgH_2 /Li cell [11], which is in good agreement with the theoretical value of 0.560 V versus Li^+ /Li^0 obtained from Nernst law.

Regarding AB_x intermetallic compounds, typical discharge curves obtained from ternary hydrides $LaNi_4MnH_5$, TiNiH [12], $Ti_{0.20}V_{0.78}Fe_{0.02}H_{1.55}$ and $Mg_{0.65}Sc_{0.35}H_{2.25}$ are presented in Figure 3b. Lengths of discharge curves of x = 4.8, 1.2, 1.3, and 2.2 are recorded for LaNi4MnH5, TiNiH, bcc $Ti_{0.20}V_{0.78}Fe_{0.02}H_{1.55}$ and $Mg_{0.65}Sc_{0.35}H_{2.25}$ hydrides, respectively. It must be noted that a nice flat plateau is obtained for the AB5 compound.

Conversion reactions with lithium ions were also carried out with different complex hydrides based on Mg or Al, especially Mg_2FeH_6, Mg_2CoH_5 and Mg_2NiH_4. These complex hydrides were prepared by reactive grinding [13,14]. They react with lithium ions at average potentials of 0.25, 0.24 and 0.27 V and give discharge capacities of 6.6, 5.5 and 3.6 Li, respectively. Using AlH_3 [15], Li_3AlH_6 [16] or more recently $LiAlH_4$, $NaAlH_4$ and Na_3AlH_6 [17,18] as negative electrode of Li-ion batteries was also reported. It demonstrates the possibility to extend the conversion process to numerous versatile complex hydrides. For these last cases, the discharge curves involve conversion process and alloying reaction, in relation to the close potential of both reaction types.

CONVERSION PROCESS REACTION MECHANISMS FOR HYDRIDES

Reactivity of hydrides with lithium ions predicted from thermodynamic rules and experimentally confirmed for different hydrides is assumed to be a conversion reaction as $MH_x + xLi \rightarrow M + xLiH$.

While this general mechanism is obvious, more complex reactions path involving the formation of alloys, solid solutions, metastable or amorphous phases can also be noticed. In the following paragraph reaction mechanisms occurring with MgH_2 [5,11], TiH_2 [10], $Mg_{0.85}Sc_{0.65}H_2$ and Mg_2TMH_x (TM = Fe, Co, Ni, x = 6, 5 , 4) [14] hydrides will be described.

Reaction of MgH_2 with lithium

The reaction of Mg hydride with lithium ions is the first example reported in the literature of a Li-driven conversion reaction with hydrides [5,11]. The electrochemical curve recorded at a low cycling rate (one equivalent of lithium in 100 h) during the reaction of MgH_2 with Li (inset of Figure 4a) shows that the full discharge (length x = 2.5 Li) involves two plateaus at 0.44 V and 0.095 V. The XRD patterns, collected at different discharge steps, are presented in Figure 4a. The XRD patterns corresponding to the first plateau (until x = 1.8 Li) show a decrease of the intensity of MgH_2 (tetragonal and orthorhombic phases) Bragg peaks and the appearance of hcp Mg and bcc Li peaks ((100, 002, 101) and (111, 200), respectively). Above x = 1.8 (Figure 4b) the first slope observed in the discharge curve corresponds to a shift of the Mg XRD lines in agreement with the formation of a Mg-type solid solution (Figure 4b, no. 6 and Table 1). The last slope corresponds to the formation of a bcc Li-type solid solution (Figure 4b, no. 8), and the plateau to the coexistence of both solid solution types (Li and Mg) (Figure 4b, no. 7). In short, MgH_2 reacts with Li ions to form Mg and LiH within the conversion process $MgH_2 + 2Li^+ + 2e^- \rightarrow Mg + 2LiH$ at around 0.44 V vs Li^+/Li^0 . Then the freshly formed Mg can react with Li ions at a low potential to form alloys (hcp Mg-type and bcc Li-type solid solutions).

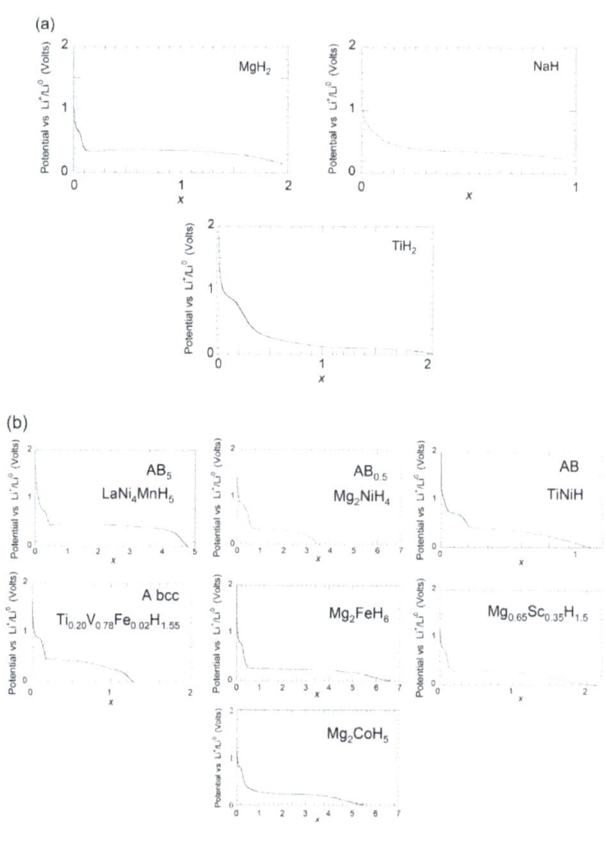

Figure 3: Potentials vs Li^+/Li^0 of MH_x/Li cells (V) as a function of the mole fraction of Li (x) recorded between 3 and 0.005 V; (a) MgH_2, NaH and TiH_2, (b) $LaNi_4MnH_5$, $Ti_{0.20}V_{0.78}Fe_{0.02}H_{1.55}$, Mg_2NiH_4, Mg_2FeH_6, Mg_2CoH_5, TiNiH, and $Mg_{0.65}Sc_{0.35}H_{2.25}$ [11, 79].

Li-Mg alloying reactions can be avoided by limiting the discharge curve to x = 2 (Figure 5). In this case a reversible capacity of 1500 mA·h·g^{-1} (irreversible loss of 25%) can be obtained while a reversible capacity of 2700 mA·h·g−1 (irreversible loss 33%) is measured for both processes (Figure 6).

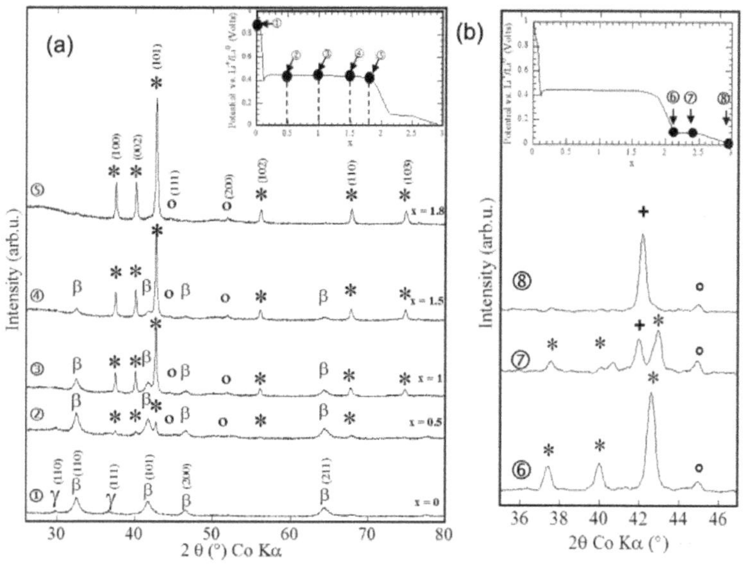

Figure 4: Potential profile and XRD patterns of MgH_2 electrode at different stages of the conversion reaction. (a) Inset: Evolution of the potential (V) as a function of x (mole fraction of Li) for a MgH_2 electrode cycled between 3 and 0.005 V at a rate of one equivalent of lithium in 100 h. The recorded XRD patterns are associated with the various stages of the discharge, denoted by the numbers (1), (2), (3), (4) and (5) in panel and correspond to x = 0 (starting electrode material), x = 0.5, 1, 1.5 and 1.8, respectively. The X-ray peaks marked by an asterisk, β, γ and circle correspond to Mg, β-MgH_2, γ-MgH_2 and LiH, respectively. b) Inset: Inset: Evolution of the potential (V) as a function of x (mole fraction of Li) for a MgH_2 electrode cycled between 3 and 0.005 V at a rate of one equivalent of lithium in 100 h. The XRD patterns (6), (7), and (8) corresponding to x = 2.1, 2.35 and 2.9. The Bragg peaks marked by an asterisk and plus sign correspond to Mg (hcp) and Li (bcc) solid solutions, respectively. Adapted from [5] (copyright 2008 Nature Publishing Group) and [11].

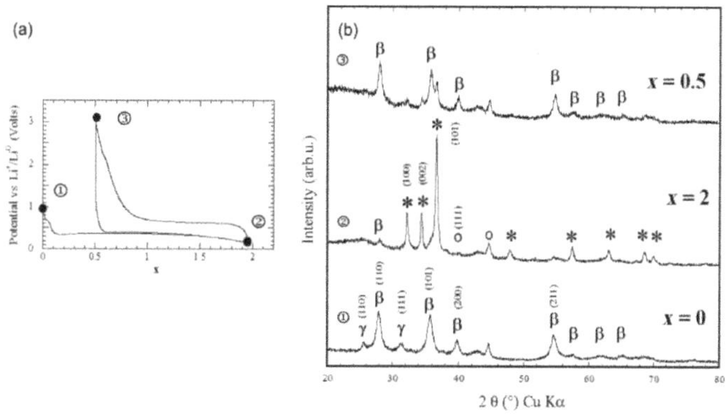

Figure 5: Potential profile and XRD patterns of MgH_2 electrode at various stages of the conversion reaction. a) Evolution of the potential (V) as a function of x for a Li/MgH_2 cell that was cycled down to x = 2 at a rate of one equivalent of lithium in 10 h. b) XRD patterns taken at various stages of the discharge, (1) x = 0 (starting electrode material), (2) x = 2 (end of the first discharge) and (3) x = 0.5 (end of the first charge). The Bragg peaks marked by an asterisk, β, γ, and circle correspond to Mg, β-MgH_2, γ-MgH_2 and LiH, respectively. Adapted from [5] (copyright 2008 Nature Publishing Group) and [11].

Table 1: Lattice parameters for Mg and Mg-type solid solution formed during the reaction of MgH_2 with Li.

	Mg	Mg hcp solid solution
lattice parameters	*a* = 3.2090(3) Å *c* = 5.2100(4) Å	*a* = 3.1970(2) Å *c* = 5.1410(6) Å

Figure 6: Potential profile of a MgH_2 electrode at various stages of the conversion reaction. a) Evolution of the potential (V) as a function of x for a Li/MgH_2 cell that was cycled down to x = 2 at a rate of one equivalent of lithium in 10 h. b) Evolution of the potential (V) as a function of x for a Li/MgH_2 cell that was cycled down to x = 4 at a rate of one equivalent of lithium in 10 h. Adapted from [5] (copyright 2008 Nature Publishing Group) and [11].

Reaction of TiH_2 with Lithium

The study of the reactions of titanium hydride with lithium is motivated by the chemical and structural properties of TiH_2 [10]. As shown on Figure 1, TiH_2 is attractive regarding its high theoretical capacities (like all binary hydrides). In addition, an improvement of the conversion process kinetics is expected because of the good electrode conductivity due to the metallic properties of titanium hydride. This reaction can be studied without any parasite reaction as Ti does not form alloys with lithium.

The TiH_2 discharge capacity, presented in Figure 7, drastically differs from that of MgH_2 through the presence of two slopes prior to a pseudo plateau. XRD analyses of the electrode during electrochemical discharge show that the reaction of TiH_2 with Li involves three steps. The two first slopes correspond to the formation of an fcc δ-TiH_{2-x} solid solution until x = 0.34 (first slope) that transforms partially from x = 0.34 to 1 in a distorted face-centered orthorhombic phase δ-TiH (fco) (second slope). Note that from x = 0.34 to 1, the peritectic transformation: hcp α-Ti(H) → fcc δ-TiH_{2-x} + δ-TiH explains hexagonal close packed (hcp) α-formation absence. Finally, conversion process occurs in the pseudo plateau with the formation of hcp α-Ti and LiH

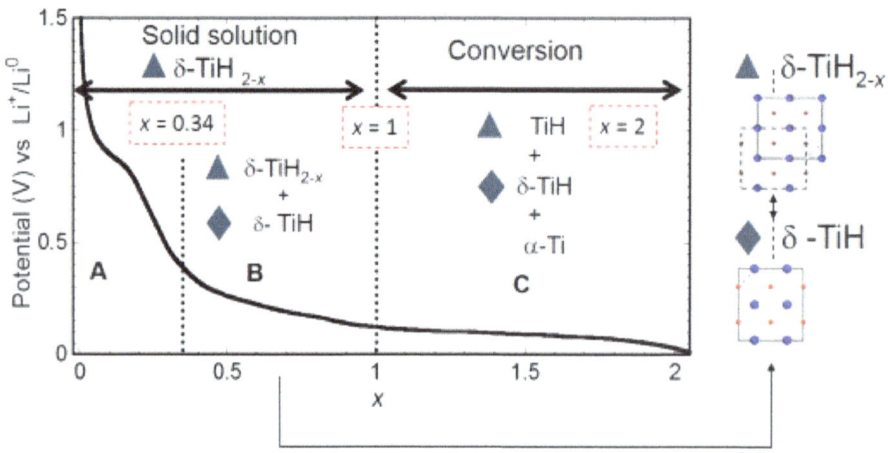

Figure 7: Summary of the dehydrogenation process of TiH_2 electrode δ-TiH_{2-x} (fcc): black triangles, TiH (fco): black diamonds, projection along the b-axis of δ-TiH_{2-x} (fcc): black triangles and TiH (fco): black diamonds showing the relationship between fcc (black triangles) and fco (black triangles) structures. Adapted from [10] (copyright 2012 Elsevier) and [11].

Reaction of Mg_2MHx with lithium

After studying the reaction of titanium hydride with lithium, during which a reaction path involving the formation of the metastable fco δ-TiH phase occurs, the complex hydrides Mg_2FeH_6, Mg_2CoH_5, Mg_2NiH_4 were chosen as models system for a conversion process with high energy storage capacities and unusual thermodynamics properties [13,19]. In fact, the decomposition of Mg_2FeH_6 and Mg_2CoH_5, which is expected during their electrochemical reaction with lithium, can be used for the formation of a conductive Fe or Co matrix, which is helpful to reverse the reaction between Mg and LiH. In addition, the far-from-equilibrium electrochemical process is an interesting tool to search for new intermetallic compounds consisting of Mg and Fe or Co [20, 21]. As shown in Section I.3, Mg_2MH_x (M = Fe, Co, Ni , x = 6,5,4) react with lithium at roughly the same potential (around 0.25 V vs Li^+ /Li^0) and the capacities measured during the discharge are close to the theoretically obtained values (Figure 8). From a structural point of view, a common behavior can be noticed for the reaction of lithium with all complex hydrides from the XRD characterizations of the electrodes. A complete (for the case of Mg_2FeH_6) or partial disappearance of the parent phases is observed, which occurs without any formation of metals (Figure 9). This loss of crystallinity suggests the formation of an electrode with nanocrystalline or amorphous structure. The formation of nanoscale Fe during the decomposition of Mg_2FeH_6 is confirmed by XAS and Mössbauer spectroscopy [22]. Ex situ XAS spectroscopy of the Mg_2CoH_5 and Mg_2NiH_4 electrodes revealed the formation of disordered MgCo and Mg_2Ni intermetallic compounds.

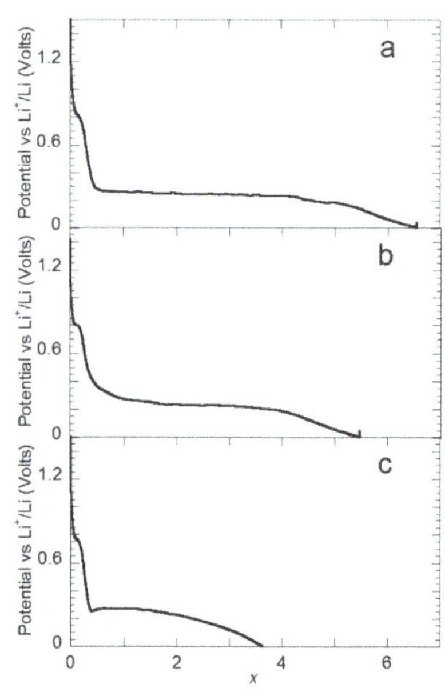

Figure 8: Discharge curves for a) Mg_2FeH_6, b) Mg_2CoH_5 and c) Mg_2NiH_4 electrodes prepared by reactive grinding as a function of Li (x) mole fraction, recorded at a current rate of one equivalent of Li in 10 h. Adapted from [13] (copyright 2013 Elsevier) and [19].

The intensity reduction of the XRD lines, which occurs without broadening, involves shifts of the lattice parameters. For instance, theMg2CoH5 lattice parameters a and c rise until x = 1 (from 4.4940(3) to 4.517(2) Å and from 6.582(1) to 6.608(1) Å, respectively; a cell volume expansion of 1.44%), evoking a phase transformation. For Mg2NiH4 the lattice parameter c decreases from 6.538(1) to 6.477(2) Å, which is concomitant with a phase transformation involving a low-hydrogen-content hydride for $x \leq 3$: $Mg_2NiH_4 \rightarrow Mg_2NiH$ [23,24].

The reaction mechanism can therefore be summarized as follows:

$$Mg_2FeH_6 + 6Li \rightarrow 2Mg + \text{amorphous Fe} + 6LiH\ ,$$

$$Mg_2CoH_5 + 5Li \rightarrow Mg + \text{disordered MgCo} + 5LiH\ ,$$

$$Mg_2NiH_4 + 3Li \rightarrow Mg_2NiH + 3LiH + Li$$

$$\rightarrow \text{disordered } Mg_2Ni\ + 4LiH\ .$$

The reaction with Mg_2FeH_6 during the conversion process with lithium ions is the first example for the production of an amorphous phase "2Mg + Fe".

Reaction of other Mg-based Hydrides with Lithium

The reactions of $2MgH_2$ + M (M = Cu, Si) and 0.65Mg + 0.35M (M = ScH_2, Ti) mixtures prepared by reactive grinding under 90 bar of hydrogen pressure with lithium ions were also studied [19, 25-27]. The electrochemical behavior of MgH_2 is not affected by the presence of a second element, Cu or Si, and significant reversible capacities for the conversion process (>1000 $mA{\cdot}h{\cdot}g^{-1}$) are obtained. In the case of the mixture $2MgH_2$ + Si, an additional capacity below 0.2 V vs Li^+ /Li^0 due to the alloying reaction of Si with Li is observed. This combined conversion (MgH_2)/alloying (Si) system presents the highest theoretical capacity anode with the possibility to reach approximately 6000 $mA{\cdot}h{\cdot}g^{-1}$

Theproductionofpure68%of$Mg_{0.65}Sc_{0.35}H_{2.25}$ (1900 $mA{\cdot}h{\cdot}g^{-1}$) by reactive grinding from magnesium and scandium hydride (MgH, + ScH,) is not possible. Instead, a mixture of $Mg_{0.65}Sc_{0.35}H_{2.25}$, 20% of MgH_2 and 12% Mg_2FeH_6 is obtained. The formation of Mg_2FeH_6 is due to the strong abrasion of the grinding tools through scandium hydride. The discharge curves of this hydride mixture involves 2.25Li at 0.32 V vs Li^+ /Li^0 and the corresponding X-ray diffraction pattern obtained at the end of the discharge shows the presence of both $Mg_{0.65}Sc_{0.35}H_{0.8}$ and $Mg_{0.65}S_{0.35}H_{1.5}$ cubic phases in agreement with the results obtained in KOH electrolyte and reported by Notten et al. [28].

Such desirable electrochemical behavior is also obtained with the $Mg_{0.65}Ti_{0.35}H_{2.25}$ mixture prepared by reactive grinding with a reversible conversion process involving both MgH_2 and TiH_2 without the addition of carbon. As expected, TiH_2 increases the conductivity of the electrode and a full discharge process is obtained without carbon. The development of such Mg-based system appears to be a promising opportunity

KINETICS LIMITATIONS OF HYDRIDES FOR CONVERSION REACTIONS: EXAMPLE OF MGH_2

The huge numbers of hydrides reported in the literature exhibit different structures, electronic properties and thermodynamics stabilities. Using the hydrides classification proposed by Libowitz et al. [29,30] metallic, covalent and ionic hydrides can be identified.

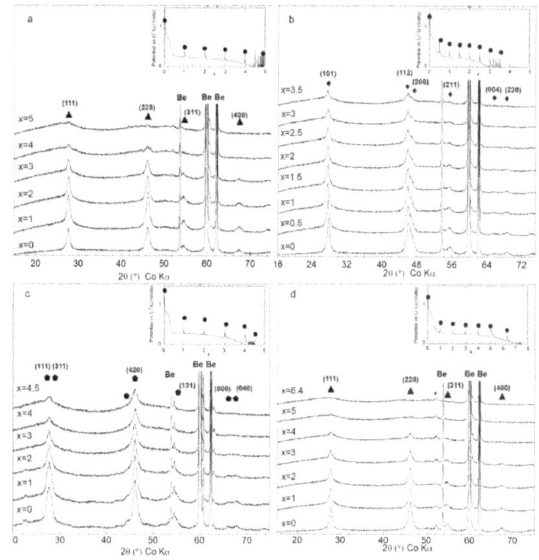

Figure 9: In situ XRD patterns of a) Mg_2FeH_6, b) Mg_2CoH_5 and c) Mg_2NiH_4 and d) Mg_2FeH_6-10% Ct, z electrodes prepared by reactive grinding as a function of the mole fraction of Li (x) Inset: Potential profile of a) b) c) and d) obtained during GITT (rate: one equivalent of Li in 10 h, relaxation time: 10 h). Adapted from [13] (copyright 2013 Elsevier) and [19].

Given the fact that the electric behavior is an important parameter for the electrochemical reaction, the issue of the poor electric conductivity of ionic and iono-covalent hydrides must be solved. For instance, the ionic hydride MgH_2 exist as α, β and γ, with the space groups P42 /mnm (tetragonal, a = b = 4.516 Å, c = 3.020 Å) Pa–3 (cubic) and Pbcn (orthorhombic, a = 4.526 Å, b = 5.448 Å, c = 4.936 Å), respectively. The tetragonal phase is the more stable phase. These hydrides exhibit band gap energies of 5.3, 5.6 and 4.2 eV, respectively, and are not electronic conductors but insulators [31].

Figure 10 shows the poor electrochemical reactivity of commercially available tetragonal MgH_2 vs Li^+ /Li^0, with no electro chemical capacity during the first discharge. The addition of an electronic conductive material, such as the graphite Super P (electronic conductivity: 10^3 $S{\cdot}cm^{-1}$) increases the electronic conductivity of the electrode. As presented in Figure 10a, the discharge capacity of the hydride increases with the amount of graphite. The addition of 25% of graphite gives a discharge capacity of 1.4Li for 2 h for a current of one mole of electron in 100 h. The contribution of Super P carbon to the total capacity remains very small and the maximum of the contribution that can be reached is less or equal to x = 0.25.

Note that the conductivity of the electrode is not only influenced by MgH_2 but also by conducting metallic Mg and insulating LiH, which are generated during the reaction. The improvement of the poor electronic conductivity of the active material in powder form has been widely addressed in the literature for different electrode materials (in aqueous and nonaqueous electrolytes). A solution is carbon-coating through chemical or physical methods. For electrochemical reactions carried out in thin films in KOH electrolyte with non-conductive hydrides such as MgH_2, or Mg_2NiH_4, the presence of a small amount of non-hydrogenated compound (few percent), as Mg or Mg_2Ni, in the starting material is sufficient to produce a satisfactory electronic conductivity inside the electrode and the addition of carbon is not necessary. Another issue consists in the production of ternary hydrides films with Mg, Ti and H [32,33]. The metallic behavior TiH_2 counters the insulating influence of MgH_2.

Influence of the Particle size on the Reversibility of the Conversion Process

Regarding the reversibility of the conversion reaction, the poor capacities obtained during the first charge for electrodes composed of commercial MgH_2 (0Li and 0.25Li for a current rate of one equivalent of Li in 10 and 100 h, respectively) clearly show than the benefit of the addition of Super P carbon to the electronic conductivity of the active material seems completely lost when the reaction is inversed (Figure 10b,c).

Other parameters than the conductivity of the active material also govern the efficiency of the conversion process. In fact, the volume variation of the electrode during the conversion process $MH_x + xLi^+ + xe^- \rightarrow M^0 + xLiH$ drastically affects the conductivity between the particles. For MgH_2 the volume variation is 83% between MgH2 (61.59 Å^3) and Mg (46.46 Å^3) + 2LiH (2 × 33.3 Å^3). Thus, during the discharge the electrode volume increases with the lithium transfer and decreases during the lithium extraction. As a consequence, voids are created inside the electrode and disconnect the particles from each other and from the current collector at the same time. Decreasing the particle size is one way to accommodate for the volume variation of the electrode and to maintain cohesion of the interfaces and the connection between particles and the current collector. Reducing the length of diffusion for Li and H can be helpful for volume accommodation and preserving interfaces. Different approaches to reduce the particles sizes and to accommodate the volume variation of the electrode can be considered and will be presented in the following paragraph: Effect of grinding on commercial hydride, effect of three hydrogen sorption, effect of grinding commercial hydride with carbon, effect of activation by sorption cycles follow by the grinding with carbon, reactive milling.

Effect of Mechanical Grinding on Commercial MgH_2

Grinding of commercial MgH_2 enables a faster hydrogen desorption with a desorption maximum at 372 °C compared to 445 °C for the untreated commercial material (Figure 11a). As expected, discharge efficiency increases with the amount of Super P carbon added to the electrode (Figure 11b). In this case, a full discharge capacity of x = 1.95Li can be achieved when 25% of graphite is added. The improved performance of the ground sample is due to a reduction of the crystallite size down to few nanometers, which facilitates the diffusion of hydrogen and lithium by increasing the number of diffusion paths. However, despite an improvement of the discharge kinetic, the reversibility of the hydride is hardly changed (Figure 12a). The reversible capacity of ground commercial MgH_2 after 30 h of grinding is actually x = 0.26Li.

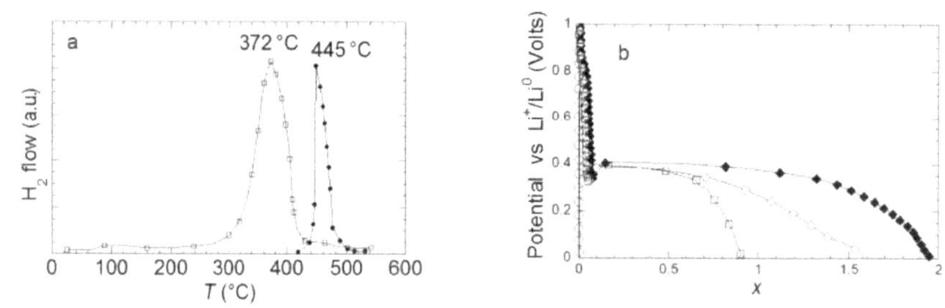

Figure 11: a) DSC traces of commercial MgH_2 unground and ground for 20 h. b) Evolution of the potential (V) as a function of x (mole fraction of Li) for a MgH_2 electrode prepared with commercial hydride ground for 20 h in the Spex mixer mill and cycled between 3 and 0.005 V at a rate of one equivalent of lithium in 100 h. Super P carbon content (%) of the electrode: open circles: 0%, open squares: 15%, open diamonds: 20%, black diamonds: 25% [11].

The particle size cannot be reduced below 0.1μm through grinding, because immediate agglomeration of smaller particles occurs (Figure 12b,c). So, even if crystallite size of few nanometers can be reached during grinding, the formation of agglomerates of 5 to 30μm (consisting of primary particles of 0.1–5μm, Figure 12b) limits the reversibility of the conversion process.

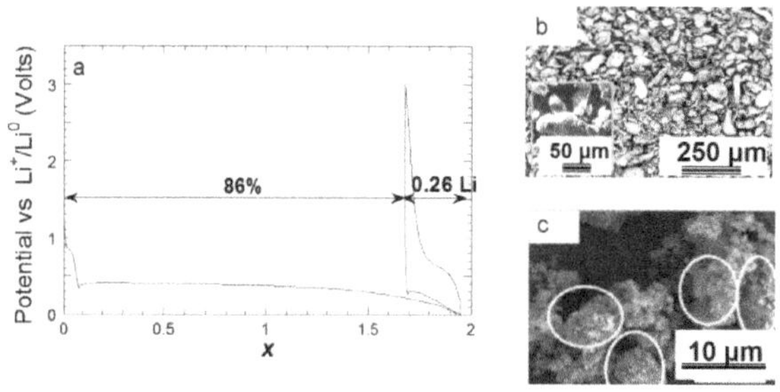

Figure 12: a) Evolution of the potential (V) as a function of x (mole fraction of Li) for a MgH_2 electrode prepared with commercial hydride, ground 30 h in a Spex mixer mill and cycled between 3 and 0.005 V at a rate of one equivalent of lithium in 100 h. b) SEM image of a MgH_2 commercial powder ground 30 h in the Spex mixer mill. c) SEM image of MgH_2 commercial powder ground 30 h in the Spex mixer mill showing the agglomeration of particles [11].

Effect of Hydrogen Sorption Cycles on MgH_2

The particle size of the hydride is also reduced through decrepitation during hydrogen desorption–absorption cycles. This solid–gas reaction not only reduces the particle size of the hydride but also enhances its reactivity vs Li-ions. In the case of the system Mg/MgH_2, 5% of Super P carbon was mixed with the Mg powder in order to increase the thermal conductivity of the powder and to prevent the necking of particles during sorption cycles.

Figure 13a and Figure 13b show the hydrogen sorption kinetics at 350 °C for Mg/MgH_2 for the first and the third cycles (uptake and loss in wt % hydrogen). Hydrogen sorption kinetics and capacities increase from the first to the third cycle and then are constant for the subsequent cycles (not shown here). This activation process observed during the first three cycles can be correlated with a particle size reduction (Figure 13c and Figure 13d) and an increase of the BET surface area from 7 to 14 $m^2\ g^{-1}$ while no change of the hydride crystallinity is observed (Figure 13e).

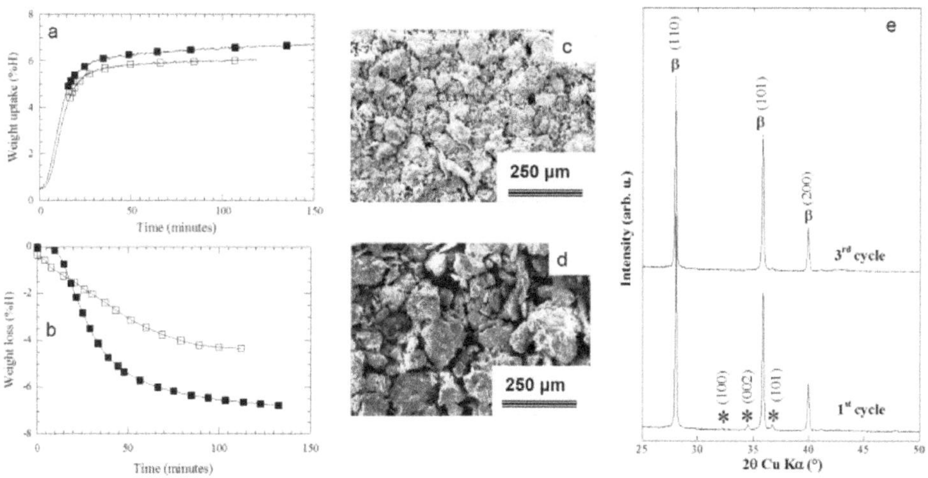

Figure 13: a) Absorption and b) desorption kinetics at 350 °C for Mg–10%C10,320 composite (open squares): first cycle, black squares: third cycle; c) SEM image of the MgH_2 powder after one absorption; d) SEM image of the MgH_2 powder after three absorptions; e) XRD patterns of the MgH_2 powder after one and after three absorptions [11].

The effect of three sorption cycles on the reaction of MgH_2 with lithium (Figure 14) shows that reversible capacity drastically increases compared to that of ground commercial hydride and reaches 0.88Li after three sorption cycles (Figure 14c). Despite an attractive increase of reversible capacity, an irreversible loss (48%) can be noticed on potential–capacity curves.

Effect of Grinding of MgH_2 with Carbon

Grinding of commercial MgH_2 with a pre-ground $C_{t,z}$ carbon, where t refers to the pre-grinding time and z to the carbon BET surface area, was used to enhance the efficiency of the conversion reaction. Grinding of commercial MgH_2 with Ct, z carbon is supposed to create of a porosity volume inside the electrode, corresponding to a volume increase due to matter transfer. This porous volume is then recovered during the lithium extraction and the total volume change of the electrode is then minimized. In addition, the carbon also acts as conductive additive and a coating agent, which prevents the agglomeration of the hydride particles during grinding. A detailed study of the effect of mechanical milling on the physical/chemical and electrochemical properties compared to AB_5 alloys is available in [34].

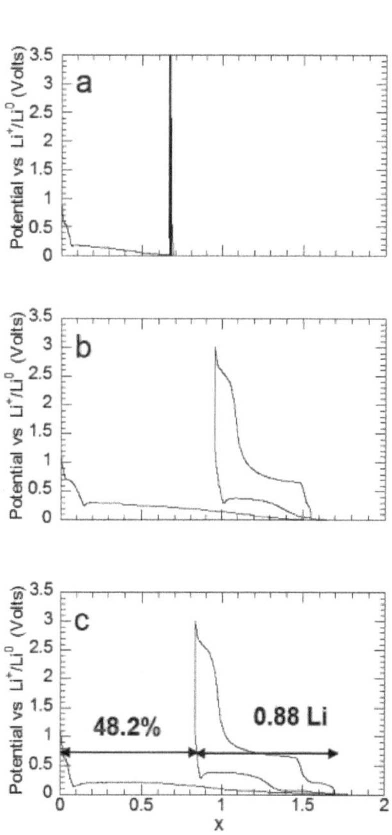

Figure 14: Evolution of the potential (V) as a function of x (mole fraction of Li) for MgH_2 electrodes cycled between 3 and 0.005 V at a rate of one equivalent of lithium in 10 h. a) MgH_2 commercial hydride; b) MgH_2 obtained after one absorption of hydrogen; c) MgH_2 obtained after three absorptions of hydrogen [11].

Figure 15 shows the evolution of the BET surface z (in $m^2\ g^{-1}$) and of the d(002) interplanar spacing of carbon $C_{t,z}$, as a function of the milling time t (in h). Two main grinding steps can be noticed in Figure 15. First the BET surface area increase while d(002) remains almost constant (step A) and then the BET surface area decrease and d(002) drastically increases (step B). Firstly, The mechanical energy transferred to the carbon produces an exfoliation of the graphene layer. Then, the cumulated mechanical energy coming from the grinding is sufficient to promote fissure propagation within the graphene layer, resulting in the fracture of the C–C covalent bonds, leading to the formation of very reactive edge carbon atoms and unstable particles which agglomerate together. The degree of disorder for carbonaceous materials increases with increased milling time and is proportional to the d(002) distance, as previously established [35,36]. Note that the C-free bonds created during the fracture of the graphene layer serve as oxygen scavengers, and their agglomeration and coating of the alloy particles enable a better chemical/physical protection against oxidation [34].

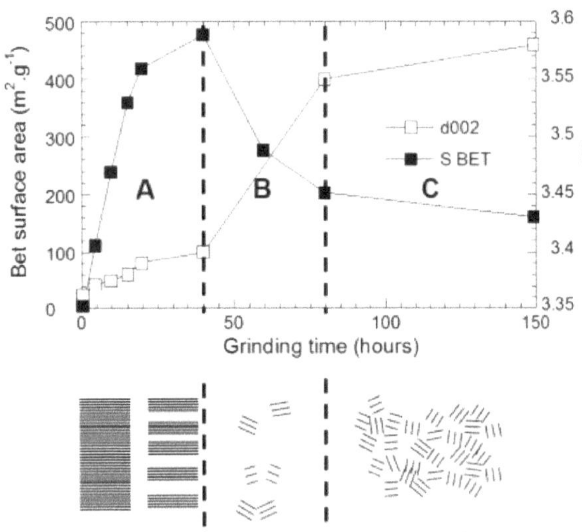

Figure 15: Graphite BET surface area ($m^2\ g^{-1}$) and d(002) interlayer distance (Å) as a function of grinding time (h) [11].

Based on the milling behavior of carbonaceous material [34,35], MgH_2 is ground using a carbon having the maximum BET surface area in order to agglomerate carbon particles on MgH_2 particles. DSC traces of MgH_2–10% $C_{t,z}$ composite obtained after 4 h of grinding shows a decrease of 48°C of the desorption peak maximum of hydride carbon composite compared to the commercial hydride, as expected (Figure 16).

The dispersion of the hydride particles into carbon increases the thermal conductivity of the powder and helps the hydrogen release. With regard to the electrochemical properties, the potential–capacity curves of an electrode composite of MgH_2–10% $C_{t,z}$ obtained after 4 h of grinding shows a spectacular enhancement of the reversible capacity with 0.96Li (= 1000 mA·h·g_{-1}) for an irreversible loss of 48% (Figure 17b).

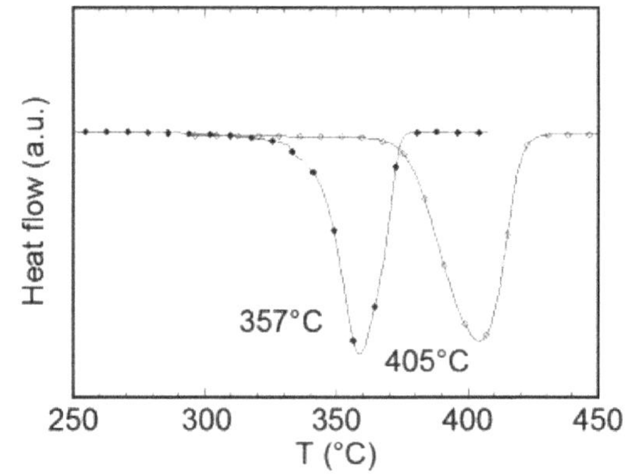

Figure 16: Thermodesorption of commercial MgH_2 (black diamonds) and commercial MgH_2 ground 4 h with 10% of $C_{t,z}$ carbon (open diamonds) [11].

The grinding of MgH_2 with carbon was also carried out with MgH_2 that was activated through three sorption cycles. A synergic effect, involving both the hydride activation by solid–gas reaction and grinding with a $C_{t,z}$ carbon to enhance the reversibility was noticed (Figure 17c,d). For instance, a capacity of 1500 mA·h·g^{-1} for an irreversible loss of 35% after three sorption cycles and 4 h of grinding with 10% of $C_{t,z}$ carbon was obtained (Figure 17d). This reactivity enhancement enables to obtain interesting reversibilities, free of alloying reaction (with a cut of voltage of 0.15 V vs Li^+ /Li^0). When the grinding time of the activated MgH_2 + $C_{t,z}$ mixture varied from 4 to 6 h, a reversible capacity of 1480 mA·h·g^{-1} for an irreversible loss of 25% is thus obtained at 0.15 V vs Li^+ /Li^0 (Figure 18).

Effect of Reactive Milling under Hydrogen

Reactive milling under hydrogen constitutes a powerful method for the synthesis of hydrides with the advantage to grind and to hydrogenate the sample in one single step. Applied to MgH_2, a subsequent grinding step with carbon $C_{t,z}$ is needed to obtain similar electrochemical performances than for the sample prepared by three hydrogen sorption steps followed by grinding with the carbon $C_{t,z}$ (1600–1700 mA·h·g^{-1} at 0.005 V vs Li^+ /Li^0 for 6 h of grinding with carbon).

Effect of Metal Catalyst Addition to the MgH_2 Carbon Composite

The improvement of the sorption kinetics of MgH_2 through catalyst addition (i.e., transition metals [37, 38] transition metal oxides [39-41] and halides [42]) has been widely studied in the literature. Nb_2O_5 is one of the most efficient catalysts [43] enabling fast hydrogen desorption kinetics with 7.6 wt % of hydrogen desorbed in 100 s at 300°C. To improve the conversion reaction of MgH_2 with lithium, the addition of a metal catalyst, which is more suitable than oxides in relation with the electrode conductivity, has been reported by Nakayama et al. in a patent [44,45]. A charge–discharge efficiency of 93.9% for the conversion reaction with MgH^2 is reported by the grinding addition of 3 atom % of a nickel catalyst (particle size 20 nm) in the hydride MCMB carbon mixture. In this last case the irreversible loss can be drastically reduced to 7% for a reversibility capacity of 2608 mA·h·g^{-1} at 0.01 V vs Li^+ /Li^0 and a current rate of C/50.

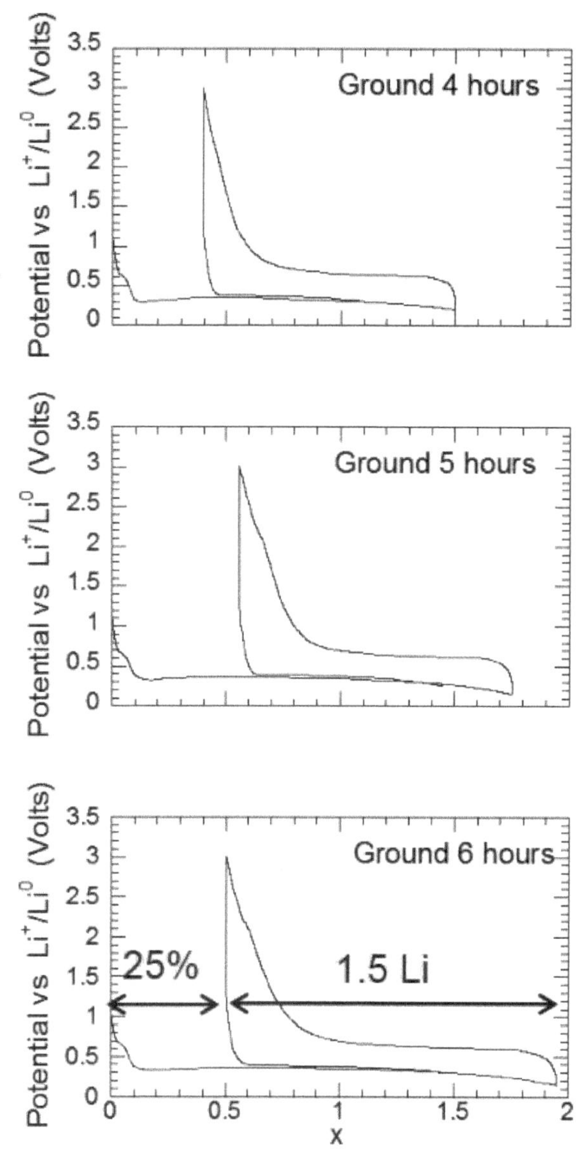

Figure 18: Evolution of the potential (V) as a function of x for a Li/MgH_2 cell that was cycled down to 0.15 V vs Li^+ /Li^0 at a rate of one equivalent of lithium in 10 h. (MgH_2 is obtained after three absorptions of hydrogen and then ground 4, 5 or 6 h with 10% of $C_{t,z}$ carbon) [11].

Performance Improvements of Hydridebased Electrodes

Different samples preparation methods for MgH_2–18% $C_{t,z}$ (activation by three hydrogen sorption steps or reactive milling followed by grinding with Ct,z carbon) are able to produce an hydride carbon composite electrode with full discharge capacity and 75% reversibility (1500 mA·h·g_{-1} at 0.15 V vs Li^+ /Li^0) during the first charge for a current of one equivalent of electrons in 10 h. Despite this improvement of the electrode reversibility of the MgH_2 carbon composite, the cycle life is however limited due to the 83% volume variation, leading to an electronic interparticular conduction loss. Moreover, at a high current rate, the slow hydrogen motion leads to a limitation of the reversible capacity. The influence of the current rate and of the number of electrochemical cycles on the reactivity of the MgH_2 carbon composite will be described in the two following paragraphs.

Influence of the Current rate on the Electrode Reactivity: example of MgH_2

The experimental capacity decreases when the current rate increases and, for an exchange rate of one equivalent of electrons in one hour, the capacity is close to zero. This result shows that the reactivity improvement with different sample preparation methods is still insufficient for real application requiring electrode power and fast charge (Figure 19). However, for a thin film of MgH_2 (200 nm) (prepared by R. Griessen group at the Vrije Universiteit Amsterdam) a full discharge capacity of 2Li can be obtained for a current of one equivalent of electrons in one hour. This fact definitively confirms than the reduction of the diffusion distances is the key to achieve high power electrodes with hydrides for conversion reaction.

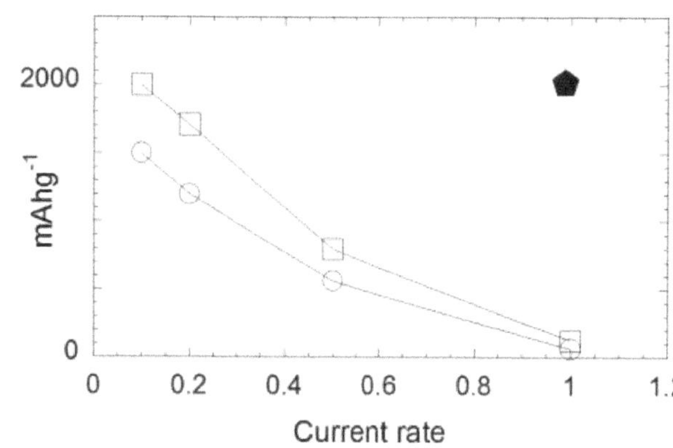

Figure 19: Capacities of a MgH_2 electrode obtained after three absorptions of hydrogen and then ground for 4 h with 10% of $C_{t,z}$ carbon as a function of the current rate (number of Li exchange per hours) open squares: discharge capacity, open circles: charge capacity; black pentagon: discharge capacity of a MgH_2 thin film of 200 nm [11].

Effect of Polymer Binders on MgH_2-Based Electrode Cycle Life

The electrochemical cycling behavior of MgH_2–18% $C_{t,z}$ / Li was studied and compared with that of MgH_2–33.3% CMC–33.3% Ct,z /Li and MgH_2–33.3% CMC-f–33.3% $C_{t,z}$ /Li cells where CMC and CMC-f are polymer binders (sodium carboxymethylcellulose and sodium carboxymethylcelluloseformate, respectively [46]). CMC [47,48], which was already widely investigated for silicon-based electrodes [49-60], surprisingly sustains the 270% volume change taking place during cycling of the Si-based electrode. Beattie et al. [54] estimated the CMC binder and carbon quantity needed to fill the holes created during lithium extraction from the Si alloy electrode to be around 66%. It was experimentally confirmed that large capacities and long cycle lives of the electrodes are obtained for a 33% Si/33% CMC/33% C mixture. This ratio of active material/binder/carbon was chosen to improve the electrode cycle life of MgH2. In addition to CMC, a derivative binder with a formyl ester group was tested (CMC-f) [61].

MgH_2–18% $C_{t,z}$/Li, MgH_2–33.3% CMC–33.3% $C_{t,z}$/Li and MgH_2–33.3% CMC-f–33.3% $C_{t,z}$/Li discharge–charge curves are shown in Figure 20. It appears there that high reversible capacities are preserved independently of the amount of active material additives in the electrode. Indeed, reversible capacities of 1700 mA·h·g^{-1}, 1800 mA·h·g^{-1} and 1900 mA·h·g^{-1} are obtained for the MgH_2–18% $C_{t,z}$/Li (Figure 20a), MgH_2–33.3% CMC–33.3% $C_{t,z}$/Li (Figure 20b) and MgH_2–33.3% CMC-f–33.3% $C_{t,z}$ /Li (Figure 20c) cells, respectively. An increase of the carbon content from 18 to 33% leads to an increase of the irreversible loss (from 25 to 39%) due to electrolyte decomposition at the carbon surface. The reversible capacities of the MgH2 electrodes as a function of the cycle number are presented in Figure 21 (MgH_2–18% $C_{t,z}$ /Li, MgH_2–33.3% CMC–33.3% $C_{t,z}$/Li and MgH_2–33.3% CMC-f–33.3% $C_{t,z}$/Li). From theses curves it is clear that the presence of the CMC-type binder enhances the cycle life of the electrode. While a capacity value of 174 mA·h·g$_{-1}$ is obtained for the electrode without binder, 240 mA·h·g$_{-1}$ and 542 mA·h·g$_{-1}$ are obtained for CMC and CMC-f binders after 40 cycles, respectively. The weak capacity, which originates from carbon (20 mAh·g$_{-1}$ after 40 cycles) is not at the origin of the better capacity retention of the MgH_2–33.3% CMC–33.3% $C_{t,z}$ /Li and MgH_2–33.3% CMCf–33.3% $C_{t,z}$ /Li electrodes. Thus, porosity created by the polymers might explain the better volume accommodation of the electrode during lithium extraction. Further studies are needed for a better understanding of the nature of the polymeric interactions with carbon and metal hydride and their role during the solid mixing of CMC-type binders with MgH_2 and Ct,z .

Role of Interface and Particle size on Reversibility of the Conversion Reaction: example of TiH_2

The reaction of titanium hydride (TiH_2) with lithium ions previously described involves a reaction path that can be summarized as follows:

$$x \leq 0.34\text{: } \delta\text{-}TiH_2(fcc) + 0.34Li \rightarrow \delta\text{-}TiH_{1.66} + 0.34LiH,$$

$$0.34 < x \leq 1\text{: } \delta\text{-}TiH_{2-x} + xLi \rightarrow \delta\text{-}TiH(fco) + xLiH,$$

$$1 < x \leq 2\text{: } \delta\text{-}TiH_{2-x}(fcc) + \delta\text{-}TiH(fco) + Li \rightarrow \alpha\text{-}Ti(hcp) + LiH.$$

Within this conversion process, a full discharge capacity of 1072 mA·h·g^{-1} is obtained for the TiH_2 ground with 10% of $C_{t,z}$ carbon for 5 h. However, while this reaction is free of any alloying reaction with lithium, no electrochemical reversibility was noticed during electrode charge (Figure 22a). To support the idea that interfaces and particles sizes play a key role on conversion reaction reversibility with titanium hydride, the electrochemical behavior of a (2LiH + Ti) composite was studied [6]. This composite was prepared through a mechanochemical reaction between TiH_2 and metallic Li. The capacity of the first charge reached 455 mA·h·g^{-1} whereas after one cycle the reversible capacity was still 428 mA·h·g^{-1}, corresponding to a capacity retention of 94% (Figure 22b). The XRD pattern collected at the end of the first charge indicated the formation of titanium hydride in agreement with the theoretical conversion reaction: 2LiH + Ti → TiH_2 + 2Li^+ + 2e^-. This is a clear example showing that when thermodynamics allow a conversion reaction to occur, interfaces and particles size constitute the pertinent parameters that governs the process reversibility.

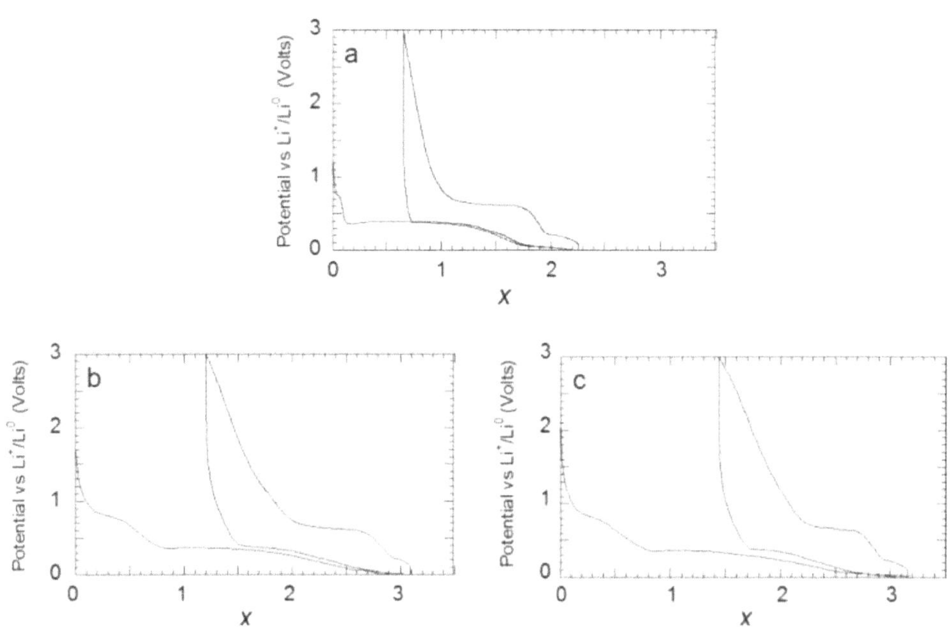

Figure 20: Evolution of the potential (V) as a function of x (mole fraction of Li) for MgH_2 electrodes cycled between 3 and 0.005 V at a rate of one equivalent of lithium in 10 h. (a) MgH_2–18% $C_{t,z}$; (b) 33.3% MgH_2–33.3% CMC–33.3% $C_{t,z}$; (c) 33.3% MgH_2–33.3% CMC-f–33.3% $C_{t,z}$. Adapted from [46]. Copyright 2011 Elsevier.

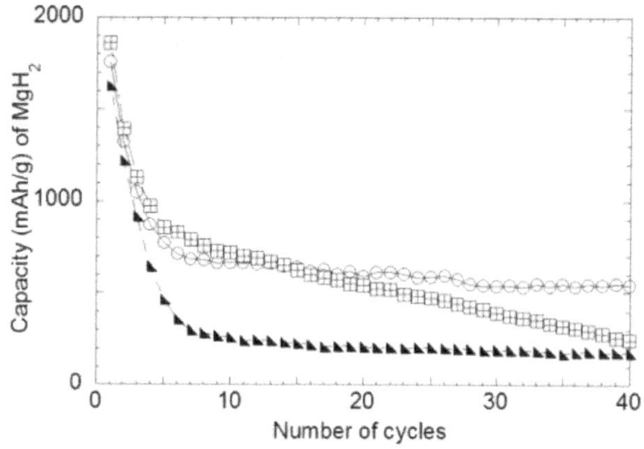

Figure 21: Electrochemical cycling performance for MgH2 composite electrodes: MgH_2–18% $C_{t,z}$ (black triangles), 33.3% MgH_2–33.3% CMC–33.3% $C_{t,z}$ (open squares) and 33.3% MgH2–33.3% CMCf–33.3% $C_{t,z}$ (open circles). Cycling between 3 and 0.005 V, rate of one equivalent of lithium in 10 h. Adapted from [46]. Copyright 2011 Elsevier.

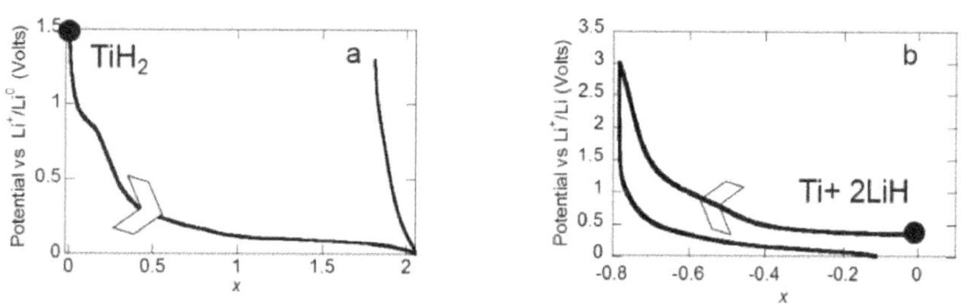

Figure 22: Potential profile of a) TiH_2 electrode ground for 5 h with 10% of $C_{t,z}$ carbon b) Ti + 2LiH electrode as a function of x, the mole fraction of Li (rate one equivalent of Li in 100 h). Adapted from [6]. Copyright 2009 Elsevier.

Synthesis of nanoscale composite hydrides: a perspective to achieve all solid state batteries

It must be emphasized first that production of hydride nanoparticles by reactive milling seems to be an unsuccessful route as long as the agglomeration of the freshly fractured particles is not controlled. Cryomilling could be a potential solution to avoid metal welding and to produce material on the laboratory scale, but this will require significant technological improvements to become economically viable for large scale material production.

Chemical methods, such as encapsulation or confinement strategies used in the design of energy storage and conversion materials, also constitute new synthetic routes that have shown promising results [62,63]. Thus, for solid-state hydrogen storage applications, very fast hydrogen absorption/desorption kinetics have been indeed confirmed for nanoscale Mg hydride (MgH_2) confined into the porosity of different carbon hosts [64,65] or chemical matrices [66,67]. Very recently, composites containing MgH_2 nanoparticles (with a narrow size distribution of 1–10 nm) which were well-dispersed into a porous carbon host have been prepared by Zlotea et al. [68]. These were produced with varying metal content up to 50 wt % and were designed to be used as a negative electrode for Li-ion batteries. These composites show interesting electrochemical behavior, especially regarding their cycle life stability (500 $mA\cdot h\cdot g^{-1}$ after 40 cycles) and have a stable nanoparticle size distribution during electrochemical cycling.

Other promising materials, produced through a physical vapor transport deposition method such as Mg/MgH_2 nanowires or nanofibers are under development for a few years now [69-71]. For instance Mg nanowire shows interesting modifications of both thermodynamics and kinetics compared to the bulk material: a decrease of the dissociation energy of about 12%, (30–50 nm nanowires: 65.3 $kJ\cdot(mol\ H_2)^{-1}$ bulk material:74 $kJ\cdot(mol\ H_2)^{-1}$). Hydriding chemical vapor deposition (HCVD) is also a powerful method to produce in situ high purity nano/microscale MgH_2 under hydrogen. This method, which enables to play with temperature and pressure, is a nice tool for the preparation of a variety of particles having controlled size and shapes (nanofibers, nanoparticles, microdendrites, irregular bulk, hexagonal microplates and microspheres). A good picture of this synthesis method, leading to the production of tailored materials, is given by a pressure–temperature diagram, similar to the diagram of Nakaya et al. [72]. Mass production and applications of such materials in the field of hydrogen storage and batteries technologies will be an interesting challenge for the next decade.

Use of computational methods to look for better materials

Computational methods can also offer interesting alternatives to help the search and development of materials for hydrogen storage and batteries. The idea is, for instance, to look for materials having high volume capacity, minimal expansion volume and high lithium mobility. A recent study by molecular dynamics simulations has revealed that the Li diffusion in MgH_2 nanocluster doped with Fe, Ni, Ti or V is independent of the presence of the transitions metal [73]. The metals improve the hydrogen desorption kinetics. Again, this example shows the mutual interest of this kind of device for hydrogen storage and batteries technologies.

The search for the existence of new stable hydrides in the Mg–Li–H system was also addressed by several groups through density functional theory approach (DFT) [74,75]. Ternary hydrides in the system Li–Mg–H, such as Li_2MgH_4 and $LiMgH_3$, are insulators dominated by ionic bonds. Their preparation from Li, Mg and H_2 is energetically favorable, but may be kinetically inhibited by separation into pure phases. The effects of various light-metal (Mg, Al, Li) and transition-metals (V, Cr, Mn, Fe, Co, Cu, Zn) dopant on the electrochemical properties of NiTiH hydrides as anodes for Li-ion batteries where theoretically studied by Qian et al. [76,77] with Al, Cr, Mn and Fe being the most promising according to the authors.

CONCLUSION

Studies of the reaction of hydrides with lithium ions started with the pioneer work with MgH2, where lithium-driven conversion reactions were firstly demonstrated for the metal hydride family. The study was subsequently extended to other metal and complex hydrides, especially TiH2 and AlH3, and is in now in progress, particularly because of the important number of different potential interesting compounds [78]. Metal hydrides present the advantage of having high capacities in a safe potential window of 0.1–1 V vs Li^+ /Li^0. Moreover, these hydrides show the lowest polarization reported to date for conversion reaction electrodes, as compared to oxides, sulfides, nitrides, phosphides and fluorides compounds, and therefore, constitute promising candidates for negative electrodes in Li-ion batteries. This research field focuses now mainly on nanocomposite synthesis in order to enhance the limited electrochemical cycling performances, the main drawback of hydrides. Electrode technology is, thus, the next challenge, considering the design of the current collector, the preparation of polymeric binders and the mixing of hydride and electrolyte together with careful studies to achieve better capacity retention performance or to scale up the nanocomposite elaboration process. On the other hand, the results obtained at the present time emphasize the importance of the knowledge of fundamentals aspects to control the conversion complex process where reaction paths, interfaces and particle size are the keys parameters. Use of hydrides as anode for Li-ion batteries needs also strong interactions between batteries and fuel-cell communities to be wholly fruitful [79]. In the final analysis, hydrides as a new concept for negative electrodes bridges Li-ion battery and hydrogen storage technologies together and can constitute a promising opportunity for the discovery and the achievement of new energy storage technology for the next 20 years.

REFERENCES

1. Cabana, J.; Monconduit, L.; Larcher, D.; Palacín, M. R. Adv. Mater. 2010, 22, E170–E192. doi:10.1002/adma.201000717
2. Bruce, P. G.; Scrosati, B.; Tarascon, J.-M. Angew. Chem., Int. Ed. 2008, 47, 2930–2946. doi:10.1002/anie.200702505
3. Breitung, B.; Reddy, M. A.; Chakravadhanula, V. S. K.; Engel, M.; Kübel, C.; Powell, A. K.; Hahn, H.; Fichtner, M. Beilstein J. Nanotechnol. 2013, 4, 705–713. doi:10.3762/bjnano.4.80
4. Nitta, N.; Yushin, G. Part. Part. Syst. Charact. 2014, 31, 317–336. doi:10.1002/ppsc.201300231
5. Oumellal, Y.; Rougier, A.; Nazri, G. A.; Tarascon, J.-M.; Aymard, L. Nat. Mater. 2008, 7, 916–921. doi:10.1038/nmat2288
6. Oumellal, Y.; Rougier, A.; Tarascon, J.-M.; Aymard, L. J. Power Sources 2009, 192, 698–702. doi:10.1016/j.jpowsour.2009.03.003
7. HSC Chemistry, v5.11; Chemistry Software Ltd.: South Nutfield, Surrey, UK.
8. Griessen, R.; Riesterer, T. Hydrogen in intermetallic compounds I. In Hydrogen in intermetallic compounds I; Schlapbach, L., Ed.; Topics in Applied Physics, Vol. 63; Springer, 1988; pp 219–284. doi:10.1007/3540183337_13
9. Sandrock, G. D.; Huston, E. L. CHEMTECH 1981, 11, 754–762.
10. Oumellal, Y.; Zaïdi, W.; Bonnet, J.-P.; Cuevas, F.; Latroche, M.; Zhang, J.; Bobet, J.-L.; Rougier, A.; Aymard, L. Int. J. Hydrogen Energy 2012, 37, 7831–7835. doi:10.1016/j.ijhydene.2012.01.107
11. Oumellal, Y. Réactivité électrochimique des hydrures métalliques vis-à-vis du lithium. Ph.D. Thesis, Université de Picardie, Jules Verne, France, 2009.
12. Bououdina, M.; Oumellal, Y.; Dupont, L.; Aymard, L.; Al-Gharni, H.; Al-Hajry, A.; Maark, T. A.; De Sarkar, A.; Ahuja, R.; Deshpande, M. D.; Qian, Z.; Rahane, A. B. Mater. Chem. Phys. 2013, 141, 348–354. doi:10.1016/j.matchemphys.2013.05.021
13. Zaïdi, W.; Bonnet, J.-P.; Zhang, J.; Cuevas, F.; Latroche, M.; Couillaud, S.; Bobet, J.-L.; Sougrati, M. T.; Jumas, J.-C.; Aymard, L. Int. J. Hydrogen Energy 2013, 38, 4798–4808. doi:10.1016/j.ijhydene.2013.01.157
14. Zhang, J.; Cuevas, F.; Zaïdi, W.; Bonnet, J.-P.; Aymard, L.; Bobet, J.-L.; Latroche, M. J. Phys. Chem. C 2011, 115, 4971–4979. doi:10.1021/jp2005293
15. Hanada, N.; Kamura, A.; Suzuki, H.; Takai, K.; Ichikawa, T.; Kojima, Y. J. Alloys Compd. 2011, 509, S584–S587. doi:10.1016/j.jallcom.2010.10.069
16. Fleischmann, S.; Bekaert, E.; Emmel, U.; Wohlfahrt-Mehrens, M., Eds. Trilithium hexahydroaluminate (Li3AlH6) as negative electrode material for lithium-ion batteries, 2nd International Conference on Materials for Energy, Karlsruhe, Germany, May 12–16, 2013; EnMat II: Karlsruhe, Germany, 2013.
17. Latroche, M.; Cuevas, F.; Zhang, J., Eds. Nanostuctured Metal Hydrides as Efficient Anode Materials for Advanced Batteries, TMS 2014, 143rd annual meeting, San Diego, California, USA, Feb 16–20, 2014; San Diego, California, USA, 2014.
18. Teprovich, J. A., Jr.; Zhang, J.; Colón-Mercado, H.; Cuevas, F.; Peters, B.; Greenway, S.; Zidan, R.; Latroche, M. J. Phys. Chem. C 2015, 119, 4666–4674. doi:10.1021/jp5129595
19. Zaidi, W. Hydrures: matériaux de conversion pour accumulateurs lithium-ion. Ph.D. Thesis, Université de Picardie, Jules Verne, France, 2013.
20. Bobet, J.-L.; Pechev, S.; Chevalier, B.; Darriet, B. J. Mater. Chem. 1999, 9, 315–318. doi:10.1039/a804971h

21. Nayeb-Hashemi, A.; Clark, J. J. Phase Equilib. 1985, 6, 238–244.
22. Zhang, J.; Zaïdi, W.; Paul-Boncour, V.; Provost, K.; Michalowicz, A.; Cuevas, F.; Latroche, M.; Belin, S.; Bonnet, J.-P.; Aymard, L. J. Mater. Chem. A 2013, 1, 4706–4717. doi:10.1039/c3ta01482g
23. Selvam, P.; Viswanathan, B.; Swamy, C. S.; Srinivasan, V. Int. J. Hydrogen Energy 1988, 13, 749–759. doi:10.1016/0360-3199(88)90036-5
24. Selvam, P.; Viswanathan, B.; Swamy, C. S.; Srinivasan, V. Thermochim. Acta 1988, 125, 1–8. doi:10.1016/0040-6031(88)87205-8
25. Mg hydride based system conversion reaction with lithium ion - Unpublished work.
26. Couillaud, S.; Kirikova, M.; Zaïdi, W.; Bonnet, J.-P.; Marre, S.; Aymonier, C.; Zhang, J.; Cuevas, F.; Latroche, M.; Aymard, L.; Bobet, J.-L. J. Alloys Compd. 2013, 574, 6–12. doi:10.1016/j.jallcom.2013.03.229
27. Braga, M. H.; El-Azab, A. Phys. Chem. Chem. Phys. 2014, 16, 23012–23025. doi:10.1039/C4CP01815J
28. Latroche, M.; Kalisvaart, P.; Notten, P. H. L. J. Solid State Chem. 2006, 179, 3024–3032. doi:10.1016/j.jssc.2006.05.032
29. Libowitz, G. G. The Solid State Chemistry of Binary Metal Hydrides; W.A. Benjamin: New York, 1965.
30. Sandia National Laboratories, Hydride Information Center. http://www.sandia.gov/ (accessed Oct 8, 2014).
31. Moysés Araùjo, C.; Ahuja, R. J. Alloys Compd. 2005, 404–406, 220–223. doi:10.1016/j.jallcom.2004.10.085
32. Gremaud, R.; Broedersz, C. P.; Borsa, D. M.; Borgschulte, A.; Mauron, P.; Schreuders, H.; Rector, J. H.; Dam, B.; Griessen, R. Adv. Mater. 2007, 19, 2813–2817. doi:10.1002/adma.200602560
33. Vermeulen, P.; Niessen, R. A. H.; Notten, P. H. L. Electrochem. Commun. 2006, 8, 27–32. doi:10.1016/j.elecom.2005.10.013
34. Aymard, L.; Lenain, C.; Courvoisier, L.; Salver-Disma, F.; Tarascon, J.-M. J. Electrochem. Soc. 1999, 146, 2015–2023. doi:10.1149/1.1391884
35. Disma, F.; Aymard, L.; Dupont, L.; Tarascon, J.-M. J. Electrochem. Soc. 1996, 143, 3959–3972. doi:10.1149/1.1837322
36. Dahn, J. R.; Sleigh, A. K.; Shi, H.; Reimers, J. N.; Zhong, Q.; Way, B. M. Electrochim. Acta 1993, 38, 1179–1191. doi:10.1016/0013-4686(93)80048-5
37. Liang, G.; Huot, J.; Boily, S.; Schulz, R. J. Alloys Compd. 2000, 305, 239–245. doi:10.1016/S0925-8388(00)00708-8
38. Liang, G.; Huot, J.; Boily, S.; Van Neste, A.; Schulz, R. J. Alloys Compd. 1999, 292, 247–252. doi:10.1016/S0925-8388(99)00442-9
39. Barkhordarian, G.; Klassen, T.; Bormann, R. J. Alloys Compd. 2004, 364, 242–246. doi:10.1016/S0925-8388(03)00530-9
40. Barkhordarian, G.; Klassen, T.; Bormann, R. Scr. Mater. 2003, 49, 213–217. doi:10.1016/S1359-6462(03)00259-8
41. Ikeda, S.; Ichikawa, T.; Yamaguchi, S.; Miyaoka, H.; Kojima, Y. J. Jpn. Inst. Energy 2014, 93, 926–930. doi:10.3775/jie.93.926
42. Bhat, V. V.; Rougier, A.; Aymard, L.; Darok, X.; Nazri, G.; Tarascon, J. M. J. Power Sources 2006, 159, 107–110. doi:10.1016/j.jpowsour.2006.04.059
43. Borgschulte, A.; Lohstroh, W.; Westerwaal, R. J.; Schreuders, H.; Rector, J. H.; Dam, B.; Griessen, R. J. Alloys Compd. 2005, 404–406, 699–705. doi:10.1016/j.jallcom.2005.01.137
44. Nakayama, H.; Matsunaga, T.; Nobuhara, K.; Kon, M. Anode material, metal secondary battery, and method for production of anode material. U.S. Patent Appl. WO 2012/007830 A1, Jan 19, 2012.
45. Nakayama, H.; Nobuhara, K.; Kon, M.; Matsunaga, T. 218th ECS Meeting Abstract 2010, 1002, 1052.
46. 46.Zaïdi, W.; Oumellal, Y.; Bonnet, J.-P.; Zhang, J.; Cuevas, F.; Latroche, M.; Bobet, J.-L.; Aymard, L. J. Power Sources 2011, 196, 2854–2857. doi:10.1016/j.jpowsour.2010.11.048
47. Heinze, T.; Koschella, A. Macromol. Symp. 2005, 223, 13–40. doi:10.1002/masy.200550502
48. Heinze, T. Macromol. Chem. Phys. 1998, 199, 2341–2364. doi:10.1002/(SICI)1521-3935(19981101)199:11<2341::AID-MACP2341>3.0.CO;2-J
49. 49. Urbonaite, S.; Baglien, I.; Ensling, D.; Edström, K. J. Power Sources 2010, 195, 5370–5373. doi:10.1016/j.jpowsour.2010.03.042
50. Guo, J.; Wang, C. Chem. Commun. 2010, 46, 1428–1430. doi:10.1039/b918727h
51. Mazouzi, D.; Lestriez, B.; Roué, L.; Guyomard, D. Electrochem. Solid-State Lett. 2009, 12, A215–A218. doi:10.1149/1.3212894
52. Bridel, J.-S.; Azaïs, T.; Morcrette, M.; Tarascon, J.-M.; Larcher, D. Chem. Mater. 2010, 22, 1229–1241. doi:10.1021/cm902688w
53. Hochgatterer, N. S.; Schweiger, M. R.; Koller, S.; Raimann, P. R.; Wöhrle, T.; Wurm, C.; Winter, M. Electrochem. Solid-State Lett. 2008, 11, A76–A80. doi:10.1149/1.2888173
54. Beattie, S. D.; Larcher, D.; Morcrette, M.; Simon, B.; Tarascon, J.-M. J. Electrochem. Soc. 2008, 155, A158–A163. doi:10.1149/1.2817828
55. Li, J.; Lewis, R. B.; Dahn, J. R. Electrochem. Solid-State Lett. 2007, 10, A17–A20. doi:10.1149/1.2398725
56. Lestriez, B.; Bahri, S.; Sandu, I.; Roué, L.; Guyomard, D.Electrochem. Commun. 2007, 9, 2801–2806. doi:10.1016/j.elecom.2007.10.001
57. Dimov, N.; Xia, Y.; Yoshio, M. J. Power Sources 2007, 171, 886–893. doi:10.1016/j.jpowsour.2007.06.026
58. Yoshio, M.; Tsumura, T.; Dimov, N. J. Power Sources 2006,163,215–218.doi:10.1016/j.jpowsour.2005.12.078
59. Buqa, H.; Holzapfel, M.; Krumeich, F.; Veit, C.; Novák, P. J. Power Sources 2006, 161, 617–622. doi:10.1016/j.jpowsour.2006.03.073

60. Liu, W.-R.; Yang, M.-H.; Wu, H.-C.; Chiao, S. M.; Wu, N.-L. Electrochem. Solid-State Lett. 2005, 8, A100–A103. doi:10.1149/1.1847685

61. Heinze, T.; Heinze, U. Macromol. Rapid Commun. 1997, 18, 1033–1040. doi:10.1002/marc.1997.030181206

62. Schüth, F. Chem. Mater. 2014, 26, 423–434. doi:10.1021/cm402791v

63. Zlotea, C.; Latroche, M. Colloids Surf., A 2013, 439, 117–130. doi:10.1016/j.colsurfa.2012.11.043

64. Zlotea, C.; Chevalier-César, C.; Léonel, E.; Leroy, E.; Cuevas, F.; Dibandjo, P.; Vix-Guterl, C.; Martens, T.; Latroche, M. Faraday Discuss. 2011, 151, 117–131. doi:10.1039/c0fd00016g

65. Nielsen, T. K.; Manickam, K.; Hirscher, M.; Besenbacher, F.; Jensen, T. R. ACS Nano 2009, 3, 3521–3528. doi:10.1021/nn901072w

66. Zhao-Karger, Z.; Hu, J.; Roth, A.; Wang, D.; Kübel, C.; Lohstroh, W.; Fichtner, M. Chem. Commun. 2010, 46, 8353–8355. doi:10.1039/c0cc03072d

67. Paskevicius, M.; Sheppard, D. A.; Buckley, C. E. J. Am. Chem. Soc. 2010, 132, 5077–5083. doi:10.1021/ja908398u

68. Oumellal, Y.; Zlotea, C.; Bastide, S.; Cachet-Vivier, C.; Léonel, E.; Sengmany, S.; Leroy, E.; Aymard, L.; Bonnet, J.-P.; Latroche, M. Nanoscale 2014, 6, 14459–14466. doi:10.1039/C4NR03444A

69. Zhu, C.; Hosokai, S.; Matsumoto, I.; Akiyama, T. Cryst. Growth Des. 2010, 10, 5123–5128. doi:10.1021/cg100856n

70. Matsumoto, I.; Akiyama, T.; Nakamura, Y.; Akiba, E. J. Alloys Compd. 2010, 507, 502–507. doi:10.1016/j.jallcom.2010.07.218

71. Zhu, C.; Hayashi, H.; Saita, I.; Akiyama, T. Int. J. Hydrogen Energy 2009, 34, 7283–7290. doi:10.1016/j.ijhydene.2009.06.080

72. Nakaya, U. Snow Crystals, Natural and Artificial; Harvard University Press: Cambridge, MA, USA, 1954. doi:10.4159/harvard.9780674182769

73. Ramzan, M.; Lebègue, S.; Ahuja, R. Int. J. Hydrogen Energy 2010, 35, 10373–10376. doi:10.1016/j.ijhydene.2010.07.024

74. Mason, T. H.; Liu, X.; Hong, J.; Graetz, J.; Majzoub, E. H. J. Phys. Chem. C 2011, 115, 16681–16687. doi:10.1021/jp203056n

75. Li, D.; Zhang, T.; Yang, S.; Tao, Z.; Chen, J. J. Alloys Compd. 2011, 509, 8228–8234. doi:10.1016/j.jallcom.2011.05.084

76. Qian, Z.; Jiang, X.; De Sarkar, A.; Adit Maark, T.; Deshpande, M. D.; Bououdina, M.; Johansson, B.; Ahuja, R. Solid State Ionics 2014, 258, 88–91. doi:10.1016/j.ssi.2014.02.007

77. Qian, Z.; De Sarkar, A.; Adit Maark, T.; Jiang, X.; Deshpande, M. D.; Bououdina, M.; Ahuja, R. Appl. Phys. ett. 2013, 103, 033902. doi:10.1063/1.4813596

78. Wietelmann, U. Chem. Ing. Tech. 2014, 86, 2190–2194. doi:10.1002/cite.201400097

79. Fichtner, M. J. Alloys Compd. 2011, 509 (Suppl. 2), S529–S534. doi:10.1016/j.jallcom.2010.12.179

CHAPTER 15

CAPACITY DEGRADATION MECHANISMS IN NICKEL/METAL HYDRIDE BATTERIES

Kwo-hsiung Young and Shigekazu Yasuoka

Keywords: Nickel/metal hydride (Ni/MH) battery, failure analysis, electrochemistry, hydrogen storage alloys, self-discharge, capacity degradation

INTRODUCTION

Nickel/metal hydride (Ni/MH) batteries are widely used in many energy storage applications. Cycle stability is one of the key criteria in judging the performance of rechargeable battery technology. The general observations regarding failed Ni/MH cells are summarized in Figure 1. In order to further investigate the mechanisms of capacity degradation and their relevant solutions to extend cycling under normal and abuse conditions, we have chosen to begin with a review of the significance of the Ni/MH battery in the overall battery market, its basic structure and chemistry, and the analytical tools used to study and characterize its performance, and to mainly focus on academic publications and reports. Patents detailing solutions for extending cycle life are reviewed in two separate articles [1,2].

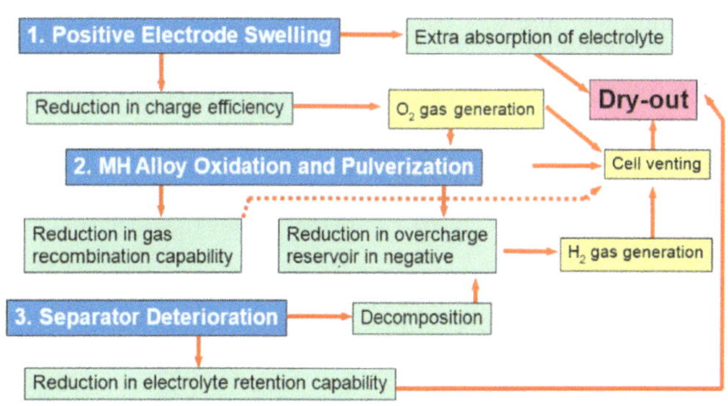

Figure 1. Schematic diagram of three key factors leading to the major failure mode of nickel/metal hydride (Ni/MH) cells—electrolyte dry-out.

Significance of Nickel/Metal Hydride Batteries

Ni/MH batteries using an alkaline KOH electrolyte have been commercialized for more than 25 years [3]. Because of its durability, abuse tolerance, compact size, and environmental friendliness, Ni/MH battery applications have steadily expanded from the traditional consumer market to include propulsion and telecommunications. However, due to its relatively low gravimetric energy density compared to the rival

Citation: Kwo-hsiung Young and Shigekazu Yasuoka, Capacity Degradation Mechanisms in Nickel/Metal Hydride Batteries, doi: 10.3390/batteries2010003.

Li-ion battery, the Ni/MH battery lost part of its market share in portable electronic devices, such as notebook computers, cell phones, and digital cameras. In the meantime, Ni/MH battery technology also invaded the primary alkaline battery market because of its voltage compatibility and low self-discharge [4,5,6], as well as the NiCd power tool market for its non-toxicity [7,8]. The success of Ni/MH in powering hybrid electric vehicles (HEV) developed by a handful of automobile manufacturers stems from its wide temperature range, abuse tolerance, superb cycle stability, high charge and discharge rate capabilities, and environmental friendliness [9]. One analyst even predicted a fourfold increase in Ni/MH battery sales for HEV and EV markets from 2014 to 2020 [10]. Although the current industries making pure battery-powered electric vehicles embrace Li-ion battery technology, a Ni/MH pouch cell developed under a five million dollar Advanced Research Projects Agency-Energy (ARPA-E) program has demonstrated a specific energy of 127 $Wh \cdot kg^{-1}$ at the cell level with an estimated target of 148 $Wh \cdot kg^{-1}$ [11]. With the recent breakthrough of high-energy Si-negative electrodes capable of storing 3635 $mAh \cdot g^{-1}$ (about ten times the current A_2B_7 alloy) [12], the future of Ni/MH in the EV application appears very bright. From the beginning of their competition, Ni/MH batteries have had higher volumetric energy density than Li-ion batteries, due to the high density active materials (rare earth metal (RE) and transition metal *versus* carbon-based products in the anode and nickel hydroxide *versus* lithiated transition metal oxides in the cathode). With the improvements in specific energy, a resurgence in Ni/MH batteries for applications that place a premium on space rather than weight, such as portable displays, wearable electronic devices, and medical devices, can be expected. As for large-scale high-power temporary energy storage applications, the GIGACELL, made by Kawasaki Heavy Industries, demonstrates superior performance using the Ni/MH chemistry [13,14]. In stationary applications, its excellent cycle stability and wide operating temperatures, combined with the low cost and easy manufacturability, have made Ni/MH the best choice [15]. The overall outlook for Ni/MH battery technology shows that it has tremendous potential in various energy storage applications following these new scientific discoveries and process improvements—a far cry from being the 25-year obsolete veteran in the battery business.

Basic Structure of Nickel/Metal Hydride Battery

There are basically seven different types of Ni/MH batteries: cylindrical with metal cases, stick (bubble gum shape), prismatic with metal cases, prismatic with plastic cases, button cell, pouch cell [16], and flooded cell [17] (Figure 2). All but button cell and pouch cell have a safety valve installed to prevent explosions from gas build-up. A simple comparison between various construction types is shown in Table 1. They all share some common parts: positive electrode, negative electrode, separator, electrolyte, case, and safety valve (except button and pouch cells). The basic electrochemistry reactions for the positive electrode, negative electrode, and full cell are:

$Ni(OH)_2 + OH^- \leftrightarrows NiOOH + H_2O + e^-$ (forward: charge, reverse: discharge) (1)

$M + H_2O + e^- \leftrightarrows MH + OH^-$ (forward: charge, reverse: discharge) (2)

$Ni(OH)_2 + M \leftrightarrows NiOOH + MH$ (forward: charge, reverse: discharge) (3)

where M is the hydrogen storage metal/alloy and MH is the hydride of metal M. During the charge process, bi-valent Ni is oxidized into the tri-valent state while metal M is reduced by the absorbed hydrogen atom. The most commonly used positive electrode in the current Ni/MH battery technology consists of active materials made of co-precipitated spherical hydroxides from Ni, Co, and Zn and some binders, pasted onto Ni-foam via a wet method. Recently, a dry application of spherical powder onto Ni-foam with no binder followed by immediate compaction has also been used to increase the energy and power density of the cell. In some high-temperature/high-rate applications, old sintered-type positive electrodes, based on fibrous Ni on stainless steel plate, are still in commission. Co-coating of the spherical particles and additives such as metallic Co and/or CoO and rare earth element (RE) oxides in the positive electrode paste are also popular. A review of the synthesis and properties of $Ni(OH)_2$ was recently reported [18].

(a) (b) (c) (d) (e) (f) (g)

Figure 2. Ni/MH batteries in: (**a**) cylindrical; (**b**) stick; (**c**) metallic prismatic; (**d**) plastic prismatic; (**e**) button; (**f**) pouch; and (**g**) flooded configurations.

Table 1. Comparison of different types of Ni/MH battery packaging

Shape	Case material	Sealed	Manufacturability	Cost	Energy density	Heat dissipation	Abuse tolerance
Cylindrical	Metal	Yes	Easy	Low	High	Easy	High
Stick	Metal	Yes	Medium	Low	High	Easy	High
Prismatic	Metal	Yes	Medium	High	Low	Easy	Med
Prismatic	Plastic	Yes	Medium	High	Low	Hard	Med
Button	Metal	Yes	Easy	Low	Low	Easy	Low
Pouch	Al foil	Yes	Easy	Low	Very high	Easy	Low
Cylindrical/prismatic	Plastic/flooded	No	Easy	Low	Low	Hard	High

The most common metal hydride (MH) alloy used in the negative electrode is a RE-based AB_5alloy. A typical atomic composition is $La_{10.5}Ce_{4.3}Pr_{0.5}Nd_{1.4}Ni_{60.0}Co_{12.7}Mn_{5.9}Al_{4.7}$. Recently, RE-based A_2B_7 MH alloys have gained popularity in high-energy and low self-discharge consumer type applications [4,5,6]. A typical atomic composition of this type is $La_{6.7}Pr_{6.3}Nd_{6.3}Ni_{72.8}Al_{4.0}$. Recent progress in MH alloys for Ni/MH battery applications can be found in the following review article [19]. The negative electrode can be prepared by dry-compacting the MH powder directly onto a Ni-mesh, Cu-mesh, expanded Ni, Ni foam, or expanded Cu substrates without the use of a binder, or by wet-pasting a slurry with MH alloy, binder, and/or additives onto nickel plated perforated stainless steel (NPPS).

A 30% KOH solution is widely used as the electrolyte for Ni/MH batteries due to the balance of conductivity and freezing point temperature. Performance comparisons for other concentrations [20] and alkaline metal hydroxides [21] are available. A small amount of LiOH (1.5 $g{\cdot}L^{-1}$), which has higher chemical reactivity, is added to boost low-temperature performance, while in high-temperature applications, part or all of the KOH is replaced by the less reactive (corrosive) NaOH to reduce corrosion. In the current standard mass production of Ni/MH cells, no other specific additive is added to the electrolyte.

Grafted polypropylene (PP)/polyethylene (PE) non-woven fabric is today's standard separator material, and an overview has been published by Kritzer and Cook [22]. While the regular separator is white in color, the sulfonated separator is brownish and offers benefits to low self-discharge due to its ability to trap redox shuttle substances, especially the nitrogen-containing compounds [23]. Both types of separators can be found in current NiMH batteries.

Experimental Methods Used in Failure Analysis

A few analytic tools are frequently used to identify the failure mode of a cycled Ni/MH battery [24,25,26]. Scanning electron microscope (SEM) with X-ray energy dispersive spectroscopy (EDS) capability is commonly used to examine the degree of pulverization, phase segregation, degree of oxidation, physical size changes, and trapping of particulates. The different features found between the secondary electron image and the backscattering electron image can be extrapolated into the changes in the average atomic weight of the area of interest. EDS mapping is especially useful in studying elemental distribution (for example, oxygen) in a relatively large area (10–100μm scale). While gas chromatography (GC) is used to identify the gas composition in the cell, inductively coupled plasma (ICP) is used to examine the metallic composition of any solid (electrode, separator, tap, *etc.*) or liquid (remaining electrolyte and solution attained through Soxhlet extraction) content from the autopsy of a cycled cell. Titration is another method to determine the content of a specific element [27]. X-ray diffraction (XRD) is an important tool to study oxide formation, phase changes, and microstructure changes in both negative [25,26,28,29] and positive [30] electrodes.

Other tools are used less frequently in failure analysis. For example, transmission electron microscope (TEM) is sometimes used to study the microstructure and composition of the surface oxide from a cycled cell [31,32,33]. Magnetic susceptibility (MS) measurements can be used to monitor the evolution of the count and size of metallic Ni-clusters embedded in the surface oxide [26,34,35]. Both X-ray photoelectron spectroscopy (XPS) [36,37,38,39,40] and Auger electron spectroscopy (AES) [41] have been used to study the surface composition, with the former being able to identify the oxidation state. The acoustic emission (AE) technique has also been used to study the volume change and pulverization of the MH alloy [42,43]. Fourier transform infrared spectroscopy (FTIR) has been used to study the OH^- ligand in $Ni(OH)_2$ [44,45,46,47]. Raman spectroscopy (RS) is another optical measurement used to characterize the changes in the separator and positive electrode [37,46,47,48]. Electrochemical impedance spectroscopy (EIS) or AC impedance measurements are usually used to isolate components with different degrees of degradation [49,50,51,52]. Polarization curves [53,54,55,56] and cyclic voltammetry (CV) [57,58,59,60] are other electrochemical tools that can be used to study the evolution of electrode surface changes. Besides experiments with real batteries, empirical capacity degradation models have also been previously developed [61,62,63].

CAPACITY DEGRADATION

Battery failure can be separated into two categories: accidental and long-term degradation. The former includes fire, electrical short-circuit, and physical damage. In Table 2, we have listed a few common symptoms and possible causes that originated the failure of the batteries. Long-term capacity loss under various test conditions is discussed in the remainder of this section.

Table 2. Common Ni/MH battery failure symptoms and possible causes

Symptom	Reasons	Possible causes
Battery short-circuit	Direct conducting path between two electrodes developed	• Separator punch-through • Conducting debris from Cu-impurities • Deformation of electrode causing direct contact between taps
Battery open-circuit	Breakage of inside connection	• Electrode breakage due to expansion/distortion • Broken tap connection • Complete electrolyte dry-out
Battery abuse	Over-discharge and overcharge	• Unbalanced capacity in positive and negative electrode • Mismatched charger
Capacity decrease	Electrode degradation	• Pulverization/oxidation of MH alloys in negative electrode • Pulverization in spherical particle due to formation of γ-NiOOH phase • Decrease in the Co-conductive network in the positive electrode
Power decrease and impedance increase	Electrolyte dry-out	• Venting from improper cell-balance • Consumption due to oxidation
	Electrode degradation	• Reduction in electrode active materials • Increase of the surface oxide of negative electrode • Loss of co-conductive network in positive electrode
	Separator degradation	• Increase in fiber diameter • Reduction in pore volume • Impurity trapped internally • Decomposition
Overheat during charge	Micro-shorting	• Conductive debris accumulation in separator
White deposits	Electrolyte leak from venting	• Improper closing of the cell • Off-balance in the remaining electrode capacity • Deterioration of gas recombination capability at the surface of MH alloy • Heavily oxidized electrode and/or electrolyte • Failure in the safety vent

Capacity Loss During Normal Cycling at Room Temperature

There are two types of capacity loss during cycling: reversible and irreversible. The reversible part is also called self-discharge, which mainly occurs through six pathways: shuttling effects from nitrogen containing compounds [64], shuttling effects from soluble ions of multi-valence transition metals [65], micro-shorts [66] from conducting/semiconducting deposits trapped in the separator [67,68], hydrogen gas desorption from MH alloys [69,70,71,72,73], direct reaction between hydrogen gas and NiOOH [69,73,74], and CoOOH protective/conductive coating breakdown due to contamination from leached MH alloys [6]. Self-discharge accelerates with rises in the environmental temperature. There is basically no self-discharge at below −5 °C [75]. Before the low self-discharge Ni/MH battery was introduced (the Eneloop cell from Sanyo using a combination of improved MH alloy, separator, and positive active materials [6]), cells initially had a monthly 20%–30% capacity reduction, which was then improved to a monthly loss of 5%–10% at room temperature [76]. Modern low-self discharge Ni/MH consumer batteries have self-discharge rates of less than 20% per year [6]. An automatically triggered re-charging algorithm may be necessary for large-scale applications [77]. Common methods used to suppress self-discharge in Ni/MH batteries are summarized in Table 3. The irreversible capacity loss, which leads to failure of the battery, covers the majority of this review.

Table 3. Summary of common methods used to suppress self-discharge in Ni/MH batteries. The star system used in the effectiveness column in Table 3, Table 4, Table 5, Table 6, Table 7, Table 8 and Table 9 was meant to show the relative strength in each method to address the problem based on authors' own experience. Interested readers are encouraged to read the original article and to form their own opinions. PP: polypropylene; PTFE: polytetrafluoroethylene; and CMC: carboxymethyl cellulose

Method	Direct impact	Environmental impact	Cost impact	Effectiveness	References
Use of a sulfonated separator	Removal of N-containing compounds	None	Modest	•••••	[22,78,79]
Use of an acrylic acid grafted PP separator	Reduction in Al- and Mn-debris formation in separator	None	None	••••	[80]
Removal of Co and Mn in A_2B_7 MH alloy	Reduction in debris formation in separator	None	None	•••••	[6,81]
Increase of the amount of electrolyte	Reduction in the hydrogen diffusion in electrolyte	None	None	••••	[82]
Removal of Cu-containing components	Reduction in micro-short	None	None	•••••	[83–85]
PTFE coating on positive electrode	Suppression of reaction between NiOOH and H_2	None	Negligible	••••	[86]

CMC solution dipping	Suppression of oxygen evolution	None	Negligible	••••	[87]
Micro-encapsulation of Cu on MH alloy	Decrease in H_2 released from MH alloy	None	Modest	•••	[88]
Ni-B alloy coating on MH alloy	Formation of a protection layer	None	Modest	•••	[89]
Alkaline treatment of negative electrode	Reduction of leach-out of Mn and Al	None	Modest	••••	[90]
Addition of LiOH and NaOH into electrolyte	Reduction in electrolyte corrosion capabilities	None	None	••••	[75]
Addition of $Al_2(SO_4)_3$ into electrolyte	Reduction in MH alloy corrosion	None	Negligible	••	[91]

Irreversible capacity losses under regular cycling conditions (temperature between 20 °C and 30 °C, rated below 2C with one or a combination of reasonable cut-off schemas during over-charge, such as those used in [26,29]) can be categorized into five main categories: degradation of negative electrode active material (MH alloy), degradation of positive electrode active material (spherical $Ni(OH)_2$), disintegration of the negative electrode, disintegration of the positive electrode, and venting of cells.

Degradation in the negative electrode includes MH alloy pulverization due to lattice expansion during hydrogenation [92,93,94,95] which results in poor electrical and protonic conduction [49,95,96], alloy surface oxidation hampering electron and proton conduction [36,52,53,54,93,94,97,98,99, 100,101], and surface fluoride formation [36]. The corrosion processes of AB_5 MH alloys have been characterized by Maurel and his coworkers using XRD, SEM, and TEM [102]. In the La-only A_2B_7 superlattice MH alloy, the pulverization due to different sequences of hydrogenation between Mg-containing A_2B_7 and Mg-free AB_5 phases dominates the failure mode [25,103].

Degradation in the positive electrode includes swelling from γ-NiOOH formation [67,104], breaking of the Co-conductive network [105], formation of less electrochemically rechargeable γ-NiOOH [25,93,106], Co dissolution and migration from the conductive network in the positive electrode [107], contamination from leach-out products (Al and Mn) in the negative electrode, deteriorating Co-conductive coating [27,104], and pulverization of positive electrode spherical particles causing detachment of active material [68,108]. The increased surface area in the positive electrode as a result of pulverization also deprives electrolyte from the separator, which increases cell resistance [109].

The mechanical disintegration of the negative electrode may include breakage of the NPPS substrate due to increased stress from electrode expansion/distortion and MH alloy powder detachment from the substrate. The mechanical disintegration of the positive electrode may include breakage of the Ni-foam substrate due to large amounts of stress from electrode expansion/distortion [110], especially in a small wounded cylindrical cell [111], and separation of spherical particles from the substrate [112]. Venting occurs when high pressure (mostly H_2) is built up inside the cell primarily from inadequate gas recombination capabilities of the MH alloy surface and/or unbalanced capacity distribution [113], which results in reduced electrolyte content [114].

Capacity Loss during Long-Term Room Temperature Storage

The irreversible capacity loss during long-term room temperature storage can be attributed to the dissolution of the surface CoOOH conducting network [115,116], corrosion/passivation of the negative electrode [23,117,118,119], decomposition of the positive electrode [115], decomposition of the separator [116], and poisoning of the positive electrode from cations that originate from the negative electrode [68,80,115].

Capacity Loss during High-Temperature Storage

Temperature is one of the key factors affecting cycle stability [120]. In addition to the regular capacity losses described in Section 2.1, high-temperature environments (≥45 °C) will accelerate the cell degradation through the following pathways: oxidation rate increases at the surface of the MH alloy particles [121,122], dissolution of Co-compounds in the Co-conductive network [113,123], higher self-discharge rates that lower the cell voltage and result in further alloy oxidation, and separator degradation [124]. The charging method used in the high-temperature range has to be specially designed. First, the cell voltage tends to be lower at higher temperature, which demands that a lower cut-off voltage be adopted during charging to prevent over-charge [122] as it can be directly correlated to capacity degradation [125]. Next, the oxygen gas evolution potential in the positive electrode tends to decrease with increased temperature, which forces the positive electrode to finish charging prematurely and for which the $-\Delta V$ cut-off method is less effective [122,126,127]. Ni/MH batteries are also more sensitive to over-charge at elevated temperatures. Ni/MH batteries overcharged at rates of 0.2C, 0.5C, and 1.0C for one month show irreversible capacity losses of 12%, 30%, and 40%, respectively [126]. Different from the irreversible capacity losses during high-temperature cycling, losses in capacity observed during low-temperature cycling are recoverable when returned to room temperature [74].

Capacity Loss Due to Low-Temperature Cycling

As stated above, low-temperature storage of Ni/MH batteries causes no apparent damage to performance. However, Chen *et al.* [128] reported capacity degradation during a −20 °C cycling experiment with MH alloy pulverization, but the alloy corrosion was less serious compared to results from room temperature and high temperature. At low temperatures, a special "surface icing" appears to form on the MH alloy, further hindering electrochemical reactions and then disappearing at higher temperature [129].

Capacity Loss Due to High-Rate Cycling

Fast charge acceptance is controlled by solid-state hydrogen diffusion [130], and the diffusion coefficient of hydrogen decreases with increasing current density [131]. The increase in the degradation rate with fast charging typically originates from an improper termination method for detecting the end of charge, which leads to a large degree of over-charge especially within an aged cell [132]. The heat generated from the internal resistance of the cell and the hydrogen-oxygen recombination reaction cannot be dissipated quickly enough, and this results in an increase in the cell temperature. Both the high rate and the high temperature conditions reduce charging efficiency [133] and therefore both conditions facilitate similar failure mechanisms, except that a high-rate cycled cell also shows electrode disintegration from extraordinarily fast gas release [134] (mostly H_2 [135]) as well as gas venting due to the insufficient time for hydrogen-oxygen recombination [136,137]. As such, fast charging of a large-sized Ni/MH battery is not recommended unless special temperature monitoring devices are installed [138,139].

Capacity Loss in a Multi-Cell Module

Thus far, the discussion in this section has focused on the cell-level where most of the capacity degradation occurs. In a single Ni/MH cell, both the over-charge (with a state-of-charge (SOC) greater than 100%) and over-discharge (depth of discharge (DOD) greater than 100%) conditions can be avoided by the proper monitoring of the cell voltage. Because of the low risk of operating Ni/MH cells under disadvantageous conditions, a multi-cell module or pack does not require voltage monitoring at the cell-level whereas the Li-ion battery does. With the proper design of the negative-to-positive capacity (*n*/*p*) ratio, the size of the over-discharge reservoir [113] and the anticipated rates of capacity degradation in both electrodes, the over-charge or the over-discharge of the cells only results in small amounts of oxygen gas or hydrogen gas evolution, respectively, in the positive electrode [17]. The small amounts of generated oxygen gas can be recombined with the hydrogen stored in the negative electrode in case of over-charge, and the small amounts of generated hydrogen gas can be stored in the negative electrode in case of over-discharge [113]. Repetitive gas evolutions from the positive electrode can result in both mechanical disintegration of the electrode and cell venting to relieve the pressure, causing a loss in capacity and an increase in cell impedance. The DOD in a multi-cell pack also plays an important role in the cycle life performance. For example, an increase of DOD from 10% to 90% in a HEV Ni/MH pack can reduce cycle life from 5000 cycles to 500 cycles [61]. For high-rate operation, as in a HEV, large swings in the SOC can result in premature MH alloy pulverization.

Methods to Improve Cycle Stability

There are many academic publications and issued patents offering, at least, partial solutions to the capacity loss problem during cycling. While patents addressing cycle stability are reviewed in two other papers [1,2], the strategies issued from the academic research community reviewed here fall under six general categories: (1) cell designs guided mainly by *n*/*p* ratio, electrolyte loading, and electrode thickness parameters; (2) active binder and additive material designs in the negative electrode; (3) composition, coating, and paste additives in the positive electrode; (4) choice of separator; (5) electrolyte; and (6) other components. Other systematic maintenance protocols for battery packs using Ni/MH cells were reported by Zhu and his coworkers [140].

Cell Design

In good Ni/MH cell design, an appropriate *n*/*p* ratio is critical to the balance of the various performance requirements in a specific application. For instance, a high energy consumer cell, a general purpose cell, and a high-rate cell may have *n*/*p* ranges of 1.05–1.2, 1.4–1.6, and 1.8–2.2, respectively. Adequate distribution of the extra negative electrode capacity into the over-charge-reservoir (OCR) and the over-discharge-reservoir (ODR) to avoid cell-venting is especially critical, particularly near the end of service life [116]. Nearly all cases of venting are due to short-circuits in the OCR that arises from material oxidation and γ-NiOOH formation that overwhelm the ODR. Other important design parameters that impact cycle life performance are electrolyte loading and electrode thickness. The amount of electrolyte added to the cell is proportional to cycle life, but too much electrolyte will eliminate the gas recombination centers and cause venting during formation. Optimal electrolyte loading is about 1.7–1.9 $g{\cdot}A^{-1}{\cdot}h^{-1}$ [141]. Thicker electrodes can improve the gravimetric and volumetric energy densities of the battery at the expense of high-rate discharge capability and mechanical integrity of the electrode. Methods for improving cycle performance through cell design are summarized in Table 4.

Table 4. Summary of cycle stability improvement methods related to cell design. ODR: over-discharge-reservoir; and *n*/*p*: negative-to-positive capacity.

Method	Direct impact	Environmental impact	Cost impact	Effectiveness	References
Pre-charge of the positive electrode	Reduction of ODR	None	Negligible	*****	[142,143]
Increase in the *n*/*p* ratio	Trade-off of capacity for longer life	None	None	*****	[134,144]
Optimization of electrolyte loading	Balance between cycle life and production yield	None	None	****	[141]
Optimization of positive electrode thickness	Reduction in electrode breakage	None	None	****	[145]
Pre-charge during the formation process	Protection of MH alloy	None	Negligible	***	[146]

Negative Electrode

While studies of degradation in MH alloys such as AB_2 [147,148,149,150,151,152,153], Mg-Ni [154,155,156,157], and V-based body-center-cubic [158] are available, we will singularly focus on the discussion of misch-metal based AB_5 and A_2B_7 superlattice MH alloys and their related electrode properties. In this section, improvements in cycle stability related to the negative electrodes are summarized in Table 5 and are categorized by alloy formula, alloy preparation, alloy post-treatment, electrode additives, and different electrode types.

Table 5. Summary of cycle stability improvement methods related to negative electrode. PVA: polyvinel alcohol; HEC: hydroxyethyl cellulose; and RE: rare earth metal.

Method		Direct impact	Environmental impact	Cost impact	Effectiveness	References
A. Alloy formula	Increase in Al-content	Increase in unit cell volume and reduction in lattice expansion during hydrogenation. Formation of Al_2O_3 protection layer on MH alloy.	None	None	•••••	[159–162]
	Increase in Co-content	Reduction in hardness and prevention of La-migration onto surface	None	Modest	•••••	[163]
	Use of misch-metal instead of pure La	Increase in degree of disorder	None	Reduction	•••••	[164,165]
	Increase in Ce and Nd content	Increase in oxidation resistance	None	Modest	•••••	[166]
	Zr addition	Decrease in pulverization rate	None	Negligible	•••••	[167,168]
	Ti addition	Decrease in pulverization rate	None	Negligible	••••	[168,169]
	Use of hyper-stoichiometry	Reduction in pressure-concentration-temperature hysteresis and pulverization	None	None	••••	[92,163]
B. Alloy preparation	Fast quenching-gas atomization	Distribution of stress from lattice expansion	None	Modest	•••••	[40,170–173]
	Fast quenching-melt spin	Improvement in alloy homogeneity	None	Modest	•••••	[174,175]
C. Surface treatment	Ni surface plating	Protection of alloy surface from oxidation and reduction in inner pressure	None	Modest	•••••	[176,177]
	Cu coating	Protection of alloy surface from oxidation	None	Modest	••••	[178–181]
	Co coating	Protection of alloy surface from oxidation	None	Modest	••••	[182]
	Pd coating	Protection of alloy surface from oxidation	None	High	••••	[183]
	Ni-B alloy coating	Protection of alloy surface from oxidation	None	Modest	••••	[89]
	Ni-P alloy coating	Protection of alloy surface from oxidation	None	Modest	••••	[184]
	Ni-S alloy coating	Protection of alloy surface from oxidation	None	Modest	••••	[185]
	Ni-Cu alloy coating	Protection of alloy surface from oxidation	None	Modest	••••	[186]
	Alkaline pre-activation	Formation of a Ni-rich surface	None	Modest	•••••	[187]
	KBH_4 treatment	Formation of a Ni-rich surface	Toxic in contact with skin	Modest	•••••	[187,188]
	Surface fluorination	Protection of alloy surface from oxidation	None	Modest	•••••	[189–192]
	Cu and HF surface treatment	Formation of CuF_2 protective layer on the surface	None	Modest	•••	[193]
D. Other treatments	AB_5 annealing	Improvement in Mn homogeneity and reduction in inner pressure	None	Modest	•••••	[166,194–196]
	La-A_2B_7 annealing	Improvement in phase homogeneity	None	Modest	•••••	[197]
	Magnetization	Improvement in mechanical integrity	None	Modest	•••	[198]
	Ultrasound treatment	Reduction in pulverization	None	Modest	•••	[128]
E. Additives	Ni fine powder	Increase in mechanical integrity	None	Negligible	••••	[199]
	Cu fine powder	Increase in mechanical integrity	None	Negligible	•••	[200]
	Co-compounds	Increase in oxidation resistance	None	Modest	••••	[60,201–203]
	CMC:PVA (3:2)	Increase in mechanical integrity	None	Negligible	••••	[204]
	Ratio of binder to conductive additives	Increase in mechanical integrity	None	None	••••	[205]
	PTFE	Improvement in hydrogen gas absorption capability to reduce pressure	None	Negligible	••••	[206]
	Teflonized carbon	Creation of 3D conductive network	None	Negligible	••••	[207]
	HEC	Improvement in hydrogen gas absorption capability to reduce pressure	Very low toxicity if swallowed	Negligible	••••	[127,195]
	BC-1 (irigenin)	Improvement in gas recombination rate	None	Negligible	••••	[208]
	Carbon nanotube	Increase in mechanical integrity	None	Modest	••••	[209,210]
	Y_2O_3	Improvement in corrosion resistance	None	Modest	••••	[211]
	Oxides of light RE	Improvement in corrosion resistance	None	Modest	••••	[212]
	Oxides of heavy RE	Improvement in corrosion resistance	None	Modest	••••	[213,214]
F. Electrode type	Use of a pellet electrode	Increase in mechanical integrity	None	Reduction	•••	[215]
	Use of a sintered type electrode	Increase in mechanical integrity	None	Reduction	••••	[216]

Positive Electrode

Currently, the most commonly used positive active material is a spherical hydroxide co-precipitated from sulphates [45] of Ni, Co, and Zn [217]. Ni has been in active use for more than one hundred years due to the chemical reversibility between Ni^{2+} and Ni^{3+} and a voltage slightly above the oxygen gas evolution potential that maximizes energy density for aqueous chemistries. Both Zn and Cd [218] are good suppressors of γ-NiOOH formation, which causes swelling of the positive electrode and consequently premature failure, but Cd is highly toxic to the environment. The element Co is interesting in that it has oxides with different oxidation states (CoO, Co_2O_3, Co_3O_4, β-CoOOH, β-$H_{0.5}CoO_2$ [219], and Co^{4+} [220]). The mechanism of reaction for Co in alkaline solution is rather complicated [38,221], but a simplified version for electrochemical engineers can be used as a guideline. Co in a +2 state is not a good conductor for electrons or protons, and it is only slightly soluble in 30% KOH. Co can be oxidized into the +3 state through solid-state reaction [222], and it is a good conductor for both electrons and protons due to the half-filled proton plane between two Co-O layers in the CoOOH crystal structure; however, the reaction is not easily reversible. The presence of Co^{4+} through a solid-state reaction can be detected at charge rates greater than C/5, and it can be reduced back to Co^{3+} at a potential of 1.05 V *versus* Cd-electrode [220]. There are three general methods for incorporating Co into the positive electrode of Ni/MH batteries, which leverage the irreversible oxidation of Co^{2+} in the normal operation voltage range (>0.63 V *versus* Cd-electrode [220]). First, Co co-precipitated with the spherical hydroxide particles form Co^{3+} to enhance the electron and proton conductivities for $Ni(OH)_2$. Second, the addition of Co, CoO, or other Co-compound into the electrode paste allows the formation of a CoOOH conductive network that surround the spherical $Ni(OH)_2$ particles. This Co-conductive network is crucial for the operation of Ni/MH batteries, especially at high rate conditions, but they can have issues with distribution, thickness uniformity, and severe degradation at high-temperatures [223]. A third method involves adding a pre-coating of CoOOH onto the spherical particles prior to making the slurry for the electrode paste, which can involve a wet-precipitation [224, 225, 226], a mud-slurry [227], or a dry mixing method. The use of Co in the pre-coating form is the most effective and economical method, and thus is indispensable in high-end Ni/MH consumer products. Suggestions to improve cycle stability related to the positive electrode are summarized in Table 6 and are categorized by spherical particle composition and size, coatings, additives, fabrication process, and substrates.

Table 6. Summary of cycle stability improvement methods related to the positive electrode. NPPS: nickel plated perforated stainless steel.

Method		Direct impact	Environmental impact	Cost impact	Effectiveness	References
A. Composition and particle size	Co-precipitation of Co	Increase in intrinsic conductivity	None	Modest	•••••	[228]
	Co-precipitation of Zn	Prevention of γ-NiOOH formation	None	Negligible	•••••	[93,229]
	Co-precipitation of Mg and/or Ca	Improvement in high-temperature performance	None	Negligible	•••	[230]
	New type of Ni-Al double layered hydroxide	High capacity α-$Ni(OH)_2$/γ-NiOOH	None	Negligible	••••	[58]
	Increase in $Ni(OH)_2$ crystallite size	Trade-off in activation	None	None	••••	[231]
B. Surface coating	CoOOH coating	Enhancement in survival rate after long-term storage	None	Modest	•••••	[121,139,223]
	$Yb(OH)_3$ coating	Improvement in high-temperature performance	None	Modest	••••	[232]
	Electrode-less plating of Co	Improvement in Co-conductive network	None	Modest	••••	[233]
	Co/Yb hydroxide coating	Improvement in high-temperature performance	None	Modest	••••	[234]
C. Additives	Nano-sized $Ni(OH)_2$	Increase in electrochemical reaction reversibility	None	None	••••	[235]
	Nano-sized ZnO	Increase in the flexibility of the electrode	None	None	••••	[236]
	Co in paste	Formation of conductive Co-network	None	Modest	••••	[237-239]
	CoO in paste	Formation of conductive Co-network	None	Modest	••••	[110,240]
	$Co(OH)_2$ in paste	Formation of conductive Co-network	None	Modest	•••••	[195,241,242]
	CoOOH in paste	Formation of conductive Co-network	None	Modest	•••••	[243,244]
	$CoSO_4$ in paste	Formation of conductive Co-network	None	Modest	••••	[245]
	Co_3O_4 in paste	Formation of conductive Co-network	None	Modest	••••	[246]
	Co and $CaCo_3$	Prevention of oxygen evolution	None	Modest	••••	[247,248]
	CuO in paste	Uniform dispersion of Co-conductive network	None	Negligible	•••	[249]
	ZnO in paste	Prevention of oxygen evolution	None	Negligible	•••	[250,251]
	$Zn(OH)_2$ in paste	Prevention of electrode swelling	None	Negligible	•••	[252]
	$Na_{0.6}CoO_2$	Formation of better conductive Co-network	None	Modest	••••	[253-255]
	RE	Decrease in oxidation rate of MH alloy	None	Modest	•••••	[256-259]
	Y_2O_3	Decrease in oxidation rate of MH alloy	None	Modest	•••••	[23,250,260-262]
	$Y(OH)_3$	Decrease in oxidation rate of MH alloy	None	Modest	•••••	[263,264]
	Oxides of heavy RE	Improvement in corrosion resistance	None	Modest	••••	[213,265,266]
	Calcium metal borate	Prevention of oxygen evolution	None	Negligible	••••	[267]
	CaF_2	Improvement in high-temperature performance	None	Negligible	•••	[116]
	$Ca(OH)_2$	Improvement in high-temperature performance	None	Negligible	•••	[268,269]
	CaS	Improvement in high-temperature performance	Reacts with acid and releases toxic H_2S gas	Negligible	•••	[270]
	$Ca_3(PO_4)_2$	Improvement in high-temperature performance	None	Negligible	•••	[271]
D. Electrode process	Use of sintered electrode	Enhancement in survival rate after long-term storage	None	Reduction	•••••	[272]
	Use of pasted electrode on NPPS	Increase in mechanical integrity	None	Reduction	•••	[273]
	Use of granulated particles	Suppression of electrode swelling	None	None	•••••	[274]
E. Substrate	Use of 3D Ni-plated steel sheet	Increase in power and cycle stability	None	Modest	••••	[275]
	Use of Ni fiber felt	Increase in surface area and flexibility	None	Modest	••••	[276]
	Pre-coating of Co-Ce alloy	Increase in contact area between substrate and $Ni(OH)_2$	None	Modest	•••	[277]

Separator

The selection of the separator has a strong impact on the discharge capacity, voltage, and cycle stability [278]. Degradation related to the separator under storage and cycling conditions includes: (1) lower rates of electrolyte permeation in the separator; (2) lower electrolyte holding capability; (3) reduction in the separator volume due to electrode expansion; and (4) reduced gas recombination abilities [96]. Degradations (1) and (2) can be attributed to the debris formed in the separator as precipitation products ($ZnMn_2O_4$) of ions leached from the negative and positive electrodes [23,67]. These deposits not only offer a path for self-discharge, but also reduce the ionic conductivity and electrolyte holding capacity by filling the fine pores in the separator [26,29]. Degradation (3) can be traced to swelling of the positive electrode active material that accompanies over-charging, converting β-NiOOH to γ-NiOOH with Al-contamination leached from the negative electrode [29]. Methods to address separator degradation are listed in Table 7.

Table 7. Summary of cycle stability improvement methods related to the separator. EVOH: ethylene-vinyl alcohol copolymer; and AMPE: alkaline microporous polymer electrolyte.

Method	Direct impact	Environmental impact	Cost impact	Effectiveness	References
Sulfonated separator	Reduction in N-compound shuttling effects	None	Modest	*****	[23,279–282]
Grafted acrylic acid/PP	Improvement in electrolyte holding capability	None	Negligible	***	[283]
Polymer gel-type	Improvement in durability	None	Negligible	***	[284,285]
Hydroentangled CMC composite	Improvement in integrity	None	Negligible	***	[286]
EVOH	Improvement in integrity	Cytotoxic	Modest	***	[287,288]
AMPE	Improvement in voltage window	None	Modest	**	[289]
Addition of a K-conducting solid oxide film	Elimination of cross-contamination from the negative electrode	None	High	**	New idea

Electrolyte

The earliest indication of performance degradation is a decrease in cell voltage, which can be traced to a reduction in the amount of electrolyte stored in the separator [50]. Electrolyte losses can be traced to: (1) electrode active material expansion and pulverization, causing an increase in surface area and the wicking of electrolyte away from the separator [54,290]; (2) venting of the cell; and (3) oxidation of metal [98,290]. The contamination in/through the electrolyte is also crucial for the life of both electrodes. Strategies involving the modification of electrolyte that can enhance cycle life performance in Ni/MH batteries are listed in Table 8.

Table 8. Summary of cycle stability improvement methods related to the electrolyte

Method	Direct impact	Environmental impact	Cost impact	Effectiveness	References
Reduction in KOH concentration	Slow-down in alloy oxidation	None	None	****	[291]
Replacement with NaOH	Slow-down in alloy oxidation	None	Negligible	*****	[292]
ZnO additives	Slow-down in alloy oxidation	None	Negligible	***	[293]
LiOH additives	Prevention of K^+ migrating into $Ni(OH)_2$ and suppression of Fe-poisoning	None	Negligible	***	[93]
$Al_2(SO_4)_3$ additives	Slow-down in alloy oxidation	None	Negligible	***	[91]
NaH_2PO_4 additives	Formation of a Ni-rich surface on MH alloy	None	Negligible	***	[294]
$NaBO_2$ additives	Improvement of high-temperature cycle stability	None	Negligible	***	[295]
Na_2WO_4 additives	Increase in oxygen evolutionary potential	None	Negligible	***	[296]
$K_4Fe(CN)_6$ additives	Prevention of electrolyte decomposition	Highly toxic	Modest	***	[297]
Use of gel-type electrolyte	Reduction in corrosion and pulverization in the positive electrode	None	Modest	***	[298,299]
Use of polymer electrolyte	Wide voltage window and better mechanical integrity	None	Modest	****	[300–310]

Other Components

Strategies to improve cycle stability not covered in Section 3.1, Section 3.2, Section 3.3, Section 3.4and Section 3.5 are summarized in Table 9, which include charging processes, formation processes, storage conditions, and hardware modifications.

Table 9. Summary of cycle stability improvement methods related to other components. OCV: open-circuit voltage

Method	Direct impact	Environmental Impact	Cost impact	Effectiveness	References
Install super water absorbing material at cell bottom	Reservoir for additional electrolyte	None	Negligible	*****	[104]
Maintain cell OCV above 1.0 V	Prevention of Co dissolution and migration from the conductive network in the positive electrode	None	None	****	[107,311,312]
Maintain cell OCV above 1.1 V	Prevention of Co dissolution and migration from the conductive network in the positive electrode	None	None	*****	[313]
Reduction of depth of discharge	Prevention of swelling in the positive electrode	None	None	*****	[110]
Reduction of number of shallow depth discharge	Prevention of memory effect	None	None	*****	[314]
Implementation of an improved battery management system	Prevention of abuse	None	Modest	*****	[315]
Pulse charging	Reduction in heat generated	None	Negligible	***	[316]
Optimization of formation parameters	Reduction in cell performance variation	None	None	****	[317,318]
Battery sealing under vacuum	Reduction in inner pressure	None	Modest	**	[319]
Improvement in sealing technology	Prevention of electrolyte leak	None	Negligible	**	[320,321]

REVIVAL OF DEGRADED/FAILED BATTERY

After long-term storage, a few small-current charge/discharge cycles can bring back some of the lost capacity in Ni/MH batteries [322,323]. A more complicated method proposed by Li and Meng [324] involves 33% SOC small-current charge, high-temperature storage (45–60 °C for 20–24 h), and a small current charge/discharge cycle to restore at least part of the lost capacity. Its strategy is to redistribute the Co-conductive network that was destroyed during storage. An alternative method uses ultrasound to disperse active materials from both electrodes in order to create a fresh surface and increase the capacity and power of the used cells [325].

Since the most common failure mode for Ni/MH batteries is electrolyte dry-out, opening cycled cells and refilling with fresh electrolyte can restore the capacity almost to the level before cycling [204]. For re-activation of MH alloy, a patent describes a method of recycling a deteriorated nickel-hydrogen battery by cleaning the cells with a concentrated sulfuric acid containing at least one type of Ni ion, Co ions, and La ions [326]. The concentrated sulfuric acid is poured into the deteriorated nickel-hydrogen battery and maintained at a temperature of 60 ± 10 °C while an electric current is applied to charge the nickel-hydrogen battery. After cleaning, the interior of the nickel-hydrogen battery is filled with an alkaline electrolyte containing a reducing agent. Consequently, γ-NiOOH converts to β-NiOOH, which restores the capacity of the positive electrode, and $RE(OH)_3$, $Al(OH)_3$, $Mn(OH)_2$, and $Co(OH)_2$ dissolve in the concentrated sulfuric acid to activate the negative electrode surface. In addition, the hydrophilic properties of the separator are restored following this method. Recycling used negative electrodes is also possible through the removal of oxide by acetic acid [327]. At the end of usable cycle life, procedures of dismantling, recovery, and reuse of spent Ni/MH batteries have been reported by Nan *et al.* [328,329], Tenorio and Espinosa [330], Bertuol *et al.* [331], Zhang *et al.* [332], Rodrigues and Mansur [333], Muller and Friedrich [334], Rabah *et al.*[335], Santos *et al.* [336], and Larsson *et al.* [337]. U.S. Patents regarding recycling Ni/MH batteries are reviewed in a separate article [2].

CONCLUSIONS

Various failure modes and capacity degradation mechanisms are reviewed here. Solutions to enhance the cycle stability have been summarized in seven tables covering cell design, negative and positive electrodes, separator, electrolyte, and other hardware. After investigating the capacity-fade issue in a single cell, the next step is to study the consistency in the capacity degradation in a battery module composed of multiple cells.

ACKNOWLEDGMENTS

The authors would like to thank the following people for their help: from Wayne State University (Simon Ng, Shuli Yan, and Ms. Shiuan Chang), BASF-Ovonic (Michael A. Fetcenko, Taihei Ouchi, Benjamin Reichman, Cristian A. Fierro, John Koch, Mr. Michael Zelinsky, Benjamin Chao, Jean Nei, Baoquan Huang, Tiejun Meng, Lixin Wang, Jun Im, Haoting Shen, Diana F. Wong, David Pawlik, Allen Chan, Ryan J. Blankenship, Nathan English, and Su Cronogue), FDK (Masazumi Tsukada, Hiroaki Yanagawa, Hirohito Teraoka, Kazuta Takeno, and Jun Ishida), Japan Metals & Chemicals Company (Jin Nakamura), and Shenzhen High Power Battery Company (Wenliang Li, Lingkun Kong, and Xingqun Liao).

REFERENCES

1. Ouchi, T.; Young, K. Reviews on the Japanese Patents about nickel/metal hydride batteries. *Batteries* **2016**. submitted for publication.
2. Chang, S.; Nei, J.; Young, K. Reviews on the U.S. Patents about nickel/metal hydride batteries. *Batteries* **2016**. submitted for publication.
3. Young, K. Metal Hydrides. In *Reference Module in Chemistry, Molecular Sciences and Chemical Engineering*; Reedijk, J., Ed.; Elsevier: Waltham, MA, USA, 2012.
4. Yasuoka, S.; Magari, Y.; Murata, T.; Tanaka, T.; Ishida, J.; Nakamura, H.; Nohma, T.; Kihara, M.; Baba, Y.; Teraoka, H. Development of high-capacity nickel-metal hydride batteries using superlattice hydrogen-absorbing alloys. *J. Power Sources* **2006**, *156*, 662–666.
5. Watada, M.; Kuzuhara, M.; Oshitani, M. Development trend of rechargeable nickel-metal hydride battery for replacement of dry cell. *GS Yuasa Tech. Rep.* **2006**, *3*, 46–53.
6. Teraoka, H. Development of Low Self-Discharge Nickel-Metal Hydride Battery. Available online: http://www.scribd.com/doc/9704685/Teraoka-Article-En (accessed on 30 September 2015).
7. Noréus, D. Substitution of Rechargeable NiCd Batteries. Available online: http://ec.europa.eu/environment/waste/studies/batteries/nicd.pdf (accessed on 30 September 2015).
8. Yonezu, I.; Yasuoka, S.; Kitaoka, K.; Nagae, T.; Takee, M. Advanced Ni-MH Batteries for Hybrid Electric Vehicles Using Super-Lattice Alloys as A Negative Electrode Materials. In Proceedings of the 8th International Advanced Automotive Battery and Ultracapacitor Conference (AABC-08), Tampa, FL, USA, 14–16 May 2008.
9. Young, K.; Wang, C.; Wang, L.Y.; Strunz, K. Electric Vehicle Battery Technologies. In *Electric Vehicle Integration into Modern Power Networks*; Garcia-Valle, R., Pe□as Lopes, J.A., Eds.; Springer: New York, NY, USA, 2013; pp. 15–56.
10. Yano Economic Research Institute Ltd. The Investigation Results from World Market of Batteries for Vehicle Use. Available online: https://www.yano.co.jp/press/pdf/1346.pdf(accessed on 30 September 2015).
11. Young, K.; Ng, K.Y.S.; Bendersky, L. A Technical Report of the Robust Affordable Next Generation Energy Storage System-BASF Program. *Batteries* **2016**, *2*.
12. Meng, T.; Young, K.; Beglau, D.; Yan, S.; Zeng, P.; Cheng, M.M. Hydrogenated Amorphous Silicon Thin Film Anode for Proton Conducting Batteries. *J. Power Sources* **2016**, *302*, 31–38.
13. Takasaki, T.; Nishimura, K.; Saito, M.; Fukunaga, H.; Iwaki, T.; Sakai, T. Cobalt-free nickel-metal

hydride battery for industrial applications. *J. Alloys Compd.* **2013**, *580*, S378–S381.

14. Nishimura, K.; Takasaki, T.; Sakai, T. Introduction of large-sized nickel-metal hydride battery GIGACELL® for industrial applications. *J. Alloys Compd.* **2013**, *580*, S353–S358.
15. Zelinsky, M.; Fetcenko, M.A.; Kusay, J.; Koch, J. Storage-Integrated PV Systems Using Advanced NiMH Battery Technology. In Proceedings of the 5th International Renewable Energy Storage Conference, Berlin, Germany, 22–24 November 2010.
16. Wendler, M.; Hübner, G.; Krebs, M. Development of Printed Thin and Flexible Batteries. Available online: https://www.hdmstuttgart.de/international_circle/circular/issues/11_01/ICJ_04_32_wendler_huebner_krebs.pdf (accessed on 4 October 2015).
17. Fetcenko, M.A.; Koch, J. Nickel-Metal Hydride Batteries. In *Linden's Handbook of Batteries*, 4th ed.; Reddy, T.B., Ed.; McGraw-Hill: New York, NY, USA, 2011; pp. 22.1–22.50.
18. Hall, D.S.; Lockwood, D.J.; Bock, C.; MacDougall, B.R. Nickel Hydroxides and Related Materials: A Review of Their Structures, Synthesis and Properties. *Proc. Math. Phys. Eng. Sci.* **2015**, *471*.
19. Young, K.; Nei, J. The current status of hydrogen storage alloy development for electrochemical applications. *Materials* **2013**, *6*, 4574–4608.
20. Ruiz, F.C.; Martinez, P.S.; Castro, E.B.; Humana, R.; Peretti, H.A.; Visintin, A. Effect of electrolyte concentration on the electrochemical properties of an AB_5-type alloy for Ni/MH batteries. *Int. J. Hydrog. Energy* **2013**, *38*, 240–245.
21. Karwowska, M.; Jarson, T.; Fijalkowski, K.J.; Leszczynski, P.J.; Rogilski, Z.; Czerwinski, A. Influence of electrolyte composition and temperature on behaviour of AB_5 hydrogen storage alloy used as negative electrode in Ni/MH Batteries. *J. Power Sources* **2014**, *263*, 304–309.
22. Li, X.; Chen, B.; Li, X.; Liu, J. Study on the storage performance improvement of Ni-MH cell. *Chin. Battery Ind.* **2001**, *6*, 18–20.
23. Shinyama, K.; Harada, Y.; Maeda, R.; Nakamura, H.; Matsuta, S.; Nohma, T.; Yonezu, I. Effect of separators on the self-discharge reaction in nickel-metal hydride batteries. *Res. Chem. Intermed.* **2006**, *32*, 447–452.
24. Kong, L.; Chen, B.; Young, K.; Koch, J.; Chan, A.; Li, W. Effects of Al- and Mn-contents in the negative MH alloy on the self-discharge and long-term storage properties of Ni/MH battery. *J. Power Sources* **2012**, *213*, 128–139.
25. Zhou, X.; Young, K.; West, J.; Regalado, J.; Cherisol, K. Degradation mechanisms of high-energy bipolar nickel metal hydride battery with AB_5 and A_2B_7 alloys. *J. Alloys Compd.* **2013**, *580*, S373–S377.
26. Young, K.; Koch, J.; Yasuoka, S.; Shen, H.; Bendersky, L.A. Mn in misch-metal based superlattice metal hydride alloy—Part 2 Ni/MH battery performance and failure mechanism. *J. Power Sources* **2015**, *277*, 433–442.
27. Bernard, P. Effects on the positive electrode of the corrosion of AB_5 alloys in nickel-metal-hydride batteries. *J. Electrochem. Soc.* **1998**, *145*, 456–458.
28. Maurel, F.; Knosp, B.; Backhaus-Ricoult, M. Characterization of corrosion products of AB_5-type hydrogen storage alloys for nickel-metal hydride batteries. *J. Electrochem. Soc.* **2000**, *147*, 78–86.
29. Wang, L.; Young, K.; Meng, T.; English, N.; Yasuoka, S. Partial substitution of cobalt for nickel in mixed rare earth metal based superlattice hydrogen absorbing alloy—Part 2 battery performance and failure mechanism. *J. Alloys Compd.* **2016**, *664*, 417–427.
30. Rajamathi, A.; Kamath, V.P.; Seshadri, R. Polymorphism in nickel hydroxide: Role of interstratification. *J. Mater. Chem.* **2000**, *10*, 503–506.
31. Young, K.; Chao, B.; Liu, Y.; Nei, J. Microstructures of the oxides on the activated AB_2 and AB_5 metal hydride alloys surface. *J. Alloys Compd.* **2014**, *606*, 97–104.
32. Young, K.; Chao, B.; Pawlik, D.; Shen, H. Transmission electron microscope studies in the surface oxide on the La-containing AB_2 metal hydride alloy. *J. Alloys Compd.* **2016**. submitted for publication.
33. Young, K.; Chang, S.; Chao, B.; Nei, J. Microstructures of the activated Si-containing AB_2 metal hydride alloy surface by transmission electron microscope. *Batteries* **2016**, *2*.
34. Young, K.; Huang, B.; Regmi, R.K.; Lawes, G.; Liu, Y. Comparisons of metallic clusters imbedded in the surface of AB_2, AB_5, and A_2B_7 alloys. *J. Alloys Compd.* **2010**, *506*, 831–840.
35. Ayari, M.; Paul-Boncour, V.; Lamloumi, J.; Percheron-Guégan, A.; Guillot, M. Study of the aging of $LaNi_{3.55}Mn_{0.4}Al_{0.3}(Co_{1-x}Fe_x)_{0.75}$ ($0 \leq x \leq 1$) compounds in Ni-MH batteries by SEM and magnetic measurements. *J. Magn. Magn. Mater.* **2005**, *288*, 374–383.
36. Qi, D.; Yang, Y.; Lin, Z. Investigation on degradation of storage characteristics of Ni-MH batteries. *Chin. J. Power Sources* **1998**, *22*, 236–239.
37. Watanabe, N.; Arakawa, T.; Sasaki, Y.; Yamashita, T.; Koiwa, I. Influence of the memory effect on X-ray photoelectron spectroscopy and Raman scattering in positive electrode of Ni-MH batteries. *J. Electrochem. Soc.* **2012**, *159*, A1949–A1953.
38. Higuchi, E.; Otsuka, H.; Chiku, M.; Inoue, H. Effect of pretreatment on the surface structure of a $Co(OH)_2$ electrode. *J. Power Sources* **2014**, *248*, 762–768.
39. Yan, D.; Suda, S. Electrochemical characteristics of $LaNi_{4.7}Al_{0.3}$ alloy activated by alkaline solution containing hydrazine. *J. Alloys Compd.* **1995**, *223*, 28–31.
40. Imoto, T.; Kato, K.; Higashiyama, N.; Kimoto, M.; Itoh, Y.; Nishio, K. Influence of surface treatment by HCL aqueous solution on electrochemical characteristics of a $Mm(Ni\text{-}Co\text{-}Al\text{-}Mn)_{4.76}$ alloy for nickel-metal hydride battery. *J. Alloys Compd.* **1999**, *282*, 274–278.
41. Young, K.; Ouchi, T.; Banik, A.; Koch, J.; Fetcenko, M.A. Improvement in the electrochemical properties of gas atomized AB_2 metal hydride alloys by hydrogen annealing. *Int. J. Hydrog. Energy* **2011**, *36*, 3547–3555.

42. Etiemble, A.; Idrissi, H.; Meille, S.; Roué, L. *In situ* investigation of the volume change and pulverization of hydride materials for Ni-MH batteries by concomitant generated force and acoustic emission measurements. *J. Power Sources* **2012**, *205*, 500–505.

43. Etiemble, A.; Roue, L.; Idrissi, H. Study of Metal Electrodes for Ni-MH Batteries by Acoustic Emission. In Proceedings of the European Working Group on Acoustic Emission (EWAGE 2010), Vienna, Austria, 8–10 September 2010.

44. Oliva, P.; Leonardi, J.; Laurent, J.F.; Delmas, C.; Braconnier, J.J.; Figlarz, M.; Fievet, F.; Guibert, A. Review of the structure and the electrochemistry of nickel hydroxides and oxy-hydroxides. *J. Power Sources* **1982**, *8*, 229–255.

45. Yang, S.; Yin, Y.; Chen, H.; Jia, J.; Zhang, M.; Ding, L. Charge/discharge characteristics and structural changing of spherical β-$Ni(OH)_2$ prepared from different nickel salt. *J. Electrochem.* **2001**, *7*, 310–315.

46. Bantignies, J.L.; Deabate, S.; Righi, A.; Rols, S.; Hermet, P.; Sauvajol, J.L.; Henn, F. New insight into the vibration behavior of nickel hydroxide and oxyhydroxide using inelastic neutron scattering, far/mid-infrared and Raman spectroscopies. *J. Phys. Chem. C* **2008**, *112*, 2193–2201.

47. Hall, D.S.; Lockwood, D.J.; Poirier, S.; Bock, C.; McDougall, B.R. Raman and infrared spectroscopy of α and β phases of thin nickel hydroxide films electrochemically formed on nickel. *J. Phys. Chem. A* **2012**, *116*, 6771–6784.

48. Hall, D.S.; Lockwood, D.J.; Poirier, S.; Book, C.; MacDougall, B.R. Applications of *in situ*Raman spectroscopy for identifying nickel hydroxide materials and surface layers during chemical aging. *ACS Appl. Mater. Interface* **2014**, *6*, 3141–3149.

49. Kuriyama, N.; Sakai, T.; Miyamura, H.; Uehara, I.; Ishikawa, H. Electrochemical impedance spectra and deterioration mechanism of metal hydride electrodes. *J. Electrochem. Soc.* **1992**, *139*, L72–L73.

50. Cheng, S.; Zhang, J.; Liu, H.; Leng, Y.; Yuan, A.; Cao, C. Study on cycling deterioration of Ni-MH battery by electrochemical impedance spectroscopy (EIS). *Chin. J. Power Sources* **1999**, *23*, 62–63.

51. Mo, Z.; Xu, P.; Han, X.; Zhao, L. Electrochemical impedance behavior of Ni-MH battery under different depth of discharge. *Chin. J. Power Sources* **2006**, *30*, 388–390.

52. Guenne, L.L.; Benard, P. Life duration of Ni-MH cells for high power applications. *J. Power Sources* **2002**, *105*, 134–138.

53. Geng, M.; Han, J.; Feng, F.; Northwood, D.O. Charging/discharging stability of a metal hydride battery electrode. *J. Electrochem. Soc.* **1999**, *146*, 2371–2375.

54. Yuan, J.; Qi, J.; Tu, J.; Feng, H.; Chen, C.; Chen, W. Electrochemical analysis of failed Ni-MH batteries. *Chin. J. Power Sources* **2001**, *25*, 279–282.

55. Yuan, X.; Xu, N. Electrochemical and hydrogen transport kinetic performance of $MlNi_{3.75}Co_{0.63}Mn_{0.4}Al_{0.2}$ metal hydride electrodes at various temperatures. *J. Electrochem. Soc.* **2002**, *149*, A407–A413.

56. Sequeira, C.A.C.; Chen, Y.; Santos, D.M.F. Effects of temperature on the performance of the $MmNi_{3.6}Co_{0.7}Mn_{0.4}Al_{0.3}$ metal hydride electrode in alkaline solution. *J. Electrochem. Soc.***2006**, *153*, A1863–A1867.

57. Gamboa, S.A.; Sebastian, P.J.; Feng, F.; Geng, M.; Northwood, D.O. Cyclic voltammetry investigation of a metal hydride electrode for nickel metal hydride batteries. *J. Electrochem. Soc.* **2002**, *149*, A137–A139.

58. Béléké, A.B.; Higuchi, E.; Inoue, H.; Mizuhata, M. Durability of nickel-metal hydride (Ni-MH) battery cathode using nickel-aluminum layered double hydroxide/carbon (Ni-AL LDH/C) composite. *J. Power Sources* **2014**, *247*, 572–578.

59. Lyons, M.E.G.; Doyle, R.L.; Godwin, I.; O'Brien, M.; Russell, L. Hydrous nickel oxide: Redox switching and the oxygen evolution reaction in aqueous alkaline solution. *J. Electrochem. Soc.***2012**, *159*, H932–H944.

60. Zhang, Q.; Su, G.; Li, A.; Liu, K. Electrochemical performance of AB_5-type hydrogen storage alloy modified with Co_3O_4. *Trans. Nonferrous Met. Soc. Chin.* **2011**, *21*, 1428–1434.

61. Serrao, L.; Chehab, Z.; Guezennec, Y.; Rizzoni, G. An Aging Model of Ni-MH Batteries for Hybrid Electric Vehicles. In Proceedings of the IEEE Vehicle Power and Propulsion Conference, Chicago, IL, USA, 7–9 September 2005; pp. 78–85.

62. Somogye, R.H. An Aging Model of Ni-MH Batteries for Use in Hybrid-Electric Vehicles. Master's Thesis, Ohio State University, Columbus, OH, USA, 2004.

63. Picciano, N. Battery Aging and Characterization of Nickel Metal Hydride and Lead Acid Batteries. Bachelor's Thesis, Ohio State University, Columbus, OH, USA, 2007.

64. Ikoma, M.; Hoshina, Y.; Matsumoto, I. Self-discharge mechanism of sealed-type nickel/metal-hydride battery. *J. Electrochem. Soc.* **1996**, *143*, 1904–1907.

65. Young, K.; Ouchi, T.; Koch, J.; Fetcenko, M.A. Compositional optimization of vanadium-free hypo-stoichiometric AB_2 metal hydride alloy for Ni/MH battery application. *J. Alloys Compd.***2012**, *510*, 97–106.

66. Guo, H.; Qiao, Y.; Zhang, H. Fast determination of micro short circuit in sintered MH-Ni battery. *Chin. J. Power Sources* **2010**, *34*, 608–609.

67. Shinyama, K.; Magari, Y.; Kumagae, K.; Nakamura, H.; Nohma, T.; Takee, M.; Ishiwa, K. Deterioration mechanism of nickel metal-hydride batteries for hybrid electric vehicles. *J. Power Sources* **2005**, *141*, 193–197.

68. Zhu, W.H.; Zhu, Y.; Tatarchuk, B.J. Self-discharge characteristics and performance degradation of Ni-MH batteries for storage applications. *Int. J. Hydrog. Energy* **2014**, *39*, 19789–19798.

69. Lee, J.-H.; Lee, K.-Y.; Lee, J.-Y. Self-discharge behaviour of sealed Ni-MH batteries using $MmNi_{3.3+x}Co_{0.7}Al_{1.0-x}$ anodes. *J. Alloys Compd.* **1996**, *232*, 197–203.

70. Wang, C.; Marrero-Rivera, M.; Serafini, D.A.;

Baricuatro, J.H.; Soriaga, M.P.; Srinivasan, S. The self-discharge mechanism of AB_5-type hydride electrodes in Ni/MH batteries. *Int. J. Hydrog. Energy* **2006**, *31*, 603–611.

71. Lee, J.; Lee, K.; Lee, S.; Lee, J. Self-discharge characteristics of sealed Ni-MH batteries using $Zr_{1-x}Ti_xV_{0.8}Ni_{1.6}$ anodes. *J. Alloys Compd.* **1995**, *221*, 174–179.
72. Jang, K.; Jung, J.; Kim, D.; Yu, J.; Lee, J. Self-discharge mechanism of vanadium-titanium metal hydride electrodes for Ni-MH rechargeable battery. *J. Alloys Compd.* **1998**, *268*, 290–294.
73. Yang, X.; Liaw, B.Y. Self-discharge and charge retention in AB_2-based nickel metal hydride batteries. *J. Electrochem. Soc.* **2004**, *151*, A137–A143.
74. Fan, M.; Liao, W.; Wu, B.; Chen, H.; Jian, X. Temperature characteristic of Ni-MH battery used in EVs. *Chin. Battery Ind.* **2004**, *9*, 287–289.
75. Huang, J.; Li, X. Factors affecting the self-discharge of Ni-MH batteries. *Chin. Battery Ind.***2005**, *10*, 290–294.
76. Zhu, W.H.; Zhu, Y.; Davis, Z.; Tatarchuk, B.J. Energy efficiency and capacity retention of Ni-MH batteries for storage applications. *Appl. Energy* **2013**, *106*, 307–313.
77. NiMH Battery Charging Algorithm. Available online: http://www.schaffler.com/whitepapers/NiMH Battery Algorithm.pdf (accessed on 4 October 2015).
78. Kritzer, P. Separators for nickel metal hydride and nickel cadmium batteries designed to reduce self-discharge rates. *J. Power Sources* **2004**, *137*, 317–321.
79. Liu, Y.; Tang, Z.; Xu, Q.; Zhang, X.; Liu, Y. Assessment of separators in high rate discharge performance of Ni–MH battery. *Chin. J. Process Eng.* **2006**, *6*, 114–119.
80. Li, X.; Song, Y.; Wang, L.; Xia, T.; Li, S. Self-discharge mechanism of Ni-MH battery by using acrylic acid grafted polypropylene separator. *Int. J. Hydrog. Energy* **2010**, *35*, 3798–3801.
81. Takasaki, T.; Nishimura, K.; Saito, M.; Iwaki, T.; Sakai, T. Cobalt-free materials for nickel-metal hydride battery: Self-discharge suppression and overdischarge resistance improvement. *Electrochemistry* **2013**, *81*, 553–558.
82. Li, X.; Li, J.; Tan, P.; Li, Y. Effects of electrolyte weight on self-discharge of Ni-MH battery. *Chin. Battery Ind.* **2008**, *13*, 299–302.
83. Zai, Y.; Han, J.; Li, Y.; Lai, X.; Duan, Q. Influence of copper on the micro-short in Ni-MH battery. *Chin. Battery Ind.* **2012**, *17*, 78–80.
84. Zha, Y.; Chen, X.; Li, L.; Yang, Z.; Wang, L. Effects of the negative current collector on the performance of the Ni-MH battery. *Chin. Battery Ind.* **2006**, *11*, 91–93.
85. Cheng, S.; Zhang, J.; Yuan, A.; Ding, W.; Cao, C. Effects of separator on discharge capacity and cycle-life of Ni-MH battery. *Chin. J. Power Sources* **1999**, *23*, 10–12.
86. Li, X.; Wang, X.; Dong, H.; Xia, T.; Wang, L.; Song, Y. Self-discharge performance of Ni-MH battery by using electrodes with hydrophilic/hydrophobic surface. *J. Phys. Chem. Solids* **2013**, *74*, 1756–1760.
87. Li, X.; Li, J. Effects of the surface treatment of formed positive electrode on self-discharge performance of Ni-MH batteries. *Chin. Battery Ind.* **2009**, *14*, 298–301.
88. Feng, F.; Northwood, D.O. Self-discharge characteristics of a metal hydride electrode for Ni-MH rechargeable batteries. *Int. J. Hydrog. Energy* **2005**, *30*, 1367–1370.
89. Lv, Y.; Zhang, J. Study on the performance of metal-hydride electrodes coated with Ni-B alloy. *Contemp. Chem. Ind.* **2014**, *43*, 1453–1455, 1463.
90. Deng, Z. Approach on influence factors of discharge oneself MH-Ni battery. *Jiangxi Metal.***2003**, *23*, 137–139.
91. Wang, Z.M.; Tsai, P.; Chan, S.L.I.; Zhou, H.Y.; Lin, K.S. Effects of electrolytes and temperature on high-rate discharge behavior of $MmNi_5$-based hydrogen storage alloys. *Int. J. Hydrog. Energy* **2010**, *35*, 2033–2039.
92. Notten, P.H.L.; Einerhand, R.E.F.; Daams, J.L.C. How to achieve long-term electrochemical cycling stability with hydride-forming electrode materials. *J. Alloys Compd.* **1995**, *231*, 604–610.
93. Dong, Q.; Yan, G.; Yu, C. Research on the cycle life of Ni-MH battery. *Chin. J. Power Sources***2000**, *24*, 306–310.
94. Yan, J.; Zhou, Z.; Li, Y.; Song, D.; Zhang, Y. Structure and property changes of positive and negative electrodes in Ni/MH batteries during charge/discharge cycles. *J. Inorg. Chem.* **1998**, *14*, 74–78.
95. Tliha, M.; Mathlouthi, H.; Lamloumi, J.; Percheron-Guegan, A. AB_5-type hydrogen storage alloy used as anodic materials in Ni-MH batteries. *J. Alloys Compd.* **2007**, *436*, 221–225.
96. Lin, J.; Cheng, Y.; Liang, F.; Sun, L.; Yin, D.; Wu, Y.; Wang, L. High temperature performance of $La_{0.6}Ce_{0.4}Ni_{3.45}Co_{0.75}Mn_{0.7}Al_{0.1}$ hydrogen storage alloy for nickel/metal hydride batteries. *Int. J. Hydrog. Energy* **2014**, *39*, 13231–13239.
97. Meli, F.; Schlapbach, L. Surface analysis of AB_5-type electrodes. *J. Less Common Met.* **1991**, *172*, 1252–1259.
98. Li, L.; Chen, Y.; Wu, F.; Chen, S. Development of recycling and rusing process for nickel metal hydride batteries for HEV. *J. Funct. Mater.* **2007**, *38*, 1928–1932.
99. Geng, M.; Hah, J.; Feng, F.; Northwood, D.O. Characteristics of the high-rate discharge capability of a nickel/metal hydride battery electrode. *J. Electrochem. Soc.* **1999**, *146*, 3591–3595.
100. Wang, C.; Jin, H.; Li, G.; Wang, R. Study on microstructures of $La_{0.8}Nd_{0.2}Ni_{3.55}Co_{0.75}Mn_{0.4}Al_{0.3}$ alloy during the charge-discharge cycle process. *Chin. J. Power Sources* **1998**, *22*, 65–67.
101. Chen, Y.; Wei, J.; Gao, F.; Zhou, W.; Zhang, Y.; Zhou, Z. Discussing of invalidation factors of Ni-MH battery at high charge and discharge rate. *Chin. J. Power Sources* **2001**, *25*, 142–145.
102. Maurel, F.; Hytch, M.J.; Knosp, B.; Backhaus-Ricoult, M. Formation of rare earth hydroxide nanotubes and whiskers as corrosion product of $LaNi_5$-type alloys in aqueous KOH. *Eur. Phys. J. Appl. Phys.* **2000**, *9*, 205–213.
103. Liu, Y.; Pan, H.; Gao, M.; Lei, Y.; Wang, Q. Degradation

mechanism of the La-Mg-Ni-based metal hydride electrode $La_{0.7}Mg_{0.3}Ni_{3.4}Mn_{0.1}$. *J. Electrochem. Soc.* **2005**, *152*, A1089–A1095.

104. Xie, D.; Cheng, H. Study on the fading of MH-Ni batteries. *Chin. Battery Ind.* **2000**, *5*, 7–10.

105. Xie, J.; Wang, S.; Xia, B.; Zhang, Q.; Shi, P. Research on the degradation of Ni-MH batteries (I)—Swelling of nickel electrode. *Chin. J. Power Sources* **1997**, *21*, 22–27.

106. Singh, D. Characteristics and Effects of γ-NiOOH on cell performance and a method to quantify it in nickel electrodes. *J. Electrochem. Soc.* **1998**, *145*, 116–120.

107. Deng, X.; Huang, S.; Wang, Y.; Wang, J. Study on the shelf performance of Ni-MH batteries. *Chin. Battery Ind.* **2006**, *11*, 172–174.

108. Durairajan, A.; Haran, B.S.; White, R.E.; Popov, B.N. Pulverization and corrosion studies of bare and cobalt-encapsulated metal hydride electrodes. *J. Power Sources* **2000**, *87*, 84–91.

109. Li, X.; Li, X.; He, X.; Liu, J.; Chen, B.; Li, C. Investigation on internal resistance of Ni-MH battery during cycle life. *Chin. Battery Ind.* **2001**, *6*, 69–71.

110. Wu, J.; Li, W.; Zai, Y. Relations of specific capacity of Ni-MH battery with self discharge and cycle life. *Chin. Battery Ind.* **2004**, *9*, 197–199.

111. Sastry, A.M.; Choi, S.B.; Cheng, X. Damage in composite NiMH positive electrodes. *J. Eng. Mater. Technol.* **1998**, *120*, 280–283.

112. Zhou, Z.Q.; Lin, G.W.; Zhang, J.L.; Ge, J.S.; Shen, J.R. Degradation behavior of foamed nickel positive electrodes of Ni-MH batteries. *J. Alloys Compd.* **1999**, *293–295*, 795–798.

113. Young, K.; Wu, A.; Qiu, Z.; Tan, J.; Mays, W. Effects of H_2O_2 addition to the cell balance and self-discharge of Ni/MH batteries with AB_5 and A_2B_7 alloys. *Int. J. Hydrog. Energy* **2012**, *37*, 9882–9891.

114. Chen, W.; Xu, Z.; Tu, J. Electrochemical investigations of activation and degradation of hydrogen storage alloy electrodes in sealed Ni/MH battery. *Int. J. Hydrog. Energy* **2007**, *27*, 439–444.

115. Singh, P.; Wu, T.; Wendung, M.; Bendale, P.; Ware, J.; Ritter, D.; Zhang, L. Mechanisms causing capacity loss on long term storage in NiMH system. *Mater. Res. Soc. Symp. Proc.* **1998**, *496*, 25–36.

116. Zhang, X.; Gong, Z.; Zhao, S.; Geng, M.; Wang, Y.; Northwood, D.O. High-temperature characteristics of advanced Ni-MH batteries using nickel electrodes containing CaF_2. *J. Power Sources* **2008**, *175*, 630–634.

117. Iwakura, C.; Kajiya, Y.; Yoneyama, H.; Sakai, T.; Oguro, K.; Ishikawa, H. Self-discharge mechanism of nickel-hydrogen batteries using metal hydride anodes. *J. Electrochem. Soc.* **1989**, *136*, 1351–1355.

118. Leblanc, P.; Blanchard, P.; Senyarich, S. Self-discharge of sealed nickel-metal hydride batteries. *J. Electrochem. Soc.* **1998**, *145*, 844–847.

119. Wang, C.S.; Marrero-Rivera, M.; Baricuatro, J.H.; Soriaga, M.P.; Serafini, D.; Srinivasan, S. Corrosion behavior of AB_5-type hydride electrodes in alkaline electrolyte solution. *J. Appl. Electrochem.* **2003**, *33*, 325–331.

120. Rukpakavong, W.; Phillips, I.; Guan, L. Lifetime Estimation of Sensor Device with AA NiMH Batteries. Available online: http://www.ipcsit.com/vol55/018-ICICM2012-Contents.pdf(accessed on 04 October 2015).

121. Bäuerlein, P.; Antonius, A.; Löffler, J.; Kümpers, J. Progress in high-power nickel–metal hydride batteries. *J. Power Sources* **2008**, *176*, 547–554.

122. Shi, F. The research of voltage's accuracy and temperature effect on the MH-Ni battery group. *Microcomput. Appl.* **2013**, *32*, 60–62.

123. Li, X.; Ma, L.; Lou, Y.; Li, F.; Xia, B. Storage performance of Ni-MH battery at state of discharge. *Chin. J. Power Sources* **2004**, *28*, 364–368.

124. Notten, P.H.L.; Van Beek, J.R.G. Nickel-metal hydride batteries: From concept to characteristics. *Chem. Ind.* **2000**, *54*, 102–115.

125. Li, L.; Wu, F.; Yang, K.; Wang, J.; Chen, S. The effects of overcharge on electrochemical performance of MH-Ni batteries. *Mater. Rev.* **2004**, *18*, 101–102.

126. Hu, W.K.; Geng, M.M.; Gao, X.P.; Burchardt, T.; Gong, Z.X.; Noréus, D.; Nakstad, N.K. Effect of long-term overcharge and operated temperature on performance of rechargeable NiMH cells. *J. Power Sources* **2006**, *159*, 1478–1483.

127. Ye,Z.; Noréus, D. Oxygen and hydrogen gas recombination in NiMH cells. *J. Power Sources***2012**, *208*, 232–236.

128. Chen, R.; Li, L.; Wu, F.; Qiu, X.; Chen, S. Effects of low temperature on performance of hydrogen-storage alloys and electrolyte. *Min. Metal. Eng.* **2007**, *27*, 44–46.

129. Guo, H.; Chao, M.; Chen, X.; Zhang, B.; Zhang, J. Study on the low temperature performance of MH-Ni battery. *J. Henan Normal Univ. Nat. Sci.* **2003**, *31*, 76–78.

130. Liaw, B.Y.; Yang, X. Limiting mechanism on rapid charging Ni-MH batteries. *Electrochim. Acta* **2001**, *47*, 875–884.

131. Raju, M.; Ananth, M.V.; Vijayaraghavan, L. Rapid charging characterization of $MmNi_{3.03}Si_{0.85}Co_{0.60}Mn_{0.31}Al_{0.08}$ alloy used as anodes in Ni-MH batteries. *Int. J. Hydrog. Energy* **2009**, *34*, 3500–3505.

132. Chen, S.; Zheng, Q.; Wang, F.; Hu, D. Effects of fast charging on efficiency of Ni-MH batteries. *J. Beijing Technol. Bus. Univ. Nat. Sci. Ed.* **2006**, *24*, 5–8.

133. Taheri, P.; Yazdanpour, M.; Bahrami, M. Analytical assessment of the thermal behavior of nickel-metal hydride batteries during fast charging. *J. Power Sources* **2014**, *245*, 712–720.

134. Hou, X.; Xue, J.; Nan, J.; Xia, X.; Yang, M.; Cui, Y. The cycle life performance of MH-Ni batteries for high power applications. *Acta Sci. Nat. Univ. Sunyats.* **2005**, *44*, 77–80.

135. Ayeb, A.; Otten, W.M.; Mank, J.G.; Notten, P.H.L. The hydrogen evolution and oxidation kinetics during overdischarging of sealed nickel-metal hydride batteries. *J. Electrochem. Soc.***2006**, *153*, A2055–A2065.

136. Cha, C.; Yu, J.; Zhang, J. Comparative experimental study of gas evolution and gas consumption reactions in sealed Ni-Cd and Ni-MH cells. *J. Power Sources* **2004**, *129*, 347–357.

137. Cuscueta, D.J.; Salva, H.R.; Ghilarducci, A.A. Inner pressure characterization of a sealed nickel-metal hydride cell. *J. Power Sources* **2011**, *196*, 4067–4071.

138. Li, N. Methods to prevent explosion in MH-Ni battery. *Chin. Battery Ind.* **2003**, *8*, 277.

139. Li, N.; Xu, Y. Discussion on explosion of MH-Ni batteries. *Chin. Battery Ind.* **2008**, *13*, 372–373.

140. Zhu, X.; Ni, J.; Xu, W.; Huang, L.; Feng, S. Analysis of common malfunctions and maintenance of nickel metal hydride battery used in AGV. *Log. Mater. Handl.* **2013**, *10*, 156–158.

141. Liu, C.; Li, J.; Chen, H.; Liu, J. The development of high performance SubC MH-Ni battery. *J. Hunan Univ. Technol.* **2012**, *26*, 64–67.

142. Fetcenko, M.A.; Young, K.; Fierro, C. Hydrogen Storage Battery; Positive Nickel Electrode; Positive Electrode Active Material and Methods of Making. U.S. Patent 7,396,379, 8 July 2008.

143. Li, X.; Tian, J.; Xia, T. A new way to reduce the consumption of hydrogen-absorbing alloy in Ni-MH battery. *Chin. Battery Ind.* **2011**, *16*, 74–78.

144. Hu, W.; Shan, Z.; Tian, J. Impact of ratio of anode/cathode active materials on internal pressure of MH-Ni batteries. *Chem. Ind. Eng.* **2006**, *23*, 95–97.

145. Sastry, A.M.; Cheng, X.; Wang, C. Mechanics of stochastic fibrous networks. *J. Thermoplast. Compos. Mater.* **1998**, *3*, 288–296.

146. Yu, T.; Zhai, Y.; Zhou, S.; Duan, Q. Effect of precharge on performance of Ni-MH battery. *Chin. Battery Ind.* **2005**, *10*, 74–76.

147. Guiose, B.; Cuevas, F.; Décamps, B.; Leroy, E.; Percheron-Guégan, A. Microstructural analysis of the ageing of pseudo-binary (Ti,Zr)Ni intermetallic compounds as negative electrodes of Ni-MH batteries. *Electrochim. Acta* **2009**, *54*, 2781–2789.

148. Liu, B.H.; Li, Z.P.; Matsuyama, Y.; Kitani, R.; Suda, S. Corrosion and degradation behavior of Zr-based AB_2 alloy electrodes during electrochemical cycling. *J. Alloys Compd.* **2000**, *296*, 201–208.

149. Liu, B.H.; Li, Z.P.; Kitani, R.; Suda, S. Improvement of electrochemical cyclic durability of Zr-based AB_2 alloy electrodes. *J. Alloys Compd.* **2002**, *330–332*, 825–830.

150. Lee, H.; Lee, K.; Lee, J. Degradation mechanism of Ti-Zr-V-Mn-Ni metal hydride electrodes. *J. Alloys Compd.* **1997**, *260*, 201–207.

151. Yu, J.; Lee, S.; Cho, K.; Lee, J. The cycle life of $Ti_{0.8}Zr_{0.2}V_{0.5}Mn_{0.5-x}Cr_xNi_{0.8}$ ($x = 0$ to 0.5) alloys for metal hydride electrodes of Ni-metal hydride rechargeable battery. *J. Electrochem. Soc.* **2000**, *147*, 2013–2017.

152. Liu, B.H.; Li, Z.P.; Suda, S. Electrochemical cycle life of Zr-based Laves phase alloys influenced by alloy stoichiometry and composition. *J. Electrochem. Soc.* **2002**, *149*, A537–A542.

153. Martíez, P.S.; Ruiz, F.C.; Visintin, A. Influence of different electrolyte concentrations on the performance of an AB_2-type alloy. *J. Electrochem. Soc.* **2014**, *161*, A326–A329.

154. Rongeat, C.; Roué, L. On the cycle life improvement of amorphous MgNi-based alloy for Ni-MH batteries. *J. Alloys Compd.* **2005**, *404–406*, 679–681.

155. Ruggeri, S.; Roué, L. Correlation between charge input and cycle life of MgNi electrode for Ni-MH batteries. *J. Power Sources* **2003**, *117*, 260–266.

156. Rongeat, C.; Grosjean, M.-H.; Ruggeri, S.; Dehmas, M.; Bourlot, S.; Marcotte, S.; Roué, L. Evaluation of different approaches for improving the cycle life of MgNi-based electrodes for Ni-MH batteries. *J. Power Sources* **2006**, *158*, 747–753.

157. Goo, N.H.; Woo, J.H.; Lee, K.S. Mechanism of rapid degradation of nanostructured Mg_2Ni hydrogen storage alloy electrode synthesized by mechanical alloying and the effect of mechanically coating with nickel. *J. Alloys Compd.* **1999**, *288*, 286–293.

158. Tsukahara, M.; Takahashi, K.; Isomura, A.; Sakai, T. Improvement of the cycle stability of vanadium-based alloy for nickel-metal hydride (Ni-MH) battery. *J. Alloys Compd.* **1999**, *287*, 215–220.

159. Bliznakov, S.; Lefterova, E.; Dimitrov, N.; Petrov, K.; Popov, A. A study of the Al content impact on the properties of $MmNi_{4.4-x}Co_{0.6}Al_x$ alloys as precursors for negative electrodes in NiMH batteries. *J. Power Sources* **2008**, *176*, 381–386.

160. Zhou, W.; Ma, Z.; Wu, C.; Zhu, D.; Huang, L.; Chen, Y. The mechanism of suppressing capacity degradation of high-Al-AB_5-type hydrogen storage alloys at 60 °C. *Int. J. Hydrog. Energy* **2016**, *41*, 1801–1810.

161. Nakamura, Y.; Oguro, K.; Uehara, I.; Akiba, E. X-ray diffraction peak broadening and degradation in $LaNi_5$-based alloys. *Int. J. Hydrog. Energy* **2000**, *25*, 531–537.

162. Wang, X.; Liu, B.; Yuan, H.; Zhang, Y. Effect of Al in $Mm(NiCoMnAl)_5$ on the performance of the sealed MH/Ni cell. *Chin. J. Appl. Chem.* **1999**, *16*, 54–57.

163. Zhang, Y.; Wang, X.; Chen, M.; Li, P.; Lin, Y.; Li, R. Design of chemical compositions on AB_5-type hydrogen storage alloy with high quality. *Met. Funct. Mater.* **2002**, *9*, 1–6.

164. Chandra, D.; Chien, W.; Talekar, A. Metal Hydrides for NiMH Battery Applications. Available online: http://www.sigmaaldrich.com/technical-documents/articles/material-matters/metal-hydrides-for.html (accessed on 1 October 2015).

165. Qingxue, Z.; Joubert, J.-M.; Latroche, M.; Jun, D.; Percheron-Guégan, A. Influence of the rare earth composition on the properties of Ni-MH electrodes. *J. Alloys Compd.* **2003**, *360*, 290–293.

166. Guo, J.H.; Jiang, Z.T.; Duan, Q.S.; Liu, G.Z.; Zhang, J.H.; Xia, C.M.; Guo, S.Y. Influence of Ce and Nd contents of $Ml(NiCoMnAl)_5$ alloy on the performances of Ni-MH batteries. *J. Alloys Compd.* **1999**, *293–295*, 821–824.

167. Sakai, T.; Miyamura, H.; Kuriyama, N.; Kato, A.; Oguro, K.; Ishikawa, H. Metal hydride anodes for nickel-hydrogen secondary battery. *J. Electrochem. Soc.* **1990**, *137*, 795–799.

168. Seo, C.; Choi, S.; Choi, J.; Park, C.; Lee, P.S.; Lee, J. Effect of Ti and Zr additions on the characteristics of AB_5-type hydride electrode for Ni-MH secondary battery. *Int. J. Hydrog. Energy* **2003**, *28*, 317–327.

169. Srivastava, S.; Upadhyay, R.K. Investigations of AB_5-type negative electrode for nickel-metal hydride cell with regard to electrochemical and microstructural characteristics. *J. Power Sources* **2010**, *195*, 2996–3001.

170. Young, K.; Ouchi, T.; Banik, A.; Koch, J.; Fetcenko, M.A.; Bendersky, L.A.; Wang, K.; Vaudin, M. Gas atomization of Cu-modified AB_5 metal hydride alloys. *J. Alloys Compd.* **2011**, *509*, 4896–4904.

171. Spodaryk, M.; Shcherbakova, L.; Sameljuk, A.; Wichser, A.; Zakaznova-Herzog, V.; Holzer, M.; Braem, B.; Khyzhun, O.; Mauron, P.; Remhof, A.; *et al.* Description of the capacity degradation mechanism in $LaNi_5$-based alloy electrodes. *J. Alloys Compd.* **2015**, *621*, 225–231.

172. Anderson, M.L.; Anderson, I.E. Gas atomization of metal hydrides for Ni-MH battery applications. *J. Alloys Compd.* **2000**, *313*, 47–52.

173. Bowman, R.C., Jr.; Witham, C.; Fultz, B.; Ratnakumar, B.V.; Ellis, T.W.; Anderson, I.E. Hydriding behavior of gas-atomized AB_5 alloys. *J. Alloys Compd.* **1997**, *253–254*, 613–616.

174. Li, C.; Wang, F.; Cheng, W.; Li, W.; Zhao, W. The influence of high-rite quenching on the cycle stability and the structure of the AB_5-type hydrogen storage alloys with different Co content. *J. Alloys Compd.* **2001**, *315*, 218–223.

175. Li, C.; Wang, X.; Wang, C. Investigations on $ML(NiCoMnTi)_5$ alloys prepared with different solidification rates. *J. Alloys Compd.* **1998**, *266*, 300–306.

176. Zou, J.; Gao, J.; Zhong, F.; Wang, X. Effect of nickel electroplating on negative electrode surface on cycle life and inner pressure of nickel hydride battery. *Electroplat. Finish.* **2009**, *28*, 9–13.

177. Yang, K.; Wu, F.; Chen, S.; Zhang, C. Influences of Ni plated on electrodes on the cycle life and inner pressure of batteries. *Min. Metal. Eng.* **2006**, *26*, 74–76.

178. Zhang, D.; Yuan, H.; Wang, G.; Zhou, Z.; Zhang, Y. Study on electrochemical characteristic of Cu-coated hydrogen storage alloy. *J. Funct. Mater. Dev.* **1998**, *4*, 167–174.

179. Feng, F.; Northwood, D.O. Effect of surface modification on the performance of negative electrodes in Ni/MH batteries. *Int. J. Hydrog. Energy* **2004**, *29*, 955–960.

180. Huang, J.; Huang, H. Effect of Cu addition on discharge performance of metal hydride electrode for Ni-MH battery. *Rare Met. Lett.* **2000**, *3*, 18–19.

181. Raju, M.; Ananth, M.V.; Vijayaraghavan, L. Influence of electroless coatings of Cu, Ni-P and Co-P on $MmNi_{3.25}Al_{0.35}Mn_{0.25}Co_{0.66}$ alloy used as anodes in Ni-MH batteries. *J. Alloys Compd.* **2009**, *475*, 664–671.

182. Manimaran, K.; Ananth, M.V.; Raju, M.; Renganathan, N.G.; Ganesan, M.; Nithya, G. Electrochemical investigations on cobalt-microencapsulated $MmNi_{2.38}Al_{0.82}Co_{0.66}Si_{0.77}Fe_{0.13}Mn_{0.24}$ hydrogen storage alloys for Ni-MH batteries. *Int. J. Hydrog. Energy* **2010**, *35*, 4630–4637.

183. Visintin, A.; Tori, C.A.; Garaventta, G.; Triaca, W.E. The effect of palladium coating on hydrogen storage alloy electrodes for nickel/metal hydride batteries. *J. Braz. Chem. Soc.* **1997**, *8*, 125–129.

184. Lv, Y. A study on decorative electrochemical coating of hydrogen storage alloy powder with nickel-phosphorus alloy. *J. Tangshan Coll.* **2014**, *27*, 42–44.

185. Bi, X.; Shan, Z.; Tian, J. Study on the performance of metal-hydride electrodes coated with Ni-S alloy in Ni-MH battery. *Chin. J. Power Sources* **2005**, *29*, 746–749.

186. Casella, I.G.; Gatta, M. Electrodeposition and characterization of nickel-copper alloy films as electrode material in alkaline media. *J. Electrochem. Soc.* **2002**, *149*, B465–B471.

187. Chen, W.; Chen, Y.; Pan, H.; Chen, C. Surface treatments of hydrogen storage alloy improve charge-discharge performances of Ni/MH battery. *Acta Phys. Chim. Sin.* **1998**, *14*, 742–746.

188. Lee, H.; Yang, D.; Park, C.; Park, C.; Jang, H. Effects of surface modifications of the $LMNi_{3.9}Co_{0.6}Mn_{0.3}Al_{0.2}$ alloy in a $KOH/NaBH_4$ solution upon its electrode characteristics within a Ni-MH secondary battery. *Int. J. Hydrog. Energy* **2009**, *34*, 481–486.

189. Li, C. Hydrogen storage alloy and fluorinated hydrating alloy batteries. *Chin. J. Power Sources***1995**, *19*, 34–37.

190. Sakashita, M.; Li, Z.P.; Suda, S. Fluorination mechanism and its effects on the electrochemical properties of metal hydrides. *J. Alloys Compd.* **1997**, *253–254*, 500–505.

191. Sun, Y.; Iwata, K.; Chiba, S.; Matsuyama, Y.; Suda, S. Studies on the properties and characteristics of the fluorinated AB_5 hydrogen-absorbing electrode alloys. *J. Alloys Compd.***1997**, *253–254*, 520–524.

192. Sun, Y.; Suda, S. Studies on the fluorination method for improving surface properties and characteristics of AB_5-types of hydrides. *J. Alloys Compd.* **2002**, *330–332*, 627–631.

193. Chao, D.; Shi, P.; Zhang, S. Effect of surface modification on the electrochemical performances of $LaNi_5$ hydrogen storage alloy in Ni/MH batteries. *Mater. Chem. Phys.* **2006**, *98*, 514–518.

194. Lou, Y.; Ma, L.; Xia, B.; Yang, C. XRD study on microstructure of rare earth containing hydrogen storage alloy in Ni-MH battery. *Rare Met. Mater. Eng.* **2006**, *35*, 412–417.

195. Pang, L.; Ye, P.; Guo, J. Methods of reducing MH-Ni battery internal pressure and increasing cycle life. *Liaoning Chem. Ind.* **1998**, *27*, 228–230.

196. Sakai, T.; Miyamura, H.; Kuriyama, N.; Ishikawa, H.; Uehara, I. Rare-earth-based hydrogen storage alloys for rechargeable nickel-metal hydride batteries. *J. Alloys Compd.* **1993**, *192*, 155–157.

197. Gao, J.; Yan, X.L.; Zhao, Z.Y.; Chai, Y.J.; Hou, D.L.

Effect of annealed treatment on microstructure and cyclic stability for La-Mg-Ni hydrogen storage alloys. *J. Power Sources***2012**, *209*, 257–261.

198. Ma, J.; Pan, H.; Zhu, Y.; Li, S.; Chen, C. A novel method to improve the electrochemical properties of hydrogen storage electrode alloy. *Acta Metal. Sin.* **2001**, *37*, 57–60.
199. Zhou, C.; Tao, M.; Yan, K.; Chen, Y.; Wu, C. Effects of nickel additive on the performance of the negative electrode for Ni-MH battery. *Chin. J. Power Sources* **2003**, *27*, 301–304.
200. Jung, J.; Lee, S.; Kim, D.; Jang, K.; Lee, J. Effect of Cu powder as compacting material on the discharge characteristics of negative electrodes in Ni–MH batteries. *J. Alloys Compd.* **1998**, *266*, 271–275.
201. Mao, L.; Tong, J.; Shan, Z.; Yin, S.; Wu, F. Effect of Co additives on the cycle life of Ni-MH batteries. *J. Alloys Compd.* **1999**, *293–295*, 829–832.
202. Di, L.; Chen, L. Influence of the CoO in MH electrode on the performance of battery. *Bull. Chin. Ceram. Soc.* **2003**, *3*, 88–90.
203. Chen, L.; Tian, Y.; Zhang, F. Influence of CoO addition on the MH electrode's performance. *J. Tianjin Inst. Technol.* **2003**, *19*, 54–56.
204. Meng, M.; Liu, M.; Deng, X.; Huang, S.; Zhan, J. Research on the cycle life of Ni-MH batteries. *Chin. J. Power Sources* **1998**, *22*, 155–157.
205. Guo, J.; Li, W.; Meng, M. Study on effect of technology factors on the properties of Ni-MH battery. *Chin. J. Power Sources* **2000**, *24*, 319–321.
206. Pang, L.; Zhang, J.; Guo, J.; Wang, B.; Li, K.; Duan, Q. Method of reducing internal pressure of Ni-MH battery and increasing cycle life. *Chin. J. Power Sources* **1999**, *23*, 158–160.
207. Petrov, K.; Rostami, A.A.; Visintin, A.; Srinivasan, S. Optimization of composition and structure of metal-hydride electrodes. *J. Electrochem. Soc.* **1994**, *141*, 1747–1750.
208. Guo, W.; Xue, J.; Xu, Q.; Tang, Z. Effect of additive BC-1 on the performance of Ni-MH battery. *Chin. J. Power Sources* **2007**, *31*, 151–153.
209. Cheng, Y.; Zhang, H.; Zhang, G.; Chen, Y.; Zhu, Q. Influence of addition of carbon nanotubes in the hydrogen storage alloy electrode on the performance of SC high power batteries. *Acta Phys. Chim. Sin.* **2008**, *24*, 527–532.
210. Xie, F.; Dong, C.; Qian, W.; Zhai, Y.; Li, L.; Li, D. Electrochemical properties of nickel-metal hydride battery based on directly grown multiwalled carbon nanotubes. *Int. J. Hydrog. Energy* **2015**, *40*, 8935–8940.
211. Arnaud, O.; Le Guenne, L.; Audry, C.; Bernard, P. Effect of yttria content on corrosion of AB_5-type alloys for nickel-metal hydride batteries. *J. Electrochem. Soc.* **2005**, *152*, A611–A616.
212. Tao, M.; Chen, Y.; Yan, K.; Zhou, C.; Wu, C.; Tu, M. Effect of rare earth oxides on the electrochemical performance of hydrogen storage electrode. *Rare Met. Mater. Eng.* **2005**, *34*, 552–556.
213. Tanaka, T.; Kuzuhara, M.; Watada, M.; Oshitani, M. Effect of rare earth oxide additives on the performance of NiMH batteries. *J. Alloys Compd.* **2006**, *408–412*, 323–326.
214. Shinyama, K.; Nakamura, H.; Nohma, T.; Yonezu, I. Improvement of high- and low-temperature characteristics of nickel-metal hydride secondary batteries using rare-earth compounds. *J. Alloys Compd.* **2006**, *408–412*, 288–293.
215. Gamboa, S.A.; Sebastian, P.J.; Geng, M.; Northwood, D.O. Temperature, cycling, discharge current and self-discharge electrochemical studies to evaluate the performance of a pellet metal-hydride electrode. *Int. J. Hydrog. Energy* **2001**, *26*, 1315–1318.
216. Ma, G.; Guo, J.; Yan, Y.; Zhang, J.; Duan, Q.; Liu, G. Application of sintered metal hydride in Ni-MH battery. *Chin. Battery Ind.* **2003**, *8*, 113–115.
217. Oshitani, M.; Yufu, H.; Takashima, K.; Tsuji, S.; Matsumaru, Y. Development of a pasted nickel electrode with high active material utilization. *J. Electrochem. Soc.* **1989**, *136*, 1590–1593.
218. Oshitani, M.; Sasaki, Y.; Takashima, K. Development of a nickel electrode having stable performance at various charge and discharge rates over a wide temperature range. *J. Power Sources* **1984**, *12*, 219–231.
219. Friebel, D.; Bajdich, M.; Yao, B.S.; Louie, M.W.; Miller, D.J.; Casalongue, H.S.; Mbuga, F.; Weng, T.; Nordlund, D.; Sokaras, D.; *et al.* On the chemical state of Co oxide electrocatalysts during alkaline water splitting. *Phys. Chem. Chem. Phys.* **2013**, *15*, 17460–17467.
220. Pralong, V.; Delahaye-Vidal, A.; Beaudoin, B.; Leriche, J.-B.; Tarascon, J.-M. Electrochemical behavior of cobalt hydroxide used as additive in the nickel hydroxide electrode. *J. Eletrochem. Soc.* **2000**, *147*, 1306–1313.
221. Otsuka, H.; Chiku, M.; Higuchi, E.; Inoue, H. Characterization of pretreated $Co(OH)_2$-coated $Ni(OH)_2$ positive electrode for Ni-MH batteries. *ECS Trans.* **2012**, *41*, 7–12.
222. Pralong, P.; Chabre, Y.; Delahaye-Vidal, A.; Tarascon, J.-M. Study of the contribution of cobalt additive to the behavior of the nickel oxy-hydroxide electrode by potentiodynamic techniques. *Solid State Ion.* **2002**, *147*, 73–84.
223. Li, X.; Xia, T.; Dong, H.; Wie, Y. Study on the reduction behavior of CoOOH during the storage of nickel/metal-hydride battery. *Mater. Chem. Phys.* **2006**, *100*, 486–489.
224. Fu, Z.; Jiang, W.; Yu, L. Preparation of nickel hydroxide coated by cobalt hydroxide. *Chin. J. Nonferr. Met.* **2005**, *15*, 1775–1779.
225. Chang, Z.; Tang, H.; Peng, P. Studies of electrochemical characteristics of nickel hydroxide with surface deposited cobalt. *J. Funct. Mater.* **2001**, *32*, 487–489.
226. Yin, T. Surface modification of nickel hydroxide particles by micro-sized cobalt oxide hydroxide and properties as electrode materials. *Surf. Coat. Technol.* **2005**, *200*, 2376–2379.
227. Ovshinsky, S.R.; Fetcenko, M.A.; Fierro, C.; Gifford, P.R.; Corrigan, D.A.; Benson, P.; Martin, F.J. Enhanced Nickel Hydroxide Positive Electrode Materials for

Alkaline Rechargeable Electrochemical Cells. U.S. Patent 5,523,182, 4 June 1996.

228. Yin, X.; Li, J.; Fu, J.; Gong, L.; Xia, X. Preparation, characterization and electrochemical performance of cobalt-substituted nickel hydroxide. *Chem. Res. Appl.* **2002**, *14*, 405–408.

229. Tessier, C.; Faure, C.; Guerlou-Demourgues, L.; Denage, C.; Nabias, G.; Delmas, C. Electrochemical study of zinc-substituted nickel hydroxide. *J. Electrochem. Soc.* **2002**, *149*, A1136–A1145.

230. Fierro, C.; Zallen, A.; Koch, J.; Fetcenko, M.A. The influence of nickel-hydroxide composition and microstructure on the high-temperature performance of nickel metal hydride batteries. *J. Electrochem. Soc.* **2006**, *153*, A492–A496.

231. Zhang, S.; Han, Z. Influence of average grain size on the electrochemical properties of nickel hydroxide. *Battery Ind.* **1999**, *4*, 133–135.

232. Li, W.; Jiang, C.; Wan, C. High-temperature performances of spherical nickel hydroxide coated with $Yb(OH)_3$. *J. Inorg. Mater.* **2006**, *21*, 121–127.

233. Ortiz, M.G.; Real, S.G.; Castro, E.B. Preparation and characterization of positive electrode of Ni-MH batteries with cobalt additives. *Int. J. Hydrog. Energy* **2014**, *39*, 8661–8666.

234. He, X.M.; Jiang, C.Y.; Li, W.; Wan, C.R. Co/Yb hydroxide coating of spherical $Ni(OH)_2$cathode materials for Ni-MH batteries at elevated temperatures. *J. Electrochem. Soc.* **2006**, *153*, A566–A569.

235. Han, E.; Xu, H.; Kang, H.; Feng, Z. Performance comparison between nano-sized and micro-sized spherical nickel hydroxide. *J. Dalian Marit. Univ.* **2008**, *34*, 87–90.

236. Chen, Q.; Yi, S.; Zhao, M. Influence of nanometric zinc oxide on the electrochemical performance of nickel metal hydride batteries. *Rare Met. Mater. Eng.* **2010**, *39*, 39–42.

237. Yuan, A.; Cheng, S.; Zhang, J.; Cao, C. Effects of metallic cobalt addition on the performance of pasted nickel electrodes. *J. Power Sources* **1999**, *77*, 178–182.

238. Wang, X.; Yan, J.; Yuan, H.; Zhou, Z.; Song, D.; Zhang, Y.; Zhu, L. Surface modification and electrochemical studies of spherical nickel hydroxide. *J. Power Sources* **1998**, *72*, 221–225.

239. Wang, X.; Leo, H.; Yang, H.; Sebastian, P.J.; Gamboa, S.A. Oxygen catalytic evolution reaction on nickel hydroxide electrode modified by electroless cobalt coating. *Int. J. Hydrog. Energy***2004**, *29*, 967–972.

240. Li, Q. Study on capacity loss of Ni-MH batteries after storage at low potential. *Chin. Battery Ind.* **2006**, *11*, 303–306.

241. Chang, Z.R.; Tang, H.; Chen, J.G. Surface modification of spherical nickel hydroxide for nickel electrodes. *Electrochem. Commun.* **1999**, *1*, 513–516.

242. Cheng, S.; Wenhau, L.; Zhang, J.; Cao, C. Electrochemical properties of the pasted nickel electrode using surface modified $Ni(OH)_2$ powder as active material. *J. Power Sources* **2001**, *101*, 248–252.

243. Hu, W.K.; Gao, X.P.; Gong, Z.X. Synthesis of CoOOH nanorods and application as coating materials of nickel hydroxide for high temperature Ni-MH cells. *J. Phys. Chem. B* **2005**, *109*, 5392–5394.

244. Hu, Q.D.; Gao, X.P.; Li, G.R.; Pan, G.L.; Yan, T.Y. Microstructure and electrochemical properties of Al-substituted nickel hydroxides modified with CoOOH nanoparticles. *J. Phys. Chem. C* **2007**, *111*, 17082–17087.

245. Xia, X.; Zhang, W.; Huang, H.; Gan, Y.; Zhang, L. Effects of $CoSO_4$ as additive on the electrochemical properties of nickel hydroxide electrode. *Zhejiang Chem. Ind.* **2007**, *38*, 1–4.

246. Su, G.; He, Y.; Liu, K. Effects of Co_3O_4 as additive on the performance of metal hydride electrode and Ni–MH battery. *Int. J. Hydrog. Energy* **2012**, *37*, 11994–12002.

247. Nathira Begum, S.; Muralidharan, V.S.; Ahmed Basha, C. The influences of some additives on electrochemical behaviour of nickel electrodes. *Int. J. Hydrog. Energy* **2009**, *34*, 1548–1555.

248. Yuan, A.; Cheng, S.; Zhang, J.; Cao, C. The influence of calcium compounds on the behaviour of the nickel electrode. *J. Power Sources* **1998**, *76*, 36–40.

249. Li, J.; Li, R.; Wu, J.; Su, H. Effect of cupric oxide addition on the performance of nickel electrode. *J. Power Sources* **1999**, *79*, 86–90.

250. Jung, K.; Yang, D.; Park, C.; Park, C.; Choi, J. Effects of the addition of ZnO and Y_2O_3 on the electrochemical characteristics of a $Ni(OH)_2$ electrode in nickel-metal hydride secondary batteries. *Int. J. Hydrog. Energy* **2010**, *35*, 13073–13077.

251. Yuan, X.; Wang, Y.; Zhan, F. Study on the additives for inhibiting swelling of nickel electrode. *Chin. J. Power Sources* **2000**, *24*, 192–196.

252. Chen, J.; Bradhurst, D.H.; Dou, S.X.; Liu, H.K. Nickel hydroxide as an active material for the positive electrode in rechargeable alkaline batteries. *J. Electrochem. Soc.* **1999**, *146*, 3606–3612.

253. Douin, M.; Guerlou-Demourgues, L.; Goubault, L.; Bernard, P.; Delmas, C. Effect of deep discharge on the electrochemical behavior of cobalt oxides and oxyhydroxides used as conductive additives in Ni-MH cells. *J. Power Sources* **2009**, *193*, 864–870.

254. Douin, M.; Guerlou-Demourgues, L.; Goubault, L.; Bernard, P.; Delmas, C. Influence of the electrolyte alkaline ions on the efficiency of the $Na_{0.6}CoO_2$ conductive additive in Ni-MH cells. *J. Electrochem. Soc.* **2008**, *155*, A945–A951.

255. Tronel, F.; Guerlou-Demourgues, L.; Basterreix, M.; Delmas, C. The $Na_{0.60}CoO_2$ phase, a potential conductive additive for the positive electrode of Ni–MH cells. *J. Power Sources* **2006**, *158*, 722–729.

256. Morioka, Y.; Narukawa, S.; Itou, T. State-of-the-art of alkaline rechargeable batteries. *J. Power Sources* **2001**, *100*, 107–116.

257. Ye, X.; Zhu, Y.; Wu, S.; Zhang, Z.; Zhou, Z.; Zheng, H.; Lin, X. Study on the preparation and electrochemical performance of rare earth doped nano-$Ni(OH)_2$. *J. Rare Earths* **2011**, *29*, 787–792.

258. Ren, J.X.; Zhou, Z.; Gao, X.P.; Yan, J. Preparation of porous spherical α-$Ni(OH)_2$ and enhancement of high-temperature electrochemical performances through yttrium addition. *Eletrochim. Acta* **2006**, *52*, 1120–1126.

259. Mi, X.; Gao, X.P.; Jiang, C.Y.; Geng, M.M.; Yan, Y.; Wan, C.R. High temperature performances of yttrium-doped spherical nickel hydroxide. *Eletrochim. Acta* **2004**, *49*, 3361–3366.

260. Li, F.; Yang, Y. Effects of nano-sized Y_2O_3 additive on performance of Ni-MH battery. *Chin. Battery Ind.* **2008**, *13*, 226–228.

261. Kaiya, H.; Ookawa, T. Improvement in cycle life performance of high capacity nickel-metal hydride battery. *J. Alloys Compd.* **1995**, *231*, 598–603.

262. Mi, X.; Ye, M.; Yan, J.; Wei, J.; Gao, X. High temperature performance of spherical nickel hydroxide with additive Y_2O_3. *J. Rare Earths* **2004**, *22*, 422–425.

263. Wu, Q.D.; Liu, S.; Li, L.; Yan, T.Y.; Gao, X.P. High-temperature electrochemical performance of Al-α-nickel hydroxides modified by metallic cobalt or $Y(OH)_3$. *J. Power Sources* **2009**, *186*, 521–527.

264. Fan, J.; Yang, Y.; Yu, P.; Chen, W.; Shao, H. Effects of surface coating of $Y(OH)_3$ on the electrochemical performance of spherical $Ni(OH)_2$. *J. Power Sources* **2007**, *171*, 981–989.

265. Oshitani, M.; Watada, M.; Shodai, K.; Kodama, M. Effect of lanthanide oxide additives on the high-temperature charge acceptance characteristics of pasted nickel electrodes. *J. Electrochem. Soc.* **2001**, *148*, A67–A73.

266. Ren, J.; Wang, X.; Li, Y.; Gao, X.; Yan, J. Effect of Lu_2O_3 on charge/discharge performances of spherical nickel hydroxide at high temperature. *J. Rare Earths* **2005**, *23*, 732–736.

267. Li, J.; Shangguan, E.; Guo, D.; Li, Q.; Chang, Z.; Yuan, X.; Wang, H. Calcium metaborate as a cathode additive to improve the high-temperature properties of nickel hydroxide electrodes for nickel-metal hydride batteries. *J. Power Sources* **2014**, *263*, 110–117.

268. Li, W.; Zhang, S.; Chen, J. Synthesis, characterization, end electrochemical application of $Ca(OH)_2$-, $Co(OH)_2$- and $Y(OH)_3$-coated $Ni(OH)_2$ tubes. *J. Phys. Chem. B* **2005**, *109*, 14025–14032.

269. Hu, M.; Lei, L.; Chen, J.; Sun, Y. Improving the high-temperature performances of a layered double hydroxide, $[Ni_4Al(OH)_{10}]NO_3$, through calcium hydroxide coatings. *Eletrochim. Acta***2011**, *56*, 2862–2869.

270. Matsuda, H.; Ikoma, M. Nickel/metal-hydride battery for electric vehicles. *Electrochemistry***1997**, *65*, 96–100.

271. He, X.; Ren, J.; Li, W.; Jiang, C.; Wan, C. $Ca_3(PO_4)_2$ coating of spherical $Ni(OH)_2$ cathode materials for Ni–MH batteries at elevated temperature. *Eletrochim. Acta* **2011**, *56*, 4533–4536.

272. Guo, H.; Wang, W.; Zhang, W.; Wang, J. Effects of long-time storage in discharging state on the performance of Ni-MH batteries. *Chin. Battery Ind.* **2009**, *14*, 166–168.

273. Fukunaga, H.; Kishimi, M.; Igarashi, N.; Ozaki, T.; Sakai, T. Non-foam-type nickel electrodes using various binders for Ni-MH batteries. *J. Electrochem. Soc.* **2005**, *152*, A42–A46.

274. Nishimura, K.; Takasaki, T.; Sakai, T. Development of high-capacity $Ni(OH)_2$ electrode using granulation process. *Stud. Sci. Technol.* **2013**, *2*, 107–112.

275. Fukunaga, H.; Kishimi, M.; Matsumoto, N.; Ozaki, T.; Sakai, T.; Tanaka, T.; Kishimoto, T. A nickel electrode with Ni-coated 3D steel sheet for hybrid electric vehicle applications. *J. Electrochem. Soc.* **2005**, *152*, A905–A912.

276. Xi, Z.; Zhang, J.; Wu, L.; Li, J.; Zhu, R.; Yang, Y.; Feng, P.; Zuo, C.; Shi, D. New type of anode material for the MH-Ni battery. *Rare Met. Mater. Eng.* **1999**, *28*, 371–374.

277. Wang, D.; Wang, C.; Dai, C.; Sun, D. Study of Co-Ce coating and surface on pasted nickel electrodes substrate. *Rare Met.* **2006**, *25*, 47–50.

278. Kritzer, P.; Cook, J.A. Nonwoven as separators for alkaline batteries—An overview. *J. Electrochem. Soc.* **2007**, *154*, A481–A494.

279. Yu, T.; Zhai, Y.; Li, Q.; Duan, Q. Study on storage of Ni-MH battery. *Chin. Battery Ind.* **2005**, *10*, 149–152.

280. Ni, J.; Wan, D.; Zhang, L. Influences of using separators made in China on Ni-MH battery performance in electric bicycle. *Electr. Bicycle* **2013**, *6*, 14–20.

281. Cao, S.; Wang, H.; Jin, X. Study on the Sulfonation Treatment of ES Nonwoven Fabrics as Battery Separator. In Proceedings of the 7th National Conference of Functional Materials and Applications, Changsha, China, 15–18 October 2010; Zhao, G.-M., Ed.; Scientific Research Publishing: Merced, CA, USA; pp. 2075–2078.

282. Furukawa, N. Development and commercialization of nickel-metal hydride secondary batteries. *J. Power Sources* **1994**, *51*, 45–59.

283. Du, H.; Liu, D.; Kong, Y.; Yang, J. Hydrophilic modification of polypropylene MH-Ni battery separator and its characterizations. *Polym. Mater. Sci. Eng.* **2013**, *29*, 101–104.

284. Wang, F.; Wu, F.; Mao, L.; Zou, L.; Liu, Y.; Chen, S. Research on polymer-gel electrolyte bipolar nicker/metal hydride batteries. *Mater. Rev.* **2005**, *19*, 124–126.

285. Cai, Z.; Gong, Z.; Geng, M.; Tang, Z. Study on a novel polymer gel improved separator in nickel/metal hydride battery. *Chin. J. Power Sources* **2005**, *29*, 142–145.

286. Tian, Y.; Gao, H.; Wang, J.; Jin, X.; Wang, H. Preparation of hydroentangled CMC composite nonwoven fabrics as high performance separator for nickel metal hydride battery. *Electrochim. Acta* **2015**, *177*, 321–326.

287. Shigematsu, T.; Oku, Y.; Ishikawa, T. Nonwoven Fabric for Alkaline Battery Separator and Manufacture Thereof. Jpn. Patent Application H06–251760, 9 September 1994.

288. Yao, Y.; Hu, J.; Yang, J.; Wang, Y.; Liang, Y. Preliminary study of making MH-Ni battery separator. *Pap. Sci. Technol.* **2009**, *28*, 76–80.

289. Chen, Z.; Ju, Y.; Zhang, W.; Zhu, Y.; Li, L. Research in preparation and performance of a novel alkaline microporous polymer electrolyte. *Mater. Rev.* **2009**, *23*, 316–318, 334.

290. Leblanc, P.; Jordy, C.; Knosp, B.; Blanchard, P.H. Mechanism of alloy corrosion and consequences on sealed nickel-metal hydride battery performance. *J. Electrochem. Soc.* **1998**, *145*, 860–863.

291. Khaldi, C.; Mathlouthi, H.; Lamloumi, J. A comparative study, of 1 M and 8 M KOH electrolyte concentrations, used in Ni-MH batteries. *J. Alloys Compd.* **2009**, *469*, 464–471.

292. Pei, L.; Yi, S.; He, Y.; Chen, Q. Effect of electrolyte formula on the self-discharge properties of nickel-metal hydride batteries. *J. Guangdong Univ. Technol.* **2008**, *25*, 10–12.

293. Wang, C.; Marrero-Rivera, M.; Soriaga, M.P.; Serafini, D.; Srinivasan, S. Improvement in the cycle life of LaB_5 metal hydride electrodes by addition of ZnO to alkaline electrolyte. *Eletrochim. Acta* **2002**, *47*, 1069–1078.

294. Matsuoka, M.; Asai, K.; Fukumoto, Y.; Iwakura, C. Electrochemical characterization of surface-modified negative electrodes consisting of hydrogen storage alloys. *J. Alloys Compd.***1993**, *192*, 149–151.

295. Shangguan, E.; Wang, J.; Li, J.; Dan, G.; Chang, Z.; Yuan, X.; Wang, H. Enhancement of the high-temperature performance of advanced nickel-metal hydride batteries with NaOH electrolyte containing $NaBO_2$. *Int. J. Hydrog. Energy* **2013**, *38*, 10616–10624.

296. Shangguan, E.; Li, J.; Guo, D.; Chang, Z.; Yuan, X.; Wang, H. Effects of different electrolytes containing Na_2WO_4 on the electrochemical performance of nickel hydroxide electrodes for nickel–metal hydride batteries. *Int. J. Hydrog. Energy* **2014**, *39*, 3412–3422.

297. Zhu, X.; Yang, H.; Ai, X. Possible use of ferrocyanide as a redox additive for prevention of electrolyte decomposition in overcharged nickel batteries. *Electrochim. Acta* **2003**, *48*, 4033–4037.

298. Mao, L.; Wu, F.; Chen, S.; Wang, F.; Liu, Y. Application of gel polymer electrolytes in Ni/MH battery. *J. Funct. Mater.* **2005**, *36*, 1389–1390.

299. Sun, S.; Song, J.; Shan, Z.; Hu, M.; Lei, L.; Chen, J.; Sun, Y. Electrochemical properties of a low molecular weight gel electrolyte for nickel/metal hydride cell. *Eletrochim. Acta* **2014**, *130*, 689–692.

300. Ju, Y.; Chen, Z.; Zhu, Y.; Li, L. Modification mechanism and research progress of polymer electrolytes in Ni/MH batteries. *Mater. Rev.* **2009**, *23*, 54–57.

301. Yang, C. Polymer Ni-MH battery based on PEO-PVA-KOH polymer electrolyte. *J. Power Sources* **2002**, *109*, 22–31.

302. Vassal, N.; Salmon, E.; Fauvarque, J.-F. Nickel/metal hydride secondary batteries using an alkaline solid polymer electrolyte. *J. Electrochem. Soc.* **1999**, *146*, 20–26.

303. Iwakura, C.; Ikoma, K.; Nohara, S.; Furukawa, N. Charge-discharge and capacity retention characteristics of new type Ni/MH batteries using polymer hydrogel electrolyte. *J. Electrochem. Soc.* **2003**, *150*, A1623–A1627.

304. Wang, C.; Wallace, G.G. Flexible electrodes and electrolytes for energy storage. *Electrochim. Acta* **2015**, *175*, 87–95.

305. Zhang, Y.; Pang, H.; Yang, H.; Zhou, Z. Electrochemical stability of PAA alkaline polymer electrolyte and its application in Ni/MH secondary battery. *Acta Sci. Nat. Univ. Nankaiensis***2008**, *41*, 57–61.

306. Tang, Z.; Li, C.; Liu, Y. Research and development of polymer Ni/MH battery. *Chem. Ind. Eng. Prog.* **2006**, *25*, 1251–1255.

307. Wu, R.; Lu, X.; Zhu, Y.; Li, L. Preparation and properties of PVA/PAAK alkaline polymer electrolytes for nickel metal hydride batteries. *Polym. Mater. Sci. Eng.* **2013**, *29*, 171–174.

308. Lu, X.; Wu, R.; Zhu, Y.; Li, L. Preparation and properties of PVA-PAA-KOH alkaline polymer electrolyte membrane. *J. Funct. Mater.* **2013**, *44*, 590–594.

309. Lu, X.; Wu, R.; Li, B.; Zhu, Y.; Li, L. Novel PVA/SiO_2 alkaline micro-porous polymer electrolytes for polymer Ni-MH batteries. *Acta Chim. Sin.* **2013**, *71*, 427–432.

310. Vignarooban, K.; Dissanayake, M.A.K.L.; Albinsson, I.; Mellander, B.-E. Ionic conductivity enhancement in PEO:CuSCN solid polymer electrolyte by the incorporation of nickel-chloride. *Solid State Ion.* **2015**, *278*, 177–180.

311. Li, X.; Xia, B. Corrosion of AB_5 alloy during the storage on the performance of nickel-metal-hydride battery. *J. Alloys Compd.* **2005**, *391*, 190–193.

312. Li, X.; Xia, B. Effect of charged-state storage on Ni-MH battery performance. *Chin. J. Power Sources* **2004**, *28*, 737–739.

313. Zhang, Z.; Jia, C.; Xing, Z.; Li, L.; Ma, Y. Factors on storage performance of MH-Ni battery. *J. Rare Earths* **2004**, *22*, 347–348.

314. Cao, M.; Zhang, J.; Du, P.; Wang, J. Investigation on performance of Ni-MH battery. *Chin. J. Power Sources* **2006**, *30*, 852–855.

315. Jung, D.Y.; Lee, B.H.; Kim, S.W. Development of battery management system for nickel-metal hydride batteries in electric vehicle applications. *J. Power Sources* **2002**, *109*, 1–10.

316. Darcy, E.C. Investigation of the Response of NiMH Cells to Burp Charging. Ph.D. Thesis, University of Houston, Houston, TX, USA, 1998.

317. Ma, S.; Yang, K.; Lui, Y.; Lan, Z. The effect of charge/discharge manner on properties of hydrogen storage alloy electrode. *J. Guangxi Univ. Nat. Sci. Ed.* **2006**, *31*, 205–207.

318. Cao, S.; Ma, Y. Influences of the formation on the storage performance of high power MH/Ni battery. *Sci. Technol. Baotou Steel Group Corp.* **2008**, *34*, 50–52.

319. Chen, B.; Li, X.; Li, X.; Zhang, Q. Method of reducing internal pressure of alkaline rechargeable battery. *Chin. Battery Ind.* **2000**, *5*, 211–213.

320. Bai, Z.; Yu, X.; Lu, F. Sealing technology about cylindrical alkaline battery. *Chin. J. Power Sources* **2003**, *27*, 381–384.

321. Yuan, J.; Chen, W. Research analysis on design optimization of cell sealing structure. *Electr. Bicycl.* **2015**, *3*, 10–11.

322. Song, Q. Performance recovery of Ni-MH battery after long storing. *Chin. Battery Ind.* **2005**, *10*, 77–79.

323. Xie, D.; Li, Q.; Wu, T. Discussion on some aspects of MH-Ni battery. *Chin. Battery Ind.* **2001**, *6*, 114–117.

324. Li, N.; Meng, J. Capacity recovery of Ni-MH batteries after long-term storage. *Chin. Battery Ind.* **2008**, *13*, 95–96.

325. Li, L.; Wu, F.; Chen, R.; Chen, S.; Wang, G. Regeneration of electrochemical performance for MH-Ni batteries. *J. Funct. Mater.* **2006**, *37*, 587–590.

326. Minohara, T. Recycling Method of Nickel-Hydrogen Secondary Battery. U.S. Patent 6,077,622A, 20 June 2000.

327. Wang, R.; Yan, J.; Zhou, Z.; Deng, B.; Gao, X.; Song, D. Regeneration of anode power from waste metal hydride/Ni rechargeable batteries. *Chin. J. Appl. Chem.* **2001**, *18*, 979–982.

328. Nan, J.; Han, D.; Yang, M.; Cui, M. Dismantling, recovery, and reuse of spent nickel-metal hydride batteries. *J. Electrochem. Soc.* **2006**, *153*, A101–A105.

329. Nan, J.; Han, D.; Yang, M.; Cui, M.; Hou, X.L. Recovery of metal values from a mixture of spent lithium-ion batteries and nickel-metal hydride batteries. *Hydrometallurgy* **2006**, *84*, 75–80.

330. Tenorio, J.A.S.; Espinosa, D.C.R. Recovery of Ni-based alloys from spent NiMH battery. *J. Power Sources* **2002**, *108*, 70–73.

331. Bertuol, D.A.; Bernardes, A.M.; Tenorio, J.A.S. Spent NiMH batteries: Characterization and metal recovery through mechanical processing. *J. Power Sources* **2006**, *160*, 1465–1470.

332. Zhang, P.; Yokoyama, T.; Itabashi, O.; Wakui, Y.; Suzuki, T.M. Recovery of metal values from spent nickel-metal hydride rechargeable batteries. *J. Power Sources* **1999**, *77*, 116–122.

333. Rodrigues, L.; Mansur, M.B. Hydrometallurgical separation of rare earth elements, cobalt and nickel from spent nickel-metal-hydride batteries. *J. Power Sources* **2010**, *195*, 3735–3741.

334. Muller, T.; Friedrich, B. Development of a recycling process for nickel-metal hydride batteries. *J. Power Sources* **2006**, *158*, 1498–1509.

335. Rabah, M.A.; Farghaly, F.E.; Abd-El Motaleb, M.A. Recovery of nickel, cobalt and some salts from spent Ni-MH batteries. *Waste Manag.* **2008**, *28*, 1159–1167.

336. Santos, V.E.O.; Celante, V.G.; Lelis, M.F.F.; Freitas, M.B.J.G. Chemical and electrochemical recycling of the nickel, cobalt, zinc and manganese from the positives electrodes of spent Ni–MH batteries from mobile phones. *J. Power Sources* **2012**, *218*, 435–444.

337. Larsson, K.; Ekberg, C.; Ødegaard-Jensen, A. Dissolution and characterization of HEV NiMH batteries. *Waste Manag.* **2013**, *33*, 689–698.

CHAPTER 16 SIMULATION OF NI-MH BATTERIES VIA AN EQUIVALENT CIRCUIT MODEL FOR ENERGY STORAGE APPLICATIONS

Ying Zhu, Wenhua H. Zhu, Zenda Davis, and Bruce J. Tatarchuk

INTRODUCTION

The prosperity of rechargeable Ni-MH and Li-ion batteries rapidly commercializes their application to plug-in hybrid electric vehicles (PHEVs), hybrid electric vehicles (HEVs), and battery electric vehicles (BEVs). Li-ion batteries are in general 50% lighter in weight and 25%–30% smaller in size [1], making an easier assembly in vehicle, generating more power to boost acceleration, and increasing energy efficiency through on-board battery energy storage via regenerative braking and mechanical-electric energy conversion. A gradual capacity reduction and performance decay are inevitable based on the current knowledge of rechargeable battery technology. It is essential to monitor and control the level of state-of-charge (SoC) in consideration of operating lifetime for the rechargeable batteries. Although the Li-ion battery power packs have remarkable advantages, the on-board energy storage using Ni-MH battery packs is currently powering many of today's hybrid vehicles on the road. This nickel-based metal-hydride battery is lighter than the traditional lead-acid one and approximately delivers twice the power output for the weight as the Pd-acid battery. The accuracy for SoC prediction is a challenge for the requirements of both battery management systems (BMS) and various EV charging infrastructure. Battery impedance is a valid tool to diagnose the cell health conditions and to assist the understanding of electrochemical reaction mechanisms. The limitation for practical applications comes from the difficulties of measurement at load and impedance data interpretation for related physical parameters and electrochemical processes. Electrochemical impedance spectroscopy (EIS) is a valid in situ nondestructive technique for battery evaluation to assess the dynamic responses at loads. It has been experimentally proven by its wide applications to various electrochemical systems, including electrodes-aqueous electrolyte interfaces [2], conducting polymers [3, 4], supercapacitors [5], solar cells [6], fuel cells [7, 8], and rechargeable batteries [9–12]. As a whole, it is valuable for further product improvement through deep understanding of the battery deterioration mechanisms during degradation process. The interpretation of impedance data can be achieved by equivalent circuit (EC) simulation and mechanistic analysis.

Characterizations through EIS impedance for NiOOH/$Ni(OH)_2$ electrodes [13, 14] delivered valuable contributions to the development of nickel-based alkaline rechargeable batteries. The modification of Ni electrodes [15, 16] is experienced with the abundant assistance of the EIS technique for performance improvement of Ni-MH batteries. Previously, Kuriyama et al. [17] fully developed a deterioration study of metal-hydride (MH) electrodes with impedance measurement and EC simulation. Recently, researches focused more on characterization of MH electrodes with different structures [18–20]. At the era of lead-acid (Pb-acid) and nickel cadmium (Ni-Cd) batteries, EIS had been proposed as a promising technique to monitor and determine the SoC levels of full batteries [21]. The study of Bundy et al. [22] was recognized as one of the few thorough impedance studies of full Ni-MH battery with attempt to predict the SoC values. Hammouche

Citation: Ying Zhu, Wenhua H. Zhu, Zenda Davis, and Bruce J. Tatarchuk, "Simulation of Ni-MH Batteries via an Equivalent Circuit Model for Energy Storage Applications," Advances in Physical Chemistry, vol. 2016, Article ID 4584781, 11 pages, 2016. doi:10.1155/2016/4584781.

et al. [23] reported the impedance measurement of a 42 Ah Ni-MH pack. Half-cell impedance was measured against cell case. Later, Thele et al. [24] followed up the research on large sealed Ni-MH battery pack. Pulse load simulation and hysteresis phenomenon were utilized to validate the Ni-MH battery model. This work provided a valid reference for simulation of the Ni-MH systems by impedance analysis. Castro et al. [25] published a mathematical simulation of their lab-made Ni-MH battery based on physicochemical model. Diffusion process was considered in this model, but no fitting curves of impedance spectra were illustrated. Slepski et al. [26] used additional reference electrode to separate impedance measurement of positive and negative electrodes. An EC model of "two time constants" was suggested for the entire battery, but neither simulation process nor fitting curve was reported in detail. In this work, impedance measurement is conducted at the whole cell level for studying of the Ni-MH rechargeable batteries. An improved equivalent circuit model considering diffusion process is proposed for battery impedance data simulation at different charge input levels. The model-based impedance analysis is expected to particularly understand the battery performance deterioration at different cell health conditions and assist the mechanism diagnostics for cell capacity degradation.

EXPERIMENTAL

Rechargeable Ni-MH Cells for Testing and Evaluation

Two Ni-MH rechargeable batteries were selected for experimental measurements in this work, named as Cell A and Cell B, respectively. The batteries were sealed D-size cylindrical cells with a height of 60 mm and a diameter of 32 mm. The Ni-MH cells were originally purchased from a Radioshack® store (a division of Tandy Corp., Fort Worth, TX) in a brand new condition. The nominal voltage is labeled as 1.2 V and the nominal cell capacity is marked as 4500 mAh (Radioshack #23-519). Sealed Ni-MH batteries can generally achieve a cycle life of more than 500 charge-discharge cycles [27] with regular operation. Performance degradation was gradually occurring in the Ni-MH Cell A and Cell B due to years of charge/discharge service cycles and shelf storage. The Arbin Battery Tester (Abts 4.0 with 8 channels) was applied for battery capacity measurement and rate performance evaluation. The cell capacities of the Ni-MH Cell A and Cell B were measured at 3702 mAh and 4362 mAh at a 0.2 C rate after years of services for various laboratory duties and overcharge endurance, respectively. This means the batteries are under different state-of-health (SoH) conditions. In this work, we simply the study by using the nonidentical battery capacity at different cell health status for imitating individual cell position in the real battery power systems after long-term cycling tasks. Impedance measurement is then conducted to study the cell deterioration and capacity degradation mechanisms.

Battery Characterization by AC Impedance Spectroscopy

AC impedance is a type of EIS technique that measures the impedance of electrochemical systems by employing sinusoidal signals as electrical stimulus. To measure cell impedance at a certain capacity level, the specified amount of charge was input to the fully discharged cells at a rate of 0.2 C. As detailed in our previous work [28], the ratio of the charge input to the rated cell capacity (RoC) is defined as

$$\text{RoC} = \%\frac{\text{Amount of Charge Input}}{\text{Rated Cell Capacity}}. \tag{1}$$

The value of RoC has some difference in comparison with the value of SoC after the charge input is more than *ca.* 85% SoC at a 0.2 C rate. When the specified RoC level was reached during the tests, the impedance measurement was applied to the cells under *dc* current loads. The loaded *dc* current was 370mA for the Ni-MH Cell A and 436mA for Cell B. The Gamry impedance monitor and TDI-Dynaload® programmable load were employed to obtain impedance spectra. The frequency swept from 10 kHz down to 0.01Hz in each measurement. Impedance data were recorded at a rate of 5 points per decade frequency. When finishing the impedance measurement, the cells were continuously discharged to the cut-off voltage of 1.0V at a rate of 0.2 C. In other words, the Ni-MH cells were fully discharged to a SoC level of 0% after each impedance measurement. The impedance data at a higher RoC level was measured by repeating this procedure. The proposed equivalent circuit model was utilized to simulate all measured impedance spectra through the software called Gamry Echem Analyst.

The mode of hybrid EIS was selected in this work for Ni-MH cell measurement. The EIS mode was firstly put forward as variable-amplitude Galvanostatic method when studying H-cells with copper/sea water system [29]. The Galvanostatic EIS mode is a perfectly valid technique for impedance data collection but it possibly produces a voltage outside the linear range due to the significant impedance increase. This results in the unacceptable impedance data.

The hybrid EIS mode is able to overcome this problem by continually adjusting the applied excitation current during experiment. The electrochemical systems were studied under constant current density. However, the current signal applied to the testing system is regulated in the hybrid EIS mode before measurement of each data point.The signal regulation is able to yield an *ac* voltage at desired magnitude to ensure does not extend beyond the linear range. The validation and accuracy of impedance measurement highly rely on the pseudolinear behavior of electrochemical systems.

During the impedance measurement in this work, a desired voltage perturbation and an estimated impedance magnitude were set before the measurement. The initial current perturbation [30] was calculated by

$$\Delta I_1 = \frac{\Delta V}{|Z_{\text{est}}|}, \tag{2}$$

where ΔI_1 was the amplitude of the current perturbation superimposed on the *dc* load for measurement at the first frequency point. ΔV was the amplitude of the desired voltage perturbation, chosen as 5mVfor this specific application. The value of $|Z_{est}|$ was the magnitude of the system impedance estimated before the measurement. The measured impedance value at this frequency point was then recorded as Z_1. The current perturbation employed for each following frequency point was calculated by [29, 30]

$$\Delta I_n = \frac{\Delta V}{|Z_{n-1}|}, \tag{3}$$

where ΔIn was the amplitude of the current perturbation used at the nth measuring point and $|Z_{n-1}|$ was the magnitude of the system impedance measured at the $(n-1)$th point.

Electrode Materials Composition by Energy Dispersive X-Ray Spectroscopy

One of these types of nickel metal-hydride cell was cut and broken down to get small amount of samples fromthe negative electrodes for surface analysis.The electrode composition was analyzed by energy dispersive Xray spectroscopy (EDX or EDS). The EDX surface technique is a bulk analytical technique used to qualitatively and quantitatively detect element composition. It is able to probe 3-4 μm below the surface of the powder and has a sensitivity of about 1000 ppm for elements from beryllium (Be) with atomic number 4 to uranium(U)with atomic number 92.The electrode samples were powdered and pressed into doublesided C tape on the EDX specimen mount. The composition of metal alloy used for the negative electrodes was detected by identifying the metal elements and their percentage.

RESULTS AND DISCUSSION

Electrode Compositions

The formulas of metal-hydride alloys used for the negative electrode of the Ni-MH batteries basically follow the disordered AB_5 type, A_2B_7 type, or disordered AB_2 type [31]. The typical elements on "A" sites and "B" sites are different in various alloy types; however, "A" generally refers to rare earth mixture and "B" refers to transition metals. According to Ovshinsky's valuable work on the MH negative electrodes for Ni-MH batteries [32], the disordered AB_5-type mischmetal (Mm) alloy [18–20] has been developed to be available for the most commercial use performing better cycle ability than other alloy structures. In this work, EDX reported that the electrode samples approximately follow the atomic ratio of AB5 as shown in Figure 1. The "A" type elements include lanthanum (La) and cerium (Ce). The "B" type elements include nickel (Ni), cobalt (Co), aluminum (Al), and manganese (Mn) with the primary component of Ni. Small amount of potassium (K) was also detected in the analytic process. This results from the use of KOH electrolyte during assembling of the Ni-MH cell. The element of Mn is able to adjust the metal-hydrogen bond strength [32], and the existence of Al and Co significantly promoted electrode kinetics [31,

32]. Based on the EDX report and analysis, the formula of this AB5-type hydride alloy can be roughly expressed as $(La_{9.70}Ce_{4.39})(Ni_{60.75}Co_{6.66}Al_{3.59}Mn_{4.45})$, according to Fetcenko et al.'s work [31] on the composition of metal-hydride alloys.

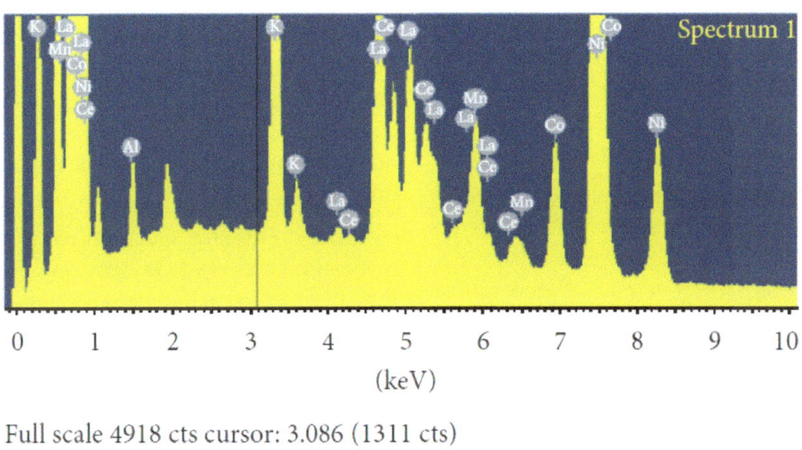

Figure 1: EDX spectrum for a sample of the Ni-MH negative electrode. The sample approximately matches the AB_5-type alloy with the primary components including nickel (B) and lanthanum (A) elements, based on the surface analysis tool.

Impedance Spectra and Equivalent Circuit Simulation

The Ni-MH cells became aged states after various operating tasks, and the cell capacities were measured as 3702mAh and 4362mAh, less than the nominal capacity of 4500mAh as specified by the manufacturer. The capacity reduction reflects cell degradation during numerous cycling loads at ambient shelf storage conditions. This represents the state-ofhealth (SoH) change for individual cells. Impedance spectra of the Ni-MH cells measured at different capacity levels are illustrated in the same Nyquist plots as shown in Figure 2.The impedance arcs of both cells at higher frequency region have the arc shapes similar to depressed semicircles. Comparing the magnitudes of these impedance arcs at X direction and Y direction, these "semicircles" are depressed to a large extent.

A change of curvature can be observed clearly at higher frequency region from the impedance spectra of the Ni-MH cell (Cell A), which implies the possibility of two overlapped impedance arcs. One straight tail dominates the behavior of cell impedance at lower frequency region of both cells at all SoC levels. Despite the similar shape of impedance spectra, the total impedance of the Ni-MH Cell B is slightly larger than the value of Cell A. The impedance of Cell B also executes larger changes with increasing the charge input than the value of Cell A. The altered magnitude and behavior of impedance spectra illustrate different amount of capacity decay in two Ni-MH rechargeable cells.

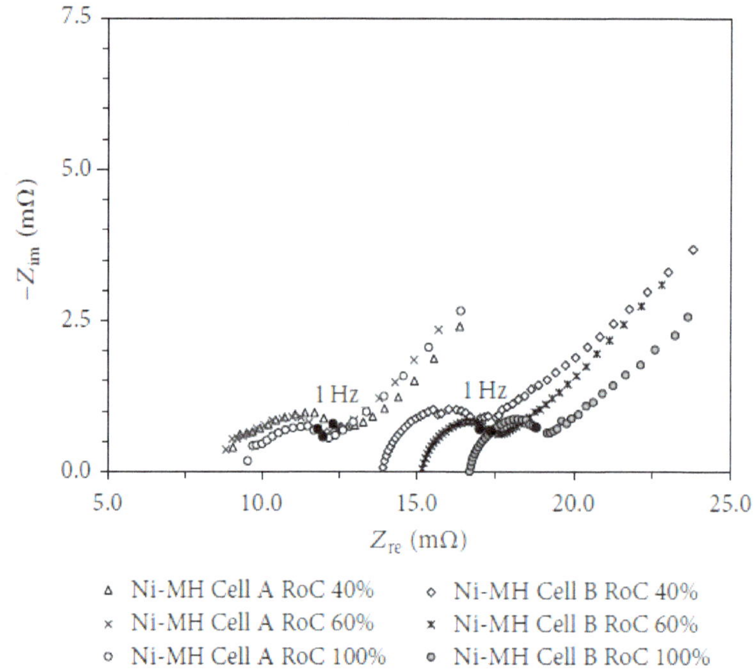

Figure 2: Impedance spectra of the Ni-MH Cell A and Cell B measured after charging to the RoC levels of 40%, 60%, and 100%. The impedance data measured at the frequency of 1Hz are highlighted in the plot (·).

A typical Randles circuit as shown in Figure 3(a) consists of one ohmic resistance, R_Ω, one parallel ($C_{dl}R_{ct}$) subcircuit behaving as a semicircle in Nyquist plot, and one Warburg element (symbolized by$W$) behaving as a unit slope line at the lowest frequency region [33]. This model considers faradaic impedance contributed by both kinetic reaction processes and diffusion processes. Since the depressed semicircles were observed from the impedance spectra of the Ni-MH cells, a modification to Randles circuit is necessary to be considered for the circuit model expansion and improvement.

One explanation to the depression is the existence of nonideal behaviors, usually caused by the surface roughness and varying thickness of electrodes, unevenly distributed current, and nonhomogeneous reaction rate. A generalized element, constant phase element (CPE, symbolized by Q), can be used to replace ideal physical elements in the case of nonideal behaviors. The phase angle of CPE remains constant through all the frequency range [34].The impedance of a CPE [33] can be written as

$$Z_Q = A\left(j\omega\right)^{-n} = \frac{1}{Y_0\left(j\omega\right)^n}. \quad (4)$$

The exponential number n reflects the degree of nonideality. When the value of n in (4) equals 1, 0, −1, and 0.5, this equation turns out to be the expression of pure capacitor C, pure resistor R, pure inductor L, and Warburg element, respectively. To simulate the nonideal capacitive behavior, the value of n needs to be limited between 0.8 and 1.0 [33]. With this valid restriction, one ($Q_{dl}R_{ct}$) failed to simulate the impedance arcs at higher frequency region. In this case, the overlap of two semicircles is preferred to explain the large extent depression. One more parallel (CR) subcircuit (Figure 3(b)) was added into a series with the Randles circuit to derive the ideal equivalent circuit model (Figure 3(c)) for studying of the full Ni-MH cells. To improve the fitness of the equivalent circuit model to the experimental data, nonideal processes are essential to be considered for the Ni-MH cells.

In this work, the ideal equivalent circuit model in Figure 3(c) was further modified to the nonideal equivalent circuit model as shown in Figure 4. It is an improved equivalent circuit model in this work proposed for the following simulation efforts. Three CPEs were applied to replace two ideal capacitors and one Warburg element for nonideal capacitive behaviors and nonideal diffusion behavior. The simulation fitness of this nonideal improved model to the experimental data measured in this work is illustrated as shown in Figure 5.

The impedance spectrum measured from the Ni-MH cell at a capacity level of *ca*. 30% SoC (dots) was plotted to detail the fitness of simulation curve (solid line). The pure resistor R_Ω refers to the smaller intercept of impedance spectra in real axis. Two parallel subcircuits ($Q_{dl,1}R_{ct,1}$ and $Q_{dl,2}R_{ct,2}$) are applied to simulate two overlapping impedance arcs in higher frequency range. For nonideal capacitive behavior, the exponential numbers of $Q_{dl,1}$ and $Q_{dl,2}$ are expected to stay between 0.8 and 1.0.The element Q_{diff} simulates the straight lines in the low frequency range for the nonideal Warburg behavior with an exponential number slightly less than 0.5. The simulation result from the fitting curves presents good consistency with measurement data. Kuriyama et al. [17] proposed an equivalent circuit model of "four time constants" for the MH electrodes. The model consisted of three (CR) parallel circuits and one Warburg element. The fitting values of model elements were utilized to study the deterioration mechanism of electrodes without showing the fitting curves. However, equivalent circuit models having less time constants were preferred to avoid redundancy. An equivalent circuit model with diffusion element of "three time constants" [20] was developed by Tliha et al. to simulate their AB5-typeMHelectrodes at altered SoC levels. The mature research on NiOOH/$Ni(OH)_2$ electrodes confirmed that an equivalent circuit model consisted of one

(CR) parallel circuit and one diffusion circuit [13]/diffusion element [35] was capable of providing excellent electrode simulation. Slepski et al. [26] suggested an equivalent circuit model of "one time constant" for their positive electrode and an equivalent circuit model of "two time constants" for their negative electrode. However, the suggested equivalent circuit model only had two time constants. No diffusion element was considered either in electrode models or in full battery model. In this work, the proposed nonideal equivalent circuit model has three time constants with consideration of diffusion processes. The impedance spectra of positive and negative electrode overlapped to some extent at higher frequency region; however, two (QR) subcircuits were preferred to distinguish contributions from each electrode.

The simulation results of the Ni-MH Cell A and Cell B at different RoC levels are shown in Figures 6 and 7. In higher frequency range, the two parallel (QR) circuits simulate the measured impedance spectra quite well under all cell status. However, the fitting curves gradually deviate from the impedance data when the frequency decreases to lower than 1Hz. This deviation becomes more significant at higher charge input levels, especially at 100% RoC level. Mathematical models likely implement better simulation in this frequency region than semi-empirical EC models [25].

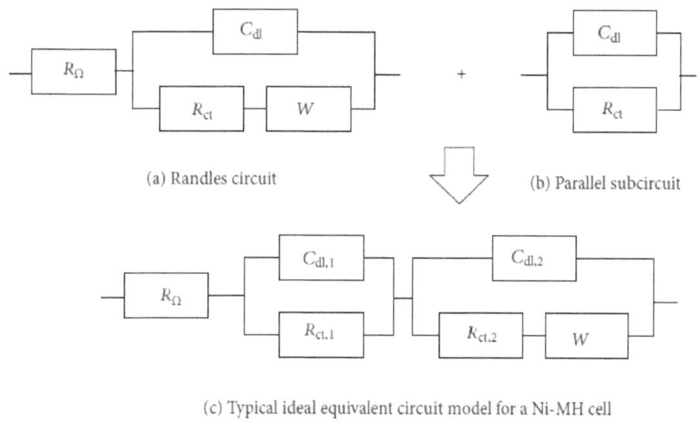

Figure 3: (a) Randles circuit. (b) Paralleled (CR) subcircuit. (c) Typical ideal equivalent circuit model for a full Ni-MH cell.

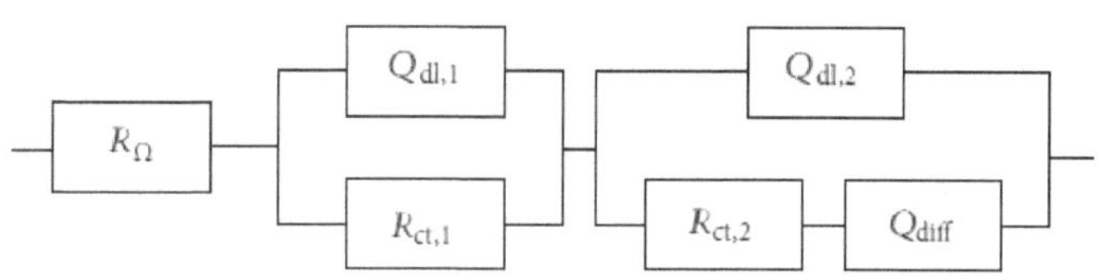

Figure 4: An improved nonideal equivalent circuitmodel proposed for impedance simulation of the Ni-MH cells.

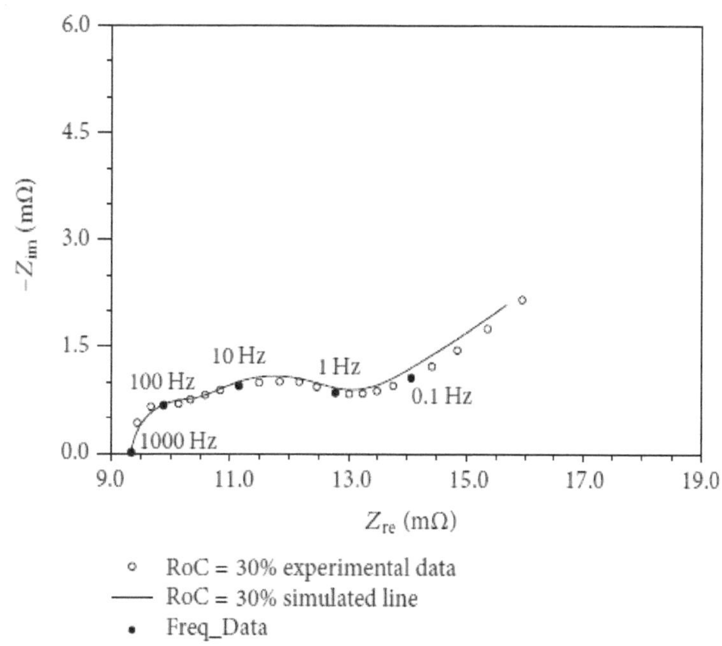

Figure 5: Impedance spectra measured from the Ni-MH cell (Cell A) at a 30% RoC (*ca.* 30% SoC) level, simulated by the improved equivalent circuit model.

Other impedance studies of full Ni-MH cells [22, 23, 26] paid more attention to measurement data and SoC studies. Valoen et al. [36] reported an improvement of simulation at low frequency region by considering hydrogen evolution process; however, this work was done merely for MH electrodes but not for full Ni-MH cells. As a whole, the simulation curves are basically identical with the measured experimental data.

Interpretation for Equivalent Circuit Elements

The understanding of cell mechanisms can be obtained through equivalent circuit simulation in addition to proper interpretation for physical elements and electrochemical processes. The electrolyte of Ni-MH rechargeable batteries is usually a concentrated potassium hydroxide (KOH) solution. The whole discharge process [27] can be described by the following electrochemical reaction as

$$NiOOH + MH \xrightarrow{\text{discharge}} Ni(OH)_2 + M$$

$$E^0_{cell} = 1.35\ V \qquad (5)$$

The reactant of nickel oxyhydroxide (NiOOH) refers to the active material of positive electrodes. Hydrogen is absorbed into metal alloys during the cell charge process and the reactant of MH refers to the metal hydride alloy at the negative electrodes after charging. The overall reaction of the charging process is simply the reversing process of (5). However, Karden et al. [37] demonstrated that the electrode kinetics is an asymmetric process during charge and discharge for a 12V Pb-acid battery. Batteries potentially undertake different impedance spectra during charging process in comparison to discharging process under the same operating condition. The mechanisms discussed below for the Ni-MH cells are for better understanding of the discharging processes at different RoC levels.

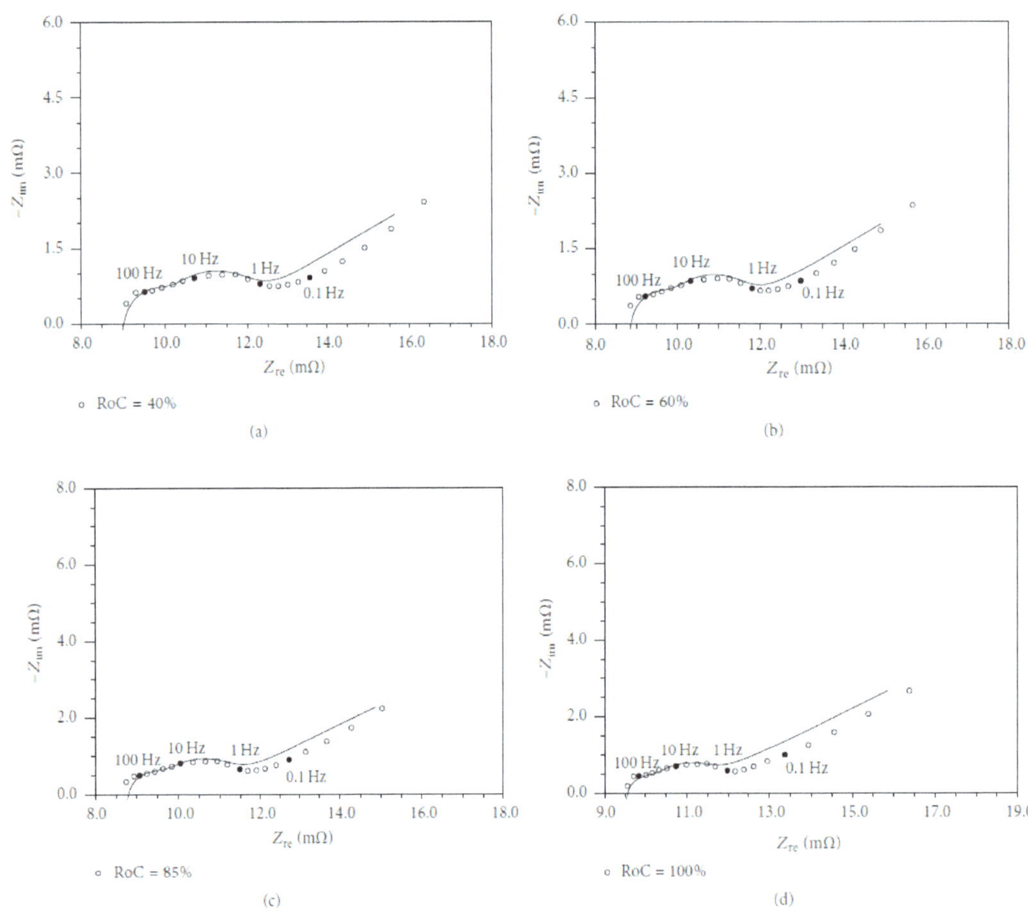

Figure 6: Impedance spectra measured from the Ni-MH cell (Cell A) at different RoC levels with fitting curves simulated from the improved equivalent circuit model.

Mechanism on Negative Electrodes

The half-cell process on the negative electrode is called the dehydriding reaction [38]. It can be expressed as follows:

(a) Hydrogen atoms adsorbed in metal alloy transfer from bulk sites to surface sites. These are metal alloy sites at the interface of electrode and electrolyte, described as

$$M_B \cdots H_{ad} + M_S \longrightarrow M_S \cdots H_{ad} + M_B \qquad (6)$$

where H_{ad} refers to the hydrogen atom adsorbed in the metal alloy; M_S and M_B refer to the empty sites of metal alloy on surface and in bulk, respectively.

Vitanen [39] explained this process as hydrogen released from the hydride phase to form adsorbed hydrogen atoms.

b) Hydrogen oxidation reaction (HOR) at the electrode/electrolyte interface is a charge transfer step coupled by the diffusion process of hydroxide ions from bulk electrolyte to the electrode/electrolyte interface. It is represented by

$$OH_B^- \longrightarrow OH_S^- \quad (7)$$

$$M_S \cdots H_{ad,S} + OH_S^- \longrightarrow M_S + H_2O_S + e^- \quad (8)$$

where OH_B^- and OH_S^- refer to the hydroxide ions in the bulk of electrolyte and at the interface of electrode/electrolyte, respectively; H_2O_S is water atoms produced by the HOR process at the electrode/ electrolyte interface.

(c) Transport of water molecules away from electrode/ electrolyte interface to bulk electrolyte can be expressed with the following reaction

$$H_2O_S \longrightarrow H_2O_B \quad (9)$$

where H_2O_B is water molecules in bulk electrolyte.

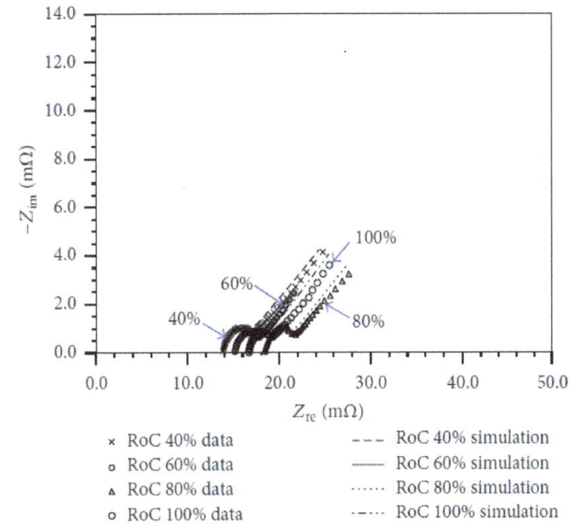

Figure 7: Impedance spectra measured from the Ni-MH cell (Cell B) at different RoC levels with fitting curves simulated from the improved equivalent circuit model.

The rate-limiting step on the negative electrode is the one-electron transfer kinetic process (see (8)) when there are sufficient hydrogen atoms adsorbed in the metal alloy sites on electrode surface. Diffusion of hydrogen atoms from bulk electrode to surface sites (see (6)) is a semi-infinite diffusion process featured by Warburg behavior at low frequency. This process does not have significant effect on cell impedance under common conditions [18, 38] but will become the rate-limiting process at low temperature [40] due to decrease of diffusion coefficient. The mass transfer process of hydroxide ions (see (7)) from bulk electrolyte to electrolyte/electrode interface is a porous bounded diffusion process. Its impedance effect can be negligible under general conditions but becomes significant under a large *dc* current load. The rate of hydroxide ion diffusion process lags behind the electrode kinetics at large current loads. The transfer process of water molecules (see (9)) can be neglected when compared to other processes, because the hydroxide ion concentration in electrolyte is assumed to be large enough [38] and globally constant during battery operation [23].

Mechanism on Positive Electrodes

Zimmerman and Effa [14] proposed a mechanism for the discharge process of NiOOH/Ni(OH)$_2$ electrodes. Following this recommended mechanism, the step-by-step reaction can be summarized as follows: (a) Transport process of water molecules from bulk electrolyte to the interface of electrolyte and electrode, followed by the formation of proton at catalytic site at the interface, is described by the following reactions:

$$H_2O_B \longrightarrow H_2O_S \quad (10)$$

$$H_2O_S \longrightarrow OH_S^- + H_S^+ \quad (11)$$

(b) Transport process of hydroxide ions from the interface back to the electrolyte bulk is written as

$$OH_S^- \longrightarrow OH_B^- \quad (12)$$

(c) Electrode reduction reaction of NiOOH active material, which is a charge transfer process coupled by the diffusion process of proton from the surface site to the charge transfer site in the electrode bulk, is expressed with the following reaction steps:

$$H_S^+ \longrightarrow H_B^+ \quad (13)$$

$$NiOOH + H_B^+ + e^- \longrightarrow Ni(OH)_2 \quad (14)$$

The processes on positive electrodes are different from what is discussed above for the negative electrodes. Diffusion process is considered as the rate-limiting step. Its effect on NiOOH/Ni(OH)$_2$ electrode can be observed under general battery operating conditions [14].This means that the charge transfer process expressed as (14) is not the rate-limiting process. And the diffusion process of proton in positive electrode (see (13)) is considered as the rate-limiting step under normal discharge process. The impedance of proton diffusion process can be applied to simulate the Warburg behavior.

Full Battery Impedance

The impedance arcs of semi-infinite diffusion processes were observed from impedance spectra of both MH electrodes [17–20] and NiOOH/Ni(OH)$_2$ electrode studies [13, 15, 16, 35]. However, only one Warburg behavior arc was observed at low frequency region in this work. Thus, the difficulty of interpretation lies in the diffusion process. Karden et al. [37] measured the half-cell impedance separately and compared the sum of the half-cell impedance to the total cell impedance. The comparison clearly illustrated that the negative electrode impedance dominated higher frequency region and the positive electrode contributed to the diffusion impedance in lower frequency region. Hammouche et al. [23] also reported that the nickel hydroxide electrode contributed the diffusion impedance to the full cell under normal discharge conditions. Based on the discussion of half-cell mechanism in the previous section, the diffusion element Q_{diff} simulates the rate-limiting diffusion process of positive electrode. It is explained as the solid-state diffusion process of proton in nickel hydroxide electrode expressed as (13). The parallel subciruit ($Q_{2,dl}R_{2,ct}$), connecting to the diffusion element, simulates the charge transfer process occurring on the electrolyte/positive electrode interface coupling with the reduction reaction of the active material NiOOH (see (14)). And the parallel subcircuit ($Q_{1,dl}R_{1,ct}$) is ascribed to the one-electron transfer process of HOR occurring on the surface of the negative electrode. The pure resistor R_{Ω}, connecting in series to other equivalent circuit elements, simulates the ohmic resistance. It consists of the resistance of cell components (electrodes, electrolyte, and conducting substrates) and resistance introduced by contact and connection.

Changes of Ohmic and Charge Transfer Resistances

The correlation between RoC and SoC was studied on the Ni-MH cell (Cell B) [28]. When certain amount of charge input was applied to Cell B, its value of RoC was calculated based on the amount of charge. The SoC level was determined when Cell B was discharged to 1.00V cut-off voltage at a 0.2 C rate. Based on the test and calculation results, the values of RoC are identical to the SoC parameter when the capacity level is below *ca.* 85% SoC. As the value of RoC continues to linearly increase with the amount of charge input, the rate of the SoC increase gradually slows down. The values of SoC deviate from RoC at a higher capacity level, but the difference starts at the beginning with a small magnitude. The reason is that the side reaction of oxygen reduction reaction (ORR) which occurred on the positive electrode (see (15)) competes for the amount of charge input. To protect Ni-MH batteries under the overcharge condition, the capacity of negative electrodes is usually designed as 1.5 to 2 times that of positive electrodes [27]. After evolving on the positive electrode, oxygen diffuses through the separators with unsaturated electrolyte and reaches the negative electrode to recombine with water molecules (see (16)). The side reactions on both electrodes during overcharge [27] can be expressed as follows:

Overcharge at positive electrodes:

$$4OH^- \longrightarrow 2H_2O + O_2 + 4e^- \quad (15)$$

Overcharge at negative electrodes:

$$O_2 + 2H_2O + 4e^- \longrightarrow 4OH^- \quad (16)$$

Side reactions also compete with cell reactions when the operation falls into over discharge conditions at low capacity levels and are simply described as follows:

Over discharge at positive electrodes:

$$2H_2O + 2e^- \longrightarrow H_2 + 2OH^- \quad (17)$$

Over discharge at negative electrodes:

$$H_2 + 2OH^- \longrightarrow 2H_2O + 2e^- \quad (18)$$

Hydrogen transfers from the positive electrode to the negative electrode through the separators with unsaturated electrolyte. Side reactions consume the charge energy without changing electrolyte and electrodes under overcharge and over discharge conditions. In this way, the Ni-MH cells are balanced but cell energy efficiency drops to a lower level. However, it is preferred to avoid these overcharge and over discharge operating windows because the oxidation reactions are harmful to the battery material durability and shorten the cell operating lifetime. Nelson pointed out that the realistic operation window for hybrid driving duty cycle of HEVs was from30% SoC to 70% SoC [41].When studying the battery mechanism and capacity decay, this work chooses more tests for the impedance measured at the operating window between 30% and 70% SoC, equivalent to the same values of RoC.

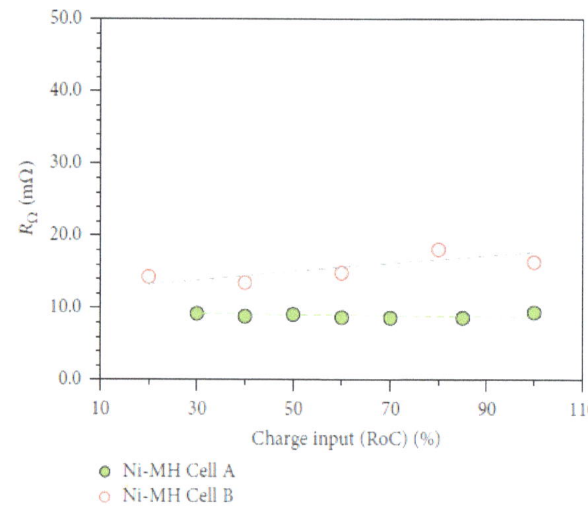

Figure 8: Ohmic resistance (R_{Ω}) of the Ni-MH Cell A and Cell B at different RoC levels. Values are calculated from the equivalent circuit simulation.

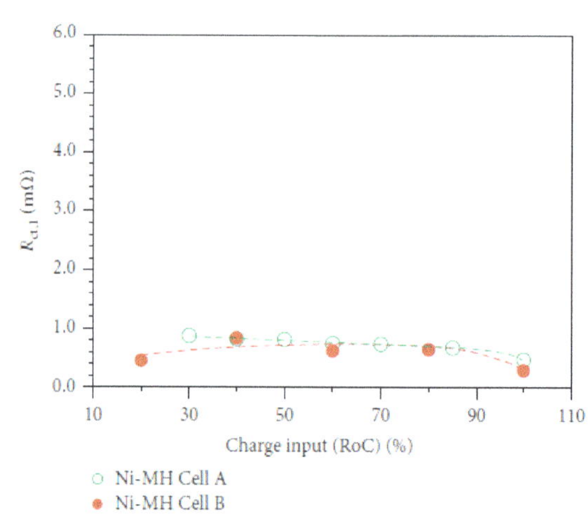

Figure 9: Charge transfer resistance contributed by HOR process on the negative electrodes of the Ni-MH Cell A and Cell B at different RoC levels. Values are calculated from the equivalent circuit simulation.

The change of ohmic resistance and charge transfer resistance with charge input levels is shown in Figures 8–10.

The resistances of the Ni-MH Cell A and Cell B contributed by the same processes are plotted on the same figure for comparison. The ohmic resistance R_Ω of Cell A stays at a lower and more stable level than Cell B (Figure 8). Between the capacity levels from 20% to 100% RoC, the value of R_Ω of Cell B slightly increases with increasing the charge input level likely due to temporary water formation inside the positive electrodes. This potentially results in the changes of local electrolyte concentration and the battery equivalent series resistance R_Ω. However, this expected variation is not obviously shown in the Ni-MH Cell A. More concentrated

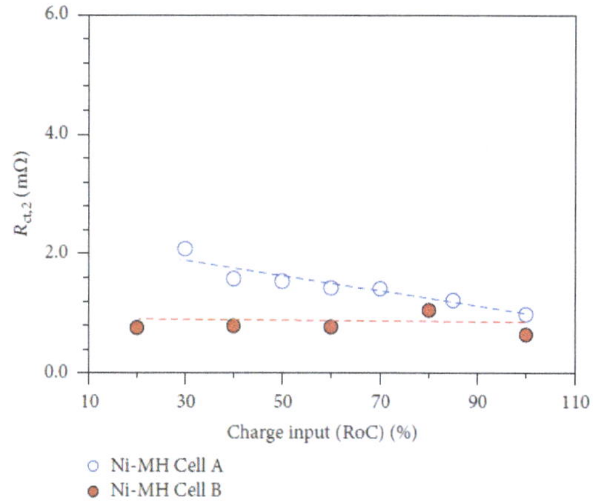

Figure 10: Charge transfer resistance contributed by NiOOH reduction process on the positive electrodes of the Ni-MH Cell A and Cell B at different RoC levels. Values are calculated from the equivalent circuit simulation.

KOH electrolyte likely contributes to this lower equivalent series resistance R_Ω for Cell A. The status of electrolyte and any other auxiliary components such as electrode material status for hydrogen storage capability also likely contribute to the value change of R_Ω parameter. The charge transfer resistance contributed by the HOR process on negative electrodes ($R_{ct,1}$) is plotted against the charge input levels as shown in Figure 9. The cell parameter of $R_{ct,1}$ is relatively stable with increasing charge input levels. $R_{ct,1}$ of both cells stay at a similar value, which means the electrode kinetics of HOR process on negative electrodes of both cells are equivalent to each other. Although the cell capacities of two cells are different after aging processes, the capacity of negative electrodes is designed at a sufficient level for cell protection. The reduction of cell capacity available is not contributed by the negative electrodes due to a relative stable charge transfer resistance showing the identical negative HOR kinetics.

The Ni-MH cell capacity available is dominated by the amount of capacity at the positive electrode. A visible difference exists between the charge transfer resistances contributed by the positive electrodes ($R_{ct,2}$) of two Ni- MH cells at different state-of-health levels after years of providing electrical power services through numerous charge and discharge cycles (Figure 10). The Ni-MH Cell B has a relatively more stable $R_{ct,2}$ in comparison with that of Cell A, because more active materials available for cell discharge are kept at the positive electrode of Cell B, corresponding to faster kinetics and higher electrode capacity. The more cell capacity available for Cell B can be reflected by smaller charge transfer resistance at its positive electrode ($R_{ct,2}$). As a result of experimental measurement, the cell capacity of theNi-MH Cell A after aging process is measured at about 3702mAh, much smaller than the amount of 4362mAh capacity for Cell B. The positive electrode of Cell A is more degraded compared to Cell B, likely due to limitation of insufficient amount of aqueous KOH supply, that is, starved electrolyte for the electrochemical reaction of active materials inside the battery. Additionally, the product of $Ni(OH)_2$ at a low level of SoC is estimated to have more impeding effects on electrode kinetic process for energy conversion and delivery (see (14)).

Thus, the value of $R_{ct,2}$ of the Ni-MH Cell A presents a slightly more increase with decreasing the charge input level (RoC) as expected. The discrimination and interpretation of impedance spectra illustrate that the impedance difference between the Ni-MH cells at different health status resulted firstly from the difference of ohmic resistances and secondly from the active material difference available for NiOOH reduction reactions on the positive electrodes inside the Ni-MH cells.

CONCLUSION

In this work, measurement using electrochemical impedance spectroscopy (EIS), simulation through an improved equivalent circuit model (ECM), and energy dispersive X-ray spectroscopy (EDX) were applied together to characterize Ni-MH rechargeable batteries via an improved equivalent circuit model to examine and analyze chemical reaction mechanisms in the Ni-MH cells. The cell performance degraded due to altered load conditions and various power cycle services. A nonideal equivalent circuit model was derived to simulate impedance spectra measured from both cells at different capacity levels. The significant performance differences were observed from their impedance spectra and equivalent circuit simulation after the cell-aging process. The simulation through the equivalent circuit model broke down the total cell impedance into several parts, corresponding to the real physical elements and chemical reaction steps. Based on the analysis of battery chemistry, the simulation results are interpreted as the ohmic resistance, the impedance of electron transfer process of hydrogen oxidation reaction (HOR) on the negative electrode, the impedance of charge transfer process of NiOOH reduction process on the positive electrode, and the impedance of solid-state diffusion process on the positive electrode.

The Ni-MH cells studied in this work were at different health status. Their impedance spectra were measured at different charge input levels during discharging process. The ohmic resistance of the Ni-MH Cell A stayed at a slightly lower and more stable level. It is likely contributed by more concentrated KOH electrolyte. The activation resistances on the negative MH electrodes of both Ni-MH cells are basically identical in spite of the amount of charge input increase for the battery. The Ni-MH Cell A had more capacity degradation than that of Cell B, prominently reflected by the activation resistance of the positive electrode inside the cell structure. This work reveals that the impedance tool combined with equivalent circuit simulation is an effective technique to study the effect of the capacity degradation on the battery energy storage and perform battery diagnostics through the electrode reactions and mechanism analysis. It is valuable for further understanding of the Ni-MH power system failure after the battery years of service at different loads, power levels, and ambient temperature conditions, highly beneficial to the extensive product improvement, enhancement of battery management systems, and charging infrastructures for better energy harvest on-board with extended service-life cycles.

ACKNOWLEDGMENTS

This work was performed under a US Army contract at Auburn University (W56HZV-05-C0686) administered through TARDEC. The authors would like to thank Mr. Dwight Cahela, Mrs. Kimberly Dennis, and all CM3 group members for their assistance and contribution to this work.

REFERENCES

1. C. Lambertson, "Lithium-ion battery simulation for greener ford vehicles," COMSOL News, Comsol, Inc., Burlington, Mass, USA, pp. 8–11, 2011.
2. A. S. Bandarenka, "Exploring the interfaces between metal electrodes and aqueous electrolytes with electrochemical impedance spectroscopy," Analyst, vol. 138, no. 19, pp. 5540–5554, 2013.
3. A. K. Arof, S. Amirudin, S. Z. Yusof, and I. M. Noor, "A method based on impedance spectroscopy to determine transport properties of polymer electrolytes," Physical Chemistry Chemical Physics, vol. 16, no. 5, pp. 1856–1867, 2014.
4. J. F. Rubinson and Y. P. Kayinamura, "Charge transport in conducting polymers: insights from impedance spectroscopy," Chemical Society Reviews, vol. 38, no. 12, pp. 3339–3347, 2009.
5. C. Z. Yang, M. Zhou, and Q. Xu, "Three-dimensional ordered macroporous MnO_2/carbon nanocomposites as high-performance electrodes for asymmetric supercapacitors," Physical Chemistry Chemical Physics, vol. 15, no. 45, pp. 19730–19740, 2013.
6. I. Mora-Seró, G. Garcia-Belmonte, P. P. Boix, M. A. Vázquez, and J. Bisquert, "Impedance spectroscopy characterisation of highly efficient silicon solar cells under different light illumination intensities," Energy & Environmental Science, vol. 2, no. 6, pp. 678–686, 2009.
7. R. U. Payne, Y. Zhu, W. H. Zhu, M. S. Timper, S. Elangovan, and B. J. Tatarchuk, "Diffusion and gas conversion analysis of solid oxide fuel cells at loads via AC impedance," International Journal of Electrochemistry, vol. 2011, Article ID 465452, 11 pages, 2011.
8. Y. Zhu, W. H. Zhu, and B. J. Tatarchuk, "Performance comparison between high temperature and traditional proton exchange membrane fuel cell stacks using electrochemical impedance spectroscopy," Journal of Power Sources, vol. 256, pp. 250–257, 2014.
9. Y. F. Liu, H. G. Pan, M. X. Gao, and Q. D. Wang, "Advanced hydrogen storage alloys for Ni/MH rechargeable batteries," Journal of Materials Chemistry, vol. 21, no. 13, pp. 4743–4755, 2011.
10. M. R. Palacín, "Recent advances in rechargeable battery materials: a chemist's perspective," Chemical Society Reviews, vol. 38, no. 9, pp. 2565–2575, 2009.
11. J. G. Zhu, Z. C. Sun, X. Z. Wei, and H. F. Dai, "A new electrochemical impedance spectroscopy model of a high-power lithium-ion battery," RSC Advances, vol. 4, no. 57, pp. 29988–29998, 2014.
12. W. H. Zhu, Y. Zhu, and B. J. Tatarchuk, "A simplified equivalent circuit model for simulation of Pb-acid batteries at load for energy storage application," Energy Conversion and Management, vol. 52, no. 8-9, pp. 2794–2799, 2011.
13. D. D. Macdonald, B. G. Pound, and S. J. Lenhart, "The application of electrochemical impedance spectroscopy for characterizing the degradation of $Ni(OH)_2$/NiOOH electrodes," Journal of Power Sources, vol. 29, no. 3-4, pp. 477–502, 1990.
14. A. H. Zimmerman and P. K. Effa, "Discharge kinetics of the nickel electrode," Journal of the Electrochemical Society, vol. 131, no. 4, pp. 709–713, 1984.
15. Y. W. Li, J. H. Yao, Y. X. Zhu, Z. G. Zou, and H. B. Wang, "Synthesis and electrochemical performance of mixed phase α/β nickel hydroxide," Journal of Power Sources, vol. 203, pp. 177–183, 2012.
16. M. Ortiz, D. Becker, G. Garaventta, A. Visintin, E. B. Castro, and S. G. Real, "Dynamic monitoring of structural changes in nickel hydroxide electrodes during discharge in batteries," Electrochimica Acta, vol. 56, no. 23, pp. 7946–7954, 2011.
17. N. Kuriyama, T. Sakai, H. Miyamura, I. Uehara, H. Ishikawa, and T. Iwasaki, "Electrochemical impedance and deterioration behavior of metal hydride electrodes," Journal of Alloys and Compounds, vol. 202, no. 1-2, pp. 183–197, 1993.
18. R. M. Humana, J. E. Thomas, F. Ruiz, S. G. Real, E. B. Castro, and A. Visintin, "Electrochemical behavior of metal hydride electrode with different particle size," International Journal of Hydrogen Energy, vol. 37, no. 19, pp. 14966–14971, 2012.
19. F. C. Ruiz, P. S. Martínez, E. B. Castro, R. Humana, H. A. Peretti, and A. Visintin, "Effect of electrolyte concentration on the electrochemical properties of an AB_5-type alloy for Ni/MH batteries," International Journal of Hydrogen Energy, vol. 38, no. 1, pp. 240–245, 2013.
20. M. Tliha, S. Boussami, H. Mathlouthi, J. Lamloumi, and A. Percheron-Guégan, "Electrochemical characteristics of AB5-type hydrogen storage alloys," Journal of Alloys and Compounds, vol. 506, no. 2, pp. 559–564, 2010.
21. F. Huet, "A review of impedance measurements for determination of the state-of-charge or state-of-health of secondary batteries," Journal of Power Sources, vol. 70, no. 1, pp. 59–69, 1998.
22. K. Bundy, M. Karlsson, G. Lindbergh, and A. Lundqvist, "An electrochemical impedance spectroscopy method for prediction of the state of charge of a nickel-metal hydride battery at open circuit and during discharge," Journal of Power Sources, vol. 72, no. 2, pp. 118–125, 1998.

23. A. Hammouche, E. Karden, and R. W. De Doncker, "Monitoring state-of-charge of Ni-MH and Ni-Cd batteries using impedance spectroscopy," Journal of Power Sources, vol. 127, no. 1-2, pp. 105–111, 2004.
24. M. Thele, O. Bohlen, D. U. Sauer, and E. Karden, "Development of a voltage-behavior model for NiMH batteries using an impedance-based modeling concept," Journal of Power Sources, vol. 175, no. 1, pp. 635–643, 2008.
25. E. B. Castro, D. J. Cuscueta, R. H. Milocco, A. A. Ghilarducci, and H. R. Salva, "An EIS based study of a Ni-MH battery prototype. Modeling and identification analysis," International Journal of Hydrogen Energy, vol. 35, no. 11, pp. 5991–5998, 2010.
26. P. Slepski, K. Darowicki, E. Janicka, and A. Sierczynska, "Application of electrochemical impedance spectroscopy to monitoring discharging process of nickel/metal hydride battery," Journal of Power Sources, vol. 241, pp. 121–126, 2013.
27. D. Linden and T. B. Reddy, Handbook of Batteries, McGraw-Hill, 2002.
28. W. H. Zhu, Y. Zhu, Z. Davis, and B. J. Tatarchuk, "Energy efficiency and capacity retention of Ni-MH batteries for storage applications," Applied Energy, vol. 106, pp. 307–313, 2013.
29. P. T. Wojcik, P. Agarwal, and M. E. Orazem, "A method for maintaining a constant potential variation during galvanostatic regulation of electrochemical impedance measurements," Electrochimica Acta, vol. 41, no. 7-8, pp. 977–983, 1996.
30. Y. Zhu, W. H. Zhu, and B. J. Tatarchuk, "In-situ dynamic characterization of energy storage and conversion systems," in Energy Storage—Technologies and Applications, A. F. Zobaa, Ed., chapter 10, pp. 239–270, InTech, Rijeka, Croatia, 2013.
31. M. A. Fetcenko, S. R. Ovshinsky, B. Reichman et al., "Recent advances in NiMH battery technology," Journal of Power Sources, vol. 165, no. 2, pp. 544–551, 2007.
32. S. R. Ovshinsky, M. A. Fetcenko, and J. Ross, "A nickel metal hydride battery for electric vehicles," Science, vol. 260, no. 5105, pp. 176–181, 1993.
33. E. Barsoukov and J. R. Macdonald, Eds., Impedance Spectroscopy: Theory, Experiment, and Applications, Wiley-Interscience, 2nd edition, 2005.
34. A. Lasia, "Modeling of impedance of porous electrodes," in Modeling and Numerical Simulations, M. Schlesinger, Ed., vol. 43 of Modern Aspects of Electrochemistry, chapter 3, pp. 67–137, Springer, New York, NY, USA, 2009.
35. K. A. Murugesamoorthi, S. Srinivasan, and A. J. Appleby, "AC impedance studies on nickel oxide electrodes to determine self-discharge characteristics of Ni-H_2 batteries," Journal of Applied Electrochemistry, vol. 21, no. 2, pp. 95–98, 1991.
36. L. O. Valoen, A. Lasia, J. O. Jensen, and R. Tunold, "The electrochemical impedance of metal hydride electrodes," Electrochimica Acta, vol. 47, no. 18, pp. 2871–2884, 2002.
37. E. Karden, S. Buller, and R. W. De Doncker, "A method for measurement and interpretation of impedance spectra for industrial batteries," Journal of Power Sources, vol. 85, no. 1, pp. 72–78, 2000.
38. W. L. Zhang, M. P. S. Kumar, S. Srinivasan, and H. J. Ploehn, "AC impedance studies on metal hydride electrodes," Journal of the Electrochemical Society, vol. 142, no. 9, pp. 2935–2943, 1995.
39. M. Vitanen, "A mathematical model for metal hydride electrodes," Journal of the Electrochemical Society, vol. 140, no. 4, pp. 936–942, 1993.
40. H. Wenzl, "BATTERIES: capacity," in Encyclopedia of Electrochemical Power Sources, C. K. Dyer, P. T. Moseley, Z. Ogumi, D. A. J. Rand, B. Scrosati, and J. Garche, Eds., vol. 1, pp. 395–400, Elsevier B.V., Madrid, Spain, 2009.
41. R. F. Nelson, "Power requirements for batteries in hybrid electric vehicles," Journal of Power Sources, vol. 91, no. 1, pp. 2–26, 2000.

CHAPTER 17

PREDICTION OF STATE-OF-HEALTH FOR NICKEL-METAL HYDRIDE BATTERIES BY A CURVE MODEL BASED ON CHARGE-DISCHARGE TESTS

Huan Yang, Yubing Qiu and Xingpeng Guo

Keywords: Nickel-metal hydride (Ni-MH) battery, state-of-health (SOH), curve fitting, nonlinear least-squares method, prediction

INTRODUCTION

Nickel-metal hydride (Ni-MH) batteries have been applied in portable electronics and electric or hybrid-electric vehicles (EVs and HEVs) owing to their relatively good storage and power, higher safety, and excellent environmental acceptability [1,2,3,4]. With regard to these applications, the failure of Ni-MH batteries, resulting from irreversible capacity degradation and loss of performance in cycle service [5,6], is a key issue that warrants close attention. The state-of-health (SOH) status of a battery, which defines the current battery performance relative to its unused condition, is a powerful indicator of battery performance [7] and is usually used to predict the end-of-life and aging of batteries [8,9]. However, the definition of battery SOH is still somewhat equivocal [10] because different battery parameters can be used as the indicators of battery performance. This ambiguity makes the determination of battery SOH a difficult task.

Generally, three definitions of battery SOH have been reported, including SOH values based on battery impedance [11,12], battery capacity [9], and comprehensive battery parameters such as impedance, capacity, open circuit potential (OCV), charging or discharging current, and temperature (*T*) [13,14,15,16,17]. Apparently, the SOH based on comprehensive battery parameters reflects the present battery performance more accurately, but it also makes the estimation more complicated. The SOH values based on battery impedance and capacity are considered to reflect the capability of the battery to provide a certain power and to store energy, respectively. Both of the SOH values are widely applied in EVs and HEVs. In this case, all the methods for the estimation of battery impedance and battery capacity can be used as a basis for the SOH estimation. However, the estimations of battery impedance and battery capacity are also not easy to do, especially online estimation [18,19].

The reported methods for the estimation of battery SOH can be generally divided into two categories: physics-based model estimations and non-physics-based model estimations. Physics-based model estimations are based on electrical or electrochemical cell models [20,21,22,23,24,25,26,27]. In order to improve the estimation accuracy, some algorithms, such as adaptive observers [28], extended Kalman filter (EKF) [29,30], and relevance vector machines (RVMs)—particle filters (PFs) [31], are usually employed. The main problem for this kind of estimation is that the battery electrical or electrochemical models may not be unique and it is difficult to verify their validity. If the used cell model is not appropriate, the estimation accuracy may be lower and difficult to improve. In addition, the used algorithms are relatively complicated, which make it difficult to use them for online estimation.

Citation: Huan Yang, Yubing Qiu and Xingpeng Guo, Prediction of State-of-Health for Nickel-Metal Hydride Batteries by a Curve Model Based on Charge-Discharge Tests, doi: 10.3390/en81112322.

In the non-physics-based model estimations, some learning algorithms, such as neural networks [32,33,34], support vector machines (SVMs) [35,36], RVMs [37], and fuzzy logic [38,39], are applied, in which the measured impedance parameters [12,32] and other characteristic battery parameters (e.g., discharge current [33], temperature, and state-of-charge (SOC) [34]) are employed as input variables. These methods can learn the battery behavior based on monitored data and thus do not demand battery physics models, but they need lots of training data and depend on the availability of a historic data set. In this case, it is difficult to use them for online estimations. Another kind of non-physics-based model estimation uses various curve equations between the practicable capacity and aging cycles [40,41,42,43,44], in which the parameters of the equations can be obtained by data fitting algorithms, and further they can be adjusted online by using a particle filtering (PF) approach [43] or by combining sets of training data based on Dempster-Shafer theory (DST) and the Bayesian Monte Carlo (BMC) method [44]. This method also needs a lot of accelerated aging test data to determine the curve equations, but it does not need complicated mathematic computations. In addition, the number of aging cycles of a used battery may be unknown in practical applications.

The purpose of this work is to find a simple model for the SOH prediction of Ni-MH batteries to avoid the problems mentioned above, such as the uncertainty of battery physics models, complicated algorithms, comprehensive parameters, and unknown cycle numbers. Therefore we tried to construct a simple curve model without electrical or electrochemical battery models and cycle numbers.

In this work, the SOH is defined as Equation (1) due to the presence of charge-discharge efficiency:

$$SOH = C_{discharge}/C_{rated} \quad (1)$$

where $C_{discharge}$ (Ah) is the discharging capacity of fully charged Ni-MH batteries and C_{rated} (Ah) is the rated capacity. As the C_{rated} for one type of Ni-MH battery is a constant value, the SOH prediction is reduced to the prediction of $C_{discharge}$. Based on the analysis of a lot of charge-discharge cycle test data for commercial Ni-MH batteries, the voltage change in 1 s occurring at the start of the charging process (ΔV_{charge}, mV) was selected as the characteristic parameter, and a curve equation between $C_{discharge}$ and ΔV_{charge} was determined to construct a curve model for the $C_{discharge}$ prediction. Then the prediction precision of this curve model was verified by using some typical charge and discharge data for the Ni-MH batteries.

EXPERIMENTAL

Commercial AA-type Ni-MH cells (Pisen®, Sichuan, China, C_{rated} = 1.8 Ah and rated voltage = 1.2 V) were used in this study. All charge-discharge tests were conducted using a computer-controlled charge-discharge instrument (BTS-3008-5V3A, Xinwei, Guangzhou, China) at room temperature (20 ± 5 °C). The charge-discharge data, including current (I), potential (V), time (t), and cycle number (n), were recorded automatically with the sampling frequency of 1 Hz. The charge-discharge rates have significant influence on battery performance, so three different charge-discharge rates (0.5C, 1.0C, and 1.6C) were selected to simulate low, medium, and high charge-discharge rate conditions. The charge-discharge test program is described as follows:

The Ni-MH cells with initial SOC at 0% were fully charged at a constant current (I_{charge}: 0.5C, 1.0C, and 1.6C), then after an interval of 600 s, they were discharged at the same constant current ($I_{discharge}$: 0.5C, 1.0C, and 1.6C) to the cutoff voltage (0.9 V). After an interval of 1800 s, the next cycle of charge-discharge test was started. When the $C_{discharge}$ of the test cells fell to 80% of the initial capacity, the charge-discharge tests were terminated [45]. The failure capacity is still defined as $0.8C_{rated}$ in this paper for convenient applications.

RESULTS AND DISCUSSION

Curve Modeling for $C_{discharge}$

Figure 1 gives the typical I-t and V-t curves of two cycles for a Ni-MH battery in the charge-discharge test at 1.0C. In each cycle there are four abrupt voltage changes (ΔV) occurring at the start/end of the charging and discharging processes, respectively. According to the equivalent circuit of the test cell [8,46], these four ΔV values in each cycle generally reflect the internal resistance ($R_{internal}$) of the test cell in different states, and there is little difference among their values. However, the measurement of ΔV is much easier than the test of $R_{internal}$, so it is selected as studied parameter. Before the start of the charging process, the completely discharged test cell was in a relatively stable state after the recovery of 1800 s, so in this paper we select the ΔV in 1 s at the start of the charging process (ΔV_{charge}, mV) as the characteristic parameter to verify its relationship with the aging state of the test cell, as shown in Figure 1.

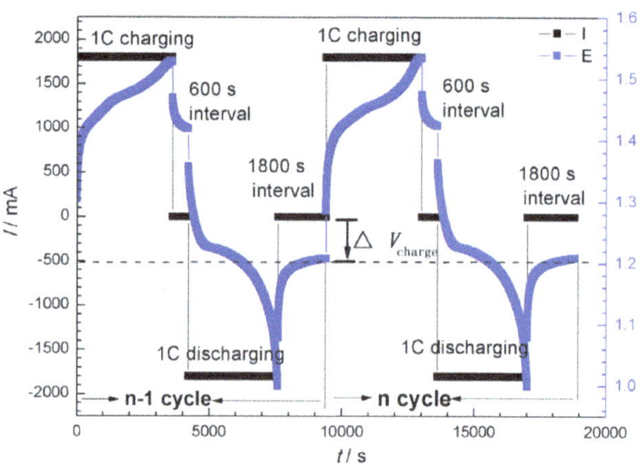

Figure 1. Typical I-t and V-t curves in the charge-discharge test at 1.0C for Ni-MH batteries (AA-type, 1.8 Ah, 1.2 V).

Through the analysis of the I-t and V-t data for each test battery, we obtained $C_{discharge}$ and ΔV_{charge} in each cycle. In order to observe the relationship of these three parameters, we made a three-dimensional (3D) diagram of ($C_{discharge}$-ΔV_{charge}-n) for all the test Ni-MH batteries. Figure 2a shows a typical 3D diagram for a Ni-MH battery in the charge-discharge test at 1.0C and its two-dimensional (2D) projection diagrams ($C_{discharge}$-n, ΔV_{charge}-n, $C_{discharge}$-ΔV_{charge}).

From Figure 2a it can be seen that ($C_{discharge}$; ΔV_{charge}; n) data points form a curve in the 3D space without obvious discrete point; suggesting that there would be nonlinear relationships between these three parameters. From the different 2D diagrams in Figure 2b–d; it can be seen more clearly that $C_{discharge}$-n; ΔV_{charge}-n and $C_{discharge}$-ΔV_{charge} curves all display good nonlinear relationships. The results in Figure 2b,c can be explained appropriately [41] and the nonlinear relationships between $C_{discharge}$-n and ΔV_{charge}-n can be used to predict the state of Ni-MH batteries. However; sometimes the number of cycles for a Ni-MH battery is unknown. In this case; the nonlinear relationship between $C_{discharge}$ and ΔV_{charge}; as shown in Figure 2d; is important for a curve model without cycle numbers. If we can obtain the relationship between $C_{discharge}$ and ΔV_{charge} for a Ni-MH battery; we can obtain its $C_{discharge}$ through the measurement of its ΔV_{charge} and then estimate its SOH with Equation (1).

Finally, we identified that Equation (2) can be used to describe the nonlinear relationship between ΔV_{charge} (mV) and $C_{discharge}$ (Ah):

$$C_{discharge} = \frac{a}{1 + \exp[-k \times (\Delta V_{charge} - c)]} \tag{2}$$

where a (Ah), k (mV^{-1}) and c (mV) are constants related to test batteries. We found that k was negative with a small absolute value ($|k|$) and related to the charge-discharge rate. In the initial cycles, $\Delta V_{charge} << c$ and therefore, if $|k|$ is not very small, there is, $\exp[-k \times (\Delta V_{charge} - c)] \rightarrow 0$ and $a \approx C_{discharge}$. So a can be considered as a parameter related to the initial $C_{discharge}$ of the test battery when $|k|$ is not very small, which may be near its rated capacity. When $\Delta V_{charge} = c$, $C_{discharge} = 0.5a$. Thus, c can be considered as a parameter related to the ΔV_{charge} when $C_{discharge}$ is at its 50% initial value.

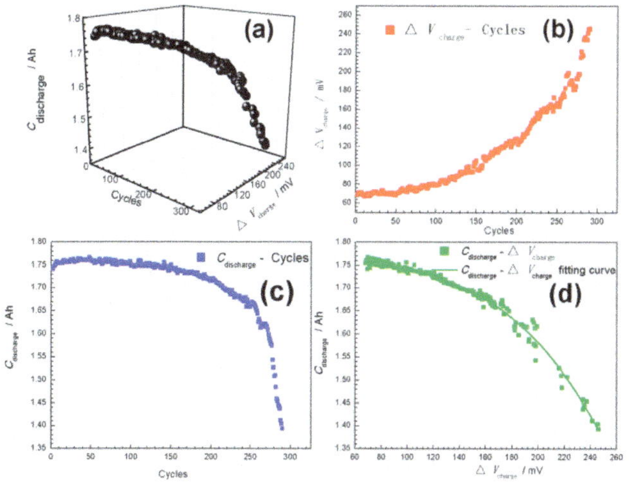

Figure 2. 3D diagram for a Ni-MH battery in the charge-discharge test at 1.0C (AA-type, 1.8 Ah, 1.2 V): (**a**) 3D diagram and (**b**), (**c**) and (**d**) projection of the 3D diagram.

The raw data in Figure 2d was fitted with Equation (2) by a nonlinear least-squares method, and the fitted curve indicates that it fits the raw data well. In order to verify the validity of Equation (2), different ($C_{discharge}$, ΔV_{charge}) raw data obtained from different Ni-MH batteries with different charge-discharge rates were fitted with Equation (2) by the nonlinear least-squares method. Figure 3 shows some typical ($C_{discharge}$, ΔV_{charge}) raw data at different charge-discharge rates and their fitted curves with Equation (2). Table 1 lists the fitting parameters of a, c, and k for different Ni-MH cells at different rates, in which R^2 is a correlation coefficient and P_{range} is the selection range of initial values for each fitting parameter when using the nonlinear least-squares method to solve their values.

The results in Figure 3 and Table 1 prove that Equation (2) can well describe the relationship between $C_{discharge}$ and ΔV_{charge}. From Table 1, it can be seen that at 0.5C and 1.0C the parameter $a$is approximate to the rated capacity of the test Ni-MH batteries (1.8 Ah), while at 1.6C it becomes larger than the rated capacity because of the too small $|k|$ value. It should be noted that at 1.6C the too small $|k|$ value makes $|-k \times (\Delta V_{charge} - c)| << 1$ and then, there is $\exp[-k \times (\Delta V_{charge} - c)] \approx 1 - [k \times (\Delta V_{charge} - c)]$. In this case, the nonlinear relationship between $C_{discharge}$ and ΔV_{charge}described with Equation (2) becomes an approximate linear relationship especially with an increase in ΔV_{charge}, as shown in Figure 3.

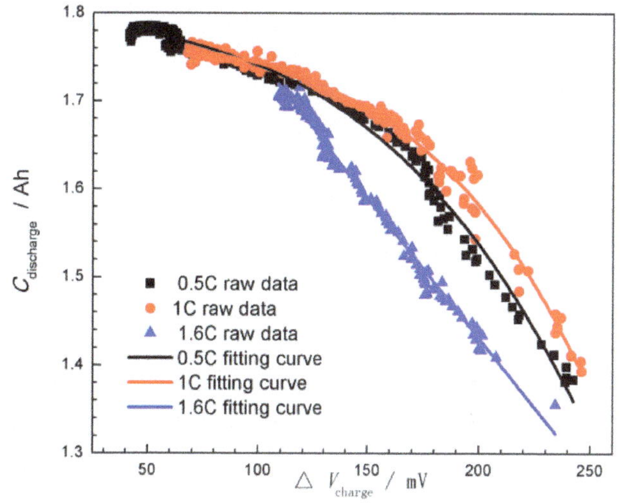

Figure 3. ($C_{discharge}$, ΔV_{charge}) raw data at different rates and their fitting curves with Equation (2).

Table 1. Fitting parameters of a, c, and k for different Ni-MH cells at different rates

Rates		a/Ah	c/mV	k/mV^{-1}	R^2
0.5C	P_{range}	1.5–2.0	200–400	−0.02–0	-
		1.8131	317.58	−0.0146	0.9862
1.0C	P_{range}	1.5–2.0	200–400	−0.03–0	-
		1.7828	320.26	−0.0172	0.9882
		1.7420	310.20	−0.0164	0.9851
		1.7680	223.30	−0.0188	0.9796
		1.7839	286.11	−0.0166	0.9800
		1.9433	313.77	−0.0078	0.9865
		1.7962	381.17	−0.0078	0.9869
1.6C	P_{range}	2.0–3.0	200–400	−0.015–0	-
		2.4479	258.21	−0.0057	0.9898
		2.6365	239.15	−0.0048	0.9890

Apparently, the values of the a, c, and k change with the charge-discharge rate. Table 1 shows that when the charge-discharge rate increases from 0.5C to 1.0C, the change of these three parameters is small, while at 1.6C the change of a and k is obvious. It may be speculated that the values of the three parameters just fluctuate slightly when the charge-discharge rate is in a certain range, such as 0.5C–1.0C. If so, this curve model should also be used in practical fields that the charge-discharge rates are not constant during applications. Further research will be conducted to verify the application of this curve model in practical fields. In this paper we just focus on the curve model itself and its SOH prediction accuracy.

State-of-Health (SOH) Prediction of Ni-MH Batteries Based on the Curve Model for $C_{discharge}$

When the relationship between $C_{discharge}$ and ΔV_{charge} of one battery is built as Equation (2), the $C_{discharge}$ can be predicted by measuring the ΔV_{charge} of the battery and then the SOH can be obtained with Equation (1). So we used the ($C_{discharge}$, ΔV_{charge}) raw date in Table 1 to check whether the curve model for $C_{discharge}$ can be used for the SOH prediction.

Theoretically, only three groups of ($C_{discharge}$, ΔV_{charge}) data are needed to solve the three parameters in Equation (2). However, the measured raw data usually have some discrete points, as shown in Figure 2 and Figure 3, so more groups of ($C_{discharge}$, ΔV_{charge}) data are needed and the prediction error is inevitable. In order to evaluate the prediction error, the root-mean-square error (*RMSE*) is defined as Equation (3):

$$RMSE = \sqrt{\sum_{1}^{n} [C_{discharge,j} - C_{discharge,j}^{test}]^2 / n}, \ j = 1, \ldots, n \tag{3}$$

where Ctestdischarge,j is the test value for sample *j*; Cdischarge,j is the predicted value for sample *j*; and *n* is the total number of samples.

The prediction method is described as follows:

(1) A certain number of ($C_{discharge}$, ΔV_{charge}) data groups in continuous charge-discharge test cycles were selected to fit the parameters in Equation (2) using a nonlinear least-squares method to obtain Equation (2) for the test Ni-MH battery;

(2) The predicted $C_{discharge}$ values were calculated with Equation (2) using the tested ΔV_{charge} values apart from those used in Step 1; and (3) the predicted *RMSE* was checked with Equation (3) using the predicted $C_{discharge}$ values and the corresponding tested $C_{discharge}$ values. Through the predicted *RMSE* we can judge how large the prediction error is and whether the prediction is valid.

Using different numbers of ($C_{discharge}$, ΔV_{charge}) data groups in different parts of the $C_{discharge}$-ΔV_{charge} curve, as shown in Figure 4, may result in different prediction results. In this paper, the first cycle $C_{discharge}$ of a selected ($C_{discharge}$, ΔV_{charge}) data group (C0discharge) is described with the SOH value of the test battery, and a relative capacity drop (RCd) is defined to describe the variation range of $C_{discharge}$ in the selected ($C_{discharge}$, ΔV_{charge}) data group, as presented in Equation (4):

$$RCd = \frac{C_{discharge}^{0} - C_{discharge}^{n}}{C_{discharge}^{0}} \tag{4}$$

where Cndischarge is the last cycle $C_{discharge}$ of the selected ($C_{discharge}$, ΔV_{charge}) data group. Apparently, the RCd value determines the last ($C_{discharge}$, ΔV_{charge}) data of the selected data group. Because we used all the ($C_{discharge}$, ΔV_{charge}) data in continuous test cycles, a larger RCd value means more data in the selected data group. In this case, the SOH and RCd values for a selected data group can determine its position on the $C_{discharge}$–ΔV_{charge} curve and also indicate the corresponding battery states.

Firstly, the effect of the number of used ($C_{discharge}$, ΔV_{charge}) data groups, which is described by RCd, on the prediction of $C_{discharge}$ is investigated. Figure 4 shows a 3D diagram of ($C_{discharge}$-ΔV_{charge}-*n*) for a Ni-MH battery in the charge-discharge test at 1.0C and the predicted results obtained with SOH = 97.38% and different RCds (0.5%, 1.0%, and 1.5%), in which the corresponding cycle numbers of the selected data range and the prediction *RMSE* values are also listed. In Figure 4, the three data groups are selected from 90–138 cycles (48 data), 90–166 cycles (77 data), and 90–184 cycles (95 data), respectively. The first cycle of the three date groups is the same, so they have the same SOH value (97.38%). Their RCd values are calculated with Equation (4) using the $C_{discharge}$ of the 90th cycle (*i.e.*, C0discharge) and C138discharge, C166discharge, and C184discharge, respectively. It is seen that the prediction *RMSE* value decreases with an increase in RCd, suggesting that using more ($C_{discharge}$, ΔV_{charge}) data groups may decrease the prediction error.

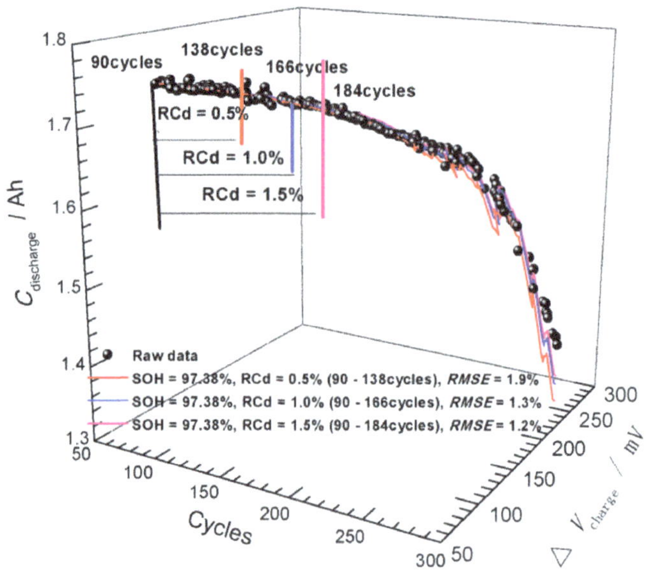

Figure 4. 3D diagram of ($C_{discharge}$-ΔV_{charge}-*n*) for a Ni-MH battery in the charge-discharge test at 1.0C and the prediction curves obtained with SOH = 97.38% and different relative capacity drops (RCds) (0.5%, 1.0%, and 1.5%).

The SOH value of the selected data group, *i.e.*, the position of the starting point of the selected data group, may also influence the prediction result. So we selected data groups at different SOH with various RCds (0.5%, 1.0%, 1.5%, and 2.0%) to fit the parameters in Equation (2) and calculated the corresponding prediction *RMSE* values, as shown in Figure 5.

In Figure 5, the average *RMSE* value at each SOH is presented and shown with a star symbol, while the general average *RMSE* value $(\overline{RMSE})$ is also given and shown with a horizontal dashed line. The *RMSE* value obtained by using data groups at SOH = 94.89% with RCd = 0.5% deviates significantly from other *RMSE* values, which may be due to the fluctuation of the raw data groups, so it was eliminated from the calculation of the average *RMSE* values.

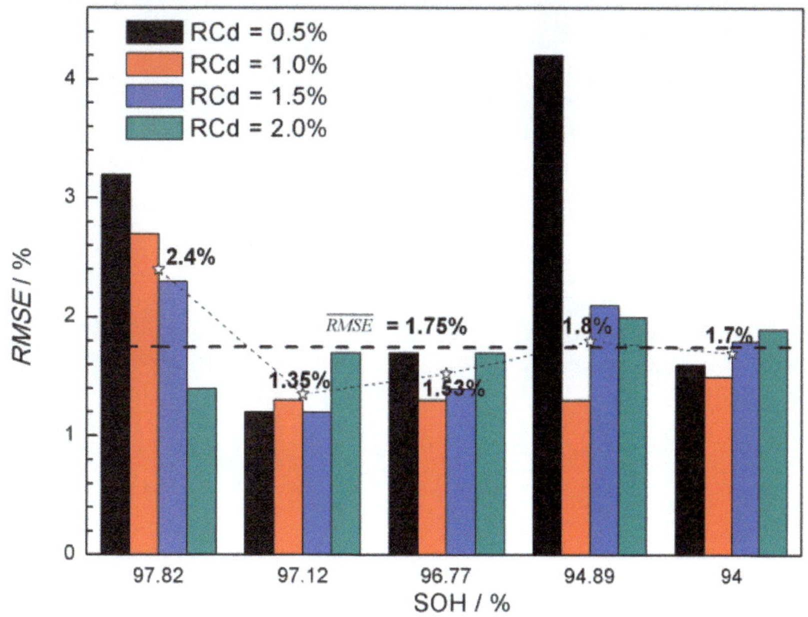

Figure 5. Prediction root-mean-square error (*RMSE*) for a Ni-MH battery in the charge-discharge test at 1.0C using ($C_{discharge}$, ΔV_{charge}) data groups at different SOH with various RCds.

From Figure 5, it can be seen that when SOH = 97.82%, the prediction *RMSE* decreases obviously with an increase in RCd, which is consistent with the result in Figure 4. But when the SOH value decreases to 97.12%, 96.77%, 94.89%, and 94%, at each SOH the prediction *RMSE* does not decrease with an increase in RCd, suggesting that using more data groups does not necessarily help improve the prediction precision. In addition, the prediction *RMSE* does not show some regular change with a decrease in the SOH, but generally it seems that when the SOH is in an intermediate range (97.12% and 96.77%), the prediction *RMSE* is relatively lower.

Figure 6 shows the selected ($C_{discharge}$, ΔV_{charge}) raw data in Figure 5 and the prediction data with the best *RMSE* (1.2%) and the *RMSE* (1.7%) near the $\overline{RMSE}$ in Figure 5 (1.75%).

From Figure 6, we can clearly see that the prediction data well accords with the raw data, especially when $C_{discharge} > 1.44$ Ah, *i.e.*, before the failure capacity ($0.8C_{rated}$), indicating that the prediction accuracy of this curve model is satisfied. It should be noted that the initial part of the raw data in Figure 6 (SOH > 97.12%) has some volatility. So using this part of the raw data for the $C_{discharge}$ prediction may result in higher prediction error (*i.e.*, large *RMSE*) and using more data groups (*i.e.*, large RCd) may improve the prediction precision, as shown in Figure 5 at SOH = 97.82%. When the SOH decreases to 97.12%–96.77%, the fluctuation of the raw data decreases, as shown in Figure 6; this may result in lower prediction *RMSE*, as shown in Figure 5.

Considering the influence of different charge-discharge rates on the prediction results, we selected the ($C_{discharge}$, ΔV_{charge}) raw data at 0.5C and 1.6C to check the curve model using the same method described above. Figure 7 shows the prediction *RMSE* for a Ni-MH battery in the charge-discharge test at 0.5C using ($C_{discharge}$, ΔV_{charge}) data groups at different SOH with various RCds, in which the average prediction *RMSE* values are also given as described in Figure 5. The *RMSE* values at SOH = 98.97% are relative large, which may also result from the fluctuation in the initial data groups, so they are eliminated from the calculation of the $\overline{RMSE}$ in Figure 7.

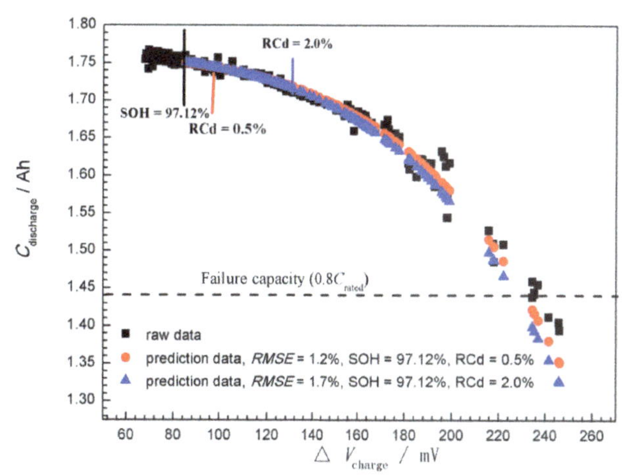

Figure 6. ($C_{discharge}$, ΔV_{charge}) raw data used in Figure 5 and the prediction data with the best *RMSE* (1.2%) and the *RMSE*(1.7%) near the $\overline{RMSE}$ in Figure 5 (1.75%).

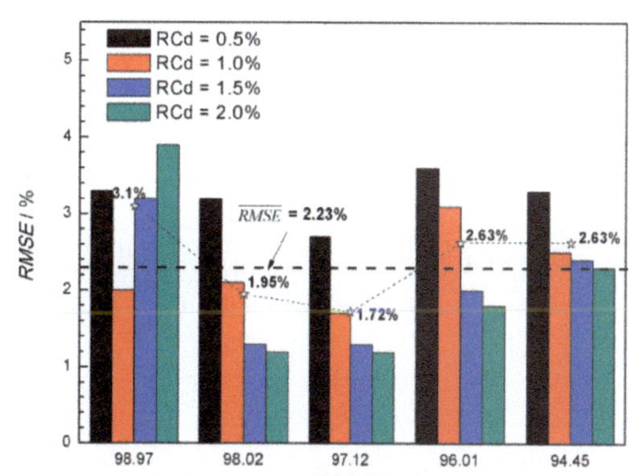

Figure 7. Prediction *RMSE* for a Ni-MH battery in the charge-discharge test at 0.5C using ($C_{discharge}$, ΔV_{charge}) data groups at different state-of-health (SOH) with different RCds.

From Figure 7, it can be seen that at SOH = 98.97% the prediction *RMSE* is higher and does not decrease with an increase in RCd, while at lower SOH values (98.02%–94.45%) the prediction *RMSE*decreases with an increase in RCd. In general, it seems that when the SOH is in an intermediate range (98.02% and 97.12%), the prediction *RMSE* is relatively lower, which is consistent with the results in Figure 5.

Figure 8 presents the ($C_{discharge}$, ΔV_{charge}) raw data used in Figure 7 and the prediction data with the best *RMSE* (1.2%) and the *RMSE* (2.1%) near the $\overline{RMSE}$ in Figure 7 (2.23%). Similarly, there is fluctuation in the initial part of the ($C_{discharge}$, ΔV_{charge}) raw data and the prediction data well accords with the raw data before the failure capacity.

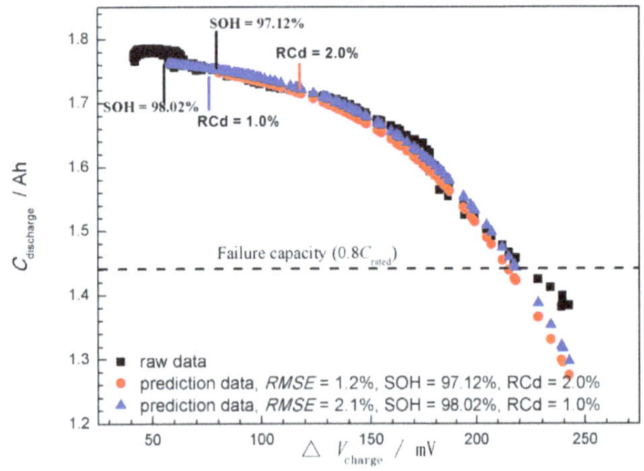

Figure 8. ($C_{discharge}$, ΔV_{charge}) raw data used in Figure 7 and the prediction data with the best *RMSE* (1.2%) and the *RMSE*(2.1%) near the $\overline{RMSE}$ in Figure 7 (2.23%).

Figure 9 shows the prediction *RMSE* for a Ni-MH battery in the charge-discharge test at 1.6C, in which the average prediction *RMSE* values are also given as described in Figure 5. Similarly, the *RMSE* values at SOH = 95% are eliminated from the calculation of the $\overline{RMSE}$ value in Figure 9.

Figure 10 presents the ($C_{discharge}$, ΔV_{charge}) raw data used in Figure 9 and the prediction data with the best *RMSE* (1.2%) and the *RMSE* (1.6%) near the $\overline{RMSE}$ in Figure 9 (1.53%). It can be seen that there is an obvious increase in the $C_{discharge}$ value in the initial charge-discharge period, as shown in Figure 10, so these data groups cannot be used for the prediction. As shown in Figure 9, only when the SOH < 95%, the prediction *RMSE* decreases to a lower value and the prediction data also accords well with the raw data before the failure capacity, as shown in Figure 10.

The results in Figure 4, Figure 5, Figure 6, Figure 7, Figure 8, Figure 9 and Figure 10 indicate that the curve model based on Equation (2) can be effectively applied to predict the SOH of the Ni-MH batteries, and the best prediction *RMSE* is around 1.2%. The selected raw data groups for the prediction significantly influence the prediction *RMSE*. Generally, the data groups in the early period of the charge-discharge test, which depends on the charge-discharge rate, are not suitable for the SOH prediction because of data fluctuations. Increasing the RCd value, *i.e.*, using more data groups for the SOH prediction, does not necessarily help improve the prediction precision, which may also be related to the data fluctuation. Generally, using data groups at the SOH in an intermediate range and with the RCd value in the range of 1.5%–2.0% may result in lower prediction *RMSE*.

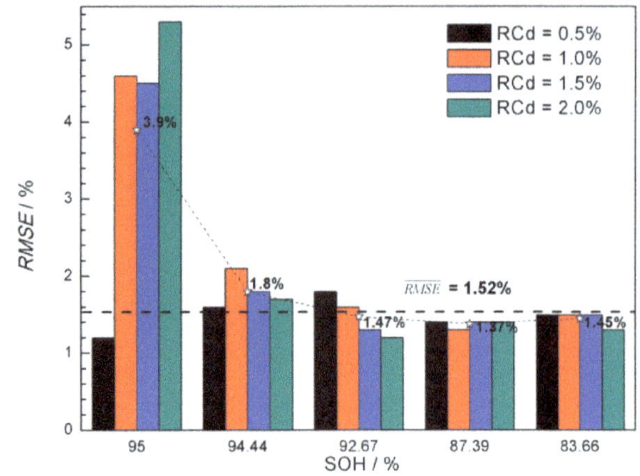

Figure 9. Prediction *RMSE* for a Ni-MH battery in the charge-discharge test at 1.6C using ($C_{discharge}$, ΔV_{charge}) data groups at different SOH with various RCds.

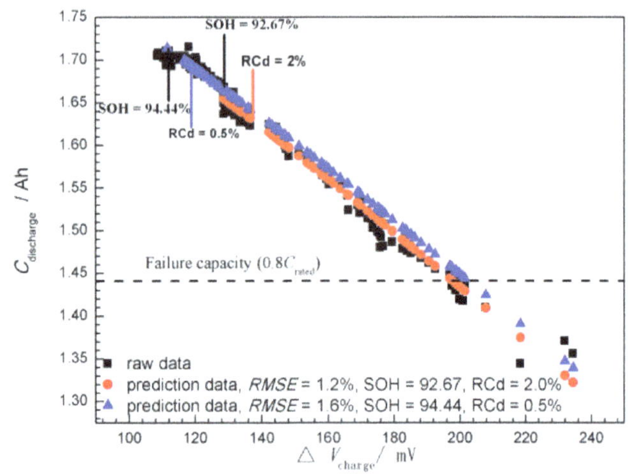

Figure 10. ($C_{discharge}$, ΔV_{charge}) raw data used in Figure 9 and the prediction data with the best *RMSE* (1.2%) and the *RMSE*(1.6%) near the $\overline{RMSE}$ in Figure 9 (1.52%).

Apart from good prediction accuracy, this curve model is also simple and easy-to-use. These advantages are important for online applications. For any type of Ni-MH battery, we can use charge-discharge cycle tests to determine the three parameters (*a*, *c*, and *k*) in the curve model at different charge-discharge rates. Then, we can easily detect the SOH of the same type Ni-MH batteries without considering their aging history (e.g., cycle numbers). As shown in Table 1, the three parameters for different batteries change a little in a certain range of charge-discharge rates, such as 0.5–1.0C. In this case, it is possible to use their average values to construct a general model for the applications with fluctuated charge-discharge rates. Moreover, if more than three ($C_{discharge}$, ΔV_{charge}) data can be obtained during applications, the new values of the three parameters (*a*, *c*, and *k*) can be calculated to modify the general curve model. Certainly, this application of the curve model is needed further verification.

One of limitations for this curve model is that it's relatively higher prediction error in a higher SOH range especially at higher charge-discharge rates. In addition, it should be noted that in this work the ($C_{discharge}$, ΔV_{charge}) data obtained from the charge-discharge cycle test were employed to construct the curve model and to verify its validity for the SOH prediction. In practical applications, batteries may undertake nonuniform charge-discharge processes, such as intermittent charge or discharge processes. Therefore, further research is needed to confirm whether this curve model can continue to be used for any Ni-MH battery in practical applications.

CONCLUSIONS

Based on charge-discharge cycle tests for commercial Ni-MH batteries (Pisen®, 1.2 V, 1.8 Ah), a curve model was constructed without battery models and cycle numbers for the SOH prediction and its prediction precision was verified. The main conclusions were drawn as follows:

(1) Based on the analysis of charge-discharge data for the Ni-MH batteries, a nonlinear relationship between the $C_{discharge}$ (Ah) and ΔV_{charge} (mV) was found and described as $C_{discharge} = a/[1+ \exp[-k \times (\Delta V_{charge} - c)]]$, where a (Ah), k (mV^{-1}) and c (mV) are constants related to charge-discharge rates. Based on this equation, the curve model for the SOH prediction of Ni-MH batteries was constructed.

(2) The ($C_{discharge}$, ΔV_{charge}) data groups obtained from the charge-discharge cycle test at different rates (0.5C, 1.0C, and 1.6C) were employed to verify the validity of the curve model for the SOH prediction. It was found that the curve model based on the nonlinear relationship between the $C_{discharge}$ and ΔV_{charge} can be effectively applied to predict the SOH of the Ni-MH batteries. The data groups used for the SOH prediction have significantly influence on the prediction accuracy, and the best prediction *RMSE* can reach 1.2%.

(3) Generally, using ($C_{discharge}$, ΔV_{charge}) data groups with the SOH in an intermediate range and with the RCd value in the range of 1.5%–2.0% may result in lower prediction *RMSE*.

ACKNOWLEDGMENTS

The authors thank the Special Funds for Basic Research of National Universities (No. 2015TS147) for the financial support of this work.

REFERENCES

1. Leblanc, P.; Blanchard, P.; Senyarich, S. Self-Discharge of Sealed Nickel-Metal Hydride Batteries Mechanisms and Improvements. *J. Electrochem. Soc.* **1998**, *145*, 844–847.
2. Yao, M.; Okuno, K.; Iwaki, T.; Tanase, S.; Harada, K.; Kato, M.; Emura, K.; Sakai, T. High-power nickel/metal-hydride battery using new micronetwork substrate: Discharge rate capability and cycle-life performance. *J. Power Sources* **2007**, *171*, 1033–1039.
3. Fetcenko, M.; Ovshinsky, S.; Reichman, B.; Young, K.; Fierro, C.; Koch, J.; Zallen, A.; Mays, W.; Ouchi, T. Recent advances in Ni-MH battery technology. *J. Power Sources* **2007**, *165*, 544–551.
4. Castro, E.; Cuscueta, D.; Milocco, R.; Ghilarducci, A.; Salva, H. An EIS based study of a Ni–MH battery prototype. Modeling and identification analysis. *Int. J. Hydrog. Energy* **2010**, *35*, 5991–5998.
5. Serrao, L.; Chehab, Z.; Guezennee, Y.; Rizzoni, G. An aging model of Ni-MH batteries for hybrid electric vehicles. In Proceedings of the 2005 IEEE Conference Vehicle Power and Propulsion, Chicago, IL, USA, 7–9 September 2005; pp. 78–85.
6. Chehab, Z.; Serrao, L.; Guezennec, Y.G.; Rizzoni, G. Aging Characterization of Nickel: Metal Hydride Batteries Using Electrochemical Impedance Spectroscopy. In Proceedings of the International Mechanical Engineering Congress and Exposition, Chicago, IL, USA, 5–10 November 2006; pp. 343–349.
7. Salkind, A.J.; Fennie, C.; Singh, P.; Atwater, T.; Reisner, D.E. Determination of state-of-charge and state-of-health of batteries by fuzzy logic methodology. *J. Power Sources* **1999**, *80*, 293–300.
8. Chiang, Y.H.; Sean, W.Y.; Ke, J.C. Online estimation of internal resistance and open-circuit voltage of lithium-ion batteries in electric vehicles. *J. Power Sources* **2011**, *196*, 3921–3932.
9. Ng, K.S.; Moo, C.S.; Chen, Y.P.; Hsieh, Y.C. Enhanced coulomb counting method for estimating state-of-charge and state-of-health of lithium-ion batteries. *Appl. Energy* **2009**, *86*, 1506–1511.
10. Waag, W.; Fleischer, C.; Sauer, D.U. Critical review of the methods for monitoring of lithium-ion batteries in electric and hybrid vehicles. *J. Power Sources* **2014**, *258*, 321–339.
11. Huet, F. A review of impedance measurements for determination of the state-of-charge or state-of-health of secondary batteries. *J. Power Sources* **1998**, *70*, 59–69.
12. Blanke, H.; Bohlen, O.; Buller, S.; de Doncker, R.W.; Fricke, B.; Hammouche, A.; Linzen, D.; Thele, M.; Sauer, D.U. Impedance measurements on lead-acid batteries for state-of-charge, state-of-health and cranking capability prognosis in electric and hybrid electric vehicles. *J. Power Sources* **2005**, *144*, 418–425.
13. Liu, G.; Ouyang, M.; Lu, L.; Li, J.; Han, X. Online estimation of lithium-ion battery remaining discharge capacity through differential voltage analysis. *J. Power Sources* **2015**, *274*, 971–989.
14. Sauer, D.U.; Wenzl, H. Comparison of different approaches for lifetime prediction of electrochemical systems—Using lead-acid batteries as example. *J. Power Sources* **2008**, *176*, 534–546.
15. Zhu, J.; Sun, Z.; Wei, X.; Dai, H. A new lithium-ion battery internal temperature on-line estimate method based on electrochemical impedance spectroscopy measurement. *J. Power Sources* **2015**, *274*, 990–1004.
16. Wenzl, H.; Baring-Gould, I.; Kaiser, R.; Liaw, B.Y.; Lundsager, P.; Manwell, J.; Ruddell, A.; Svoboda, V. Life prediction of batteries for selecting the technically most suitable and cost effective battery. *J. Power Sources* **2005**, *144*, 373–384.
17. Ramadass, P.; Haran, B.; White, R.; Popov, B.N. Capacity fade of Sony 18650 cells cycled at elevated temperatures: Part II. Capacity fade analysis. *J. Power Sources* **2002**, *112*, 614–620.

18. Bundy, K.; Karlsson, M.; Lindbergh, G.; Lundqvist, A. An electrochemical impedance spectroscopy method for prediction of the state of charge of a nickel-metal hydride battery at open circuit and during discharge. *J. Power Sources* **1998**, *72*, 118–125.
19. Meissner, E.; Richter, G. Battery monitoring and electrical energy management: Precondition for future vehicle electric power systems. *J. Power Sources* **2003**, *116*, 79–98.
20. Santhanagopalan, S.; Guo, Q.; Ramadass, P.; White, R.E. Review of models for predicting the cycling performance of lithium ion batteries. *J. Power Sources* **2006**, *156*, 620–628.
21. Gao, L.; Liu, S.; Dougal, R.A. Dynamic lithium-ion battery model for system simulation. *IEEE Trans. Compon. Packag. Technol.* **2002**, *25*, 495–505.
22. Rao, R.; Vrudhula, S.; Rakhmatov, D.N. Battery modeling for energy aware system design. *Computer* **2003**, *36*, 77–87.
23. González-Longatt, F.M. Circuit based battery models: A review. In Proceedings of the 2nd Congreso IberoAmericano De Estudiantes deIngeniería Eléctrica, Puerto la Cruz, Venezuela, 3–7 April 2006; pp. 1–5.
24. Chen, M.; Rincon-Mora, G.A. Accurate electrical battery model capable of predicting runtime and IV performance. *IEEE Trans. Energy Convers.* **2006**, *21*, 504–511.
25. Zhu, W.H.; Zhu, Y.; Tatarchuk, B.J. A simplified equivalent circuit model for simulation of Pb-acid batteries at load for energy storage application. *Energy Convers. Manag.* **2011**, *52*, 2794–2799.
26. Pascoe, P.E.; Anbuky, A.H. A unified discharge voltage characteristic for VRLA battery capacity and reserve time estimation. *Energy Convers. Manag.* **2004**, *45*, 277–302.
27. Pascoe, P.E.; Anbuky, A.H. A VRLA battery simulation model. *Energy Convers. Manag.* **2004**, *45*, 1015–1041.
28. Chiang, Y.H.; Sean, W.Y. Dynamical estimation of state-of-health of batteries by using adaptive observer. In Proceedings of the 2nd International Conference on Power Electronics and Intelligent Transportation System, 19–20 December 2009; pp. 110–115.
29. Hu, C.; Youn, B.D.; Chung, J. A multiscale framework with extended Kalman filter for lithium-ion battery SOC and capacity estimation. *Appl. Energy* **2012**, *92*, 694–704.
30. Sepasi, S.; Ghorbani, R.; Liaw, B.Y. A novel on-board state-of-charge estimation method for aged Li-ion batteries based on model adaptive extended Kalman filter. *J. Power Sources* **2014**, *245*, 337–344.
31. Saha, B.; Goebel, K.; Poll, S.; Christophersen, J. Prognostics methods for battery health monitoring using a Bayesian framework. *IEEE Trans. Instrum. Meas.* **2009**, *58*, 291–296.
32. Chang, W.Y. Estimation of the state of charge for a LFP battery using a hybrid method that combines a RBF neural network, an OLS algorithm and AGA. *Int. J. Electr. Power Energy Syst.***2013**, *53*, 603–611.
33. Gorman, C.; Ingersoll, D.; Jungst, R.; Paez, T.L. Artificial neural network simulation of battery performance. In Proceedings of the Thirty-First Hawaii International Conference, Kohala Coast, HI, USA, 6–9 January 1998; pp. 115–121.
34. Andre, D.; Nuhic, A.; Soczka-Guth, T.; Sauer, D. Comparative study of a structured neural network and an extended Kalman filter for state of health determination of lithium-ion batteries in hybrid electricvehicles. *Eng. Appl. Artif. Intell.* **2013**, *26*, 951–961.
35. Rufus, F.; Lee, S.; Thakker, A. Health monitoring algorithms for space application batteries. In Proceedings of the Prognostics and Health Management, Denver, CO, USA, 6–9 October 2008; pp. 1–8.
36. Widodo, A.; Shim, M.C.; Caesarendra, W.; Yang, B.S. Intelligent prognostics for battery health monitoring based on sample entropy. *Expert Syst. Appl.* **2011**, *38*, 11763–11769.
37. Xing, Y.; Ma, E.W.; Tsui, K.L.; Pecht, M. Battery management systems in electric and hybrid vehicles. *Energies* **2011**, *4*, 1840–1857.
38. Singh, P.; Fennie, C.; Reisner, D. Fuzzy logic modelling of state-of-charge and available capacity of nickel/metal hydride batteries. *J. Power Sources* **2004**, *136*, 322–333.
39. Singh, P.; Kaneria, S.; Broadhead, J.; Wang, X.; Burdick, J. Fuzzy logic estimation of SOH of 125Ah VRLA batteries. In Proceedings of the International Telecommunications Energy Conference, Chicago, IL, USA, 19–23 September 2004; pp. 524–531.
40. Micea, M.V.; Ungurean, L.; Carstoiu, G.N.; Groza, V. Online state-of-health assessment for battery management systems. *IEEE Instrum. Meas.* **2011**, *60*, 1997–2006.
41. Wang, F.; Fan, B.; Liu, S.; Qian, G.; Huang, X.; Han, S. Attenuation test and duplication of power battery's cycle life. *J. Automot. Saf. Energy* **2012**, *3*, 71–76.
42. Kang, L.; Zhao, X.; Ma, J. A new neural network model for the state-of-charge estimation in the battery degradation process. *Appl. Energy* **2014**, *121*, 20–27.
43. Xing, Y.; Ma, E.W.; Tsui, K.-L.; Pecht, M. An ensemble model for predicting the remaining useful performance of lithium-ion batteries. *Microelectron. Reliab.* **2013**, *53*, 811–820.
44. He, W.; Williard, N.; Osterman, M.; Pecht, M. Prognostics of lithium-ion batteries based on Dempster-Shafer theory and the Bayesian Monte Carlo method. *J. Power Sources* **2011**, *196*, 10314–10321.
45. Eom, S.W.; Kim, M.K.; Kim, I.J.; Moon, S.I.; Sun, Y.K.; Kim, H.S. Life prediction and reliability assessment of lithium secondary batteries. *J. Power Sources* **2007**, *174*, 954–958.
46. Remmlinger, J.; Buchholz, M.; Meiler, M.; Bernreuter, P.; Dietmayer, K. State-of-health monitoring of lithium-ion batteries in electric vehicles by on-board internal resistance estimation. *J. Power Sources* **2011**, *196*, 5357–5363.

CHAPTER 18

ARRHENIUS EQUATION-BASED CELL-HEALTH ASSESSMENT: APPLICATION TO THERMAL ENERGY MANAGEMENT DESIGN OF A HEV NIMH BATTERY PACK

Yalian Yang, Xiaosong Hu, Datong Qing and Fangyuan Chen

Keywords: Battery thermal management, battery health, electrified vehicle

INTRODUCTION

Energy supply uncertainty, volatile oil prices, and increasingly stringent regulations on pollutants and carbon footprint pose heavy challenges for today's fossil fuel-dominant ground transportation [1,2]. To accelerate the shift to sustainable transportation, electrified vehicles, such as battery electric vehicles (BEVs), hybrid electric vehicles (HEVs), and plug-in hybrid electric vehicles (PHEVs), are being studied and developed around the globe [3,4,5]. The traction battery packs have a significant impact on the performance, cost, and reliability of electrified vehicles. An efficient battery management system (BMS) is thus required to ensure safe and reliable battery operations in complex and highly dynamic onboard environment [6,7,8,9].

A key task of BMS is to manage cell temperature to avoid overheating and to reduce large temperature gradients among cells. Temperature has been recognized as the dominant stressing factor for battery degradation [10]. Both high temperature and large temperature gradients lead to reduced cell life. Diverse methods of battery thermal management were studied. For example, phase change materials which can absorb large amounts of heat were often used to control thermal runaways under extreme conditions [11,12,13,14]. Active methods, e.g., liquid-cooled and air-cooled systems, were also adopted [15,16]. A comparison of the effectiveness of the two types of methods was investigated in [17]. Heat pipes were also used to cool the battery system [18]. Several commonly used battery cooling techniques are compared in [19]. The vast majority of relevant literature was focused on the thermal behavior of the cooling material, and the performance of different approaches in preventing unsafe temperatures is compared. Studies on the cell temperature uniformity, however, are relatively scarce. A rule-of-thumb threshold, e.g., 5 °C, was often used as limit to the temperature nonuniformity for batteries of different chemistries, or from different manufacturers. Little research has been done to study how small the temperature gradient should be for a specific battery to ensure cell health. Since different batteries typically have considerably different aging characteristics, the same temperature deviation will induce different degradations for different batteries. A mathematical model that can explicitly capture the relationship between cell temperature and degradation would be beneficial to addressing this problem. According to the requirement for the cell health uniformity, the model can be used to assessing the acceptable level of temperature gradient, thus predicting degradation and guiding cooling system design. The main contributions of this paper are the development of such a model and its application to temperature-gradient management for cells inside a battery pack.

Citation: Yalian Yang, Xiaosong Hu, Datong Qing and Fangyuan Chen, Arrhenius Equation-Based Cell-Health Assessment: Application to Thermal Energy Management Design of a HEV NiMH Battery Pack, doi: 10.3390/en6052709.

In this paper, the effect of temperature deviation on cell degradation consistency is modeled using the Arrhenius equation, based on experimental data collected on Nickel Metal Hydride (NiMH) cells. According to the calibrated model and cell health-consistency requirement, the feasibility and applicability of a NiMH battery pack design, together with its air cooling system, is experimentally evaluated for an HEV prototype manufactured by Changan Automobile Company Limited in Chongqing city in China. The shortcomings of the current design are analyzed using computational fluid dynamics (CFD). Subsequently, an improved design is proposed. The performance of the new design is validated by means of CFD simulations and experimental analysis.

The remainder of this paper is organized as follows: in Section 2, the mathematical model describing the influence of the temperature discrepancy on the lifetime consistence of NiMH cells is presented. Based on the model, Section 3 presents the assessment of a NiMH battery pack design, along with its cooling system. The design deficiencies are also analyzed. An improved design solution is depicted in Section 4, in which simulation and experimental analysis are applied to demonstrate its performance. Finally, conclusions are drawn in Section 5.

EFFECT OF TEMPERATURE GRADIENT ON CELL DEGRADATION VARIATION

Cell temperature has a large impact on cell health degradation: the higher the temperature, the faster the rate of aging [20]. Therefore, non-uniform temperatures will result in different cell health levels within the same pack. The cell with the highest degradation usually determines the lifetime of the whole pack. To facilitate effective thermal management system design, it is important to build a mathematical model that can accurately quantify the effect of temperature gradients on cell lifetime uniformity.

Model Structure Based on Arrhenius Equation

The Arrhenius equation is a succinct, but precise characterization of the temperature dependence of the rate of a chemical reaction [21]. It was often applied to model the temperature-induced alteration of diffusion coefficients, creep rates, and other thermal processes [22]. The Arrhenius equation was also used to delineate the temperature-dependent degradation rates of electrochemical cells [23,24,25].The basic Arrhenius equation is shown as follows:

$$\frac{dC}{dn} = \Lambda e^{-\frac{\Delta E}{RT}} = \Lambda e^{-\frac{\lambda}{T}} \quad (1)$$

where dC / dn is the rate of cell capacity change with respect to aging cycle, Λ is the pre-exponential factor, and R is the universal gas constant [i.e., 8.314 J/(mol·K)]. ΔE and T are activation energy in the unit of J/mol and absolute temperature in the unit of K, respectively [21,22]. Here, we view Equation (1) as an empirical formula such that Λ and $\lambda = \Delta E / R$ are interpreted as two unknown parameters to be calibrated. Integrating both sides of Equation (1) over the cell lifespan yields:

$$C_r = -\Lambda n_c e^{-\frac{\lambda}{T}} \quad (2)$$

where C_r is the capacity reduction threshold indicative of the cell failure, and n_c is the cycle life. Assign two different temperatures, T_1 and T_2 ($T_1 > T_2$), to Equation (2) to constitute:

$$\Delta n_c = \frac{C_r}{\Lambda}\left(e^{\frac{\lambda}{T_2}} - e^{\frac{\lambda}{T_1}}\right) \quad (3)$$

where Δn_c is the lifetime deviation. It is evident that Equation (3) is a quantitative description of the effect of the temperature gradient on cell life.

Model Parameter Calibration

Linear Identifiable Form

An identifiable form of the model is obtained by taking the natural logarithm of both sides of Equation (2):

$$\ln\left(\frac{C_r}{n_c}\right) = \ln(-\Lambda) - \frac{\lambda}{T} = \underbrace{\begin{bmatrix} 1 & -\frac{1}{T} \end{bmatrix}}_{\varphi} \underbrace{\begin{bmatrix} \ln(-\Lambda) \\ \lambda \end{bmatrix}}_{\theta} \quad (4)$$

where $\boldsymbol{\varphi}$ is the regressor, and $\boldsymbol{\theta}$ is the parameters to be identified. For electrified vehicle applications, C_r is often specified as 20%. In order to extract the unknown parameters, the cell cycle life n_c under different temperatures should be determined or estimated.

HEV NiMH Cell Aging Test

NiMH cells from the same batch for an HEV prototype manufactured by Changan Company were tested. The testing system can implement diverse charging/discharging modes, monitor and record cell current, voltage, and power, as well as implement safety protection. The nominal capacity and voltage of the NiMH cells are 6.5 Ah and 1.2 V. The upper and lower cut-off voltages are 1.55 V and 1.05 V, respectively. The aging cycle was chosen to be constant-current full charge and discharge at the rate of 1 C, and the cell static capacity was calibrated every 10 cycles to record the cell degradation levels. Three aging tests were conducted under 25 °C, 40 °C, and 50 °C, respectively. In each test, five NiMH cells were cycled, and then the average capacity was calculated and used to take the cell manufacturing variance into account. Simple aging cycles are used because our goal is to identify the relationship between the cell health and operating temperature, as opposed to establishing a complete battery degradation model as a function of temperature, current rate, and depth of discharge (DOD). Furthermore, since in practice the cells in the pack undergo approximately the same current and DOD, the cell health inconsistency is mainly attributed to the cell temperature deviation. The measured capacities under the three aging tests are shown in Figure 1. The capacity reduction is shown in Figure 2.

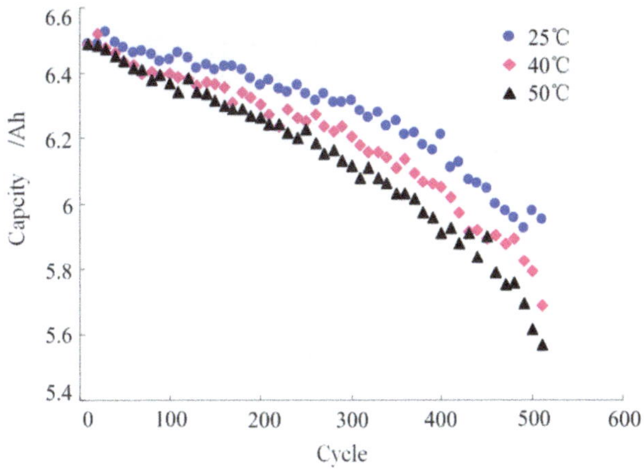

Figure 1. Cell capacities under three different temperatures.

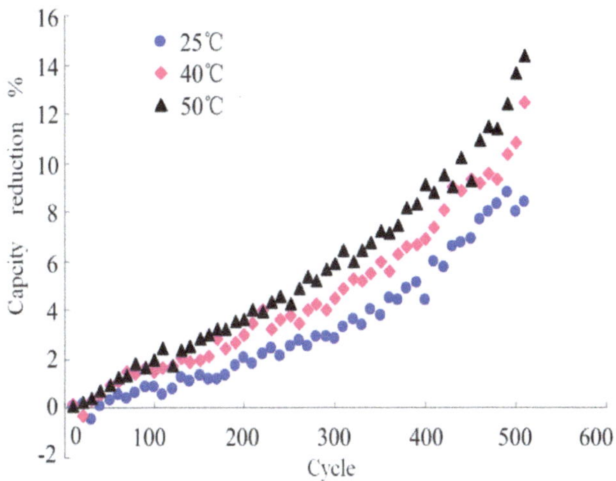

Figure 2. Capacity percent reductions.

It is clear that the cell degraded most rapidly under 50 °C. Notice that since the aging tests did not damage the cells up to failure, this necessitates estimating the cell life based on the available capacity history. From Figure 2, it can be observed that the trajectory of the capacity reduction is a quadratic-function of the number of aging cycles. We use a nonlinear least-squares optimization algorithm to identify the parameters of the quadratic function. The results for the three temperatures are shown in Table 1. For $C_d = C_r = 20\%$, the solutions are shown in Table 2.

Table 1. Quadratic functions depicting the relationship between capacity reduction and the number of aging cycles

Temperature/°C	Quadratic function/%
25	$C_d = 3.2\times10^{-5}n^2 - 9.8\times10^{-4}n + 0.21$
40	$C_d = 2.6\times10^{-5}n^2 + 6\times10^{-3}n + 0.15$
50	$C_d = 2.3\times10^{-5}n^2 + 9\times10^{-3}n + 0.13$

Table 2. Estimated cell cycle lifetime at the three temperatures

Temperature/°C	25	40	50
Cycle life	801.87	765.96	754.19

Least-Squares Algorithm Based Parameterization

The least-squares algorithm is used to estimate the optimal model parameters using the following equation:

$$\hat{\theta} = (X^T X)^{-1} X^T y \tag{5}$$

where

$$X^T = \left[\varphi(T_1)^T \quad \varphi(T_2)^T \quad \varphi(T_3)^T\right] \tag{6}$$

$$y = \left[\ln(\frac{C_r}{n_{c,T_1}}) \quad \ln(\frac{C_r}{n_{c,T_2}}) \quad \ln(\frac{C_r}{n_{c,T_3}})\right]^T. \tag{7}$$

T_1, T_2 and T_3 are the three operating temperatures. After obtaining $\hat{\theta}$, we can determine the model parameters:

$$\Lambda = -e^{\hat{\theta}(1)} \tag{8}$$

$$\lambda = \hat{\theta}(2). \tag{9}$$

The identification result is shown in Table 3, and the mean square error of the fitting is 4.04×10^{-5}. Note that thanks to the nonlinear Equation (8), the standard error of Λ is derived from Monte Carlo simulation, based on statistics (mean and standard deviation) of $\hat{\theta}(1)$.It is clear from error analysis that the fitting equation has good accuracy.

Table 3. Optimal model parameters and error bounds identified by the least squares method

Result	Λ	λ
Estimate	-5.61×10^{-4}	240.74
Standard error (1 Sigma)	6.20×10^{-5}	34.29

Assessment Criterion

HEVs are often subject to complicated traffic and climate conditions so that the on-board battery packs undergo varying loads and ambient temperatures. Because the cells suffer more severe degradation at high temperatures, leading to shorter life, the temperature deviation among cells in a hot vehicle environment is of great concern. A performance index proposed by the HEV manufacturer requires that the cell life inconsistency for the battery cells should be less than 1.5% when the temperature is higher than 35 °C [26]. According to Equation (3), the maximum allowable temperature gradient among cells can be efficiently calculated to be 6 °C, which will serve as a performance assessment criterion and a design guide for the subsequent thermal energy management system of the HEV NiMH battery pack.

FEASIBILITY ASSESSMENT OF PRIMITIVE NIMH BATTERY PACK DESIGN WITH ITS COOLING SYSTEM

Experimental Analysis

The configuration of a prototype NiMH battery pack with no consideration of the impact of temperature discrepancy on cell health consistency is shown in Figure 3. It consists of three layers. The first layer contains 36 cells arranged in 3 rows and 12 columns, and there are 42 cells in the form of 3 rows and

14 columns for each of the other two layers. Notice that all 120 cells are electrically connected in series to increase the voltage level, as the HEV with an integrated starter/generator propulsion system has a low degree of hybridization which does not require a lot of energy. The cell dimensions and main electrical specifications are listed in Table 4. Five air ducts are designed for each cell, as shown in Figure 3. The cells in each layer were cooled by two fans placed symmetrically. T-shaped thermocouples were used to collect the cell temperature data. The measurement points were selected considering the structural characteristics of each layer. There are 12 sensors in the first layer shown in Figure 4, and the remaining layers each have 15 sensors shown in Figure 5.

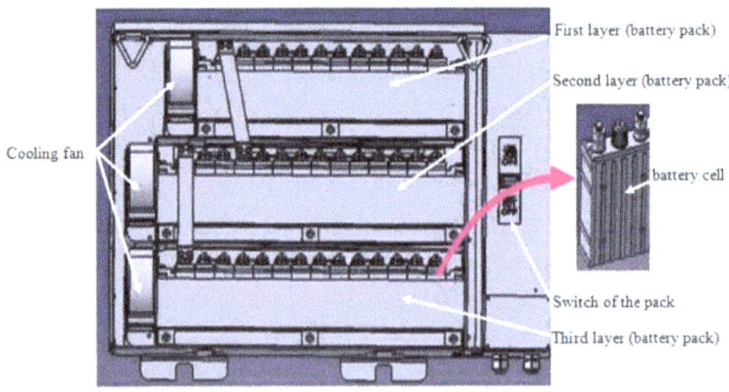

Figure 3. Configuration of the primitive HEV NiMH battery pack: 36 cells (3 rows and 12 columns) in the first (top) layer; 42 cells (3 rows and 14 columns) in each of the other layers.

Table 4. Cell dimensions and main electrical specifications

Thickness	23.5 mm
Width	60 mm
Height	84 mm
Weight	180 g
Nominal Voltage	1.2 V
Nominal Capacity	6.5 Ah
Power Density	≥ 1400 W/kg
Energy Density	≥ 44 Wh/kg
Ambient Temperature	−40 ~ 80 °C
Working Temperature	−30 ~ 60 °C
Usable SOC range	30% ~ 80%

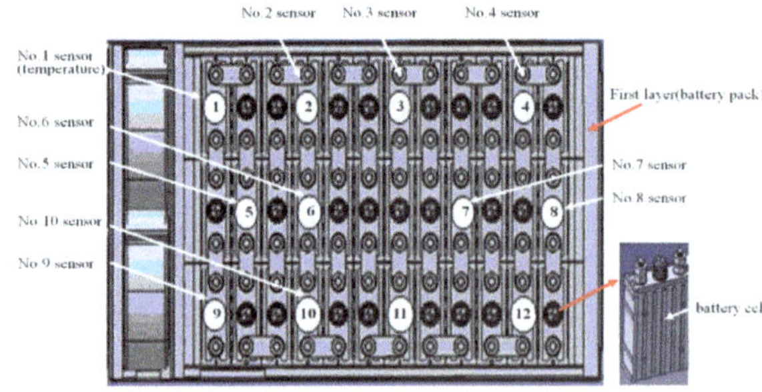

Figure 4. Sensor location in the first layer.

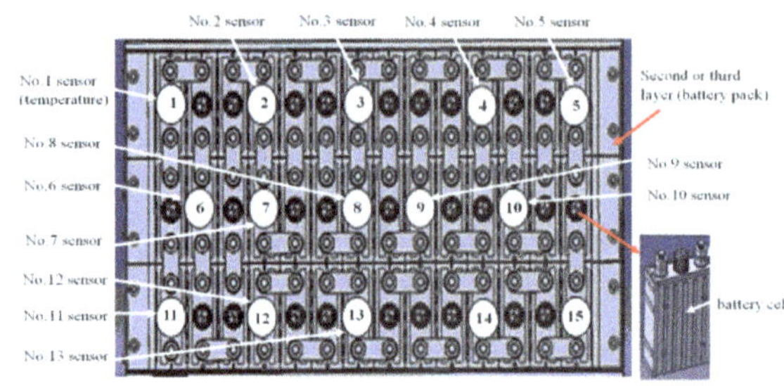

Figure 5. Sensor location in the second and third layers.

The road testing of the HEV vehicle with the prototype battery pack was conducted in a test field by DongFeng Motor Company Limited in Xiangfan city, China, during a hot summer. The test involved diverse road conditions and driving patterns. Please refer to [27] for more details on the test procedures. We measured the cell temperature in a 4.5 minute idling process intentionally designed as part of the test. The measured temperatures are shown in Figure 6, Figure 7 and Figure 8. It can be seen that the cell with the highest temperature of 45.5 °C occurred in the third layer of the pack, while the lowest one with 36 °C is in the second layer. The maximum temperature difference is thus 9.5 °C. Taking a closer examination, we can find that the largest temperature discrepancy in the first layer is around 5.1 °C, and those for the second and third layers are 8 °C and 7.5 °C. It is clear that the preliminary battery pack and its cooling system design cannot meet the performance requirement of a maximum temperature difference of 6 °C.

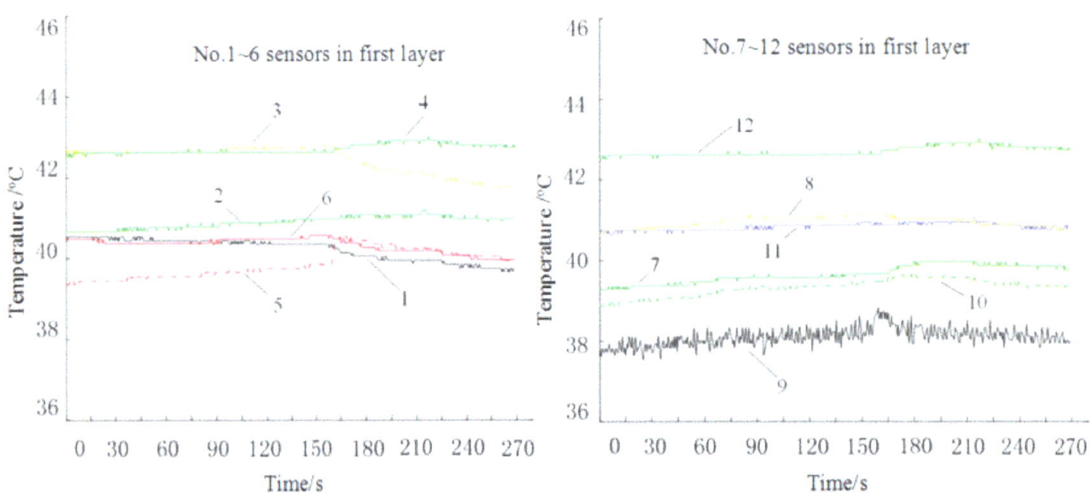

Figure 6. Temperature measurement in the first layer.

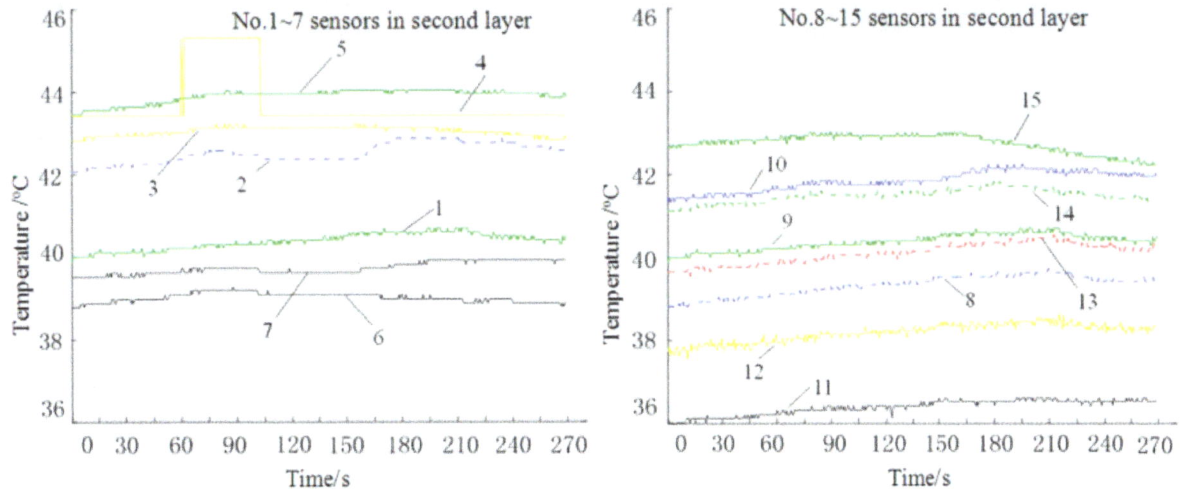

Figure 7. Temperature measurement in the second layer.

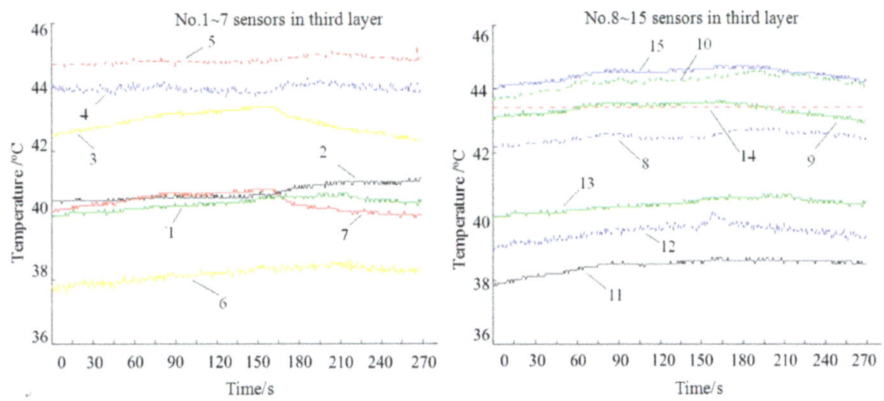

Figure 8. Temperature measurement in the third layer.

Analysis of the Shortcomings of the Preliminary Design

Since the pack structure is quite complicated, a flow field analysis was performed first to investigate the air flow allocation among the three layers. The CFD model of the pack is shown in Figure 9. The air inlet and outlet were designated as the pressure import and the pressure exportation, respectively; the corresponding pressures were 30 Pa and −220 Pa. The air inlet is assumed to be at 25 °C. The calculated result is shown in Table 5. It can be seen that the second and third layers have different air flow, leading to large temperature non-uniformity. Because of the smaller flow, the majority of cells in the third layer had higher temperatures. The flow inconsistence was mainly attributed to the interference of air fans between the two layers.

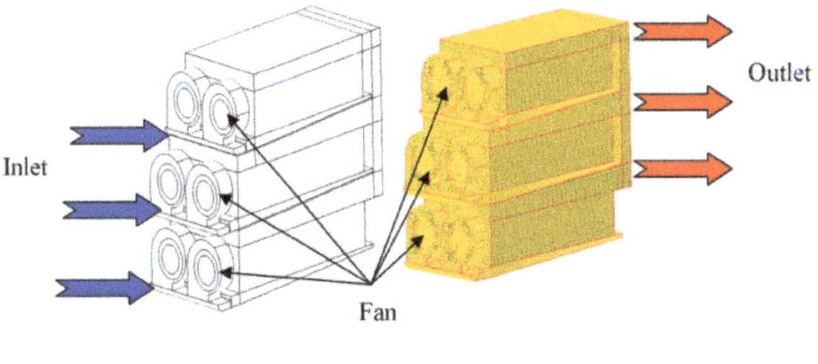

Figure 9. CFD simulation model of the primitive pack.

Table 5. Flow distribution results

Layer No.	Flow/ Kg/s	Flow percent
1	0.0130	29.23%
2	0.0170	38.11%
3	0.0145	32.66%

Based on the flow distribution analysis, more in-depth CFD simulations for each layer were conducted. The simulation model of the third layer, for example, is shown in Figure 10. The section velocity contours of two fans are shown in Figure 11. It can be noticed that the middle area assumed larger air flow velocity in that the interference between the two fans existed. The cross-section velocity contour of the third battery layer is shown in Figure 12. It is obvious that flow non-uniformity existed among cells in the layer, causing the cell temperature gradient. Similar results can be obtained for the other two layers.

According to the analysis above, we can find that the undesired performance of the preliminary pack and cooling system design were due to the non-uniformity in the flow levels of battery layers. The design of the pack structure and the six air fans were unsatisfactory; the interference among fans had a considerably negative influence on cell temperature uniformity.

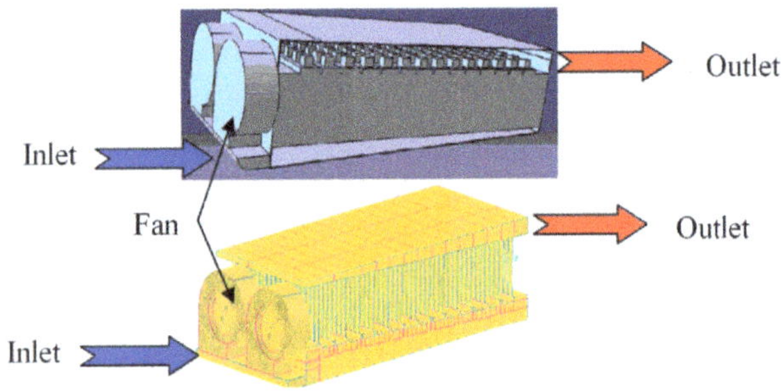

Figure 10. Simulation model of the third layer.

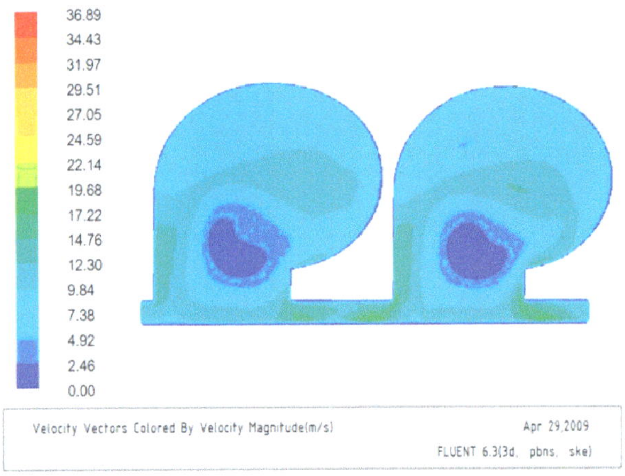

Figure 11. Section velocity contours of two fans.

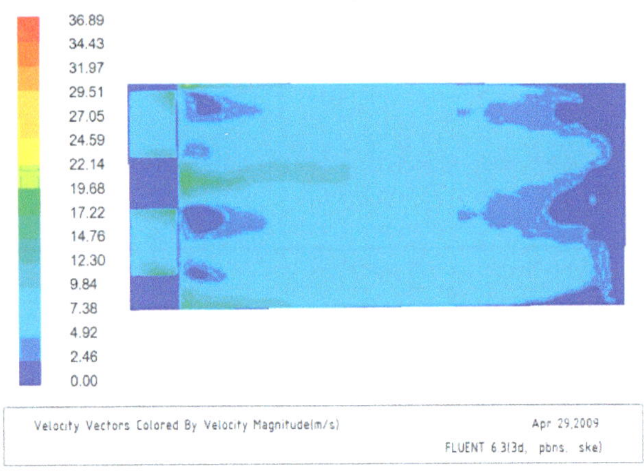

Figure 12. Cross-section velocity contour of the third layer.

IMPROVED THERMAL ENERGY MANAGEMENT

Enhanced Structure Design

In order to avoid the uneven flow distribution among layers which appeared in the original pack structure, 120 cells are placed in only one layer in the form of eight modules. Each module comprises 15 cells in series. Although the eight modules have a parallel topology, they are connected electrically in series. The configuration of the improved pack structure is shown in Figure 13. The primitive six air fans are replaced by one fan placed in the outlet to alleviate the flow non-uniformity among cells caused by the interference of multiple fans. The enhanced pack, together with its cooling system, is shown in Figure 14. The whole system consists of a box body (1), the battery pack (2) shown in Figure 13, an air intake duct (3), a DC/DC converter (4), an intelligent control unit (5), and a centrifugal fan (6). The pack dimensions and main electrical specifications are listed in Table 6. Please refer to our previous paper [28] to find more details on the structure design of the DC/DC converter and control unit for effectively dissipating the heat. The fan can provide larger air flow than those in the primitive design, with the maximum flow of 280 m^3/h. The air enters the pack from the intake duct under the suction role of the centrifugal fun, and then the convective heat transfer occurring at air ducts on the cell body and the internal gaps of the pack structure is able to cool the cells. After that, the air continues to dissipate the heat from the DC/DC converter and the controller and is finally extracted by the fan.

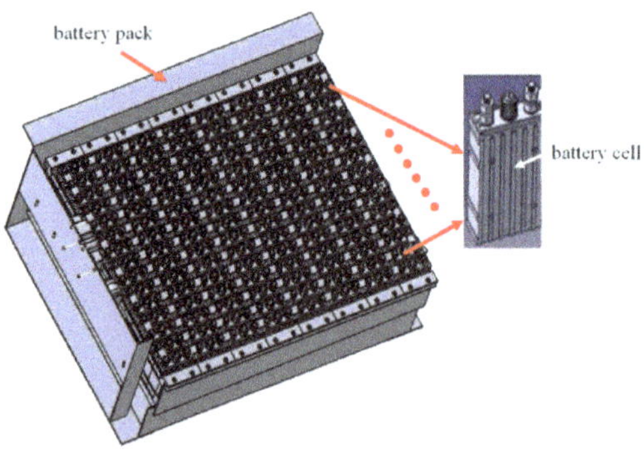

Figure 13. Improved battery pack structure.

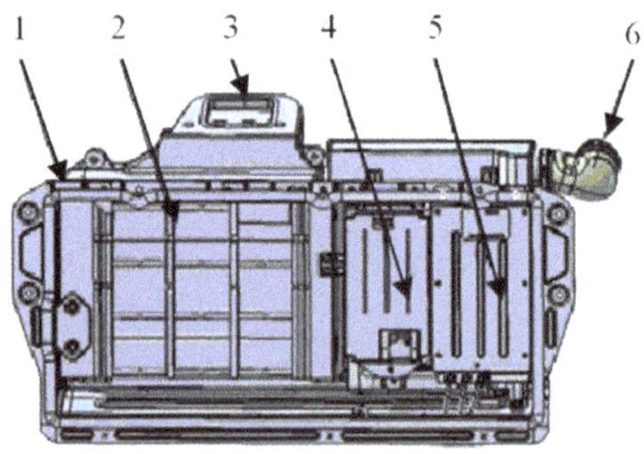

(1)Box Body (2)Battery Pack (3)Air Intake Duck (4)DC DC Converter
(5)Intelligent Control Unit (6)Centrifugal Fan and Air Outlet Duck

Figure 14. Configuration of the integrated pack and cooling system.

Table 6. Dimensions and main electrical specifications of the improved pack design

Width	444 mm
Length	542 mm
Height	198 mm
Weight	35 kg
Nominal Capacity	6.5 Ah
Nominal Voltage	144 V
Number of cells	120

Simulation Validation

In the CFD simulation of the battery pack, the air flow at the inlet boundary provided by the fan was 140 m^3/h, and its temperature was assumed to be 35 °C. Since the average heating power of the pack was measured to be 494 W in the practical HEV filed test at 40 °C, the heat flux of 718.5 W/$m^3$was specified for the battery pack system in the simulation. Please refer to [27] for more details on the boundary conditions applied. The simulation result of the first row of cells is shown in Figure 15. We can see similar temperature change of air in the ducts and gaps for eight cells, indicating relatively uniform heat dissipation. The maximum temperature of cell surface is 49.3 °C; the minimum is 45.5 °C; the maximum deviation is therefore 3.8 °C. It meets the performance index of thermal energy control depicted in Section 2. The simulation results of the other rows are similar.

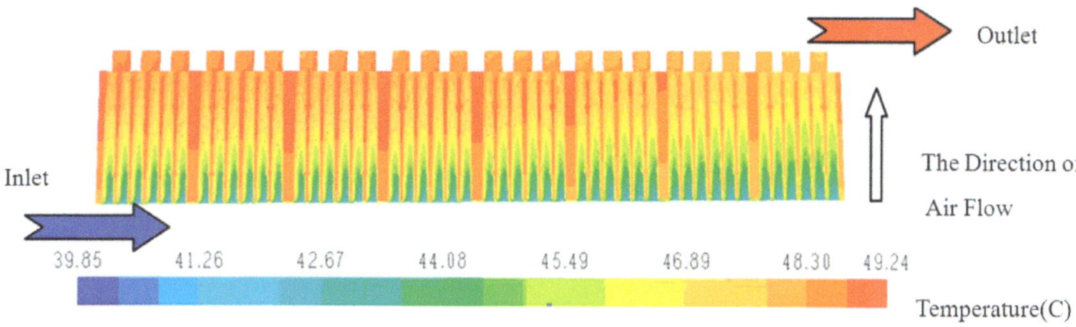

Figure 15. Temperature field of the first row of cells in the enhanced pack design.

Experimental Validation

An experimental platform for evaluating the validity of the improved battery pack mainly was composed of an environmental chamber, a chassis dynamometer, a blower, and a data collection unit. The chamber was used to simulate the actual HEV environment parameters, such as ambient temperature, humidity, and sunlight intensity, such that the onboard test can be accomplished in a well-controlled indoor condition. The dynamometer was employed to load the HEV and control the vehicle speed to realize the specified driving cycles. In order to simulate the airflow around the vehicle during running, the blower was placed in front of the vehicle. The data collection system was to measure and record the cell temperature. The experiment arrangement is shown in Figure 16. The locations of temperature sensors (white circles) in the pack are shown in Figure 17. The sensors are numbered by using the indices of the matrix of 15 rows and 8 columns, for example, the number of the sensor in the fourth row and first column is 4-1.

To better evaluate the quality of thermal energy management of the pack, driving cycles inducing serious heating of energy storage device should be adopted. Two cycles simulating climbing (10% slope) and urban traffic jam were thus chosen. Please refer to [27] for the details of the two testing cycles. The results of assessing the temperature deviation among cells in the two tests are shown in Figure 18 and Figure 19. For good readability, we only show the temperatures from a part of sensors without loss of generality and validity. It can be found that the maximum temperature difference was 4.5 °C when the maximum cell temperature of 46.8 °C appeared in the climbing cycle; the maximum deviation was 5.2 °C when simulating the urban traffic jam. Therefore, the experimental results demonstrate the validity of the improved battery pack thermal energy management design, according to the prescribed performance criterion in Section 2.

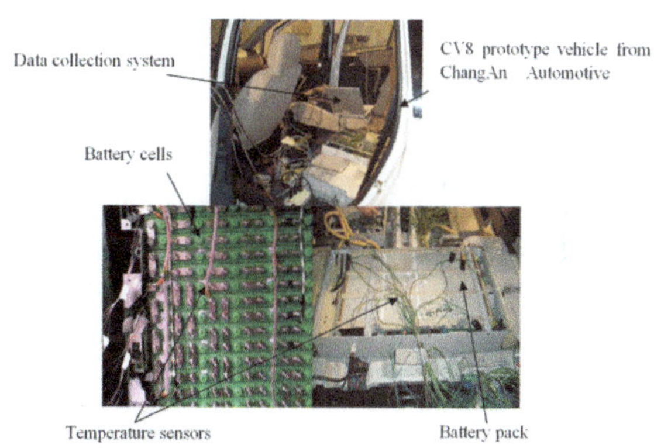

Figure 16. Experimental arrangement for performance assessment of the improved battery pack.

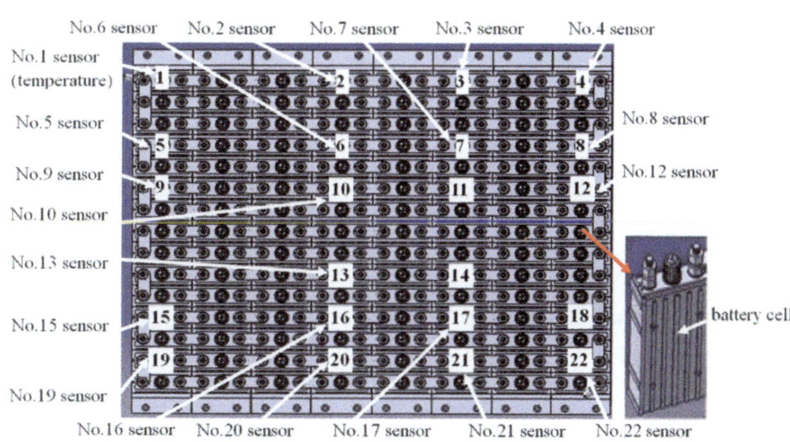

Figure 17. Sensor locations in the improved pack.

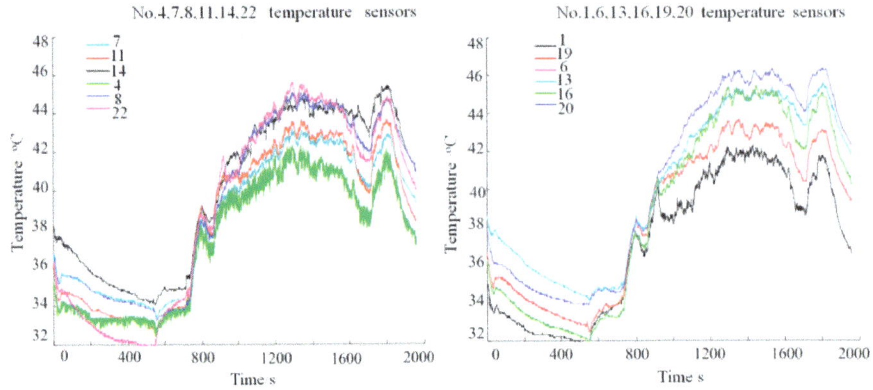

Figure 18. Temperature measurement in the climbing cycle.

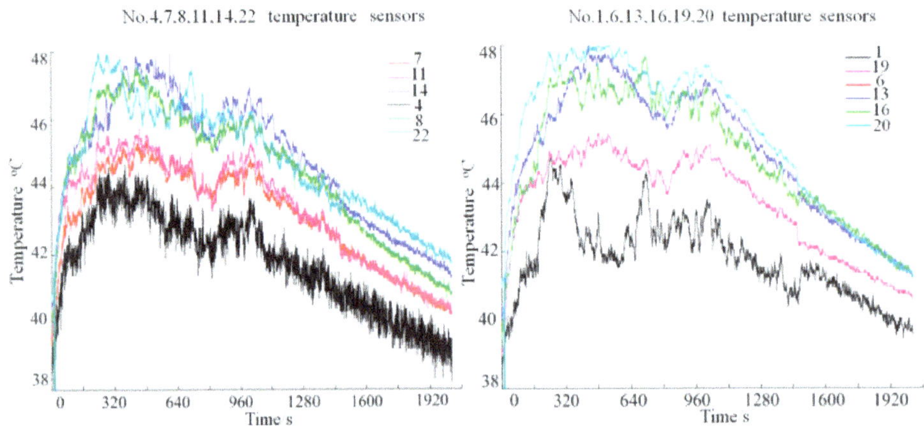

Figure 19. Temperature measurement in the urban traffic jam situation.

CONCLUSIONS

The impact of the temperature gradient on the cell lifetime consistency is first modeled for cells inside a HEV NiMH battery pack, based on Arrhenius equation, by analyzing the aging datasets collected under multiple environmental temperatures. This model, together with the requirement for the cell health consistency provided by the HEV manufacturer, is used as an evaluation criterion and guidance for the thermal energy management design of the NiMH battery pack adopted. It is found that the maximum temperature gradient should not be larger than 6 °C to guarantee the allowable cell life inconsistency. Since different types of cells typically have obviously dissimilar degradation characteristics, it is unreasonable to give a universal performance index for batteries of different chemistries or from different manufacturers, and a specific criterion is necessary for thermal energy management of the chosen battery to avoid under- or overdesign. According to the calibrated model, the applicability of a primitive NiMH battery pack design and its air cooling system without considering the actual relationship between the cell-health and temperature deviations is assessed, and its shortcomings are analyzed using computational fluid dynamics (CFD) analysis. The original three-layer structure is vulnerable to the uneven flow allocation among layers, and the large interference between two air fans in each layer also leads to serious flow non-uniformity among cells. In order to avert these deficiencies, an improved design scheme with one-layer pack structure and one potent cooling fan at the outlet is then proposed. The CFD simulation and experimental validation results show that the maximum temperature gradient among cells in the enhanced packaging and cooling design is less than 6 °C, verifying its effectiveness and superiority.

ACKNOWLEDGMENTS

The authors would like to acknowledge Huei Peng at The University of Michigan (Ann Arbor, MI, USA) for insightful and constructive comments and language polishing. The research work was also partly supported by the National Natural Science Foundation of China (51075411) and Specialized Research Fund for State Key Laboratories in China (0301002109148).

REFERENCES

1. Juul, N.; Meibom, P. Road transport and power system scenarios for Northern Europe in 2030. Appl. Energy 2012, 92, 573–582.
2. Williams, B.; Martin, E.; Lipman, T.; Kammen, D. Plug-in-hybrid vehicle use, energy consumption, and greenhouse emissions: An analysis of household vehicle placements in northern California. Energies 2011, 4, 435–457.
3. Ehsani, M.; Falahi, M.; Lotfifard, S. Vehicle to grid services: Potential and applications. Energies 2012, 5, 4076–4090.
4. Doucette, R.T.; McCulloch, M.D. Modeling the prospects of plug-in hybrid electric vehicles to reduce CO_2 emissions. Appl. Energy 2011, 88, 2315–2323.

5. Zou, Y.; Sun, F.; Hu, X.; Guzzella, L.; Peng, H. Combined optimal sizing and control for a hybrid tracked vehicle. Energies 2012, 5, 4697–4710.
6. Hu, X.; Sun, F.; Zou, Y. Estimation of state of charge of a lithium-ion battery pack for electric vehicles using an adaptive Luenberger observer. Energies 2010, 3, 1586–1603.
7. Xing, Y.; Ma, E.W.M.; Tsui, K.L.; Pecht, M. Battery management systems in electric and hybrid vehicles. Energies 2011, 4, 1840–1857.
8. Hu, X.; Li, S.; Peng, H. A comparative study of equivalent circuit models for Li-ion batteries. J. Power Sources 2012, 198, 359–367.
9. Hu, X.; Li, S.; Peng, H.; Sun, F. Robustness analysis of State-of-Charge estimation methods for two types of Li-ion batteries. J. Power Sources 2012, 217, 209–219.
10. Thomas, E.V.; Bloom, I.; Christophersen, J.P. Rate-based degradation modeling of lithium-ion cells. J. Power Sources 2012, 206, 378–382.
11. Rao, Z.; Wang, S.; Zhang, G. Simulation and experiment of thermal energy management with phase change material for ageing $LiFePO_4$ power battery. Energy Convers. Manag. 2011, 52, 3408–3414.
12. Duan, X.; Naterer, G.F. Heat transfer in phase change materials for thermal management of electric vehicle battery modules. Int. J. Heat Mass Transf. 2010, 53, 5176–5182.
13. Khateeb, S.A.; Farid, M.M.; Selman, J.R.; Al-Hallaj, S. Design and simulation of a lithium-ion battery with a phase change material thermal management system for an electric scooter. J. Power Sources 2004, 128, 292–307.
14. Khateeb, S.A.; Amiruddin, S.; Farid, M.M.; Selman, J.R.; Al-Hallaj, S. Thermal management of Li-ion battery with phase change material for electric scooters: experimental validation. J. Power Sources 2005, 142, 345–353.
15. Giuliano, M.R.; Prasad, A.K.; Advani, S.G. Experimental study of an air-cooled thermal management system for high capacity lithium-titanate batteries. J. Power Sources 2012, 216, 345–352.
16. Mahamud, R.; Park, C. Reciprocating air flow for Li-ion battery thermal management to improve temperature uniformity. J. Power Sources 2011, 196, 5685–5696.
17. Sabbah, R.; Kizilel, R.; Selman, J.R.; Al-Hallaj, S. Active (air-cooled) vs. passive (phase change material) thermal management of high power lithium-ion packs: Limitation of temperature rise and uniformity of temperature distribution. J. Power Sources 2008, 182, 630–638.
18. Rao, Z.; Wang, S.; Wu, M.; Lin, Z.; Li, F. Experimental investigation on thermal management of electric vehicle battery with heat pipe. Energy Convers. Manag. 2013, 65, 92–97.
19. Rao, Z.; Wang, S. A review of power battery thermal energy management. Renew. Sustain. Energy Rev. 2011, 15, 4554–4571.
20. Serrao, L.; Chehab, Z.; Guezenec, Y.G.; Rizzoni, G. An Aging Model of Ni-MH Batteries for Hybrid Electric Vehicles. In Proceedings of IEEE Vehicle Propulsion and Power, Chicago, IL, USA, 7–9 September 2005.
21. Laidler, K.J. The World of Physical Chemistry; Oxford University Press: Oxford, UK, 1993.
22. Laidler, K.J. Chemical Kinetics, 3rd ed.; Benjamin-Cummings: San Francisco, CA, USA, 1997.
23. Zhang, Y.; Wang, C.; Tang, X. Cycling degradation of an automotive $LiFePO_4$ lithium-ion battery. J. Power Sources 2011, 196, 1513–1520.
24. Zhou, C.; Qian, K.; Allan, M.; Zhou, W. Modeling of the cost of EV battery wear due to V2G application in power systems. IEEE Trans. Energy Convers. 2011, 26, 1041–1050.
25. Sarre, G.; Blanchard, P.; Broussely, M. Aging of lithium-ion batteries. J. Power Sources 2004, 127, 65–71.
26. Yang, H.; Piao, C. Design Criteria of HEV Battery Module; Technical Report for ChangAn Automobile Company Limited: Chongqing, China, 2007.
27. Chen, F. A Study on Temperature Uniformity of Ni-MH Batteries for Hybrid Electric Vehicle. Master's Thesis, Chonqqing University, Chongqing, China, 2010.
28. Yang, Y.; Chen, F.; Qing, D.; Liang, C. Heat dissipation optimization of DC-DC and IPU for hybrid electric vehicle. J. Chongqing Univ. 2010, 33, 1–6.

CHAPTER 19

ENVIRONMENTAL IMPACT ASSESSMENT AND END-OF-LIFE TREATMENT POLICY ANALYSIS FOR LI-ION BATTERIES AND NI-MH BATTERIES

Yajuan Yu, Bo Chen, Kai Huang, Xiang Wang and Dong Wang

Keywords: Li-ion battery, Ni-MH battery, Life Cycle Assessment, Eco-indicator 99, cycle, recycle, incineration

INTRODUCTION

Secondary batteries, also known as rechargeable batteries, are a group of batteries that can be used after discharge by charging the active substances. As they can be used repeatedly, the secondary batteries' lives are longer than those of primary batteries. As a result, the use of secondary batteries has the potential advantages of conserving resources and reducing waste. Moreover, given their high capacity and high energy density, secondary batteries are widely used throughout the World [1,2].

However, these environmentally friendly features do not imply that secondary batteries have no impact on the ecological environment. Different types of secondary batteries have different potential effects on resources, ecosystems and human health during their production, use, disposal, or recycle stages. For example, the production and disposal phases of heavy metals, which secondary batteries contain, can be hazardous to the environment [3,4,5] due to the high worldwide levels of battery consumption. Meanwhile, material production has increased resource depletion. These metal elements and other poisonous substances could become harmful to the ecosystem if a battery is arbitrarily discarded or improperly buried.

To date, the environmental effects of secondary batteries, their resource pressures, their potential risks and hazards have not attracted enough or sufficient attention. More attention should be paid to the environmental characteristics of secondary batteries as the ecological impacts of electronic products come into focus [6]. Environmentally friendly secondary batteries will lead the development of the battery industry; therefore, an appropriate method to analyse the environmental impact of secondary batteries is necessary.

Various tools have been developed to evaluate the environmental impacts of different systems [7,8,9,10,11,12]. Life Cycle Assessment (LCA) is a suitable tool for calculating the environmental impact of certain electric products [13,14,15]. As with any product, the environmental impact of an electric product over its entire life cycle is affected by many factors. For secondary batteries, LCA analysis should consider pollution emissions and the environmental impact of producing, recycling and disposing of the batteries in a life cycle.

To reveal the environmental impact of secondary batteries, Matheys compared the environmental indicators of five electric vehicle batteries using the LCA method [16]. Zackrisson *et al.* performed a LCA study of two lithium-ion batteries to optimize the design of lithium-ion batteries for plug-in hybrid electric vehicles [17]. The LCA results of

Citation: Yajuan Yu, Bo Chen, Kai Huang, Xiang Wang and Dong Wang, Environmental Impact Assessment and End-of-Life Treatment Policy Analysis for Li-Ion Batteries and Ni-MH Batteries, doi:10.3390/ijerph110303185.

three batteries (nickel metal hydride, nickel cobalt manganese lithium-ion, and iron phosphate lithium-ion batteries) for plug-in hybrid and full performance battery electric vehicles were presented in Guillaume Majeau-Bettez's study [3]. Longo *et al.* assessed the energy and environmental impacts of sodium/nickel chloride batteries [18]. In resource recovery, Carl found that the power and energy consumption in the production stage decreased from 43% to 8% while the proportion of recycled lead increased from 50% to 99%. That study also noted that resource consumption, such as land use by a single cycle, may be reduced by an increase in the number of cycles [19]. Slack [20] focused on the source, flow and disposal methods of the UK's used household batteries, while Panero [21] studied the damage that these batteries caused to human health and the ecosystem after they were buried in a landfill.

The existing research has built a good foundation for the further study of the secondary batteries' environmental impact. The extant research considers limited aspects of the relation between environmental impact and the end-of-life treatment. Dewulfa *et al.* performed a critical analysis of natural resource savings in lithium ion battery recycling [22]. However it didn't have enough details of disposal in the end-of-life treatment and how disposal acts on the environmental impact. Based on existing research, we conducted a comprehensive assessment model to include recycle and waste stages of secondary batteries in the whole life cycle. The charge-discharge cycles were also introduced in this study to confirm how the cycles influence the LCA results and the end-of-life treatment. Through the integrated use of the Eco-indicator 99 method and several developed tools such as MATLAB and some LCA software, a quantitative model for secondary batteries' environmental impact and end-of-life treatment analysis was applied, which was used on two types of secondary batteries. This model analyses the environmental impact of secondary batteries, and it offers suggestions to reduce the environmental impact of secondary batteries.

The increasing public concern about the environment has resulted in stricter regulations worldwide on spent portable batteries related to the adequate destination of hazardous residues [4]. The environmental impact of secondary batteries shown from the LCA cases can help verify the chances and possibility that the policies prompt society to improve end-of-life treatment in order to treat these types of residues since the consumption of batteries is considerable around the World.

MATERIALS AND METHODS

The analysis was carried out following the ISO 14040 [23] and ISO 14044 [24] series, which are the basic standards providing a procedure to carry out a LCA. In our framework, the battery system and environmental impact assessment will be included in a LCA following the ISO 14040/14044 guidelines.

LCA Model

The goal and scope and inventory of a LCA study depend on the intended use of the study. The LCA model was built on a Li-ion battery and a Ni-MH battery based on a theoretical design. Material needs were based on laboratory tests and literature references. The main goal of this LCA study was the assessment of the environmental impact of different secondary batteries made from laboratories, as well as the comparison and analysis of the impact factors from different batteries. Furthermore, it highlights the environmental points within end-of-life treatment analysis.

Functional Unit. The assessment objects of this study were secondary batteries. Two types of experimental batteries were selected: (1) A type of Li-ion battery (cathode material: $LiNi_{1/3}Co_{1/3}Mn_{1/3}O_2$ + 1% Fe_3O_4, which was considered to be the most promising material to replace $LiCoO_2$ as a cathode material for Li-ion batteries) and (2) a type of Ni-MH battery (anode material: $LaMg_{12}$ + 200% Ni alloy, which had a high first discharge capacity). During the batteries' production processes, the types and quantities of their raw materials were analysed and recorded. The functional unit was defined as a battery whose specific capacity is 1,000 mAh/g under sustained charge cycles. All figures for environmental impact were related to one battery under the functional unit conditions.

System boundaries. The system boundaries were based on the general rules of ISO 14040 and 14044. This study focused on secondary batteries. The product system includes production, use and recycling. Metals and other materials will be recycled as raw materials. The battery charger and other hardware were outside the system boundary. Only the battery itself and its casing were considered. Figure 1 presents the system definition. Because there were different ways of using batteries, short transports were assumed in the use phase. Long transports were assumed in battery manufacture and the end-of-life treatment.

Life cycle inventory. The life cycle inventory of the components and materials included the $LiNi_{1/3}Co_{1/3}Mn_{1/3}O_2$ + 1% Fe_3O_4 ternary cathode material Li-ion battery and the $LaMg_{12}$+200% Ni alloy anode material Ni-MH battery. The first discharge-specific capacities of the $LiNi_{1/3}Co_{1/3}Mn_{1/3}O_2$ + 1% Fe_3O_4 ternary cathode material Li-ion battery and the $LaMg_{12}$ + 200% Ni alloy anode material Ni-MH battery, both produced in the laboratory, were 170.9 mAh/g and 932.8 mAh/g, respectively. The types and mass of the main raw materials and energy consumption are shown in Table 1 and Table 2. The masses of raw materials and the charge-discharge data were primary data and some energy data in the battery manufacturing was secondary data.

Environmental impact assessment. In line with the recommendations in Eco-indicator 99 system [25], three mainly environmental damage categories were calculated: Human health damage; Ecosystem quality impact; Resource consumption.

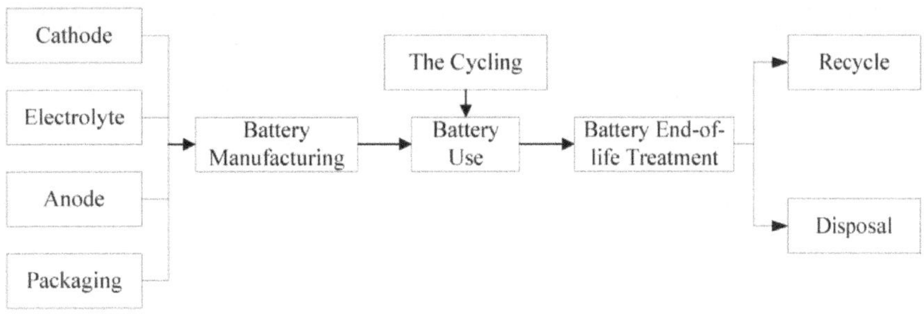

Figure 1. Flow diagram of the battery system.

Table 1. The inventory of main raw materials (kg) and energy consumption for the Li-ion battery ($LiNi_{1/3}Co_{1/3}Mn_{1/3}O_2$ + 1% Fe_3O_4, 170.9 mAh/g).

Raw material	Mass/g	w%	Energy	
Li	1.75E−05	69.63%	Fossil fuels (Heat coal in industrial furnace)	0.190 MJ
Ni	1.25E−06	4.95%	Electricity from grid	3.77 E−04 kWh
Mn	1.50E−06	5.97%		
Co	1.60E−06	6.37%		
Fe_3O_4	7.40E−08	0.29%		
Acetylene black	1.01E−06	4.02%		
PVDF	5.05E−07	2.01%		
$LiPF_6$/PC-DMC (1 mol·L^{-1})	1.70E−06	6.76%		
Total	2.51E-05	100%		

Table 2. The inventory of main raw materials (kg) and energy consumption for the Ni-MH battery ($LaMg_{12}$ + 200% Ni alloy, 932.8 mAh/g).

Raw material	Mass/g	w%	Energy	
La	3.21E−05	0.09%	Fossil fuels (Heat coal in industrial furnace)	24.6 MJ
Mg	6.75E−05	0.19%	Electricity from grid	2.95 kWh
Ni	3.70E−04	1.03%		
C	3.10E−04	0.86%		
KOH	3.36E−02	93.65%		
LiOH	1.50E−03	4.18%		
Total	3.59E−02	100%		

Choices and Assumption of the Study

The Eco-indicator methodology that is used to calculate the standard values conforms well to the ISO 14042 standard on Life Cycle Impact Assessment (LCIA), although some details will perhaps deviate. The LCA method was selected to comprehensively evaluate the environmental impact of secondary batteries.

In the Life Cycle Impact Assessment process of LCA, the Eco-indicator 99 system was chosen. The Eco-indicator 99 LCIA method is based on the principles of environmental damage. The term "damage" includes 11 aspects, which can be grouped according to three main types of damage (Table 3). Standard Eco-indicators are numbers that express the total environmental load of a product or process. The core of this method is the weighing system between the different environmental aspects. The Eco-indicator point (Pt) is representative for one thousandth of the yearly environmental load of one average European inhabitant. The indicators can be calculated with appropriate LCA software. Sima Pro 7.1.8 [26] was chosen as the LCA software in this study. The Eco-indicator of a material or process is a non-unit number indicating the environmental impact of the material or process based on data from a LCA tool, where a higher indicator means a greater environmental impact [27,28].

Table 3. The 11 "damages" grouped in three main types.

Main Types of Damage	Eco-indicator Damage
(A) human health damage	(1) carcinogens, (2) respiratory organics, (3) respiratory inorganics, (4) climate change, (5) radiation, (6) ozone layer;
(B) ecosystem quality impact	(7) eco-toxicity, (8) acid rain / eutrophication, (9) land use;
(C) resource consumption	(10) minerals, (11) fossil fuels.

Data Quality and Assumptions

Material synthesis and the assembly of batteries were implemented in the Beijing Institute of Technology under quality assurance conditions. The raw material masses and the charge-discharge data were measured in the lab. Energy consumption and environmental emission were collected from databases and data from the scientific literature [29]. The resource and energy flow data of the system were obtained from the data of experimental batteries and the industrial process information about raw material production. Data for the disposal and other end-of-life treatment originated from the Ecoinvent database v2.2 [30] as a background system. The environmental impacts of PVDF and $LiPF_6$/PC-DMC were ignored after considering the lack of data in the Ecoinvent database for the relevant components. Acetylene black was mainly composed of carbon, and its environmental impact is mainly caused by carbon. The environmental impact of carbonyl nickel was represented by nickel sediment.

Quality Standardisation of Raw Materials

For the Eco-indicator comparison of the two selected batteries, it was necessary to determine a unified functional unit and standardized inventory data. Quality standardisation of the raw materials was then performed in accordance with the standard.

In this paper, the functional unit is a battery whose specific capacity is 1,000mAh/g. The cycles are the times a consumer charges and discharges a battery in the use phase. The requirements of the use step impact the charge-discharge cycles. Each battery requires a material independently from the use step. The needs are related to the choice of the functional unit for two different batteries. Depending on the relationship between the charge-discharge cycles and the specific capacity, the material needs can be related with the use step. And this would be reflected in the life cycle inventory.

Based on the condition of battery tests in the laboratory, the total capacity of batteries can be calculated by Equation (1) in which the cycle was N_k:

$$C_{total} = \sum_{i=1}^{k} C_i \tag{1}$$

where C_i can be calculated by fitting the formula obtained through curve fitting of the cycle performance.

To facilitate the comparison of results, the mass of raw material needs (m_s) under standard specific capacity (C_s = 1,000 mAh/g) can then be calculated by Equation (2):

$$m_s = C_s \cdot m / C_{total} \tag{2}$$

where m is the battery mass under test, and m_s is the result of quality standardisation of raw materials for the functional unit. Through this calculation method, a life cycle inventory of the raw materials was assembled.

End-of-Life Treatment Policy Analysis

End-of-life treatment, such as disposal and recycle, contributes to the environmental impact of batteries [31]. Different measures for treatment and the proportion of different measures lead to changes in different environmental impact categories. With the Eco-indicator 99, the total environmental impact of a unit mass of raw materials during stages of production, use, disposal and recycle was quantified. Incineration (while mixed with other types of wastes) and landfill are the common means of disposal for batteries. Due to China's waste treatment trends, we selected incineration as the typical means of disposal for discarded batteries. Depending on the Eco-indicator points and their changing with different end-of-life treatment, it can be shown how recycle and disposal affect the environment and what level the environmental impact has.

By means of the slice function in MATLAB [32], the Eco-indicator distribution obtained above is shown in the cycles, incineration rate and recycle rate in three-dimensional space and two-dimensional space composed of any two of the above three factors. End-of-life analysis was performed based on these slice graphs. The analysis can be used to support policies concerning the end-of-life treatment.

Evaluation Procedures

Figure 2 shows the research scheme of our environmental impact and end-of-life treatment policy analysis of secondary batteries.

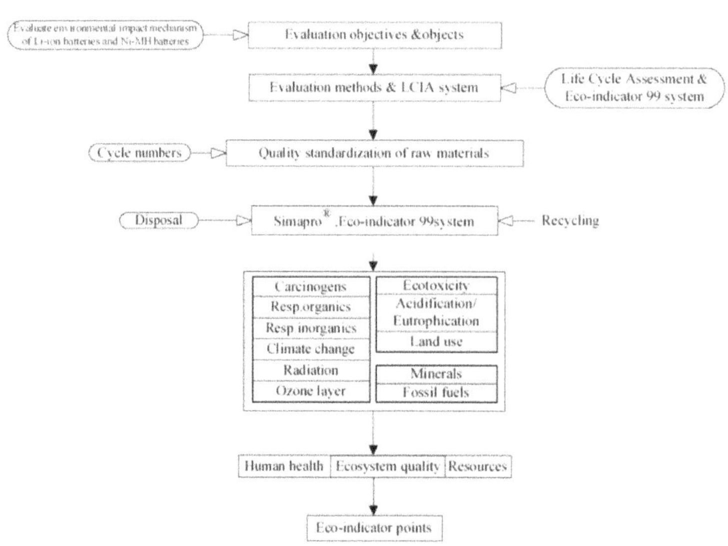

Figure 2. Research steps of the secondary battery environmental impact assessment.

The type and quantity of raw material data were analysed using the SimaPro7.1.8. Finally, with the Eco-indicators of each ingredient, the Eco-indicator distributions of secondary batteries with influencing factors (cycle number, incineration rate and recycle rate) were obtained. For discarded batteries, incineration and recycle were chosen to be the end-of-life treatment measures.

RESULTS AND DISCUSSION

Cycle Performance Fitting

The cycle performance of the selected batteries is shown in Figure 3. Exponential decay fitting was performed. The fitting equation of the red curves is as follows:

$$C_{Ni-MH} = 625.09e^{\left(-N_i/41.71\right)} + 435.85e\left(-N_i/2.60\right) + 39.64, R^2 = 0.99 \quad (3)$$

$$C_{Li-ion} = 7.20e^{\left(-N_i/6.99842\right)} + 7.20e\left(-N_i/7.0053\right) + 155.87, R^2 = 0.74 \quad (4)$$

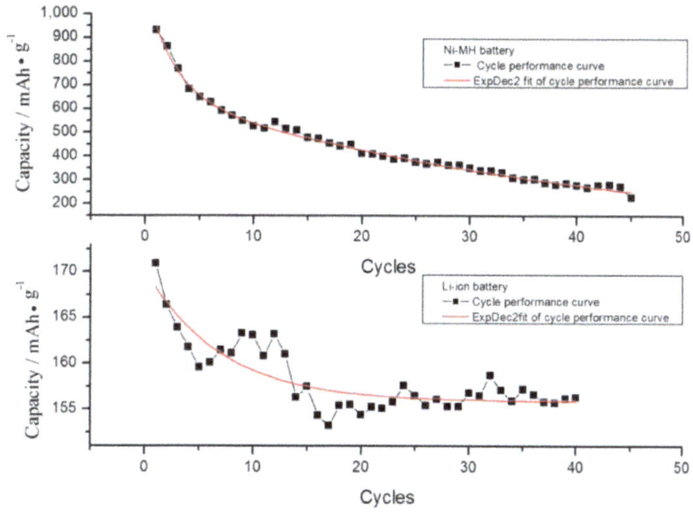

Figure 3. Cycle performance fits of the selected Ni-MH battery & Li-ion batteries.

The results were obtained in accordance with the method described in Section 2.1 and Section 2.4.

Comparison of Selected Batteries

Three-Dimensional Eco-Indicator Slices of the Two Selected Batteries

Three-dimensional Eco-indicator slices of the two selected batteries are shown in Figure 4. Here, the three axes represent the cycle number, incineration rate and recycle rate of the corresponding batteries. The Eco-indicator point of batteries is indicated by different gradient colours. The Eco-indicator point of the selected Li-ion battery is significantly lower than that of the selected Ni-MH battery. For further analysis of the influence of the three factors on the Eco-indicator of the two selected batteries, the Eco-indicator distribution in two-dimensional space composed by any two of the three factors were drawn based on Figure 4.

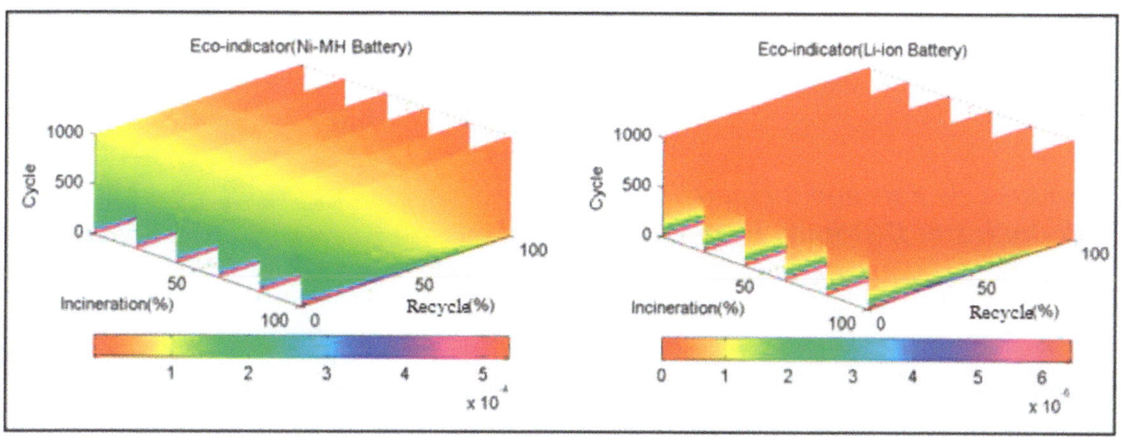

Figure 4. Three-dimensional slices of eco-indicator of the two selected batteries.

Influence of Cycles and Incineration Rate on Eco-Indicator Distribution

The influence of cycles and incineration rate on the Eco-indicator distribution of two selected batteries at a certain recycle rate (50%) is shown in Figure 5. As seen, the Eco-indicator points of the two selected batteries decrease rapidly with the increase in cycles when the number of cycles is less than 200, but its attenuation is less obvious after 200 cycles. Incineration rate has a greater impact on the Eco-indicator point of the selected Ni-MH battery, and the impact increases after the incineration rate reaches 50% or more. In Li-ion batteries, the Eco-indicator point is scarcely affected by the incineration rate, as shown in the figure.

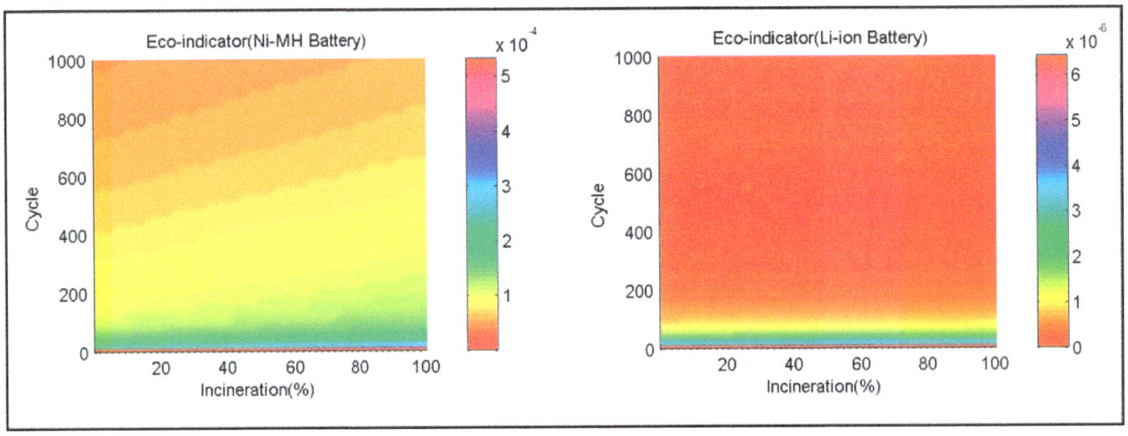

Figure 5. Influence of cycles and incineration rate on the Eco-indicator of the two selected batteries.

Influence of Cycles and Recycle Rate on Eco-Indicator Distribution

Figure 6 shows the influence of cycles and recycle rate on the Eco-indicator distributions of the two selected batteries when the incineration rate is 50%. The Eco-indicator points decrease rapidly when the cycle numbers are less than 200. The points of the two selected batteries decay with the increase of recycle rate, particularly when the rate is above 50%. The Eco-indicator point of Ni-MH battery has a larger attenuation. The influence of recycle rate is more obvious, while the influence on the Li-ion battery can be clearly observed when the recycle rate reaches 40%–50%.

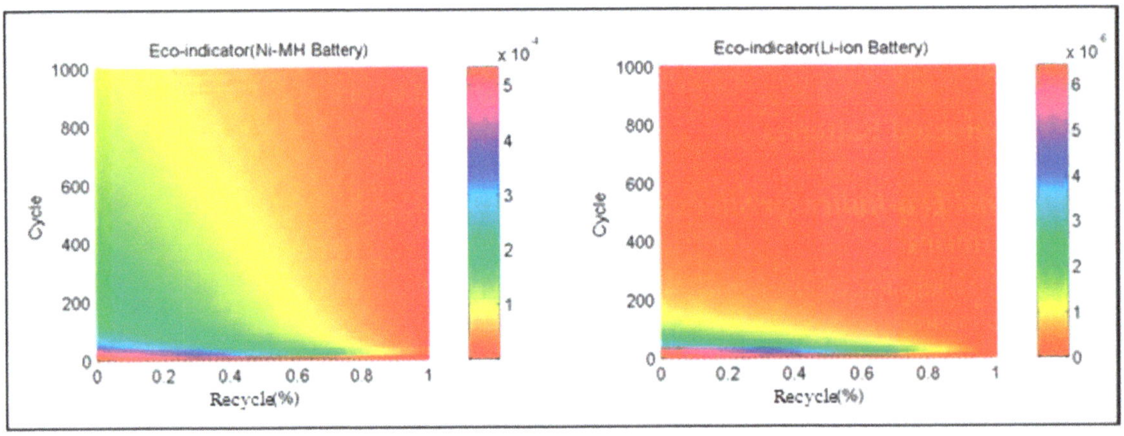

Figure 6. Influence of cycles and recycle rate on the Eco-indicator of the two selected batteries.

Influence of Recycle Rate and Incineration Rate on Eco-Indicator Distribution

Figure 7 shows the effect of recycle rate and incineration rate on the Eco-indicator of the two selected batteries when the cycle number is 100. As confirmed by this diagram, the Eco-indicator points of the two selected batteries decay with the increase of recycle rate and incineration rate. These two factors have a greater impact on the Eco-indicator of the Ni-MH battery than the Li-ion battery; in contrast, their impact on the Li-ion battery is small, and the Eco-indicator point of the Li-ion battery is nearly unaffected by the incineration rate.

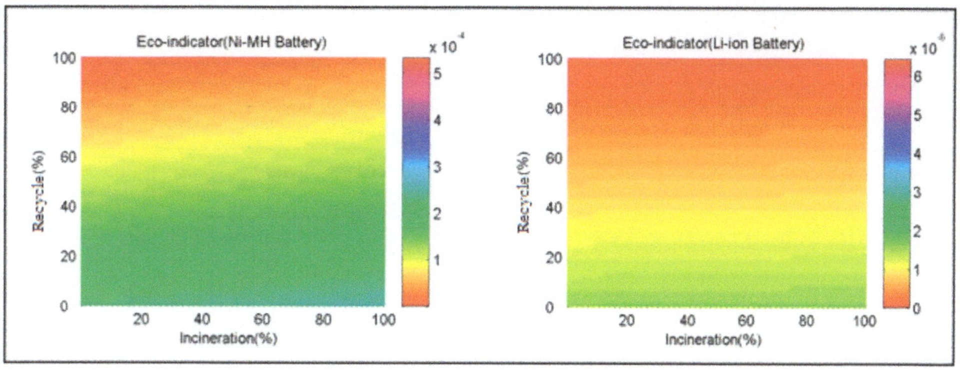

Figure 7. Influence of recycle rate and incineration rate on the Eco-indicator of the two selected batteries.

In summary, to reduce the environmental impact of selected batteries, the charge-discharge cycle number during the use stage should be at least 200. Battery incineration has a greater impact on the Eco-indicator point of the Ni-MH battery, while its impact on the Li-ion battery is small. Battery recycling can reduce the battery's environmental impact, particularly for Ni-MH batteries, and the incineration has little effect on reducing the environmental impact of Ni-MH batteries.

Uncertainty Analysis

Based on the laboratory battery test conditions, the charge-discharge cycle numbers of the batteries and the masses of raw materials were the input data of the LCA. According to the quality standardisation of the raw materials' masses, the cycle number of the batteries was assumed to be normally distributed under 95% confidence interval. The distributions presented the impact of uncertainty data. For the Ni-MH battery, the Eco-indicator point ranges from 0.5×10^{-4} Pt to 2.0×10^{-4} Pt when the charge-discharge cycle number ranges from 0 to 1,000. For the Li-ion battery, the Eco-indicator point ranges from 0.1×10^{-4} Pt to 1.0×10^{-4} Pt when the charge-discharge cycle number ranges from 0 to 1,000. According to the results, if the battery has a short use phase, the environmental impact will increase. If secondary batteries are under unsteady use conditions, they always have a short use phase and short charge-discharge cycles. Since the cycle numbers are related with energy consumption, the uncertainty of energy consumption in battery use phase has larger impacts on the Li-ion battery. Improving the cycle performance of the Li-ion battery can efficiently reduce the environmental impact.

CONCLUSIONS AND POLICY IMPLICATIONS

In this paper, an assessment framework was applied to conduct an environmental impact and end-of-life treatment analysis for secondary batteries based on the Eco-indicator 99 system. This applied model comprehensively evaluated the cycle, recycle and waste (for example, incineration) stages of secondary batteries. The analysis of the end-of-life treatment is helpful to improve some extant policies concerning secondary batteries. Two types of secondary batteries were studied by means of life cycle assessment. The results show that: (1) The Eco-indicator points of the two selected batteries decrease rapidly with the increase of the cycle number up to 200, but the attenuation is small after 200; (2) the incineration rate has greater impact on the Eco-indicator point of the Ni-MH battery than of the Li-ion battery, so policies and regulations concerning the disposal of the two kinds of batteries should be different; (3) the Eco-indicator points of the two types of batteries decay with the increase of the recycle rate, which means a strict recycle policy should be set; and (4) the influence of the recycle rate on the Ni-MH battery is more obvious, while the influence on the Li-ion battery can be clearly observed when the recycle rate is 40%–50%.

Due to the assessment of environmental loads about the selected secondary batteries, several policy implications should be considered for reducing environmental impact of secondary batteries. First, policy-makers should encourage the use of higher energy density batteries with better chemical characteristics. Additional cycles in a longer service life will result in lower environmental loads. To reduce the environmental impact of selected batteries, the charge-discharge cycle number during the use stage should be at least 200. In other words, according to the assessment, a policy should be proposed to require a minimum of 200 cycles for secondary batteries.

Second, secondary batteries should be sorted out from refuse because their contents and properties are different from those of common household garbage. For instance, most batteries contain heavy metals, such as cobalt and nickel, as well as some organic pollutants. Environmental risks will increase if secondary batteries are buried or combusted with mixed municipal waste.

Third, end-of-life treatment policies for secondary batteries should be carefully selected. For example, incineration measures should be put in place for Ni-MH batteries, and battery recycle can significantly reduce the environmental impact of Ni-MH batteries. Additionally, the recycle of Li-ion batteries may result in higher efficiency and lower environmental loads. In summary, the mechanism analysis tools proposed in this paper can also be used to evaluate the environmental impact of other secondary batteries, and the policies proposed in this paper may be useful in setting measures to reduce the environmental impact of these batteries.

Nomenclature

C_i	battery charge-discharge capacity	LCIA	life cycle inventory assessment
C_{total}	total battery capacity in use phase	m	battery mass under test
C_s	standard specific capacity per functional unit	m_s	required raw materials mass per functional unit
LCA	life cycle assessment	N_k	charge-discharge cycles in use phase

ACKNOWLEDGMENTS

The authors would like to express appreciation for these funding sources: (1) The Fundamental Research Funds for the Central Universities (No. YX2014-07); (2) National Natural Science Foundation of China (No. 51104021, 41301636); (3) Beijing Natural Science Foundation (No. 9122017); (4) National Key Technology Research and Development Program of the Ministry of Science and Technology of China (No. 2012BAD15B05). Thanks are also due to Wang Jing and Su Yue-feng for their help with the material synthesis and analysis.

AUTHOR CONTRIBUTIONS

Yajuan Yu designed the research process and conceived the whole paper. Bo Chen and Kai Huang analyzed the data and performed some of the figures. Xiang Wang programmed the codes. Dong Wang performed some of the figures. Yajuan Yu wrote and checked through the whole paper. Yajuan Yu and Kai Huang had full access to all of the data in the study and take responsibility for the integrity of the data and the accuracy of the data analysis. Both contributed equally to the study and are co-corresponding authors.

REFERENCES

1. Julien, C.M.; Mauger, A. Review of 5-V electrodes for Li-ion batteries: Status and trends. *Ionics* **2011**, *19*, 951–988.
2. Sullivan, J.L.; Gaines, L. Status of life cycle inventories for batteries. *Energ. Convers. Manage.***2012**, *58*, 134–148.
3. Majeau-Bettez, G.; Hawkins, T.R.; Stromman, A.H. Life cycle environmental assessment of lithium-ion and nickel metal hydride batteries for plug-in hybrid and battery electric vehicles. *Environ. Sci. Technol.* **2011**, *45*, 4548–4554.
4. Xu, J.Q.; Thomas, H.R.; Francis, R.W.; Lum, K.R.; Wang, J.; Liang, B. Review of processes and technologies for the recycling of lithium-ion secondary batteries. *J. Power. Sources.* **2008**, *177*, 512–527.
5. Salhofer, S.; Tesar, M. Assessment of removal of components containing hazardous substances from small WEEE in Austria. *J. Hazard. Mater.* **2011**, *186*, 1481–1488.
6. Bratt, C.; Hallstedt, S.; Robert, K.H.; Broman, G.; Oldmark, J. Assessment of eco-labelling criteria development from a strategic sustainability perspective. *J. Clean. Prod.* **2011**, *19*, 1631–1638.
7. Wrisberg, N.; Udo de Haes, H.A.; Triebswetter, U.; Eder, P. *Analytical Tools for Environmental Design and Management in a Systems Perspective*; Kluwer Academic Publishers: Dordrecht, the Netherland, 2002.
8. Chang, T.C.; You, S.J.; Yu, B.S.; Yao, K.F. A material flow of lithium batteries in Taiwan. *J. Hazard. Mater.* **2009**, *163*, 910–915.
9. Grubb, G.F.; Bakshi, B.R. Appreciating the role of thermodynamics in LCA improvement analysis via an application to titanium dioxide nanoparticles. *Environ. Sci. Technol.* **2011**, *45*, 3054–3061.
10. Sebastien, H.; Vera, I.S.; Jerome, P. Terrestrial ecotoxicity and effect factors of metals in life cycle assessment (LCA). *Chemosphere* **2007**, *68*, 1489–1496.
11. Monika, H. Perspective of the systematic and extended use of temporal and spatial aspects in LCA of long-lived products. *Chemosphere* **1998**, *37*, 265–270.
12. Gandhi, N.; Huijbregts, M.A.; Meent, D.V.; Peijnenburg, W.J.; Guinee, J.; Diamond, M.L. Implications of geographic variability on comparative toxicity potentials of Cu, Ni and Zn in freshwaters of Canadian ecoregions. *Chemosphere* **2011**, *82*, 268–277.
13. Berg, S.; Lindholm, E.L. Energy use and environmental impacts of forest operations in Sweden. *J. Clean. Prod.* **2005**, *13*, 33–42.
14. Rebitzer, G.; Ekvall, T.; Frischknecht, R.; Hunkeler, D.; Norris, G.; Ryberg, T.; Schmidt, W.P.; Suh, S.; Weidema, B.P.; Pennington, D.W. Life cycle assessment: Part 1: Framework, goal and scope definition, inventory analysis, and applications. *Environ. Int.* **2004**, *30*, 701–720.
15. Notter, D.A.; Gauch, M.; Widmer, R.; Wager, P.; Stamp, A.; Zah, R.; Althaus, H.J. Contribution of Li-ion batteries to the environmental impact of electric vehicles. *Environ. Sci. Technol.* **2010**, *44*, 6550–6556.
16. Matheys, J.; Timmermans, J.M.; Mierlo, J.V.; Mayer, S.; den Bossche, P.V. Comparison of the environmental impact of five electric vehicle battery technologies using LCA. *Int. J. Sust. Manuf.* **2009**, *1*, 318–329.
17. Zackrisson, M.; Avellan, L.; Orlenius, J. Life cycle assessment of lithium-ion batteries for plug-in hybrid electric vehicles—Critical issues. *J. Clean. Prod.* **2010**, *18*, 1519–1529.
18. Longo, S.; Antonucci, V.; Cellura, M. Life Cycle Assessment of Storage Systems: The Case Study of a Sodium/Nickel Chloride Battery. Available online: http://dx.doi.org/10.1016/j.jclepro.2013.10.004 (accessed on 10 October 2013).
19. Carl, J.R. Environmental assessment of vanadium redox and lead-acid batteries for stationary energy storage. *J. Power. Sources.* **1999**, *80*, 21–29.

20. Slack, R.J.; Bonin, M.; Gronow, J.R.; Van Santen, A.; Voulvoulis, N. Household hazardous waste data for the UK by direct sampling. *Environ. Sci. Technol.* **2007**, *41*, 2566–2571.

21. Panero, S.; Romoli, C.; Achilli, M.; Cardarelli, E.; Scrosati, B. Impact of household batteries in landfills. *J. Power Sources* **1995**, *57*, 9–12.

22. Dewulf, J.; Vorsta, G.V.; Denturcka, K.; Langenhove, H.V.; Ghyoot, W.; Tytgat, J.; Vandeputte, K. Recycling rechargeable lithium ion batteries: Critical analysis of natural resource savings. *Resour. Conserv. Recy.* **2010**, *54*, 229–234.

23. Environmental Management—Life Cycle Assessment—Principles and Framework. Available online: http://www.ce.cmu.edu/~hsm/lca2007/readings/iso14040.pdf (accessed on 3 June 1997).

24. Bureau of Indian Standards (BIS). *Environmental Management—Life Cycle Assessment—Requirements and Guidelines*; IS/ISO 14044; BIS: Delhi, India, 2006.

25. The Eco-indicator 99: A Damage Oriented Method for Life Cycle Impact Assessment—Methodology Report. Available online: http://www.pre-sustainability.com/download/EI99_methodology_v3.pdf (accessed on 22 June 2001).

26. Goedkoop, M.; Schryver, A.D.; Oele, M. *Introduction to LCA with SimaPro v2.0*; PRe Consultants: Amersfoort, the Netherlands, 2010.

27. Goedkoop, M.; Effting, S.; Collingo, M. *The Eco-Indicator'99—A Damage Oriented Method for Life Cycle Impact Assessment—Manual for Designers*; PRe Consultants: Amersfoort, the Netherlands, 2000.

28. Taoussanidis, N. Life Cycle Assessment of Combined Solar System. In Proceedings of the 4th WSEAS International Conference on Heat Transfer, Thermal Engineering and Environment, Elounda, Greece, 21–23 August 2006.

29. Sullivan, J.L.; Gaines, L. *A Review of Battery Life-Cycle Analysis: State of Knowledge and Critical Needs*; Argonne National Laboratory: Argonne, IL, USA, 2010.

30. Ecoinvent Report v2.2. Available online: http://www.ecoinvent.org/fileadmin/documents/en/00_summary_ecoinvent-v2.2.pdf (accessed on 14 April 2010).

31. Lupi, C.; Pasquali, M.; Dell'Era, A. Nickel and cobalt recycling from lithium-ion batteries by electrochemical processes. *Waste. Manage.* **2005**, *25*, 215–220.

32. Clavreul, J.; Guyonnet, D.; Tonini, D. Stochastic and epistemic uncertainty propagation in LCA. *Int. J. Life. Cycle Assess.* **2013**, *18*, 1393–1403.

CHAPTER 20

A REVISIT TO THE HYDROGEN DESORPTION/ABSORPTION BEHAVIORS OF $LIALH_4/LIBH_4$: EFFECTS OF CATALYSTS

Pattaraporn Sridechprasat, Labhatrada Phuirot, Pramoch Rangsunvigit, Boonyarach Kitiyanan and Santi Kulprathipanja

Keywords: $LiAlH_4$, $LiBH_4$, hydrogen storage

INTRODUCTION

For decades, concerns about limited fossil fuel reservoirs have increased alarmingly. In addition, increases in fossil-fuel consumption are causing global environmental issues due to CO_2 emissions. As the most promising alternative fuel, hydrogen is attractive as a clean energy source. Unlike petroleum, hydrogen can be produced from renewable sources such as water, sunlight, and biomass, thereby reducing the growing dependence on fossil fuels [1,2]. Furthermore, it can generate electricity by reacting with oxygen in PEM fuel cells with only water as the by-product without greenhouse gas emissions for hydrogen-fueled vehicles. Storage of hydrogen for on-board applications, however, is difficult because of its physical characteristics such as its very light weight. Despite the availability of a few hydrogen storage methods, solid-state hydrogen storage has been considered as a promising method to employ for on-board hydrogen storage [3,4].

Li-based complex hydrides such as lithium aluminum hydride ($LiAlH_4$) and lithium borohydride ($LiBH_4$) have been considered as promising hydrogen storage materials due to their high theoretical hydrogen capacities of about 10.5 and 18.5 wt %, respectively [5,6,7]. For $LiAlH_4$, the balling time for sample preparation and transition metals were reported to affect its decomposition behaviors [6,8,9]. Even though $LiBH_4$ has a high hydrogen desorption capacity, it is a very stable and needs a very high temperature, up to 600 °C, for any significant amount of desorbed hydrogen to be observed [10]. Jin *et al.* [5] reported that a $LiBH_4$ and $LiAlH_4$ mixture added with 3 mol % TiF_3stored hydrogen up to 7.2 wt % with a decomposition temperature between 177 and 247 °C. In addition, $LiAlH_4$+$2LiBH_4$ doped with 5 mol % TiF_3 showed the best hydrogen storage properties, with the decrease in the onset temperature of the first and second dehydrogenation steps [11]. Moreover, it could absorb 3.76 and 4.78 wt % in 1 and 14 h, respectively, at 600 °C and 4 MPa hydrogen.

In this work, $LiBH_4$ and $LiAlH_4$ were selected as the base materials for hydrogen storage. Mixtures of these hydrides were prepared by mechanical ball-milling. A small amount of Ti-based catalysts ($TiCl_3$ and TiO_2), VCl_3, and $ZrCl_4$ were added during the milling process to study their effects on the hydrogen storage capacity, decomposition temperature, kinetics, and reversibility of hydrogen desorption/absorption. Phase transformations during the hydrogen desorption/absorption were also investigated.

RESULTS AND DISCUSSION

Various molar ratios of $LiAlH_4$ and $LiBH_4$ in the mixtures were studied in order to understand effects of the hydride combination

Citation: Pattaraporn Sridechprasat, Labhatrada Phuirot, Pramoch Rangsunvigit, Boonyarach Kitiyanan and Santi Kulprathipanja, A Revisit to the Hydrogen Desorption/Absorption Behaviors of LiAlH4/LiBH4: Effects of Catalysts, doi:10.3390/en5093691.

on the amount of desorbed hydrogen and desorption temperature. As shown in Figure 1, $LiAlH_4$ (Figure 1a) decomposes in two steps starting at 145 °C and continues to 220 °C with a total hydrogen amount of 7.6 wt %. $LiBH_4$ (Figure 1g) starts to decompose a small amount of hydrogen at a low temperature of 95 °C, which is believed to occur due to the reduction in the crystallite size and hydrogen desorption distance, and the distortion of hydride structure during the high energy milling process [12]. $LiBH_4$ starts to release 1.0 wt% hydrogen at a temperature higher than 285 °C, which is the melting point of $LiBH_4$. In the case of the $LiAlH_4$ and $LiBH_4$ mixtures (Figure 1b–f), all mixtures release lower amounts of hydrogen than $LiAlH_4$. A 2:1 molar ratio of $LiAlH_4$:$LiBH_4$ ($2LiAlH_4$+$LiBH_4$) (Figure 1b) provides the highest amount of hydrogen at 6.6 wt %, and it desorbs hydrogen in the temperature range of 100–220 °C. The hydrogen desorption temperatures of the $LiAlH_4$/$LiBH_4$ mixture correspond with the data is previous works [11]. The amounts of desorbed hydrogen and temperature of each mixture are also shown in Table 1. However, the desorbed samples cannot re-absorb hydrogen under the tested conditions, as we will further discuss in the phase transformation section.

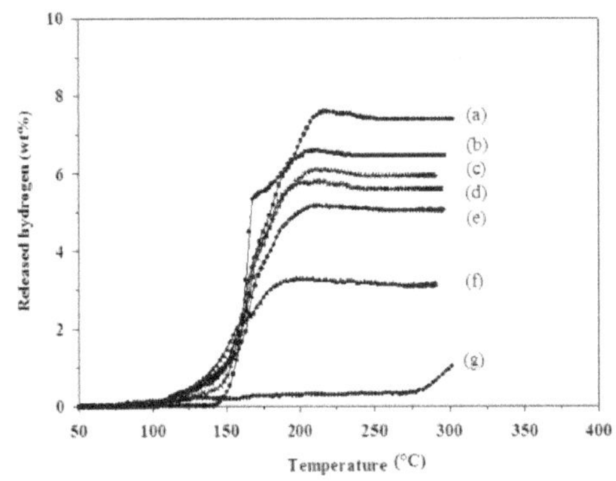

Figure 1. Hydrogen desorption profiles of: (**a**) as-milled $LiAlH_4$; (**b**) $2LiAlH_4$+$LiBH_4$; (**c**) $3LiAlH_4$+$LiBH_4$; (**d**) $4LiAlH_4$+$LiBH_4$; (**e**) $LiAlH_4$+$LiBH_4$; (**f**) $LiAlH_4$+$2LiBH_4$; and (**g**) as-milled $LiBH_4$.

Table 1. Hydrogen desorption temperature and amounts of desorbed hydrogen of $LiAlH_4$–$LiBH_4$ mixtures

Sample	Desorption temperature (°C)		Desorption amount (wt %)
	Initial	Final	
$LiAlH_4$	145	220	7.6
1:1	105	220	5.2
2:1	100	220	6.6
3:1	95	220	6.1
4:1	105	220	5.8
1:2	95	210	3.3
$LiBH_4$	75	370 *	1.0 *

One mol % of TiO_2, $TiCl_3$, VCl_3, or $ZrCl_4$ were also added to the $2LiAlH_4$+$LiBH_4$ mixture in order to investigate effects of the transition metals on the hydrogen desorption/absorption behaviors. Figure 2 shows the two-step decomposition reactions and the decrease in the desorption temperature for all doped mixtures. The mixture in the presence of $TiCl_3$ (Figure 2d) starts to decompose and release hydrogen at the lowest temperature (40 °C), which is lower than the undoped one by 60 °C. In addition, 1 mol % $TiCl_3$–$2LiAlH_4$+$LiBH_4$ provides the highest amount of hydrogen desorption among the doped samples, 6.4 wt %. A possible reason is the fact that $TiCl_3$could facilitate the crystallite size reduction of the hydride during the milling process resulting in the hydrogen storage improvement [13]. However, the hydrogen desorption capacities of all the doped mixtures are significantly lower than that of the undoped one. The results are summarized in Table 2.

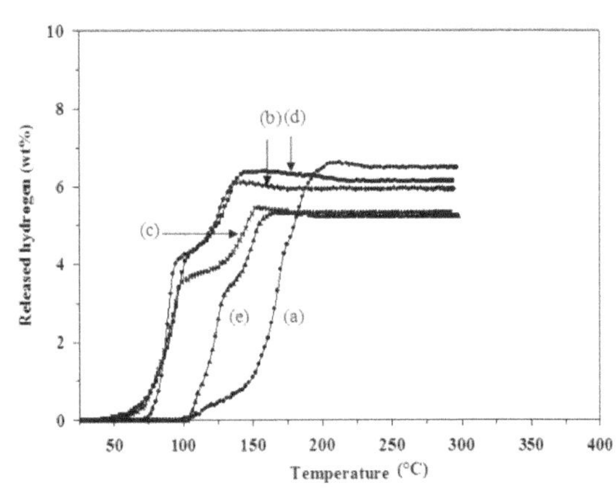

Figure 2. Hydrogen desorption profiles of: (**a**) $2LiAlH_4$+$LiBH_4$; (**b**) 1 mol % $ZrCl_4$–$2LiAlH_4$+$LiBH_4$; (**c**) 1 mol % VCl_3–$2LiAlH_4$+$LiBH_4$; (**d**) 1 mol % $TiCl_3$–$2LiAlH_4$+$LiBH_4$; and (**e**) 1 mol % TiO_2–$2LiAlH_4$+$LiBH_4$.

Table 2. Hydrogen desorption temperature and amounts of desorbed hydrogen of $2LiAlH_4$+$LiBH_4$ mixed with 1 mol % TiO_2, $TiCl_3$, VCl_3, and $ZrCl_4$ in the first (R1) and second (R2) steps.

Sample	Desorption temperature (°C)			Desorption amount (wt %)		Total (wt %)
	R1	R2	Final	R1	R2	
$2LiAlH_4$+$LiBH_4$	100	-	220	-	-	6.6
TiO_2–$2LiAlH_4$+$LiBH_4$	105	138	190	3.5	1.8	5.3
$TiCl_3$–$2LiAlH_4$+$LiBH_4$	40	110	170	4.4	2.0	6.4
VCl_3–$2LiAlH_4$+$LiBH_4$	53	110	155	3.7	1.8	5.5
$ZrCl_4$–$2LiAlH_4$+$LiBH_4$	72	105	150	4.3	1.8	6.1

To investigate effects of an amount of catalyst on the hydrogen desorption properties of the $2LiAlH_4$+$LiBH_4$ mixture, 3 and 5 mol % $TiCl_3$ were further added. The results, as shown in Table 3 and Figure 3, show that the hydrogen desorption capacities of all $TiCl_3$-doped mixtures are significantly lower than that of the undoped one. The hydrogen desorption capacity decreases with the increase in the doping amount, 6.4, 5.6, and 2.7 wt % for 1, 3, and 5 mol % doping, respectively. The desorption temperature of all $TiCl_3$-doped samples are lower than the undoped one by 40–70 °C; however, there is no significant difference in the desorption temperature among the $TiCl_3$-doped samples. Jin *et al.* [5] reported that the mixture of $LiAlH_4$+$2.2LiBH_4$ together with 3 mol % TiF_3 started to decompose between 177 °C and 247 °C and the weight loss reached up to 7.2 wt % at 387 °C. Mao *et al.* [11] also reported that $LiAlH_4$+$2LiBH_4$ doped with 5 mol % TiF_3 could decrease the onset temperature of the first and second decomposition steps by 64 and 150 °C, respectively. They further studied that doping with 5 mol % TiO_2 decreased the temperature of the first and second decomposition steps by 27 and 50 °C, respectively. In this work, no hydrogen absorption was observed for any of the doped samples at 300 °C and 8.5 MPa hydrogen for 6 h. Jin *et al.* [5] confirmed that the dehydrogenated product of 3 mol% TiF_3–$LiAlH_4$+$2.2LiBH_4$ could be hydrogenated to about 5.1 wt % H_2 at 350 °C and 70 bar hydrogen for 6 h. But the reaction kinetics was rather slow. Mao *et al.* [11] reported that 5 mol % TiF_3–$LiAlH_4$+$2LiBH_4$ absorbed 3.76 wt % and 4.78 wt % in 1 h and 14 h, respectively, at 600 °C and 4 MPa hydrogen.

Table 3. Hydrogen desorption temperature and amounts of desorbed hydrogen of $TiCl_3$-doped $2LiAlH_4$+$LiBH_4$ in the first (R1) and second (R2) steps.

Sample	Desorption temperature (°C)			Desorption amount (wt %)		Total (wt %)
	R1	R2	Final	R1	R2	
$2LiAlH_4$+$LiBH_4$	100	-	220	-	-	6.6
1 mol % $TiCl_3$–$2LiAlH_4$+$LiBH_4$	40	110	170	4.4	2.0	6.4
3 mol % $TiCl_3$–$2LiAlH_4$+$LiBH_4$	60	95	170	3.1	2.5	5.6
5 mol % $TiCl_3$–$2LiAlH_4$+$LiBH_4$	45	100	150	1.4	1.3	2.7

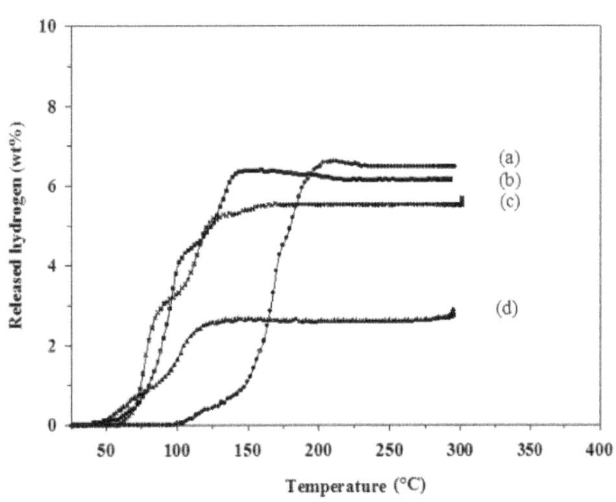

Figure 3. Hydrogen desorption profiles of: **(a)** $2LiAlH_4$+$LiBH_4$; **(b)** 1 mol % $TiCl_3$–$2LiAlH_4$+$LiBH_4$; **(c)** 3 mol % $TiCl_3$–$2LiAlH_4$+$LiBH_4$; and **(d)** 5 mol % $TiCl_3$–$2LiAlH_4$+$LiBH_4$.

Phase transformation of the 2:1 molar ratio of $LiAlH_4$:$LiBH_4$ mixture after the milling process is similar to $LiAlH_4$, without any peak of $LiBH_4$. In the case of the undoped mixture, as shown in Figure 4a, only $LiAlH_4$ is left after the 120 min milling process. In the case of the doped mixture, as shown in Figure 4b–e, Al is observed after the milling process. It is possible that the milling process and a dopant might destabilize the structure of $LiAlH_4$ [9]. After the hydrogen desorption at 300 °C (Figure 5), Al and LiH are present in both undoped and doped mixtures, without the formation of AlB_2. It is the AlB_2 phase that is claimed to facilitate the reversibility of $LiAlH_4$ and $LiBH_4$[11,14,15,16]. This could be a possible reason why the $LiAlH_4$/$LiBH_4$ mixture is not reversible. In addition, as shown in Figure 6, the XRD patterns of the desorbed samples in the presence of a higher amount of the catalyst (3 or 5 mol % $TiCl_3$) exhibit new peaks, which are those of LiCl [17]. The formation of LiCl might deteriorate the hydrogen desorption ability of the samples, consequently decreasing the hydrogen desorption capacity.

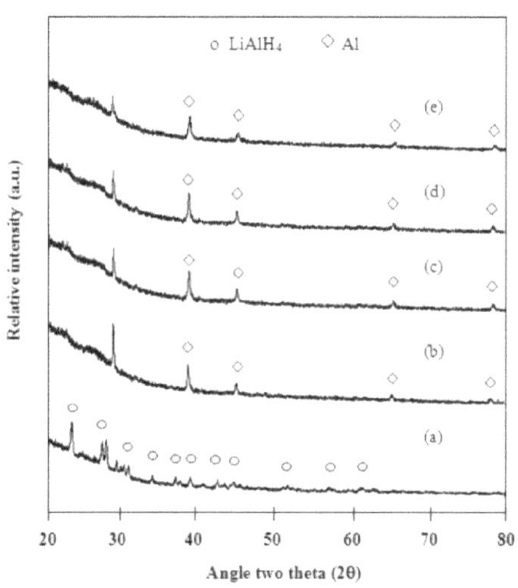

Figure 4. XRD patterns of milled samples for 120 min: (**a**) $2LiAlH_4$+$LiBH_4$; (**b**) 1 mol % $ZrCl_4$–$2LiAlH_4$+$LiBH_4$; (**c**) 1 mol % VCl_3–$2LiAlH_4$+$LiBH_4$; (**d**) 1 mol % $TiCl_3$–$2LiAlH_4$+$LiBH_4$; and (**e**) 1 mol % TiO_2–$2LiAlH_4$+$LiBH_4$.

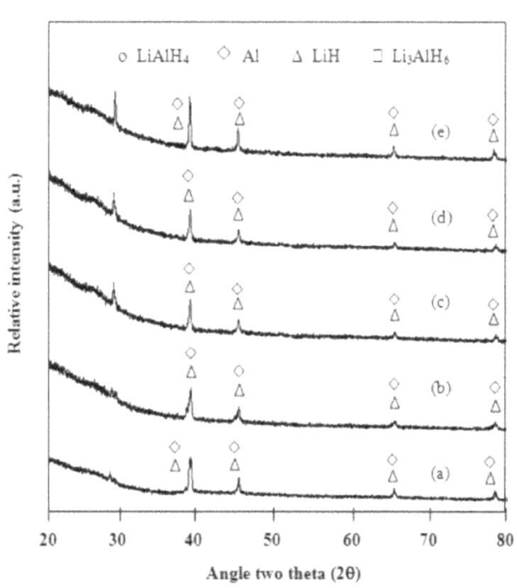

Figure 5. XRD patterns of desorbed samples at 300 °C: (**a**) $2LiAlH_4$+$LiBH_4$; (**b**) 1 mol % $ZrCl_4$–$2LiAlH_4$+$LiBH_4$; (**c**) 1 mol % VCl_3–$2LiAlH_4$+$LiBH_4$; (**d**) 1 mol % $TiCl_3$–$2LiAlH_4$+$LiBH_4$; and (**e**) 1 mol % TiO_2–$2LiAlH_4$+$LiBH_4$.

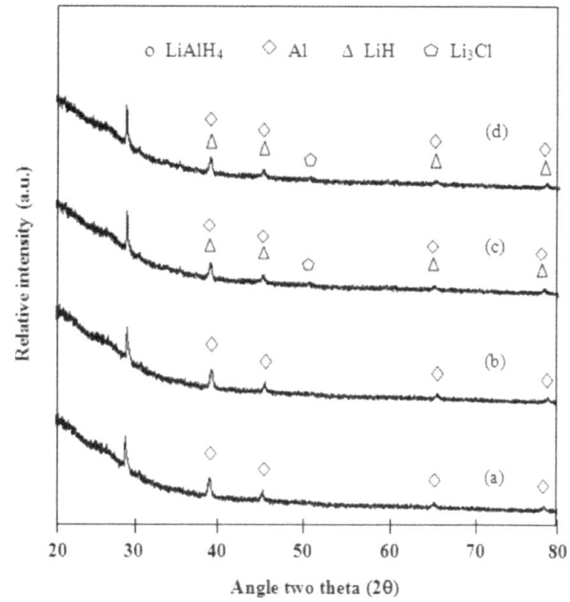

Figure 6. XRD patterns of: (**a**) milled 3 mol % $TiCl_3$–$2LiAlH_4$+$LiBH_4$; (**b**) milled 5 mol % $TiCl_3$–$2LiAlH_4$+$LiBH_4$; (**c**) desorbed 3 mol % $TiCl_3$–$2LiAlH_4$+$LiBH_4$; (**d**) desorbed 5 mol % $TiCl_3$–$2LiAlH_4$+$LiBH_4$.

The $LiAlH_4$ and $LiBH_4$ mixtures doped with TiO_2, $TiCl_3$, VCl_3, or $ZrCl_4$ partially decompose during the milling process because the structure of $LiAlH_4$ is unstable and the additives might destabilize its structure [9]. These metal catalysts in the hydride systems are reduced by the metal hydrides and transformed to TM-neutral [6,18], as shown in Reactions (1) and (2), where TM is the transition metal, M is the metal, and X can be either V or Ti:

$$TiO_2 + 4MH \rightarrow 2M_2O + Ti + H_2 \quad (1)$$

$$XCl_3 + 3MH \rightarrow 3MCl + X + 3/2H_2 \quad (2)$$

It may be this reduction that increases the hydrogen desorption capacity of $LiAlH_4$ as compared to the undoped one. Moreover, after the decomposition, these catalysts can form M_2O and MCl, which are left in the sample as stable materials and cannot further decompose [7]. In this work, MCl formation in Reaction (2) is LiCl, the presence of which might cause the decrease in the hydrogen desorption capacity with the increase in the $TiCl_3$ amount.

EXPERIMENTAL SECTION

All sample handling was carried out in a nitrogen-filled glove box. $LiBH_4$ (95%, Aldrich Chemical), $LiAlH_4$ (95%, Acros), zirconium (IV) chloride (99.5% $ZrCl_4$, Aldrich Chemicals), vanadium (III) chloride (>99% VCl_3, MERCK), and titanium dioxide (TiO_2, Degüssa) were used without further purification. Titanium (III) chloride ($TiCl_3$) powder was prepared by evaporation of 12% titanium (III) chloride solution in hydrochloric acid obtained from Riedel-de Haën. The mixing procedure was completed in a centrifugal ball mill at 300 rpm for 120 min with a balltopowder ratio of 100:1. After that, approximately 350 mg of a sample was immediately loaded into the high-pressure stainless reactor. Then, it was placed in the thermovolumetric apparatus. The desorption was performed from the room temperature to 300 °C (350 °C for $LiBH_4$) with a heating rate of 2 °C min^{-1}. The sample was held at this condition until no further desorption was observed. The absorption was carried out at 300 °C and 8.5 MPa hydrogen for 6 h. X-ray diffraction measurements (Rigaku, model: D/max-2200) were used in order to identify phase transformations during the hydrogen desorption/absorption of the hydride systems at room temperature over a range of diffraction angles from 20° to 90° with CuK-α radiation (40 kV, 30 mA). The XRD sample plate is covered with a Kapton tape to prevent oxide contamination. The tape shows a diffraction peak at around 28°.

CONCLUSIONS

The hydrogen desorption/absorption behaviors of undoped and doped $LiAlH_4/LiBH_4$ mixtures were studied. A 2:1 $LiAlH_4$:$LiBH_4$ molar ratio released the highest amount of hydrogen at 6.6 wt % in the temperature range of 100–220 °C. In the case of the doped systems, 1 mol % of metal catalysts ($TiCl_3$, TiO_2, VCl_3, or $ZrCl_4$) was doped into the mixture. The mixture with $TiCl_3$ desorbed hydrogen at the lowest temperature (40°C). Furthermore, the increase in the amount of $TiCl_3$ from 1 to 3 and 5 mol % decreased the hydrogen desorption capacity. No hydrogen absorption was observed for any of the doped $LiAlH_4$–$LiBH_4$ samples. In addition, the XRD patterns indicated the presence of Al and LiH in the mixture after the desorption corresponding to the decomposition of $LiAlH_4$.

ACKNOWLEDGMENTS

The financial support for this work was provided by the 90th Anniversary of Chulalongkorn University Fund (Ratchadaphisek Somphot Endowment); Royal Jubilee Ph.D. Program (Grant No. PHD/0249/2549), Thailand Research Fund; National Science and Technology Development Agency (Reverse Brain Drain Project); The Petroleum and Petrochemical College (PPC); The Climate Change Cluster (CC557A), National Research University Program, Office of Higher Education Commission (OHEC); Center of Excellence for Petrochemical and Materials Technology, Chulalongkorn University; and UOP, A Honeywell Company, Des Plaines, IL, USA.

REFERENCES

1. ProQuest. Available online: http://www.csa.com/discoveryguides/hydrogen/overview.php(accessed on 21 April 2009).
2. Mitch, J. Filling up with hydrogen. 22 August 2005; Chemical and Engineering News Website. Available online: http://www.cen-online.org (accessed on 22 August 2009).
3. U.S. Department of Energy. Hydrogen and Fuel Cells Program. Available online: http://www.hydrogen.energy.gov/storage (accessed on 21 April 2009).
4. Zhou, L. Progress and problems in hydrogen storage. *Renew Sustain. Energy Rev.* **2004**, *9*, 395–408.
5. Jin, S.A.; Shim, J.H.; Cho, Y.W.; Yi, K.W.; Zabarac, O.; Fichtnerc, M. Reversible hydrogen storage in $LiBH_4$–Al–LiH composite powder. *Scr. Mater.* **2008**, *58*, 963–965.
6. Blanchard, D.; Brinks, H.W.; Hauback, B.C.; Norby, P. Desorption of $LiAlH_4$ with Ti- and V-based additives. *Mater. Sci. Eng.* **2004**, *108*, 54–59.
7. Isobe, S.; Ichikawa, T.; Hanada, N.; Leng, H.Y.; Fichtner, M.; Fuhr, O.; Fujii, H. Effect of Ti catalyst with different chemical form on Li–N–H hydrogen storage properties. *J. Alloy Compd* **2005**, *404–406*, 439–442.
8. Easton, D.S.; Schneibel, J.H.; Speakman, S.A. Factors affecting hydrogen release from lithium alanate ($LiAlH_4$). *J. Alloy Compd.* **2005**, *398*, 245–248.
9. Suttisawat, Y.; Rangsanvigit, P.; Kitayanan, B.; Muangsin, N.; Kulprathipanja, S. Catalytic effect of Zr and Hf on the hydrogen desorption/absorption of $NaAlH_4$ and $LiAlH_4$. *Int. J. Hydrogen Energy* **2007**, *32*, 1277–1285.
10. Züttel, A.; Wenger, P.; Rentsch, S.; Sudan, P.; Mauron, Ph.; Emmenegger, Ch. $LiBH_4$ a new hydrogen storage material. *J. Power Sources* **2003**, *118*, 1–7.
11. Mao, J.F.; Guo, Z.P.; Liu, H.K.; Yu, X.B. Reversible hydrogen storage in titanium-catalyst $LiAlH_4$-$LiBH_4$ system. *J. Alloy Compd.* **2009**, *487*, 434–438.
12. Sridechprasat, P.; Suttisawat, Y.; Rangsunvigit, P.; Kitiyanan, B.; Kulprathipanja, S. Catalyzed $LiBH_4$ and MgH_2 mixture for hydrogen storage. *Int. J. Hydrogen Energy* **2010**, *36*, 1200–1205.
13. Varin, R.A.; Zbroniec, L.; Polanski, M.; Bystrzycki, J. A review of recent advances on the effects of microstructural refinement and nano-catalytic additives on the hydrogen storage properties of metal and complex hydrides. *Energies* **2011**, *4*, 1–25.
14. Mao, J.F.; Yu, X.B.; Guo, Z.P.; Poh, C.K.; Liu, H.K.; Wu, Z.; Ni, J. Improvement of the $LiAlH_4$–$NaBH_4$ system for reversible hydrogen storage. *Phys. Chem. C* **2009**, *113*, 10813–10818.
15. Mao, J.F.; Guo, Z.P.; Leng, H.Y.; Wu, Z.; Guo, Y.H.; Yu, X.B.; Liu, H.K. Reversible hydrogen storage in destabilized $LiAlH_4$–MgH_2–$LiBH_4$ ternary-hydride system doped with TiF_3. *Phys. Chem. C* **2010**, *114*, 11643–11649.
16. Yu, X.B.; Xia, G.L.; Guo, Z.P.; Liu, H.K. Dehydrogenation/rehydrogenation mechanism in aluminum destabilized lithium borohydride. *J. Mater. Res.* **2009**, *24*, 2720–2727.
17. Kamali, A.R.; Fray, D.J.; Schwandt, C. Thermokinetic characteristics of lithium chloride. *J. Therm. Anal. Calorim.* **2011**, *104*, 619–626
18. Siangsai, A.; Suttisawat, Y.; Sridechprasat, P.; Rangsunvigit, P.; Kitiyanan, B.; Kulprathipanja, S. Effect of Ti compounds on hydrogen desorption/absorption of $LiNH_2/LiAlH_2/MgH_2$. *J. Chem. Eng. Jpn.* **2010**, *43*, 95–98.

INDEX

N

O

P

R

S

T

X